The Produce Contamination Problem:
Causes and Solutions

Food Science and Technology International Series

Series Editor

Steve L. Taylor
University of Nebraska – Lincoln, USA

Advisory Board

Ken Buckle
The University of New South Wales, Australia

Mary Ellen Camire
University of Maine, USA

Roger Clemens
University of Southern California, USA

Hildegarde Heymann
University of California – Davis, USA

Robert Hutkins
University of Nebraska – Lincoln, USA

Ron S. Jackson
Quebec, Canada

Huub Lelieveld
Bilthoven, The Netherlands

Daryl B. Lund
University of Wisconsin, USA

Connie Weaver
Purdue University, USA

Ron Wrolstad
Oregon State University, USA

A complete list of books in this series appears at the end of this volume.

The Produce Contamination Problem: Causes and Solutions

Edited by

Gerald M. Sapers
USDA, Agricultural Research Service, Retired,
Wyndmoor, PA, USA

Ethan B. Solomon
DuPont Chemical Solutions Enterprise
Experimental Station Laboratory, Wilmington, DE, USA

Karl R. Matthews
Rutgers-The State University of New Jersey
Department of Food Science, New Brunswick, NJ, USA

AMSTERDAM • BOSTON • HEIDELBERG • LONDON
NEW YORK • OXFORD • PARIS • SAN DIEGO
SAN FRANCISCO • SINGAPORE • SYDNEY • TOKYO

Academic Press is an imprint of Elsevier

Academic Press is an imprint of Elsevier
30 Corporate Drive, Suite 400, Burlington, MA 01803, USA
525 B Street, Suite 1900, San Diego, California 92101-4495, USA
84 Theobald's Road, London WC1X 8RR, UK

Copyright © 2009, Elsevier Inc. All rights reserved.

No part of this publication may be reproduced or transmitted in any form or by any means, electronic or mechanical, including photocopy, recording, or any information storage and retrieval system, without permission in writing from the publisher.

Permissions may be sought directly from Elsevier's Science & Technology Rights Department in Oxford, UK: phone: (+44) 1865 843830, fax: (+44) 1865 853333, E-mail: permissions@elsevier.com. You may also complete your request online via the Elsevier homepage (http://elsevier.com), by selecting "Support & Contact" then "Copyright and Permission" and then "Obtaining Permissions."

Library of Congress Cataloging-in-Publication Data
APPLICATION SUBMITTED

British Library Cataloguing-in-Publication Data
A catalogue record for this book is available from the British Library.

ISBN: 978-0-12-374186-8

For information on all Academic Press publications
visit our Web site at www.elsevierdirect.com

Printed in the United States of America
09 10 9 8 7 6 5 4 3 2 1

Working together to grow
libraries in developing countries

www.elsevier.com | www.bookaid.org | www.sabre.org

ELSEVIER   BOOK AID International   Sabre Foundation

Contents

PREFACE xv
CONTRIBUTORS xix

PART 1 Introduction

CHAPTER 1 Scope of the Produce Contamination Problem 3
Gerald M. Sapers and Michael P. Doyle
Introduction 3
    Produce-Associated Outbreaks—a New Problem? 3
    Consequences of Produce-Associated Outbreaks 4
Key Aspects of the Produce Contamination Problem 5
    Characteristics of Produce-Associated Outbreaks 5
    Prevalence of Produce Contamination with Human Pathogens ... 8
    Microbial Attachment and Survival on Produce Surfaces 9
Potential Sources of Produce Contamination 9
    Preharvest Sources 9
    Contamination During Packing 10
    Contamination During Fresh-Cut Processing 11
Gaps in Our Understanding of Produce Contamination 12
    Current State of Knowledge 12
    What We Don't Know 12
    Developing Effective Interventions 13

CHAPTER 2 Microbial Attachment and Limitations of
Decontamination Methodologies 21
Ethan B. Solomon and Manan Sharma
Introduction 21
Ecological Niches and Introduction into the Plant Environment 23
Outbreak Investigations Reveal Sources and Persistence
    of Pathogens 24
The Plant Surface 25
Attachment of Pathogens to Plant Tissue 25
Attachment of *Escherichia coli* O157:H7 to Lettuce 26
Attachment of *Salmonella* to Tomatoes 28
Attachment of *Salmonella* to Cantaloupes 30

Biofilm Formation on Produce Surfaces 31
Internalization 32
Limited Efficacy of Conventional Decontamination
    Methodologies 36
Conclusion 39

PART 2 Sources of Contamination

CHAPTER 3 Identification of the Source of Contamination 49
Jeff Farrar and Jack Guzewich
Introduction 50
Overview: Phases of a Foodborne Outbreak Investigation 50
    Surveillance and Detection 50
    Epidemiologic 51
    Environmental 53
    Traceback Investigations 61
    Regulatory/Enforcement 65
    Prevention/Research 66
Training Needs for Environmental Investigators of Retail,
    Food-Processing Facilities, Packing Sheds, and Farms 66
Resuming Operations 67
Farm Investigations 68
Packinghouse Investigations 71
Vacuum Cooler/Hydrocooler Investigation 72
Fresh Cut Produce Processor Investigations 72
Intentional Contamination 73
Lessons Learned 73
Recommendations 75

CHAPTER 4 Manure Management 79
Patricia D. Millner
Introduction 79
Manure Use on Crops 82
Survival of Pathogens in Manure 84
    Bacteria 84
    Protozoan and Helminthic Parasites 87
    Viruses 89
Pastures, Lots, and Runoff 90
Manure Treatment Technologies 92
Composting 93
Summary 97

CHAPTER 5	Water Quality 105	
	Charles P. Gerba and Christopher Y. Choi	
	Introduction 105	
	Irrigation Water 106	
	Water Quality Standards for Irrigation Water 108	
	Occurrence of Pathogens in Irrigation Water 109	
	Contamination of Produce During Irrigation 111	
	Survival of Pathogens on Produce in the Field 113	
	Other Sources 113	
	Summary and Conclusions 114	
CHAPTER 6	Sapro-Zoonotic Risks Posed by Wild Birds in Agricultural Landscapes 119	
	Larry Clark	
	Introduction 120	
	Bird Species Commonly Associated with Agriculture ... 120	
	Pigeons 121	
	Gulls 123	
	Water Fowl 123	
	Passerines 123	
	Bacterial Diseases 124	
	Campylobacter 124	
	Chlamydia 125	
	Escherichia coli 125	
	Listeria 127	
	Salmonella 127	
	Fungal Diseases 128	
	Aspergillus 128	
	Cryptococcus 128	
	Histoplasma 129	
	Parasitic Diseases 130	
	Cryptosporidia 130	
	Microsporidia 130	
	Toxoplasma 130	
	Mitigation Options 131	
	Summary 132	
CHAPTER 7	Produce Contamination by Other Wildlife 143	
	Daniel H. Rice	
	Introduction 143	
	Viral Pathogens 146	
	Bacterial Pathogens 146	

Parasitic Pathogens 151
Protozoa 152
Helminths 153
Mitigating Wildlife—Crop Interactions 155
Summary 156

PART 3 Commodities Associated with Major Outbreaks and Recalls

CHAPTER 8 Leafy Vegetables 165
Karl R. Matthews
Introduction 165
Outbreaks Associated with Leafy Greens 166
Growing Conditions by Geographical Region: Link to
  Outbreaks 168
Harvesting Practices: Influence on Contamination 169
Processing Practices and Product Contamination 171
  Handling Prior to Processing 171
  Washing and Sanitizing 173
  Packaging 175
Interaction of Microbes with Leafy Greens 177
  Plant Leaf Characteristic 177
  Native Flora of Leafy Greens 178
  Microbe Characteristics 179
Influence of Cutting on Microbial Populations 181
Conclusions 181

CHAPTER 9 Melons 189
*Alejandro Castillo, Miguel A. Martínez-Téllez and
M. Ofelia Rodríguez-García*
Introduction 190
Prevalence of Human Pathogens in and on Melons 190
Outbreaks of Foodborne Disease Linked to Melons 191
  Characteristics of Outbreaks 191
  Contributing Factors 194
  Impact of Regulatory Actions 195
Potential Sources and Mechanisms of Contamination
  and Measures Recommended to Prevent Contamination 196
  Preharvest 197
  Postharvest 201
  Cutting Practices 203

        Structural Characteristics of Melons Promoting
           Microbial Survival and Growth 206
           Current Knowledge about Growth and Survival of
           Pathogens in Melons 206
           Cantaloupe Netting 207
           Biofilm Formation 208
           Microbial Infiltration and Internalization 209
        Use of Antimicrobial Treatments to Decontaminate Melons 209
           Treatments Tested on Fresh Melons 210
        Fresh-Cut Melons 212
           Treatment with Antimicrobial Agents 212
           Irradiation 213
        Conclusions 214

CHAPTER 10 Raw Tomatoes and *Salmonella* 223
        Jerry A. Bartz
        Introduction 223
        Commercial Tomato Production and Marketing 225
        Evidence that Tomatoes were The Source of Outbreaks of
           Salmonellosis 229
           Outbreak from South Carolina Tomatoes, 1990 229
           Second Outbreak Traced to South Carolina
           Tomatoes, 1993 230
           Multistate Outbreak of Salmonellosis, 1998 231
           Transplant Games Outbreak, 2002 232
           Outbreak of *S.* Newport Linked to Virginia Tomatoes, 2002 ... 232
           Multiserotype Convenience Store Outbreak, 2004 233
           Second Outbreak of Salmonellosis Caused by
           Serotype Newport Traced Back to Tomatoes Produced
           on the Delmarva Peninsula, 2005 234
           Three-State Outbreak, 2005 235
           Recurrence of Serotype Newport in Delmarva
           Tomatoes, 2006 235
           Outbreak Linked to Ohio Tomatoes, 2006 236
        Unanswered Questions 236
        Recommendations for Commercial Tomato
           Production and Handling, Farm-To-Fork 243

CHAPTER 11 Tree Fruits and Nuts: Outbreaks, Contamination
        Sources, Prevention, and Remediation 249
        Susanne E. Keller
        Introduction 249

Organisms of Concern 250
Outbreaks Associated with Tree Fruits 252
Outbreaks Associated with Tree Nuts 254
Routes of Contamination 255
Prevention 258
Remediation 259
Conclusions 260

CHAPTER 12 Berry Contamination: Outbreaks and
Contamination Issues 271
Kalmia E. Kniel and Adrienne E.H. Shearer
Introduction 271
The Impact of Major Outbreaks 273
History of Viral Contamination of Berries 274
Hepatitis A Outbreaks with Raspberries and Strawberries 275
Norovirus Associated Outbreaks with Raspberries 277
The Role of *Cyclospora cayetanensis* in Berry-Associated
   Outbreaks 279
Transmission of *Cyclospora* Oocysts and the Role of Foods 280
Bacterial Contamination of Berries 283
Contamination Reduction Strategies 285
In Summary 297

PART 4 Avoidance of Contamination

CHAPTER 13 Produce Contamination Issues in México and Central
America 309
Jorge H. Siller-Cepeda, Cristobal Chaidez-Quiroz and
Nohelia Castro-del Campo
Introduction 309
Sources of Contamination 312
   Irrigation Water 312
   Runoff 315
   Inadequate Disinfection Processes at Packinghouses 316
   Conditions for Agricultural Workers 318
Good Agricultural Practices 321
Outbreak-Related Cases in México and Central America 323
Conclusions 326

CHAPTER 14 Regulatory Issues in Europe Regarding Fresh Fruit and
Vegetable Safety 331
Gro S. Johannessen and Kofitsyo S. Cudjoe
Introduction 332

The European Union 333
    Basic Facts 333
    The European Free Trade Association (EFTA) and
    The European Economic Area (EEA) 334
    European Fruit and Vegetable Production 334
Fresh Produce Contamination Problems in Europe 335
    Foodborne Bacteria 336
    Parasites 337
    Foodborne Human Pathogenic Virus 337
    Molds and Mycotoxins 338
European Regulations 339
    EU Central Regulations 339
    European Food Safety Authority (EFSA) 340
    Rapid Alert System for Food and Feed 340
    Hygiene and Control Rules 342
    GAP and Quality Assurance in Fruit and
    Vegetable Production 344
    Import from Countries Outside EU and EEA
    (third countries) 345
GlobalGAP (formerly EurepGAP) 346
Differences from US Regulations 346
Funding of Food-Safety Research in Europe 348
Sources for Further Information 349
Acknowledgements 349

CHAPTER 15 Regulatory Issues in Japan Regarding Produce Safety 353
Kenji Isshiki, Md. Latiful Bari, Shinichi Kawamoto and
Takeo Shiina
Introduction 354
Domestic Fresh Produce Production 357
Domestic Consumption of Fresh Produce 361
Fresh Produce-Related Outbreaks in Japan 361
Domestic Food Chain Approach from Farm to Table 364
    Good Agricultural Practices (GAP) in Japan 364
    Outline of Japan GAP 365
    Dealing with GAP in the Private Sector and Producer 366
Government Initiatives for Ensuring Produce
    Safety and Gaining Consumer Confidence 367
    Initiatives in Ensuring Produce Safety and
    Stable Supply 367
    Initiatives for Gaining Consumer Confidence 367

Imports and Distribution of Fresh Produce 368
    Increasing Agricultural and Food Imports 368
    Distribution Route of Fresh Produce 373
    Vegetable Imports and Compliance 374
Safety Regulations and Their Enforcement in Japan 376
    Scandals and a Major Change of Attitude 376
    New Food Safety Policy and Public Standards 377
    Main Role of the Food Safety Commission 378
    Applicable Laws and Regulations 380
    Plant Protection Law 380
    Food Sanitation Law 380
    Product Liability Law 381
    Enforcement of Regulations and Standards in Practice 381
    Private Sector View on Standards 384
    Company Strategies and Company-Specific Quality
    Standards 384
    Company-Specific Quality Standards as a
    Differentiation Strategy 385
    Traceability 385
Conclusions 386

PART 5 Technology for Reduction of Human Pathogens in Fresh Produce

CHAPTER 16 Disinfection of Contaminated Produce with Conventional Washing and Sanitizing Technology 393
Gerald M. Sapers
Introduction 393
Washing and Sanitizing Agents 395
    Detergent Products 395
    Chlorine 395
    Alternatives to Chlorine 397
    Other Approved Sanitizing Agents for Produce 401
    Sanitizing Agents for Organic Crops 403
    Expectations for Sanitizing Agents 403
Washing Equipment 403
    Types of Washers 403
    Efficacy of Commercial Washers 404
Produce Washes for Food Service and Home Use 405
Efficacy of Washing and Sanitizing Methods for Problem
    Commodities 407

        Leafy Vegetables 407
        Tomatoes.. 409
        Cantaloupe 410
        Apples.. 412
    Conclusions 414

CHAPTER 17 Advanced Technologies for Detection and Elimination of Pathogens 425
Brendan A. Niemira and Howard Q. Zhang
    Introduction 425
    Detection Methods 426
        Immunomagnetic Beads and Biosensors: Separation and Concentration 427
        PCR-Based Methods 428
        Computer/AI Optical Scanning... 429
    Antimicrobial Intervention Technologies 430
        Cold Plasma 431
        Irradiation.. 433
        Pulsed Light 434
        High-Pressure Processing.. 436
        Sonication 436
        Biological Controls 437
    The Challenge of Technology Development for Organic Foods... 438
    Acknowledgements 439

CHAPTER 18 Conclusions and Recommendations 445
Casey J. Jacob, Benjamin J. Chapman and Douglas A. Powell
    Introduction 445
    Sources of Contamination 446
    Commodities at Risk 447
    Challenges of Produce Disinfection... 449
    Investigating Contamination on the Farm 449
    Pre-emptive Food Safety Programs 450
    The Farm-to-Fork Approach.. 451

INDEX 453
CONTENTS OF RECENT VOLUMES 465

Preface

At its inception this book was intended as an examination of the problem of disinfecting fresh produce contaminated with human pathogens. In recent years numerous outbreaks of foodborne illness have been attributed to the presence of human pathogens in such widely consumed fresh commodities as salad greens, tomatoes, apples, cantaloupes, and fresh juices. This problem has been the subject of extensive research, published in hundreds of scientific papers and reports. One might think that conventional or innovative new technology for produce disinfection might be up to the task of reducing pathogen loads on produce to levels consistent with product safety.

However, though much has been learned about produce decontamination, the reality of the situation is that with few exceptions (ionizing radiation, high pressure), existing methods of cleaning and disinfecting fresh produce are incapable of achieving reductions in pathogen levels greater than 90 to 99%, which are insufficient to assure product safety. In part this is due to certain intrinsic aspects of microbial attachment to plant surfaces—the inaccessibility of some attachment sites, internalization of microorganisms within the plant tissue, and formation of resistant biofilms. Perhaps a wiser strategy would be to take a fresh look at the problem of *avoiding* human pathogen contamination of produce rather than focusing on disinfection technologies with limited efficacy.

Much has been written about potential sources of *Escherichia coli* O157:H7 contamination, which resulted in recent widespread outbreaks of illness associated with fresh romaine and spinach, or of *Salmonella* Saintpaul contamination, which caused the very large 2008 outbreak attributed first to tomatoes but later to imported peppers, but there is little hard evidence establishing specific sources of these pathogens. Did contamination occur as the result of runoff from nearby feedlots or other animal production activities? Was flooding of fields during winter storms a factor? Were feral animals involved? How much do we really know about the contamination process?

In this book, contributors who are experts in the areas of food safety and produce production, harvesting, packing and fresh-cut processing provide a critical, problem-oriented look at produce contamination and its avoidance.

The book is organized into five sections. In the first section an introductory chapter describes the scope of the problem. This is followed by a chapter that examines microbial attachment, survival of human pathogens attached to produce, and the limitations of conventional sanitizing treatments in assuring microbiological safety. The third chapter describes methods used to identify contamination sources through epidemiological methods and environmental investigations, including traceback, strain identification, and location of the specific source, if possible.

The second section includes three chapters that focus on major sources of contamination—water, manure, and wildlife—and examines where and how during crop production, harvesting, packing, or fresh-cut processing these sources might contaminate fresh produce.

In the third section, commodities associated with major outbreaks (leafy vegetables, melons, tomatoes, apples, and berries) are each examined to determine what intrinsic characteristics or production practices make them especially vulnerable to contamination.

Chapters in the fourth section provide an international perspective on produce contamination issues, focusing on outbreak trends, marketing and distribution practices, produce imports and exports, governmental agencies and regulations concerned with produce safety, avoidance of contamination through application of Good Agricultural and Manufacturing Practices and guidance documents, and regulatory actions such as recalls and restrictions on imports.

In the fifth section, technology for reduction of human pathogens in fresh produce is examined. Current technology for produce disinfection by washing and application of sanitizing agents is described. The prospects for technological advances in rapid detection and inactivation of microbial contaminants on produce are examined. The book ends with a chapter summarizing conclusions and recommendations for reduction in the risk of human pathogen contamination of fresh produce.

At this time, I wish to express my gratitude to my coeditors, Dr. Ethan B. Solomon and Professor Karl R. Matthews, for their many contributions to the development of this book, including identification of prospective chapter authors, monitoring the progress of authors in completing their chapters (and nagging them, when required), and providing in-depth reviews of the incoming chapter manuscripts.

I wish to thank the chapter authors for sharing their expertise and insights regarding produce contamination in well-written, comprehensive, and up-to-date examinations of their respective topics.

I am grateful to Carrie Bolger and Nancy Maragioglio at Elsevier for the enthusiastic support of this project and their great patience in dealing with our difficulties in meeting major deadlines.

Finally, I wish to thank my dear wife, Ellie, for her constant support and for tolerating my avoidance of household responsibilities, recreational activities, and quiet evenings together when superceded by my editing obligations. She was great!

Gerald M. Sapers, Ph.D.
Editor

Contributors

Md. Latiful Bari (Ch. 15)
National Food Research Institute, Food Hygiene Laboratory, Tsukuba, Japan

Jerry A. Bartz (Ch. 10)
University of Florida, Department of Plant Pathology, Gainesville, FL, USA

Alejandro Castillo (Ch. 9)
Texas A&M University, Department of Animal Science, College Station, TX, USA

Nohelia Castro-del Campo (Ch. 13)
CIAD-Culiacan, Culiacán, Sinaloa, C.P., Mexico

Christobal Chaidez-Quiroz (Ch. 13)
CIAD-Culiacan, Culiacán, Sinaloa, C.P., Mexico

Benjamin J. Chapman (Ch. 18)
North Carolina State University, Department of 4-H Youth Development and Family & Consumer Sciences, Raleigh, NC, USA

Christopher Choi (Ch. 3)
University of Arizona, Department of Agricultural and Biosystems Engineering, Tucson, AZ, USA

Larry Clark (Ch. 6)
USDA, National Wildlife Research Center, Fort Collins, CO, USA

Kofitsyo S. Cudjoe (Ch. 14)
National Veterinary Institute, Section for Food Bacteriology and GMO, Oslo, Norway

Michael Doyle (Ch. 1)
University of Georgia, Center for Food Safety, Griffin, GA, USA

Jeff Farrar (Ch. 3)
California Department of Public Health, Food & Drug Branch, Sacramento, CA, USA

Charles Gerba (Ch. 5)
University of Arizona, Department of Soil, Water, and Environmental Science, Tucson, AZ, USA

Jack Guzewich (Ch. 3)
USDA, Center for Food Safety and Applied Nutrition, College Park, MD, USA

Kenji Isshiki (Ch. 15)
Hokkaido University, Division of Marine Life Science, Hakodate, Hokkaido, Japan

Casey J. Jacob (Ch. 18)
Kansas State University, Department of Diagnostic Medicine/Pathobiology, Manhattan, KS, USA

Gro S. Johannessen (Ch. 14)
National Veterinary Institute, Section for Food Bacteriology and GMO, Oslo, Norway

Shinichi Kawamoto (Ch. 15)
National Food Research Institute, Food Hygiene Laboratory, Tsukuba, Japan

Susanne E. Keller (Ch. 11)
FDA/CFSAN, National Center for Food Safety and Technology, Summit-Argo, IL, USA

Kalmia E. Kniel (Ch. 12)
University of Delaware, Department of Animal and Food Sciences, Newark, DE, USA

Miguel A. Martínez-Téllez (Ch. 9)
CIAD, Dirección de Tecnología de Alimentos de Origen Vegetal, Sonora, Mexico

Karl R. Matthews (Ch. 8)
Rutgers-The State University of New Jersey, Department of Food Science, New Brunswick, NJ, USA

Patricia Millner (Ch. 4)
USDA, Beltsville Agricultural Research Center, Beltsville, MD, USA

Brendan A. Niemira (Ch. 17)
USDA, Agricultural Research Service, Eastern Regional Research Center, Wyndmoor, PA, USA

Douglas Powell (Ch. 18)
Kansas State University, Department of Diagnostic Medicine/Pathobiology, Manhattan, KS, USA

Daniel H. Rice (Ch. 7)
NY State Department of Agriculture and Markets, Albany, NY, USA

M. Ofelia Rodríguez-García (Ch. 9)
Universidad de Guadalajara, CUCEI, Departamento de Farmacobiología, Jalisco, Mexico

Gerald M. Sapers (Ch. 1, 16)
USDA, Eastern Regional Research Center Agricultural Research Service, Wyndmoor, PA, USA

Manan Sharma (Ch. 2)
USDA, Agricultural Research Service, Environmental Microbial and Food Safety Laboratory, Beltsville, MD, USA

Adrienne E.H. Shearer (Ch. 12)
University of Delaware, Department of Animal and Food Sciences, Newark, DE, USA

Takeo Shiina (Ch. 15)
National Food Research Institute, Distribution Engineering Laboratory, Tsukuba, Japan

Jorge H. Siller-Cepeda (Ch. 13)
Desert Glory Mexico S De RL de CV, Guadalajara, Jalisco, Mexico

Ethan B. Solomon (Ch. 2)
DuPont Chemical Solutions Enterprise, Experimental Station Laboratory, Wilmington, DE, USA

Howard Q. Zhang (Ch. 17)
USDA, Agricultural Research Service, Eastern Regional Research Center, Wyndmoor, PA, USA

PART 1

Introduction

CHAPTER 1

Scope of the Produce Contamination Problem

Gerald M. Sapers, Ph.D. (Emeritus)
Eastern Regional Research Center, Agricultural Research Service, US Department of Agriculture, Wyndmoor, PA

Michael P. Doyle, Ph.D.
Regents Professor and Director, Center for Food Safety, University of Georgia, Griffin, GA

CHAPTER CONTENTS

Introduction	3
Produce-Associated Outbreaks—a New Problem?	3
Consequences of Produce-Associated Outbreaks	4
Key Aspects of the Produce Contamination Problem	5
Characteristics of Produce-Associated Outbreaks	5
Prevalence of Produce Contamination with Human Pathogens	8
Microbial Attachment and Survival on Produce Surfaces	9
Potential Sources of Produce Contamination	9
Preharvest Sources	9
Contamination During Packing	10
Contamination During Fresh-Cut Processing	11
Gaps in Our Understanding of Produce Contamination	12
Current State of Knowledge	12
What We Don't Know	12
Developing Effective Interventions	13

INTRODUCTION

Produce-Associated Outbreaks—a New Problem?

For decades, concerns regarding the microbiological safety of foods have focused largely on animal products that were responsible for outbreaks of *E. coli* O157:H7 from ground beef; salmonellosis from poultry, eggs, and dairy products; and listeriosis from soft cheeses and processed meats. Outbreaks of botulism were associated with canned vegetables, but fresh fruits and vegetables generally were considered to be safe, except in countries where the combination of endemic gastrointestinal diseases, unsafe agricultural practices, and poor sanitation resulted in traveler's diarrhea and other illnesses acquired by consumption of locally grown fresh produce. US produce packers and the fresh-cut industry have long believed that their products were made safe by the use of a triple-wash technology using chlorinated water or other approved sanitizing agents.

In recent years, however, this picture has changed dramatically due to an increase in the number of outbreaks of foodborne illnesses associated with fresh and fresh-cut fruits and vegetables. Many large outbreaks involving widely consumed commodities such as apple cider, cantaloupe, raspberries, bagged lettuce and spinach, tomatoes, green onions, and sprouts have been reported during the past decade (Brackett, 1999; Beuchat, 2002). This increase may be due in part to greater consumption of fresh produce in response to the recommendations of health and nutrition professionals. Increased consumption has translated into increased production and distribution of fresh produce, but the growth of produce packing and fresh-cut processing facilities with regional or national distribution capabilities has exposed more consumers to products that may have been contaminated on a single processing line or at a single farm. Additionally, to meet increased demand for out-of-season items, sourcing of fresh produce became a global endeavor, including growing locations where the potential of human pathogen contamination of fruits and vegetables is high. Furthermore, with better methods for identifying and tracking foodborne outbreaks, the local and state health departments and CDC have become in the past decade much better at detecting produce-associated outbreaks, many of which previously would not have been recognized, or the source not identified.

Consequences of Produce-Associated Outbreaks

Pathogen contamination of fresh produce has important public health consequences. Not only are there more cases of illness from produce-associated outbreaks, highly vulnerable population groups—the very young, the old, and the immunocompromised—are often affected. For these individuals, the severity of foodborne illnesses can be much greater, if not life-threatening, and there may be serious long-term consequences to health. An indirect health-related consequence is the reduced intake of beneficial nutrients from fruits and vegetables by individuals concerned about acquiring a foodborne illness.

The economic consequences of produce-associated outbreaks are substantial, including the medical costs and lost income of patients, the costs of damage control (disposal of unmarketable products, cost of product recalls, cleanups, and retrofitting) for the affected produce packer/processor, and lost production time. In addition, there are the costs associated with litigation, awards from successful lawsuits, and long-term damage to the company's reputation, reflected by reduced sales of fresh produce items. A history of outbreaks can be damaging to an entire segment of the produce industry (e.g., spinach, green onions, and tomatoes) or to a production area (e.g., the Salinas Valley of California), resulting in increased costs for government-mandated changes in production and processing practices and in reduced sales of products nationwide. The estimated cost to tomato growers from the 2008 multistate *Salmonella* Saintpaul outbreak (over 1400 cases reported) was approximately $200 million (Anon., 2008). This outbreak was originally attributed to contaminated tomatoes, but subsequent investigation implicated jalapeño peppers as the major vehicle, serrano peppers also as a vehicle, and tomatoes possibly as a vehicle (CDC, 2008a). The overall economic cost to the industry could be a generalized reduction in sales and consumption of fresh produce resulting from reduced confidence in their safety.

KEY ASPECTS OF THE PRODUCE CONTAMINATION PROBLEM

Characteristics of Produce-Associated Outbreaks

Data compiled by the Centers for Disease Control and Prevention (CDC) provides insight into trends in the prevalence, size, and causes of produce-related outbreaks. Between 1993 and 1997, the prevalence of outbreaks associated with fresh fruits and vegetables, as reported by the CDC in summary tables for each year (CDC, 2000), was erratic with no upward trend (Table 1.1). However, there was an abrupt increase in the prevalence of produce-associated outbreaks between 1998 and 2002, perhaps in part because of a change in surveillance and/or reporting methodology (CDC, 2006).

The number of outbreaks associated with specific human pathogens during 2003–2006 is shown in Table 1.2. *E. coli* O157:H7, *Salmonella*, and norovirus were responsible for most outbreaks; however, the number of outbreaks and cases for each agent varied from year to year, and each year, large single outbreaks were associated with other pathogens (hepatitis A in 2003, *Cryptosporidium* in 2004, and *Cyclospora* in 2005). Interestingly, no produce-associated outbreaks were attributed to *Listeria monocytogenes* during this period or in 2000–2002 (CDC, 2008b).

CHAPTER 1: Scope of the Produce Contamination Problem

Table 1.1 Number of Reported Foodborne-Disease Outbreaks and Cases Associated with Fruits and Vegetables, US[1]

Year	Outbreaks	Cases
1993	12	4213
1994	17	1311
1995	9	4307
1996	13	1807
1997	15	719
1998	44	1885
1999	62	1902
2000	62	2399
2001	58	2582
2002	53	1765

[1]Data for 1997–2002 from summary tables reported by the CDC (2000, 2006).

Table 1.2 Human Pathogens Involved in Reported Outbreaks Associated with Fruits and Vegetables[1]

Year	Pathogen	Number of Reported Outbreaks	Cases
2003	E. coli O157:H7	4	87
	Salmonella	10	719
	Shigella	2	62
	Cryptosporidium	1	144
	Hepatitis A	1	935
	Norovirus	4	110
2004	Campylobacter	2	22
	E. coli O157:H7	3	308
	Salmonella	4	164
	Cryptosporidium	1	212
	Norovirus	19	893
2005	E. coli O157:H7	4	70
	Salmonella	7	367
	Cyclospora	1	592
	Hepatitis A	1	40
	Norovirus	6	431
2006	E. coli O157:H7	7	380
	Salmonella	9	240
	Staphylococcus aureus	1	35
	Cyclospora	1	14
	Norovirus	25	770

[1]Data for 2003–2006 represent outbreaks reported by the CDC (2008) that were associated with specified single items of produce or produce combinations, in contrast to items described as "salad" or "salad bar," which were excluded from the tabulation because of the possible presence of nonproduce components (e.g., chicken) that were the actual source of the outbreak.

CDC data reported for 1998–2002 reveal that the incidence of outbreaks is greater for vegetables than for fruits (CDC, 2006). An in-depth examination of outbreak data for 2003–2006 (Table 1.3) reveals that the principal problem commodities were green salads and lettuce, other leafy vegetables or herbs, sprouts, tomatoes, melons, and fruit salad. Many of these commodities are vulnerable to contamination because they grow on or close to soil where contamination can occur. The number of cases and their distribution among commodities varies from year to year. In recent years, major produce-related outbreaks have been caused by *Salmonella* contamination of tomatoes (FDA, 2004, 2006a) and orange juice (FDA, 2005a), *E. coli* O157:H7 contamination of fresh-cut lettuce (FDA, 2006b, 2007) and

Table 1.3 Items of Fresh Produce Most Frequently Implicated in Outbreaks of Foodborne Disease[1,2]

Year	Produce Item	Outbreaks	Cases
2003	Green salads and lettuce	8	218
	Sprouts	4	62
	Melons	3	182
	Other vegetables (scallions, tomatoes, spinach)	3	962
	Other fruits (strawberries, mango)	2	30
	Juices (apple cider)	1	144
2004	Green salads and lettuce	17	630
	Tomatoes	4	671
	Other fruits (fruit salad)	3	184
	Melons	2	134
	Sprouts	2	37
	Other vegetables (cucumber salad, mixed vegetables)	2	329
	Juices (apple cider)	1	212
2005	Green salads and lettuce	5	137
	Tomatoes	4	176
	Fruit salad	4	339
	Other vegetables (parsley, onion, carrots, basil)	4	633
	Other fruits (watermelon, strawberries)	2	58
	Juices (orange juice)	1	157
2006	Green salads and lettuce	25	828
	Tomatoes	4	157
	Fruit salad	6	177
	Other vegetables (spinach, bean sprouts, cucumber salad)	4	238
	Other fruits (watermelon, berries)	4	54

[1]Data from the CDC (2008b).
[2]Entries represent outbreaks associated with individual produce items or combinations where each component is a specified item of produce. Items designated by the CDC as "salad" or "salad bar," without additional designations (e.g., lettuce-based) to exclude the presence of a major nonproduce component such as chicken or pasta, are not included in this tabulation.

bagged spinach (FDA, 2006c), *Cyclospora* contamination of basil (FDA, 2005b), and hepatitis A contamination of green scallions from Mexico (FDA, 2003b). Several of the outbreaks associated with leafy greens were traced to farms in the Central Valley and Salinas Valley regions of California.

Prevalence of Produce Contamination with Human Pathogens

The sporadic nature of produce-related outbreaks is suggestive of localized contamination events, which makes systematic study of contamination sources difficult. One approach to assessing the magnitude of the problem is to obtain data on the prevalence of produce contamination for different commodities and growing locations. Both the FDA and USDA have conducted large-scale studies of selected commodities to determine the prevalence of contamination. The FDA's testing of imported produce (FDA, 2001b) revealed a relatively high prevalence of *Salmonella* and *Shigella* contamination on culantro (50%), cilantro (9%), cantaloupe (7.3%), celery (3.6), parsley (2.4%), lettuce (1.7%), and scallions (1.7%), all of which are grown on or close to soil. Testing of domestic produce (FDA, 2003a) revealed a lower prevalence of contamination (total 1.1%) than was found with imported produce (total 4.4%). Domestically grown scallions (3.2%) and cantaloupe (3.1%) had the highest prevalence of contamination, whereas the contamination of cilantro, parsley, and lettuce was each about 1%.

A USDA survey of selected produce commodities sampled at wholesale and distribution centers (USDA, 2004) revealed a much lower prevalence of contamination. *Salmonella* spp. were detected only on lettuce (0.14%), and *E. coli* with a virulence factor was detected on Romaine lettuce (1.34%), leaf lettuce (1.25%), and on cantaloupe, celery, and tomatoes at prevalences less than 0.2%.

Other studies of fresh and fresh-cut produce, grown either organically or conventionally, revealed a very low or no prevalence of human pathogen contamination (Riordan et al., 2001; Sagoo et al., 2001; Anon., 2002; Phillips and Harrison, 2005; Johnston et al., 2006; Mukherjee, 2006; Dallaire et al., 2006; Danyluk et al., 2007; Bobe et al., 2007). However, Heisick et al. (1989) reported a high prevalence of *L. monocytogenes* contamination (26–30%) on potatoes and radishes at retail. Castillo et al. (2006) reported high prevalences of *Salmonella* (14–20%) and *Shigella* (6–17%) in freshly squeezed orange juice and on fresh oranges collected at public street markets and street booths in Guadalajara, Mexico.

These results suggest that contamination of most fresh produce by enteric pathogens is too low for data from broadly focused surveys to provide helpful guidance in identifying primary sources of contamination. This represents an important gap in our understanding of produce contamination.

Microbial Attachment and Survival on Produce Surfaces

When human pathogens come in contact with produce in the crop production environment, they can rapidly attach and strongly adhere (Sapers et al., 1999; Liao and Sapers, 2000; Ukuku and Fett, 2006). Some pathogens can also form biofilms on plant surfaces (Carmichael et al., 1999; Annous et al., 2005). These topics have been reviewed (Carmichael et al., 1999; Mandrell et al. 2006) and are further discussed in Chapter 2.

The extent to which attached human pathogens survive and proliferate on produce surfaces, both preharvest and postharvest, is dependent on the type of pathogen and characteristics of the produce (Carlin and Nguyen-The, 1994; Ukuku and Fett, 2006), temperature (Zhuang et al., 1995; Duffy, 2005a), humidity (Stine et al., 2005b; Fonseca, 2006; Iturriaga et al., 2007), the degree of protection from environmental stresses provided by the microbial attachment site (Zhuang et al., 1995; Chancellor et al., 2006), nutrient availability (Carmichael et al., 1999), and interactions with epiphytic microbes (Francis and O'Beirne, 1998; Aruscavage et al, 2006; Cooley et al., 2006) and plant pathogens (Wells and Butterfield, 1997). Pathogen survival is greater in porous or broken tissue than on smooth tissue (Wei et al., 1995), and growth can occur in wounds (Wei et al., 1995; Beuchat and Scouten, 2004). Pathogens also can become internalized within plant tissues via attachment and infiltration at pores and cut edges (Bartz and Showalter, 1981; Bartz, 1982; Seo and Frank, 1999; Solomon et al., 2002a, 2000b).

Studies with tomatoes and cantaloupe inoculated with human pathogens or surrogates have revealed that as the time interval between inoculation and washing with sanitizing agents increases from one hour to several days, the efficacy of the sanitizer treatment in reducing pathogen populations decreases significantly (Ukuku and Sapers, 2001; Ukuku et al., 2001; Sapers and Jones, 2006). Microbial internalization, and/or biofilm formation occurring between the time of contamination and washing may be contributing factors. If pathogen contamination of produce occurs preharvest or during harvest, sufficient time may elapse before washing in the packing or processing facility to enable development of these protective factors that reduce the efficacy of sanitizer treatments.

POTENTIAL SOURCES OF PRODUCE CONTAMINATION

Preharvest Sources

Foodborne outbreak investigations have helped shed some light on the identification of sources of produce contamination. Such investigations can be characterized into five phases: surveillance/detection, epidemiologic, environmental/traceback, regulatory/enforcement, and prevention/research. In theory, outbreak investigators should be able to genetically match laboratory-confirmed pathogens from ill individuals with laboratory-confirmed pathogens from epidemiologically implicated foods, and identify where and how the contamination occurred. However, this process is often complicated by delays and variability in diagnostic testing procedures, reporting of results, and in conducting epidemiologic investigations. Additionally, many perishable foods with a short shelf-life, such as fresh produce, may no longer be available for laboratory testing. Traceback can be complicated by poor record-keeping and commingling of products from different growers throughout the food chain from production to consumption. Hence, identifying the specific source of an outbreak at the farm or field level is often not possible. A more detailed presentation of the difficulties encountered in foodborne outbreak investigations is provided in Chapter 3.

Field studies conducted in crop production locations, packinghouses, and processing facilities, and studies with model systems have revealed some potential sources of produce contamination. These are described in greater detail in Chapters 4 through 7. The initial source of human enteric pathogens is usually the feces of domesticated animals, wildlife, or humans. Field studies have demonstrated potential sources of human pathogens in farm environments associated with animal production (Rodriguez et al., 2006; McAllister et al., 2006; Doane et al., 2007), fecal contamination from wildlife (Rice et al., 1995; Wallace et al., 1997; Kullus et al., 2002; Hamilton et al., 2006a; Yan et al., 2007), composted manure (Islam et al., 2004; Ingham et al., 2005), soil (Gagliardi et al., 2003; Johannessen et al., 2005), runoff (Muirhead et al., 2006), irrigation water (Steele et al., 2005; Stine et al., 2005a; Duffy et al., 2005b; Hamilton et al., 2006b; Espinoza-Medina et al., 2006), and the hands of packing workers (Espinoza-Medina et al., 2006). Whether such contamination sources represent an actual food safety hazard will depend on the extent to which human pathogens in the farm or packinghouse environment contact produce surfaces, attach, survive environmental stresses and exposure to sanitizing agents, and, if necessary, multiply to a population level sufficient to cause illness. These are questions that need to be answered by research to enable identification of the most effective intervention(s), be it at the farm, packinghouse, fresh-cut processing line, or elsewhere in the produce continuum.

Contamination During Packing

Studies by Duffy et al. (2005) indicate that packing equipment may be a source of human pathogen contamination of fresh produce. Gagliardi et al. (2003) implicated process water used for cooling and washing melons as a source of contamination. Garcia et al. (2006) attributed *E. coli* contamination of apples used for cider production to microbial buildup in dump tanks, in which the sanitizer/wash solution was not adequately replenished, and to inadequate cleaning and sanitizing of scrubbers, spray nozzles, and conveyors. Keller et al. (2002) also determined that bacteria proliferate in an apple cider mill when equipment is inadequately sanitized, thereby resulting in cider contamination.

Contamination During Fresh-Cut Processing

It is well established that conventional cleaning and sanitizing treatments applied to fresh produce generally reduce pathogen populations by only 90 to 99% (1- to 2-log reduction), likely due to the survival of microorganisms attached in protected sites or in biofilms on produce surfaces or to neutralization of the sanitizer by the organic load of the process water (see Chapters 2 and 16). This is true both for uncut and fresh-cut commodities. However, exposure of cut produce, especially leafy vegetables, to contaminated wash water increases the risk of bacterial attachment at cut surfaces, subsequent internalization (Seo and Frank, 1999; Solomon et al., 2002b), and proliferation of the human pathogens during product handling, storage, and distribution. Similarly, contamination of fresh-cut cantaloupe with human pathogens, by transfer from the rind surface to the flesh during cutting (Ukuku and Sapers, 2001), could result in extensive growth during storage and distribution of the fresh-cut product under conditions of temperature abuse.

Detection of *L. monocytogenes* in fresh-cut apples, which resulted in a product recall in 2001 (FDA, 2001a), provides evidence of a contamination risk associated with the use of browning inhibitors and other processing aids in fresh-cut processing. If not frequently refreshed, such solutions can contain a build-up of substantial amounts of nutrients, leached from cut produce, thereby making these processing aids suitable media for the proliferation of environmental contaminants such as *L. monocytogenes*. Additionally, this pathogen can grow, albeit slowly, at the low temperatures of fresh-cut processing rooms and under conditions of temperature abuse during product distribution and retailing.

Many studies have validated that human pathogens can survive and grow on fresh-cut produce (Steinbruegge et al., 1988; Escartin et al., 1989; Carlin et al., 1995). Human pathogen survival and growth on fresh-cut produce is affected by many of the factors discussed earlier, especially temperature (Kallander et al., 1991; Piagentini et al., 1997; Farber et al., 1998), interaction with the indigenous microflora (Carlin et al., 1996; Francis and O'Beirne, 1998), nutrient availability, and use of controlled or modified atmospheres for storage or packaging (Berrang et al., 1989; Abdul-Raouf et al., 1993; Omary et al., 1993; Kakiomenou et al., 1998). Improvements in plant sanitation and maintenance of the cold chain from the packing or processing plant through distribution and retailing to the consumer are important prerequisites to reducing pathogen contamination of produce.

GAPS IN OUR UNDERSTANDING OF PRODUCE CONTAMINATION

Current State of Knowledge

With produce-related outbreaks frequently in the news, and the public health and economic costs so high, why does this problem continue in spite of the large research effort carried out by government, academia, and the private sector to improve food safety? Means of detecting and tracking human pathogens in the food supply continue to improve. Much is known about the foodborne pathogens responsible for produce-related outbreaks, their ability to attach to fresh fruits and vegetables, and the efficacy of various conventional and new disinfection technologies. However, many challenges remain; for example, the low infectious dose of *E. coli* O157:H7, the limited efficacy of many approved sanitizers on produce surfaces, the resistance to cleaning and disinfection of bacteria in biofilms, and the limitations that outbreak investigators have in tracking a contamination event to a specific location and source, as discussed in Chapter 3.

What We Don't Know

We know how to identify and subtype the strain of the causative agent of an outbreak and link human isolates to food isolates, but we cannot, with frequency, readily pin down the actual source or contamination event. Perhaps this is because the source is a flock of birds or a meandering feral pig, both random and unpredictable events. Perhaps the event is a dust storm conveying desiccated manure from a distant feedlot to a produce farm, again a random occurrence, but this should be more predictable and a risk to avoid. In order to more comprehensibly address the problem of produce-associated outbreaks of foodborne illness, we need a better understanding of the contamination process, including transmission of pathogens in aerosols and water, survival of pathogens in manure and soil, mode of contact between human pathogens and produce surfaces, adhesion or entrapment characteristics at the attachment site, opportunities for internalization of the pathogen, opportunities for biofilm formation, and the role of environmental conditions such as temperature, humidity, rainfall, and wind velocity. These factors are discussed in Chapter 2.

Also to be considered is information regarding agricultural practices, hygienic behavior of farm workers, the risks associated with field packing and hydrocooling operations, proximity of potential contamination sources (exposed irrigation canals; nearby areas of animal production and flyways; presence of human pathogens in soil; and scat of local populations of deer, rodents, amphibians, and other wildlife). These contributing factors are discussed in Chapters 4 through 7. Special attention is needed for those commodities that are most frequently associated with large outbreaks—what makes them more vulnerable to contamination with human pathogens? These problem commodities are addressed in Chapters 8 through 12.

Developing Effective Interventions

Based on the foundation of an improved understanding of the major routes of produce contamination, and of the ability of pathogens to survive and grow on produce, more effective interventions must be developed to reduce the potential for produce contamination. These would be incorporated into guidance documents and HACCP (Hazard Analysis of Critical Control Points) plans and updated good agricultural and manufacturing practices. Also needed are more effective regulatory actions, not only in processing facilities but also at the farm, to reduce the risk of contamination and to exclude contaminated produce from the marketplace. Since many types of fresh produce are sourced internationally, regulation of produce safety should be addressed in global terms. These topics are discussed in Chapters 13 through 15.

Finally, we must consider the promise and limitations of technology in providing means of rapid detection of human pathogens in fresh produce, identification of contamination sources, and disinfection of contaminated produce to reduce the risk of foodborne illness. There is recognition of the limits of conventional produce disinfection technology, and therefore, more efficacious antimicrobial treatments need to be developed (see Chapter 16). The prospects for advanced technological solutions are addressed in Chapter 17.

The primary purpose of this book is to address what is known about contamination of fresh produce by human pathogens, and to present those interventions that may be applied to reduce the risk of contamination. Using this information, specific gaps in our understanding of these topics can be identified and used to set an agenda for prioritized research that will provide safer produce. Chapter 18 summarizes the state of our knowledge, provides recommendations for development of more effective interventions, and examines policy issues that can influence improvements in the microbiological safety of fresh produce.

In summary, there are deficiencies in the current state of knowledge of human pathogen contamination of fresh produce; the survival and proliferation of microbial contaminants during packing, processing, storage, distribution, and marketing of produce; and the efficacy of conventional interventions; all of which contribute to the problem of produce-associated outbreaks. Improvements in our understanding of sources of produce contamination coupled with implementation of more efficacious food safety interventions are needed to achieve greater success in reducing the occurrence of such outbreaks.

REFERENCES

Abdul-Raouf, U. M., Beuchat, L. R., and Ammar, M. S. (1993). Survival and growth of *Escherichia coli* O157:H7 on salad vegetables. *Appl. Environ. Microbiol.* **59**, 1999–2006.

Annous, B. A., Solomon, E. B., and Cooke, P. H. (2005). Biofilm formation by *Salmonella* spp. on cantaloupe melons. *J. Food Safety* **25**, 276–287.

Anon. (2002). Results of 4th Quarter National Survey 2002. European Commission Coordinated Programme for the Official Control of Foodstuffs for 2002. Bacteriological safety of pre-cut fruit & vegetables, sprouted seeds and unpasteurized fruit & vegetable juices from processing and retail premises. www.fsai.ie/surveillance/food_safety/microbiological/4thQuarter2.pdf. Accessed 6/11/07.

Anon. (2008). FDA criticized over its response to *Salmonella* outbreak. *Baltimore Sun*, August 1. www.baltimoresun.com/news/health/bal-te.fda01aug01,0,3276708.story. Accessed 10/15/08.

Aruscavage, D., Lee, K., Miller, S. et al. (2006). Interactions affecting the proliferation and control of human pathogens on edible plants. *J. Food Sci.* **71**, R89–R99.

Bartz, J. A. (1982). Infiltration of tomatoes immersed at different temperatures to different depths in suspensions of *Erwinia carotovora* subsp. *carotovora*. *Plant Dis.* **66**, 302–306.

Bartz, J. A. and Showalter, R. K. (1981). Infiltration of tomatoes by aqueous bacterial suspensions. *Phytopathology* **71**, 515–518.

Beuchat, L. R. (2002). Ecological factors influencing survival and growth of human pathogens on raw fruits and vegetables. *Microb. Infect.* **4**, 413–423.

Beuchat, L. R. and Scouten, A. J. (2004). Factors affecting survival, growth, and retrieval of *Salmonella* Poona on intact and wounded cantaloupe rind and stem scar tissue. *Food Microbiol.* **21**, 683–694.

Berrang, M. E., Brackett, R. E., and Beuchat, L. R. (1989). Growth of *Listeria monocytogenes* on fresh vegetables stored under controlled atmosphere. *J. Food Prot.* **52**, 702–705.

Bobe, G., Thede, D. J., Ten Eyck, T. A. et al. (2007). Microbial levels in Michigan apple cider and their association with manufacturing practices. *J. Food Prot.* **70**, 1187–1193.

Brackett, R. E. (1999). Incidence, contributing factors, and control of bacterial pathogens in produce. *Postharvest Biol. Technol.* **15**, 305–311.

Carlin, F. and Nguyen-The, C. (1994). Fate of *Listeria monocytogenes* on four types of minimally processed green salads. *Lett. Appl. Microbiol.* **18**, 222–226.

Carlin, F., Nguyen-The, C., and da Silva, A. A. (1995). Factors affecting the growth of *Listeria monocytogenes* on minimally processed fresh endive. *J. Appl. Bacteriol.* **78**, 636–646.

Carlin, F., Nguyen-The, C., and Morris, C. E. (1996). Influence of background microflora on *Listeria monocytogenes* on minimally processed fresh broadleaved endive (*Cichorium endivia* var. *latifolia*). *J Food Prot.* **59**, 698–703.

Carmichael, I., Harper, I. S., Coventry, M. J. et al. (1999). Bacterial colonization and biofilm development on minimally processed vegetables. *J. Applied Microbiol. Symposium Suppl.* **85**, 45S–51S.

Castillo, A., Villarruel-López, A., Navarro-Hidalgo, V. et al. (2006). *Salmonella* and *Shigella* in freshly squeezed orange juice, fresh oranges and wiping cloths collected from public markets and street booths in Guadalajara, Mexico: Incidence and comparison of analytical routes. *J. Food Prot.* **69**, 2595–2599.

CDC. (2000). Surveillance for foodborne disease outbreak—United States, 1993–1997. *MMWR* Mar. 17, 2000/**49**(SS01), 1–51.

CDC. (2006). Surveillance for foodborne disease outbreak—United States, 1998–2002. *MMWR* Nov. 10, 2006/**55**(SS10), 1–34.

CDC. (2008a). Outbreak of *Salmonella* serotype Saintpaul infections associated with multiple raw produce items—United States, 2008. *MMWR* August 29, 2008/**57**(34), 929–934.

CDC. (2008b). Outbreak surveillance data. Annual listing of foodborne disease outbreaks, United States, 1990–2006. www.cdc.gov/foodborneoutbreaks/outbreak_data.htm. Accessed 10/15/08.

Chancellor, D. D., Tyagi, S., Bazaco, M. C. et al. (2006). Green onions: Potential mechanism for hepatitis A contamination. *J. Food Prot.* **69**, 1468–1472.

Cooley, M. B., Chao, D., and Mandrell, R. E. (2006). *Escherichia coli* O157:H7 survival and growth on lettuce is altered by the presence of epiphytic bacteria. *J. Food Prot.* **69**, 2329–2335.

Dallaire, R., LeBlanc, D. I., Tranchant, C. C. et al. (2006). Monitoring the microbial populations and temperatures of fresh broccoli from harvest to retail display. *J. Food Prot.* **69**, 1118–1125.

Danyluk, M. D., Jones, T. M., Abd, S. J. et al. (2007). Prevalence and amounts of *Salmonella* found on raw California almonds. *J. Food Prot.* **70**, 820–827.

Doane, C. A., Pangloili, P., and Richards, H. A. (2007). Occurrence of *Escherichia coli* O157:H7 in diverse farm environments. *J. Food Prot.* **70**, 6–10.

Duffy, E. A., Cisneros-Zevallos, L., Castillo, A. et al. (2005a). Survival of *Salmonella* transformed to express green fluorescent protein on Italian parsley as affected by processing and storage. *J. Food Prot.* **68**, 687–695.

Duffy, E. A., Lucia, L. M., Kells, J. M. et al. (2005b). Concentrations of *Escherichia coli* and genetic diversity and antibiotic resistance profiling of *Salmonella* isolated from irrigation water, packing shed equipment, and fresh produce in Texas. *J. Food Prot.* **68,** 70–79.

Espinoza-Medina, I. E., Rodríguez-Leyva, F. J., Vargas-Arispuro, I. et al. (2006). PCR identification of *Salmonella*: Potential contamination sources from production and postharvest handling of cantaloupes. *J. Food Prot.* **69,** 1422–1425.

Farber, J. M., Wang, S. L., Cai, Y. et al. (1998). Changes in populations of *Listeria monocytogenes* inoculated on packaged fresh-cut vegetables. *J. Food Prot.* **61,** 192–195.

FDA. (2001a). Enforcement report. Recalls and field corrections: foods—Class I. Recall number F-535-1. 20 August 2001. Sliced apples in poly bags. www.fda.gov/bbs/topics/ENFORCE/2001/ENF00708.html

FDA. (2001b). FDA Survey of Imported Fresh Produce. US Food and Drug Administration Center for Food Safety and Applied Nutrition, Office of Plant and Dairy Foods and Beverages. January 30, 2001. www.cfsan.fda.gov/~dms/prodsur6.html

FDA. (2003a). FDA Survey of Domestic Fresh Produce. US Department of Health and Human Services, US Food and Drug Administration Center for Food Safety and Applied Nutrition Office of Plant and Dairy Foods and Beverages. January 2003. www.cfsan.fda.gov/~dms/prodsu10.html

FDA. (2003b). FDA Update on recent hepatitis A outbreaks associated with green onions from Mexico. *FDA Statement*. Food and Drug Administration, US Department of Health and Human Services. December 9, 2003. www.fda.gov/bbs/topics/NEWS/2003/NEW00993.html

FDA. (2004). FDA Investigates certain Roma tomatoes as source of foodborne illness outbreaks in Pennsylvania, Ohio, and Mid-Atlantic states. *FDA Statement*. Food and Drug Administration, US Department of Health and Human Services. July 23, 2004. www.fda.gov/bbs/topics/news/2004/NEW01096.html

FDA. (2005a). FDA issues nationwide health alert on Orchid Island unpasteurized orange juice products. *FDA Statement*. Food and Drug Administration, US Department of Health and Human Services. July 8, 2005. www.fda.gov/bbs/topics/news/2005/NEW01203.html

FDA. (2005b). FDA works to trace source of foodborne illness in Florida. *FDA News*. Food and Drug Administration, US Department of Health and Human Services. June 3, 2005. www.fda.gov/bbs/topics/NEWS/2005/NEW01183.html

FDA. (2006a). FDA notifies consumers that tomatoes in restaurants linked to *Salmonella* Typhimurium outbreak. *FDA News*. Food and Drug Administration, US Department of Health and Human Services. November 3, 2006. www.fda.gov/bbs/topics/NEWS/2006/NEW01504.html

FDA. (2006b). Update: FDA narrows investigation of *E. Coli* O157:H7 outbreak at Taco Bell restaurants. *FDA News*. Food and Drug Administration, US Department of Health and Human Services. December 13, 2006. www.fda.gov/bbs/topics/NEWS/2006/NEW01525.html

FDA. (2006c). Nationwide *E. Coli* O157:H7 outbreak: Questions & answers. US Department of Health and Human Services, US Food and Drug Administration Center for Food Safety and Applied Nutrition. September 16, 2006; updated October 20, 2006. www.cfsan.fda.gov/~dms/spinacqa.html

FDA. (2007). FDA and States closer to identifying source of *E. coli* contamination associated with illnesses at Taco John's restaurants. *FDA News*. Food and Drug Administration, US Department of Health and Human Services. January 12, 2007. www.fda.gov/bbs/topics/NEWS/2007/NEW01546.html

Fernandez Escartin, E., Castillo Ayala, A., and Saldana Lozano, J. (1989). Survival and growth of *Salmonella* and *Shigella* on sliced fresh fruit. *J. Food Prot.* **52**, 471–472, 483.

Fonseca, J. M. (2006). Postharvest quality and microbial population of head lettuce as affected by moisture at harvest. *J. Food Sci.* **71**, M45–M49.

Francis, G. A. and O'Beirne, D. (1998). Effects of the indigenous microflora of minimally processed lettuce on the survival and growth of *Listeria innocua*. *Int. J. Food Sci. Technol.* **33**, 477–488.

Gagliardi, J. V., Millner, P. D., Lester, G. et al. (2003). On-farm and postharvest processing sources of bacterial contamination to melon rinds. *J. Food Prot.* **66**, 82–87.

Garcia, L., Henderson, J., Fabri, M. et al. (2006). Potential sources of microbial contamination in unpasteurized apple cider. *J. Food Prot.* **69**, 137–144.

Hamilton, M. J., Yan, T., and Sadowsky, M. J. (2006a). Development of goose- and duck-specific DNA markers to determine sources of *Escherichia coli* in waterways. *Appl. Environ. Microbiol.* **72**, 4012–4019.

Hamilton, A. J., Stagnitti, F., Premier, R. et al. (2006b). Quantitative microbial risk assessment models for consumption of raw vegetables irrigated with reclaimed water. *Appl. Environ. Microbiol.* **72**, 3284–3290.

Heisick, J. E., Wagner, D. E., Nierman, M. I. et al. (1989). *Listeria* spp. found on fresh market produce. *Appl. Environ. Microbiol.* **55**, 1925–1927.

Ingham, S. C., Fanslau, M. A., Engel, R. A. et al. (2005). Evaluation of fertilization-to-planting and fertilization-to-harvest intervals for safe use of noncomposted bovine manure in Wisconsin vegetable production. *J. Food Prot.* **68**, 1134–1142.

Iturriaga, M. H., Tamplin, M. L., and Escartín, E. F. (2007). Colonization of tomatoes by *Salmonella* Montevideo is affected by relative humidity and storage temperature. *J. Food Prot.* **70**, 30–34.

Islam, M., Doyle, M. P., Phatak, S. C. et al. (2004). Persistence of enterohemorrhagic *Escherichia coli* O157:H7 in soil and leaf lettuce and parsley grown in fields treated with contaminated manure composts or irrigation water. *J. Food Prot.* **67**, 1365–1370.

Janes, M. E., Cobbs, T., Kooshesh, S. et al. (2002). Survival differences of *Escherichia coli* O157:H7 strains in apples of three varieties stored at various temperatures. *J. Food Prot.* **65**, 1075–1080.

Johannessen, G. S., Bengtsson, G. B., Heier, B. T. et al. (2005). Potential uptake of *Escherichia coli* O157:H7 from organic manure into Crisphead lettuce. *Appl. Environ. Microbiol.* **71**, 2221–2225.

Johnston, L. M., Jaykus, L.-A., Moll, D. et al. (2006). A field study of the microbiological quality of fresh produce of domestic and Mexican origin. *Int. J. Food Microbiol.* **112**, 83–95.

Kallander, K. D., Hitchins, A. D., Lancette, G. A. et al. (1991). Fate of *Listeria monocytogenes* in shredded cabbage stored at 5 and 25°C under a modified atmosphere. *J. Food Prot.* **54**, 302–304.

Kakiomenou, K., Tassou, C., and Nychas, G.-J. (1998). Survival of *Salmonella enteritidis* and *Listeria monocytogenes* on salad vegetables. *World J. Microbiol. Biotech.* **14**, 383–387.

Keller, S. E., Merker, R. I., Taylor, K. T. et al. (2002). Efficacy of sanitation and cleaning methods in a small apple cider mill. *J. Food Prot.* **65**, 911–917.

Kullas, H., Coles, M., Rhyan, J. et al. (2002). Prevalence of *Escherichia coli* serogroups and human virulence factors in feces of urban Canada geese (*Branta Canadensis*). *Int. J. Environ. Health Res.* **12**, 153–162.

Liao, C.-H. and Sapers, G. M. (2000). Attachment and growth of *Salmonella* Chester on apple fruits and in vivo response of attached bacteria to sanitizer treatments. *J. Food Prot.* **63**, 876–883.

Mandrell, R. E., Gorski, L., and Brandl, M. T. (2006). Attachment of microorganisms to fresh produce. In *Microbiology of Fruits and Vegetables* (G. M. Sapers, J. R. Gorny, and A. E. Yousef, Eds.), pp. 375–400. CRC Press.

McAllister, T. A., Bach, S. J., Stanford, K. et al. (2006). Shedding of *Escherichia coli* O157:H7 by cattle fed diets containing monensin or tylosin. *J. Food Prot.* **69**, 2075–2083.

Muirhead, R. W., Collins, R. P., and Bremer, P. J. (2006). Interaction of *Escherichia coli* and soil particles in runoff. *Appl. Environ. Microbiol.* **72**, 3406–3411.

Mukherjee, A., Speh, D., Jones, A. T. et al. (2006). Longitudinal microbiological survey of fresh produce grown by farmers in the upper Midwest. *J. Food Prot.* **69**, 1928–1936.

Omary, M. B., Testin, R. F., Barefoot, S. F. et al. (1993). Packaging effects on growth of *Listeria innocua* in shredded cabbage. *J. Food Sci.* **58**, 623–626.

Phillips, C. A. and Harrison, M. A. (2005). Comparison of the microflora on organically and conventionally grown spring mix from a California processor. *J. Food Prot.* **68**, 1143–1146.

Piagentini, A. M., Pirovani, M. E., Güemes, D. R. et al. (1997). Survival and growth of *Salmonella* hadar on minimally processed cabbage as influenced by storage abuse conditions. *J. Food Sci.* **62**, 616–618, 631.

Rice, D. H., Hancock, D. D., and Besser, T. E. (1995). Verotoxigenic *E. coli* O157 colonization of wild deer and range cattle. *Vet. Rec.* **137**, 524.

Riordan, D. C., Sapers, G. M., Hankinson et al. (2001). A study of U.S. orchards to identify potential sources of *Escherichia coli* O157:H7. *J. Food Prot.* **64**, 1320–1327.

Rodriguez, A., Pangloli, P., Richards, H. A. et al. (2006). Prevalence of *Salmonella* in diverse environmental farm samples. *J. Food Prot.* **69**, 2576–2580.

Sagoo, S. K., Little, C. L., and Mitchell, R. T. (2001). The microbiological examination of ready to eat organic vegetables from retail establishments in the United Kingdom. *Lett. Appl. Microbiol.* **33**, 434–439.

Sapers, G. M. and Jones, D. M. (2006). Improved sanitizing treatments for fresh tomatoes. *J. Food Sci.* **71**, M252–M256.

Sapers, G. M., Miller, R. L., and Mattrazzo, A. M. (1999). Effectiveness of sanitizing agents in inactivating *Escherichia coli* in Golden Delicious apples. *J. Food Sci.* **64**, 734–737.

Seo, K. H. and Frank, J. F. (1999). Attachment of *Escherichia coli* O157:H7 to lettuce leaf surface and bacterial viability in response to chlorine treatment as demonstrated by using confocal scanning laser microscopy. *J. Food Prot.* **62,** 3–9.

Solomon, E. B., Potenski, C. J., and Matthews, K. R. (2002a). Effect of irrigation method on transmission to and persistence of *Escherichia coli* O157:H7 on lettuce. *J. Food Prot.* **65,** 673–676.

Solomon, E. B., Yaron, S., and Matthews, K. R. (2002b). Transmission of *Escherichia coli* O157:H7 from contaminated manure and irrigation water to lettuce plant tissue and its subsequent internalization. *Applied Environ. Microbiol.* **68,** 397–400.

Steele, M., Mahdi, A., and Odumeru, J. (2005). Microbial assessment of irrigation water used for production of fruit and vegetables in Ontario, Canada. *J. Food Prot.* **68,** 1388–1392.

Steinbruegge, E. G., Maxcy, R. B., and Liewen, M. B. (1988). Fate of *Listeria monocytogenes* on ready to serve lettuce. *J. Food Prot.* **51,** 596–599.

Stine, S. W., Song, I., Choi, C. Y. et al. (2005a). Application of microbial risk assessment to the development of standards for enteric pathogens in water used to irrigate fresh produce. *J. Food Prot.* **68,** 913–918.

Stine, S. W., Song, I., Choi, C. Y. et al. (2005b). Effect of relative humidity on preharvest survival of bacterial and viral pathogens on the surface of cantaloupe, lettuce, and bell peppers. *J. Food Prot.* **68,** 1352–1358.

Ukuku, D. O. and Fett, W. F. (2006). Effects of cell surface charge and hydrophobicity on attachment of 16 *Salmonella* serovars to cantaloupe rind and decontamination with sanitizers. *J. Food Prot.* **69,** 1835–1843.

Ukuku, D. O. and Sapers, G. M. (2001). Effect of sanitizer treatments on *Salmonella* Stanley attached to the surface of cantaloupe and cell transfer to fresh-cut tissues during cutting practices. *J. Food Prot.* **64,** 1286–1291.

Ukuku, D. O., Pilizota, V., and Sapers, G. M. (2001). Influence of washing treatment on native microflora and *Escherichia coli* population of inoculated cantaloupes. *J. Food Safety* **21,** 31–47.

USDA. (2004). *Microbiological Data Program Progress Update and 2002 Data Summary*. www.ams.usda.gov/science/mpo/MDPSumm02.pdf

Wallace, J. S., Cheasty, T., and Jones, K. (1997). Isolation of Verocytotoxin-producing *Escherichia coli* O157 from wild birds. *J. Appl. Microbiol.* **82,** 399–404.

Wei, C. I., Huang, T. S., Kim, J. M. et al. (1995). Growth and survival of *Salmonella* Montevideo on tomatoes and disinfection with chlorinated water. *J. Food Prot.* **58,** 829–836.

Wells, J. M. and Butterfield, J. E. (1997). *Salmonella* contamination associated with bacterial soft rot of fresh fruits and vegetables in the marketplace. *Plant Dis.* **81,** 867–872.

Yan, T., Hamilton, M. J., and Sadowsky, M. J. (2007). High-throughput and quantitative procedure for determining sources of *Escherichia coli* in waterways by using host-specific DNA marker genes. *Appl. Environ. Microbiol.* **73,** 890–896.

Zhuang, R-Y, Beuchat, L. R, and Angulo, F. J. (1995). Fate of *Salmonella* Montevideo on and in raw tomatoes as affected by temperature and treatment with chlorine. *Appl. Environ. Microbiol.* **61,** 2127–2131.

CHAPTER 2

Microbial Attachment and Limitations of Decontamination Methodologies

Ethan B. Solomon

DuPont Chemical Solutions Enterprise, Experimental Station Laboratory, Wilmington, DE

Manan Sharma

Environmental Microbial and Food Safety Laboratory, United States Department of Agriculture, Beltsville, MD

CHAPTER CONTENTS

Introduction	21
Ecological Niches and Introduction into the Plant Environment	23
Outbreak Investigations Reveal Sources and Persistence of Pathogens	24
The Plant Surface	25
Attachment of Pathogens to Plant Tissue	25
Attachment of *Escherichia coli* O157:H7 to Lettuce	26
Attachment of *Salmonella* to Tomatoes	28
Attachment of *Salmonella* to Cantaloupes	30
Biofilm Formation on Produce Surfaces	31
Internalization	32
Limited Efficacy of Conventional Decontamination Methodologies	36
Conclusion	39

INTRODUCTION

The number of outbreaks of foodborne illness arising from the consumption of fresh and fresh-cut produce has risen dramatically over the last two decades (Sivapalasingam et al., 2004). From 1990 to 2005, fresh produce was associated with 713 outbreaks, resulting in 34,049 cases of illness (Anon., 2007). The number of illnesses linked to fresh produce surpassed those for all other foods, including poultry, beef, eggs, and seafood. In addition, the average number of illnesses per produce outbreak was significantly higher than those from other foods (Anon., 2007).

Although the reasons behind this increase are somewhat unclear, several factors play an important role. First, the per capita consumption of fresh produce has increased significantly. From 1982 to 1997, US consumption of raw fruits and vegetables increased by 18% and 29%, respectively (Garrett et al., 2003). Second, the produce industry has become increasingly global, with large volumes of produce being imported into the United States, making oversight difficult. Third, convenience foods such as fresh-cut fruits and bagged salads also have grown exponentially, but are more conducive to microbial growth and spoilage than the whole produce from which they are derived (Brandl, 2008). However, the increased consumption of leafy greens does not fully explain the increased incidence of outbreaks associated with these commodities. The incidence of foodborne outbreaks associated with leafy greens increased by 39% between 1996 and 2005, but leafy green consumption increased by only 9% (Herman et al., 2008). This indicates that other factors may also be responsible for the increased number of these outbreaks.

Fresh produce is grown in agricultural settings that are prone to contamination by microbial pathogens. Produce alone does not normally harbor pathogenic microorganisms, however zoonotic bacterial pathogens such as *Salmonella* and *Escherichia coli* O157:H7 are easily transferred from other sources. Preharvest sources of pathogenic microorganisms include soil, manure (or compost), irrigation water, water used for pesticide application, insects, and wild or domestic animals. Postharvest sources include human handling, harvesting and transport equipment, animals, dust, wash water, packing-shed equipment, improper storage, temperature abuse, cross-contamination, and improper consumer display.

Until recently, it was thought that enteric pathogens such as *E. coli* O157:H7 and *Salmonella* survived poorly in the harsh environment encountered on plant surfaces, where microorganisms must survive sunlight, desiccation, nutrient limitation, and drastic temperature fluctuations. Recent research has shown this not to be the case. Enteric pathogens have been demonstrated to persist in a variety of agricultural settings including water, soils, manure, the plant rhizosphere, and even on exposed (foliar) plant surfaces (Heaton and Jones, 2007; Brandl, 2006). As a result of outbreaks occurring during the mid 1990s, the survival and dissemination of foodborne pathogens in agricultural environments has been studied in detail. More recently, the intimate interactions between enteric pathogens and plant tissue have begun to be scrutinized.

Compared with research focused on the ecological behavior of foodborne pathogens, significantly more effort has been placed into finding postharvest decontamination methods capable of removing or inactivating foodborne pathogens (bacteria, parasites, and viruses) from contaminated plant tissue. A number of chemical and physical treatments have been evaluated, but, to our knowledge, none have been deemed completely effective in removing or inactivating attached bacteria without negatively impacting the organoleptic qualities of the product. In fact, few have succeeded in reducing the pathogenic bacterial load by more than 2 or 3 log units.

This chapter will discuss the attachment of foodborne pathogens to plant tissue and why decontamination of produce has proven so difficult. Although a variety of organisms (bacteria, viruses, parasites, etc.) have been implicated in outbreaks arising from produce, this chapter will focus primarily on *Escherichia coli* O157:H7 and *Salmonella*, because of the frequency of outbreaks associated with these pathogens and the depth to which they have been studied.

ECOLOGICAL NICHES AND INTRODUCTION INTO THE PLANT ENVIRONMENT

Both *Salmonella* and *E. coli* O157:H7 are normally found in the gastrointestinal tracts of warm-blooded farm animals. *E. coli* O157:H7 traditionally is associated with ruminant animals (cattle, sheep, and goats) (Erickson and Doyle, 2007), and *Salmonella* is found frequently in poultry, but also in pigs, cattle, goats, waterfowl, rodents, and insects (D'Aoust, 1998). Both of these organisms are introduced into the plant environment by dissemination from their animal hosts. The carriage rate of these organisms in their animal reservoirs is not quite clear. For example, estimates of the prevalence of *E. coli* O157:H7 in cattle range from less than 1% to upward of 25% (Elder et al., 2000). Carriage rate studies differ in the type of animals surveyed (age), feeding regimens, and type of samples obtained (swabs, grabs, fecal pats, etc.), so conclusions as to the carriage rate are difficult to draw. Regardless of the carriage rate, infected cattle are known to shed anywhere between 10^1 and 10^7 CFU of *Escherichia coli* O157:H7 per gram of feces (Besser et al., 2001). Given that typical cattle excrete 20 to 50 kg of feces per day, this provides a large "inoculum" of *E. coli* O157:H7 for the farm environment. The presence of "super-shedders," a few cattle in a herd that shed greater than 10^4 CFU/g feces, also present a significant source of *E. coli* O157:H7 in the producegrowing environment (Matthews et al., 2006).

Animal manures are used widely as fertilizers, and proper composting is essential to ensure that pathogens from manure do not directly interact with growing plants. A number of researchers have investigated the survival of both *E. coli* O157:H7 and *Salmonella* in manure from various animals, under different conditions such as temperature or aeration, presence of different manure amendments, and at a range of manure-to-soil ratios (Duffy, 2003). Kudva et al. (1998) found that *E. coli* O157:H7 survived for more than 21 months in ovine manure at levels ranging up to 10^6 CFU/g manure. Aeration of the ovine manure pile greatly reduced the survival time. Experiments with artificially inoculated bovine feces have also confirmed survival of *E. coli* O157:H7 for greater than 40 days, dependent on initial inoculum and holding temperature (Wang et al., 1996). More recent work states that *Salmonella* may be persistent for longer durations than *E. coli* in bovine manure when kept under constant temperature and moisture conditions (Himathongkham et al., 1999). Collectively, these studies indicate that enteric pathogens can survive for long periods of time in animal manures and composts and therefore remain in close proximity to growing crops.

In addition to direct contact with plants, pathogens in manure are often transferred to water either directly or through runoff. Studies have documented an increase in the levels of bacterial pathogens in water sources immediately after heavy rains (Cooley et al., 2007). Contaminated water has been implicated in two recent outbreaks arising from produce (tomato, 2005/2006; shredded lettuce, 2006). Extensive laboratory research has demonstrated that enteric pathogens originating in manure can survive for long periods of time in water (Wang and Doyle, 1998). Sterilized clear water has often been used as a model system (Wang and Doyle, 1998; Kolling and Matthews, 2001), but the utility of these studies is questionable, since the introduction of an organic load (such as manure) greatly increases the survival of *E. coli* O157:H7 in water (Hutchison et al., 2005). These findings, coupled with extensive field-based research, clearly indicate that pathogens can survive for long periods in contaminated water and that this water poses a serious threat to growing crops.

OUTBREAK INVESTIGATIONS REVEAL SOURCES AND PERSISTENCE OF PATHOGENS

Since the first large outbreaks in the early 1990s, specific pathogens have repeatedly been implicated in outbreaks arising from the same plants. For example, from 1990 to 2005, 24 outbreaks of *Salmonella* have involved seed sprouts, and 16 have involved melons (Anon., 2007). Spinach, lettuce, and other leafy vegetables have been involved in 29 outbreaks of *E. coli* O157:H7. Traceback investigations into some of these outbreaks have revealed details of the mechanisms of how these bacteria are introduced and persist onto the plant surface. For example, an intense investigation into the origin of the *E. coli* O157:H7 outbreak linked to spinach in 2006 (California Food Emergency Response Team, 2007) revealed the presence of the outbreak strain in cattle feces, surface water, and feral pigs present near the fields where the spinach was grown. These investigations underscore the complexity of the preharvest environment and the ease with which plant tissue can become contaminated with foodborne pathogens.

THE PLANT SURFACE

Pathogens introduced onto the plant via water, manure, improper handling, or any other vector must attach themselves and proliferate on plant surfaces. Most aerial plant surfaces are covered in cuticle, a hydrophobic material composed primarily of fatty acids, waxes, and polysaccharides. The cuticle prevents plant dehydration and also protects the plant from infiltration by phytopathogens (Solomon et al., 2006). Cracks in the cuticle, or other damage that exposes the epidermal cells of the plant surface are often sites at which bacteria colonize. The plant surface itself is an inhospitable environment for enteric pathogens. It is subject to large swings in temperature and relative humidity, limited water or nutrient availability, and potential exposure to UV from sunlight. It also contains a large population of foliar microorganisms in large aggregates that may compete with foodborne pathogens for nutrients in this environment (Lindow, 2006).

ATTACHMENT OF PATHOGENS TO PLANT TISSUE

In order to be transmitted by fresh produce, foodborne pathogens must attach to plant tissue and survive through consumption. Regardless of the environmental source, recent data indicates that enteric bacteria can attach to growing plant tissue in a relatively rapid fashion, colonize specific microenvironments that may be plant-species specific, coexist with epiphytic bacteria to survive and grow, and persist for significant periods of time (see earlier reviews by Brandl, 2006 and Solomon et al., 2006 for exhaustive information).

Laboratory experiments with excised plant tissue or intact whole produce indicate that bacterial pathogens attach rapidly and irreversibly. Ukuku demonstrated that *Salmonella* deposited onto cantaloupe melons could not be washed off after just four hours of incubation (Ukuku and Sapers, 2001). Experiments with *E. coli* and lettuce demonstrated irreversible attachment after just a few hours (Beuchat, 1999). Similar results were found with tomatoes inoculated with *Salmonella* (Iturriaga et al., 2003).

The authors indicated that 0.3% of *Salmonella* cells were attached to the external tomato surface immediately after introduction (Iturriaga et al., 2003). Liao and Cooke (2001) used a laboratory model consisting of green pepper slices to study attachment of *Salmonella* Chester. They concluded that 30% of the bacterial inoculum firmly attached to the pepper's cut surface within 30 seconds. These firmly attached cells could not be removed by washing or agitation. These results were confirmed by Han et al. (2000) who found that *E. coli* irreversibly attached to green pepper after a short time and were not removed by washing.

Collectively, these studies demonstrate that bacterial pathogens attach rapidly to produce surfaces. Attached pathogens are extremely difficult to remove with current washing or agitation regimens. In addition, many of these studies differentiate between intact and damaged plant tissues. Damaged plant tissues are far more susceptible to colonization than undamaged tissue of the same type (Seo et al., 1999). Fresh-cut produce is therefore at even higher risk in terms of bacterial colonization compared to the whole product from which it was derived. Because of the repeated implication of lettuce, cantaloupe, and tomatoes as sources of outbreaks, the following section will review research on the attachment of pathogens to these commodities.

ATTACHMENT OF *ESCHERICHIA COLI* O157:H7 TO LETTUCE

In the period from 1990 to 2005, leaf lettuce was linked to more outbreaks than any other singular type of produce (Anon., 2007). This distinction has much to do with the popularity of lettuce, its wide distribution, the use of cut (bagged) lettuce as a salad ingredient, and the physical nature of the plant itself. Lettuce is grown in close proximity to the ground, allowing for extensive contact with surrounding soil. Lettuce leaves are easily damaged by mechanical disruption, making them more conducive as an environment for bacterial attachment and growth. In addition, most lettuce is consumed raw as opposed to other vegetables that may be cooked. Research into the interactions between *E. coli* O157:H7 and lettuce has been ongoing for more than a decade. Surveys of farm-level and store-purchased lettuce report extremely low levels of the pathogen, yet outbreaks continue to occur (Delaquis et al., 2007).

Beuchat (1999) demonstrated that *E. coli* could attach to cut lettuce tissue when suspended in either bovine manure or 0.1% peptone. Low levels of bacteria (ca. 10^2 CFU/g) introduced onto lettuce leaves survived 15 days of storage at 4 °C. Treatment of contaminated leaves with 200 ppm of chlorine was no more effective at inactivating the bacteria than treatment with water alone. Similar results were reported by Delaquis et al. (2002) for lettuce pieces inoculated with *E. coli* O157:H7 and then immediately washed in both cold and warm water containing 100 ppm chlorine. Other studies indicate that *E. coli* O157:H7 (as well as *Salmonella* and *Listeria*) attached preferentially to cut edges of lettuce pieces (Takeuchi and Frank, 2000) compared to the intact surface. Seo and Frank (1999) utilized confocal microscopy to visualize cells of *E. coli* attached at stomata and trichomes of cut lettuce plants. These observations led the authors to conclude that attachment sites for *E. coli* were similar to those reported for phytopathogens.

The attachment of *E. coli* to growing lettuce plants under greenhouse or field conditions also has been determined. Collectively, greenhouse and laboratory-level experiments demonstrate that plants contaminated by direct contact with the organism, whether carried in manure or water, result in persistent contamination (Solomon et al., 2002; Cooley et al., 2006, Franz et al., 2007). Solomon et al. (2002) found that plants spray-irrigated once with water containing 10^7 CFU/ml of *E. coli* O157:H7 remained culture-positive for 20 days. Treatment of spray-irrigated plants with 200 ppm chlorine failed to completely inactivate the organism. Islam et al. (2004) planted lettuce seedlings into manure-fertilized soil that had been inoculated with a nonpathogenic strain of *E. coli* O157:H7. Lettuce plants harvested 77 days after planting were positive for the organism. Similar levels of persistence of *E. coli* in the lettuce rhizosphere and phyllosphere have been reported by other authors using real-time PCR to quantify these levels (Ibekwe et al., 2004).

A number of recent authors have investigated the role of cell surface charge, presence of divalent cations, hydrophobicity, capsule production, curli production, and other specific bacterial adherence mechanisms and their role in attachment of *E. coli* to lettuce tissue (Hassan and Frank, 2003, 2004; Boyer et al., 2007). Collectively, these studies have shown very little correlation between the presence of cell surface appendages, charge, or hydrophobicity on the ability of the bacteria to attach to lettuce tissue. Hassan and Frank (2004) reported that increased capsule production led to greater attachment of *E. coli* O157:H7 to lettuce. The possibility that the interaction between lettuce and *E. coli* introduced onto the lettuce surface is one of simple entrapment is supported by studies using live and glutaraldehyde-killed cells of *E. coli* and demonstrating little difference in the numbers retained on the lettuce surface (Solomon et al., 2006). Abiotic fluorescent microspheres adhered to lettuce pieces as well as *E. coli* O157:H7 (Solomon et al., 2006).

Recent studies have found significant molecular interactions between *E. coli* O157:H7 and plant tissue (Shaw et al., 2008; Matthysse et al., 2008; Torres et al., 2005). Although no genetic elements have been definitively identified as required for attachment, mutants deficient in cellulose production, colanic acid, and poly n-acetylglucosamine had greatly reduced ability to bind to alfalfa sprouts (Matthysse et al., 2008). Mutants deficient in OmpA were also unable to bind to alfalfa as strongly as wild type cells (Torres et al., 2005). Shaw et al. (2008) demonstrated that *E. coli* O157:H7 cells used the filamentous type III secretion system to attach to arugula, spinach, and lettuce leaves when leaves were immersed in high populations (10^6 CFU/ml) at 37 °C for extended periods of times.

ATTACHMENT OF *SALMONELLA* TO TOMATOES

Since 1998, at least 12 outbreaks of salmonellosis have been attributed to the consumption of tomatoes (Barak et al., 2008). These 12 outbreaks resulted in 1990 culture confirmed infections; however, since approximately 97.5% of all *Salmonella* infections are not confirmed, these outbreaks may have resulted in almost 79,600 illnesses (CDC, 2007). Outbreaks arising from tomatoes tend to be quite large and widely dispersed, indicating a point source of contamination early in distribution (Greene et al., 2008). Unlike lettuce, tomatoes are not grown in close proximity to the ground and as such, should be less likely to be contaminated with manure or compost. In addition, the external surface of a tomato should make it more difficult for bacteria to attach and grow, compared to the high surface area of leafy vegetables. Despite these differences, tomatoes continue to be identified as vehicles for outbreaks.

Much like early work with lettuce, the attachment, survival, and proliferation of bacteria on tomatoes was investigated in laboratory-scale experiments (Zhuang et al., 1995; Iturriaga et al., 2003) in response to outbreaks linked to tomatoes. Studies designed to mimic on-field conditions demonstrated that *Salmonella* survived well on tomato surfaces following immersion into contaminated water (Zhuang et al., 1995). Storage at temperatures above 10 °C resulted in rapid bacterial growth over a short period of time (Zhuang et al., 1995). The temperature differential between the tomato and the wash water was found to be extremely important since immersion of warmer tomatoes into colder water resulted in infiltration of *Salmonella* into the internal portion of the tomato core. Lastly, chopped ripe tomatoes supported significant growth of *Salmonella* at 20 and 30 °C. This raises the possibility that contamination of tomatoes with low levels of

Salmonella would result in high levels of growth after chopping and holding at higher temperatures. Iturriaga et al. (2003) used scanning electron microscopy (SEM) to visualize the interactions between *Salmonella* Montevideo and tomato surfaces. They found that attachment occurred rapidly and irreversibly and attributed this attachment to bacterial cells associating with the waxy cuticle present on tomato surfaces. These results were confirmed by Iturriaga et al. (2007), who found relative humidity as well as temperature to be important to the persistence of *Salmonella* on tomatoes. A recent study by Shi et al. (2007) raises the possibility that outbreaks of *Salmonella* arising from tomatoes may be serovar-specific since some strains grew better than others on ripe fruit.

Greenhouse-level studies also have been conducted to investigate interactions between *Salmonella* and tomato plants. Barak et al. (2008) demonstrated contamination incidences ranging from 6.3 to 61.1% of *S. enterica* in the phyllosphere of various tomato cultivars when seeds were planted in contaminated soil. The authors further hypothesized that since soil-borne *Salmonella* demonstrated poor attachment to tomato seedlings, soil may not be a common route of preharvest contamination. However, experiments conducted by Guo et al. (2002b) demonstrated that *Salmonella* present in soil in contact with the stem scars of mature green tomatoes could survive for at least 14 days, increase in population by 2.5 log CFU per tomato (dependent on storage conditions), and potentially contaminate the internal portions of the fruit. Field-level studies conducted by the same authors (Guo et al., 2001) indicated that *Salmonella* introduced onto flowers of tomato plants, both prior to or after fruit set or injected into the stem adjacent to the flowers (to mimic physical damage to the plant), could result in contaminated tomatoes at harvest. These results indicate that *Salmonella* attach well to flowers and can survive through fruit set and be present at harvest (up to 49 days). Furthermore, the harvest of contaminated tomatoes from plants that had received stem inoculation demonstrates systemic movement of the bacteria from the stem into the tissue of the fruit (Guo et al., 2001).

Molecular studies using *S. enterica* indicate that genes required for virulence in animals as well as those required for synthesis of the *Salmonella* extracellular matrix are also important for the colonization of plant tissue (Barak et al., 2005, 2007). Although these studies were conducted using sprouts as opposed to tomatoes, the results warrant discussion. Using an alfalfa sprout colonization assay, 6000 mutants created through transposon mutagenesis were screened for their ability to bind to plant tissue (Barak et al., 2005). Mutants deficient in *rpoS*, the stationary phase sigma factor, as well as *agfD*, a transcriptional regulator, were significantly reduced in their binding ability (approximately sevenfold fewer cells per sprout). Further experiments demonstrated that cellulose, thin aggregative fimbriae, and the O-antigen capsule are also important determinants in the attachment of *Salmonella* to plant tissue (Barak et al., 2007). Other work examining the colonization of *Salmonella* on alfalfa sprouts indicated that strains lacking elements of a Type III secretion system (TTSS) were able to "hypercolonize" alfalfa seedlings, indicating that the TTSS elements may be recognized by plant defenses to limit *Salmonella* colonization (Iniguez et al., 2005).

ATTACHMENT OF *SALMONELLA* TO CANTALOUPES

Cantaloupe melons have been implicated in at least 11 outbreaks of salmonellosis in the United States and Canada between 1973 and 2003 (Bowen et al., 2006). Most of the outbreaks linked to cantaloupe resulted in over 20 illnesses and covered a wide geographic area, indicating that contamination likely occurred at a single source. From this information, it can be inferred that *Salmonella* attaches to the surface of the cantaloupe, and survives well during transport and storage through sale, preparation, and consumption. The physical characteristics of cantaloupe melons and the manner in which they are grown makes it exceedingly difficult to prevent contamination and also to remove attached bacteria once in place. Melons are grown directly on the ground in constant contact with the soil surface. In addition, melon plant leaves provide a shady environment for growing melons, reducing the likelihood of large swings in temperature or direct sunlight. The melon fruit itself is covered in netting made up of large cracked pieces of cuticle. Fissures in the cantaloupe netting have been demonstrated to be infiltrated by cells of *Salmonella* (Annous et al., 2004, 2005) and likely provide attachment sites that aid bacterial survival when in contact with aqueous sanitizers.

The behavior of *Salmonella* on cantaloupe melons has been examined in detail by numerous investigators (Annous et al., 2005; Ukuku and Sapers, 2001; Ukuku and Fett, 2004). Laboratory experiments with whole melons indicate that once introduced onto the surface of the cantaloupe, cells of *Salmonella* are impossible to remove completely, regardless of the sanitizer used or the exposure time. Additional experiments demonstrate that the efficacy of sanitizer treatment is reduced when the organism is allowed to reside on the melon surface for an extended period of time (Ukuku and Sapers, 2001; Ukuku 2004). The authors postulate that attachment of cells to sites inaccessible to sanitizer action within the cantaloupe netting, and the initiation of bacterial biofilm formation, contribute to the minimal efficacy of aqueous sanitizers. These results contrast with those found for honeydew melons, which are closely related to cantaloupe, yet lack the characteristic surface netting. Five minute washing treatments in 2.5% or 5.0% hydrogen peroxide were effective at completely eliminating *Salmonella* from the surface of honeydew melons (>3 log reduction), yet were unable to produce the same results on cantaloupe (Ukuku, 2004). Alvarado-Casillas et al. (2007) reported similar results when comparing the efficacy of sanitizers on cantaloupe with efficacy on bell peppers.

One concern with the survival of *Salmonella* attached to cantaloupe surfaces is the ability of fresh-cut melon to support the growth of *Salmonella*. Cutting through the rind of a melon harboring *Salmonella* may result in contamination of fresh-cut pieces with the pathogen (Ukuku and Sapers, 2001). Fresh-cut melon on display at grocery stores generally is stored at temperatures above 4 °C and *Salmonella* grows extremely well in melon cubes stored at abusive temperatures (Ukuku, 2004).

Greenhouse-level studies with cantaloupe are far less prevalent than for lettuce or tomato. Stine et al. (2005) used a growth chamber to investigate the effects of humidity on the survival of *Salmonella* on cantaloupe. Inactivation rates of *Salmonella* under dry conditions were significantly greater than the inactivation rate under humid conditions. These results suggest that humid conditions may allow *Salmonella* to persist on a melon surface through harvest, transport, storage, and consumption. In terms of the intimate interactions between cells of *Salmonella* and the melon surface, no molecular mechanisms have been investigated; however, the cell surface charge and hydrophobicity have been found to play an important role (Ukuku and Fett, 2002, 2006).

BIOFILM FORMATION ON PRODUCE SURFACES

Following initial attachment of the pathogens onto produce surfaces, recent evidence indicates that bacterial pathogens may become entrapped in a biofilm (Annous et al., 2005). Biofilms are defined as "an assemblage of microorganisms adherent to each other and/or to a surface and embedded in a matrix of exopolymers" (Costerton et al., 1999). It is estimated that between 30 and 80% of the total bacterial population existing on plant surfaces are embedded in biofilms (Lindow and Brandl, 2003). The formation of biofilms by plant epiphytic or pathogenic bacteria has long been known (Danhorn and Fuqua, 2007); however, the discovery that enteric pathogens could establish biofilms on plant surfaces was surprising. In the last few years the number of studies on the formation of biofilms by foodborne bacteria on produce surfaces has expanded. *Salmonella, E. coli, Campylobacter,* and *Shigella* have been found to form distinct biofilms on the surfaces of produce ranging from tomatoes to melons to parsley (Itturiaga et al., 2007; Annous et al., 2005; Agle, 2003).

Bacterial cells embedded in biofilms are significantly different from their planktonic (free-floating) counterparts in terms of both physiology and genetics. The formation of biofilms by bacterial cells on plant surfaces is likely a survival strategy to withstand the harsh environment of the plant surface (wide temperature changes, desiccation, UV, oxidative stress). Similar to biofilms on food-processing surfaces, bacteria embedded within biofilms on plant tissue are more difficult to remove and more resistant to inactivation than their planktonic counterparts (Chmielewski and Frank, 2003). These differences allow biofilm-associated cells to survive the harsh environment of the plant surface in the field as well as during harvest, transport, and storage.

INTERNALIZATION

Bacterial foodborne pathogens can attach and persist on produce commodities, but several laboratory studies have reported the ability of these bacteria to survive in internal tissues of plants. Internalization can occur through several mechanisms: the immersion of fruit into bacterial suspensions, uptake of pathogen through plant roots, and the infiltration of these bacterial cells through damaged or cut tissues. It is unclear if the presence of aggregates or biofilms of foodborne pathogens on produce surfaces makes them more or less likely to internalize to plant tissues. Workers have shown that enteric bacteria, able to colonize the rhizosphere, are more likely to colonize internal tissues of alfalfa seedlings (Dong et al., 2003), but more work is needed to truly establish this correlation.

Several other factors can influence the opportunity and capability of foodborne pathogens to internalize in plant tissues. Despite differences in the experimental results, internalized pathogens are believed to pose a distinct threat. Because of their location within the tissues of otherwise healthy fruits and vegetables, internalized pathogens can transit through the food chain undetected. In addition, no method of treatment or sanitation that is currently used in the food industry has been proven capable of inactivating these internalized organisms.

The route of potential internalization and uptake of bacterial foodborne pathogens by produce differs depending on the produce commodity and stage of development of the plant or fruit. Several studies on internalization into harvested fruits have been conducted, and evidence of pathogen internalization has been observed when there has been a temperature differential between the fruit and the inoculum. Laboratory experiments that describe internalization through immersion in inoculum have been conducted with high levels of bacteria (10^6 to 10^8 CFU/ml). A positive temperature differential (Buchanan et al., 1999) occurs when a warm piece of fruit is immersed in a cooler fluid. This causes contraction of gases in internal spaces within the fruit, resulting in a partial vacuum that draws in some of the fluid through pores in the fruit surface. When the fluid is contaminated or intentionally inoculated with pathogenic bacteria, these bacteria infiltrate into the fruit (Penteado et al., 2004). Mangoes, held in 46 °C water for 90 minutes and then immersed in an inoculum of *Salmonella* Enteriditis at 22 °C, were more likely to have bacteria internalized near the stem end (83% of samples) than the blossom end (8% of samples) of the fruit (Peneteado et al., 2004). Only 2.5% and 3.0% of oranges exposed to inocula of *E. coli* O157:H7 and *Salmonella*, respectively, under a positive temperature differential contained the internalized pathogens (Eblen et al., 2004). Infiltration of *E. coli* O157:H7 was observed infrequently in cores of apples exposed to an inoculum under a positive temperature differential, and this low frequency of internalization was further decreased when cold apples were placed in a cold inoculum (Buchanan et al., 1999).

The uptake of foodborne pathogens through intact roots of plants is a topic that has received much scrutiny in the last decade. Previous internalization studies have not adequately addressed the duration of survival of these internalized populations in plants during their growing seasons. Previous studies have also varied in the methods used to grow plants (in soil, in hydroponic medium, and composted manure), the method of introduction of these pathogens to plants (on roots, in soil, in irrigation water), the growth medium, or whether the pathogens were delivered in combination with bacterial phytopathogens and soil parasites. The stage of plant development (seeds, seedlings, and mature plants) may also affect the degree and extent of internalization of foodborne pathogens. Previous internalization studies also used surface sterilization techniques that were varied in their effectiveness to ensure that bacteria recovered in these studies were from internalized tissue.

Root uptake studies mainly focused on the internalization of *E. coli* O157:H7 in spinach and lettuce plants, and *Salmonella* in tomato plants. Solomon et al. (2002) examined lettuce seedlings after exposure to high populations of *E. coli* O157:H7 in contaminated irrigation water or contaminated manure slurry in contact with roots. *E. coli* O157:H7 were recovered from internal lettuce tissue five days after exposure to 10^7 CFU *E. coli*

O157:H7/ml irrigation water, indicating that root uptake of the pathogen did occur. This work also showed that the uptake of the pathogen through the roots of lettuce seedlings was more frequent when populations of *E. coli* O157:H7 were high. Uptake and internalization of *E. coli* O157:H7 into root and leaf stem tissue in 45-day-old romaine lettuce plants after application of 10^9 CFU/ml to the soil resulted in approximately 10^2 CFU of *E. coli* O157:H7/g in lettuce tissues after five days (Solomon and Matthews, 2005). Populations did not change significantly over five days after application, indicating that cells did not grow within lettuce tissue.

The coinoculation of the phytopathogen *Pseudomonas syringae* with *E. coli* O157:H7 in six-week-old spinach plants did not enhance the survival of the human pathogen when internalized through a vacuum infiltration process applied to roots of plants (Hora et al., 2005). Populations of internalized *E. coli* O157:H7 initially reached 4.9 log CFU/g spinach when coinoculated with *P. syringae*, whereas inoculation of *E. coli* O157:H7 alone resulted in 4.6 log CFU/g spinach one week after inoculation of spinach plants (Hora et al., 2005). Mechanical disruption and infection of roots of spinach plants with nematodes, preceding inoculation of soil with *E. coli* O157:H7 (10^7 CFU), resulted in internalization of *E. coli* O157:H7 in root tissues (24/24 plants) but not in leaves of spinach plants (0/24 plants).

In another study, the populations of *E. coli* O157:H7 were determined nine and 49 days after exposure of hydroponically grown cress, lettuce, radish, and spinach plants to 10^2 CFU *E. coli* O157:H7/ml (Jablasone et al., 2005). Populations of *E. coli* O157:H7 internalized after nine days of growth in lettuce, radish, and spinach tissues, but were not detected in internal tissues of cress (Jablasone et al., 2005). Populations of *E. coli* O157:H7 in spinach were about 2.5 log CFU/g after nine days but fell to undetectable levels (< 1 log CFU/g) when assayed at 49 days. *Salmonella* was detected in internal tissues of lettuce and radishes, but not spinach and cress plants. *Salmonella* populations were also introduced to the same seedling varieties on hydroponic media, but internalization of these cells (1 to 1.6 log CFU/g) to lettuce and radish plants was observed only after nine days (Jablasone et al., 2005). No internalized *Salmonella* cells were observed in any plants examined after 49 days. The lack of persistence of *E. coli* O157:H7 or *Salmonella* cells in tissues of lettuce suggests that cells may be under physiological and nutritional stress in the vasculature of plants or that plant defenses are effective in killing enteric human pathogens. In another study, crisphead lettuce (*Lactuca sativa*) seedlings planted in soil and manure contaminated with *E. coli* O157:H7 (10^4 CFU/g) did not result in internalization of the pathogen in either three-week-old seedlings or seven-week-old lettuce plants and leaves (Johannessen et al., 2005).

Populations of *E. coli* O157:H7 in the soil after seven weeks had declined by approximately 2 log CFU/g.

The observed differences in internalization of *E. coli* O157:H7 in lettuce from different studies may be due to the growth of seedlings in hydroponic media, compared to growth in soil (Jablasone et al., 2005), and the time elapsed between contamination events and bacteriological analysis. Other studies point to the internalization of *E. coli* O157:H7 and *Salmonella* in lettuce seedlings when planted in soils containing high populations of bacterial pathogens. Franz et al. (2007) placed eight-day-old lettuce seedlings in hydroponic media inoculated with either *E. coli* O157:H7 or *Salmonella* Typhimurium at 10^7 CFU/ml and grew the plants for 18 days. *E. coli* internalized more readily (3.95 log CFU/g) when seedlings were grown in soil inoculated with high population of *E. coli* O157:H7 (7–8 log CFU/g), and at significantly higher levels than populations of *S.* Typhimurium (2.37 to 2.57 log CFU/g). When grown in hydroponic media, examination of surface-sterilized root and leaf tissues revealed that no *E. coli* O157:H7 cells were recovered from internal tissue of lettuce leaves or root tissues. In this same study, a wild-type *Salmonella* strain internalized to 4/10 lettuce root samples and 2/10 lettuce leaf samples at populations of 10^5 CFU/g. The increased internalization of *E. coli* O157:H7 in plants in soil indicates that *E. coli* O157:H7 may internalize when root tissue is damaged (more likely with plant growth in soil) compared to plants grown in hydroponic solution (Franz et al., 2007).

Other workers have shown that internalization of *E. coli* O157:H7 did not occur when mature spinach plants were exposed to 10^3 or 10^6 CFU/g in pasteurized soils, but was observed sporadically when plants were grown in hydroponic medium inoculated with 10^7 CFU/ml (Sharma and Donnenberg, 2008). From this work, internalized populations of 3.7 and 4.3 log CFU *E. coli* O157:H7/shoot were recovered after 14 and 21 days of growth in hydroponic solution, respectively, and were recovered sporadically when replanted into soil at day 28. Similarly, Klerks et al. (2007a, 2007b) demonstrated that lettuce seedlings planted in manure-amended soil inoculated with *S.* Typhimurium or *S.* Enteritidis did not show evidence of internalized bacteria when plants were analyzed after six weeks, but 3.9 and 4.2 log CFU/g of *S.* Typhimurium and *S.* Enteriditis were recovered when plants were grown in *Salmonella*-inoculated Hoagland's agar. However, *S.* Dublin did internalize to plants when lettuce seedlings were grown in manure-amended soil (2.21 log CFU/g) and in Hoagland's solutions (4.6 log CFU/g), indicating that *S.* Dublin may be more adept at surviving in internal tissues of lettuce plants than other *Salmonella* serovars.

It is possible that *E. coli* O157:H7 and *Salmonella* spp. have different capabilities to internalize to plant tissues. Klerks et al. (2007b) showed the presence of internalized *Salmonella* in lettuce plants in intracellular spaces between epidermal cells, and in a few cases, were observed in the pericycle and the vascular system. Internalization of foodborne pathogens may be more frequent in hydroponic systems because there is less competition from rhizospheric bacteria that are present in soil and may have physiological fitness advantages when colonizing root tissue, providing a platform for internalization (Klerks et al., 2007a).

Studies of tomato plants grown in *Salmonella*-inoculated hydroponic solutions showed the pathogen was associated with the stems, leaves, and the hypocotyl regions of plants, although no surface sterilization treatment of the tissue was performed (Guo et al., 2002) to definitively indicate that these bacteria were internalized. As expected, more consistent *Salmonella* populations were recovered from the hypocotyl region than the stem of leaf tissues, regardless of whether or not roots were cut before placing in hydroponic solution (Guo et al., 2002). Other authors have observed fluorescent EHEC in relatively underdeveloped vasculature of hypocotyls, which is more supportive of relatively uninhibited entry by *E. coli* O157:H7 (Wachtel et al., 2002).

Collectively, the results from the presented studies demonstrate that it is possible for plants to internalize bacterial foodborne pathogens through root uptake, but this phenomenon likely requires a constant high population of the pathogen in the vicinity of the root system. There does not seem to be a set of conditions that encourages consistent root uptake of pathogens similar to the positive temperature differential that promotes the internalization of bacteria to fruits, although the hydroponic growing environment may be more likely to encourage uptake of bacteria. However, no human pathogenic cells have been shown to internalize through uptake and then progress through the vasculature to leaf tissue.

LIMITED EFFICACY OF CONVENTIONAL DECONTAMINATION METHODOLOGIES

Efficient and complete decontamination of produce and fresh-cut produce items is made difficult by the attachment and survival strategies utilized by bacteria on fruit and vegetable surfaces. Further processing of fresh-cut produce may also make bacterial cells harder to remove or inactivate. Through direct microscopic observation, it was estimated that 10^5 CFU bacteria/cm^2 were present on healthy spinach leaves (Warner et al., 2008). Phyllobacteria have been shown to colonize at various sites in and on leaf surfaces, including the base of trichomes, at stomata, epidermal cell wall junctions, as well as in grooves along veins, depressions in the cuticle, and beneath the cuticle

(Beattie and Lindow, 1999). It is less clear if human enteric pathogens like *E. coli* O157:H7, *Salmonella*, and *Listeria monocytogenes* demonstrate similar behavior. *Salmonella* Thompson was shown to attach around stomata of spinach leaves and in cell margins, similar to where native bacterial biofilms and microcolonies were detected (Warner et al., 2008).

Other workers have shown that bacterial cells introduced to the leaf surface have a better chance of surviving when they are deposited on or in aggregates of other bacteria (Monier and Lindow, 2005). These aggregates are characterized by an exopolysaccharide matrix that contains a dense population of bacterial cells (Monier and Lindow, 2003). If foodborne pathogens are in these aggregates, this may limit the effectiveness of sanitizer treatments on produce. *Salmonella* Thompson and *Pantoea agglomerans* were shown to form aggregates on cilantro leaves (Brandl and Mandrell, 2003), and *Wausteria paucula* supported the survival of *E. coli* O157:H7 on lettuce leaves and in the rhizosphere (Cooley et al., 2006). The fungal phytopathogens *Cladosporium cladosporioides* and *Penicillium expansum* promoted the colonization and infiltration of *Salmonella* in wounded cantaloupe tissue (Richards and Beuchat, 2005). These studies demonstrate that the survival strategies utilized by foodborne bacterial pathogens on produce surfaces may promote their survival in the presence of hypochlorite or other sanitizers.

Other studies have shown that foodborne pathogens attach preferentially to cut surfaces, where more nutrients may be available for their growth and survival (Boyer et al., 2007; Takeuchi and Frank, 2000). *E. coli* O157:H7 and *L. monocytogenes* attached in greater numbers to cut lettuce leaves compared to whole leaf surfaces, whereas *Salmonella* Typhimurium attached equally well to both intact and cut surfaces (Takeuchi et al., 2000), but only *Pseudomonas fluorescens* formed microcolonies on intact leaves. *E. coli* O157:H7 cells penetrate into cut surfaces of lettuce more efficiently when stored at 4 °C compared to storage at 7, 25, or 37°C (Takeuchi and Frank, 2000), and treatment of inoculated cut surfaces with 200 μg/ml of free chlorine did reduce *E. coli* O157:H7 cells attached to cut surfaces by between 0.7 and 1.0 log CFU/g but did not eliminate them completely.

Other work has indicated that *E. coli* O157:H7 cells in stomata are only marginally protected from produce wash solutions (Takeuchi and Frank, 2001a). These same authors showed that cells that penetrated furthest into the stomata were more protected from killing by free chlorine than cells on damaged tissue surface or cells on intact surfaces of lettuce (Takeuchi and Frank, 2001b).

The oxidative capacity of free chlorine in hypochlorite solutions is inactivated in the presence of organic matter (Marriott, 1999). Organic matter released from cut tissue of produce quickly and effectively inactivates free chlorine, allowing bacterial cells in or on produce tissue to survive. The exopolysaccharide (EPS) surrounding bacterial biofilms on produce may also inactivate the free chlorine in hypochlorite solutions. Washing inoculated lettuce in chlorinated water (100 ppm) reduced *L. monocytogenes* populations by only 0.7 log CFU/g, whereas a peracetic acid wash reduced counts by 1.7 log CFU/g (Hellstrom et al., 2006). Other workers have shown that dipping inoculated lettuce in 100 ppm hypochlorite solution results in about 2 log CFU/g reduction of nonpathogenic *E. coli* (Behrsing et al., 2000). Populations of *Salmonella* Baildon on diced tomatoes and on shredded lettuce dipped in 200 ppm hypochlorite for 40 s were reduced by less than 1 log CFU/g (Weissinger et al., 2000).

Processing operations related to leafy greens may also encourage bacterial infiltration and attachment to produce surfaces. Lettuce leaves inoculated with *E. coli* O157:H7 and subsequently vacuum cooled had higher populations (ca. 1 log CFU/g) of the pathogen infiltrate into the leaf than those leaves that were not subjected to vacuum cooling (Li et al., 2008). Mechanical damage occurring during packing and transport may also result in favorable conditions for bacterial attachment. Damage of stems of lettuce plants resulted in the release of sugar-containing latex, which supported the growth and rapid increase of *E. coli* O157:H7 populations (Brandl, 2008). This latex may also prevent gaseous and liquid sanitizers from penetrating tissues and inactivating bacteria. Inoculated, shredded romaine lettuce leaves supported the more rapid growth of *E. coli* O157:H7 over inoculated intact, bruised, and cut romaine lettuce leaves (Brandl, 2008).

The relative inefficiency of liquid sanitizers has put recent focus on gaseous sanitizers, which may have more oxidative capability than liquid treatments to kill foodborne pathogens on produce. Chlorine dioxide (ClO_2) is an attractive alternative to hypochlorite solutions because of its increased oxidative capacity compared to hypochlorite, and its enhanced stability in the presence of organic matter (Marriott, 1999). Treatment of produce inoculated with *Salmonella* spp. with 100 mg of released ClO_2 gas for one hour at 23 °C resulted in a 2 log CFU reduction on bell peppers, 5 to 6 log CFU on cucumbers, and 1 to 4 log CFU reduction on strawberries (Gyun-Yuk et al., 2005). Sy et al. (2005) reported that ClO_2 gas was effective in killing populations of *Salmonella*, *E. coli* O157:H7, and *L. monocytogenes* on apples, tomatoes, and onions but not on peaches, cabbage, carrots, or lettuce.

Gaseous ozone treatments also have been assessed on produce commodities. Ozone is a promising sanitizing alternative for produce because of its rapid oxidative activity (Marriott, 1999). A combination of continuous (no pressure) and pressurized ozone (83 kPa) for 128 minutes reduced *Salmonella* and *E. coli* O157:H7 populations by 2.6 and 2.9 log CFU/g on strawberries and 3.6 and 3.8 log CFU/g on raspberries, respectively (Bialka et al., 2007).

Treatment with 5 ppm ozonated water for five minutes on lettuce inoculated with either *E. coli* O157:H7 or *L. monocytogenes* resulted in reductions of about 1 log CFU/g (Gyun Yuk et al., 2006), indicating that the gaseous forms of ozone may be more effective than the liquid form of ozone.

The generation of ozone, other reactive oxygen species and nitric oxide by atmospheric plasma, and its application for two and five minutes to apples inoculated with *Salmonella* and *E. coli* O157:H7, resulted in population reductions greater than 2 and 3 log CFU, respectively (Critzer et al., 2007). Cold plasma treatments (3 min) of inoculated apples resulted in reductions of 2.9 to 3.7 log CFU/ml and 3.4 to 3.6 log CFU/ml of *Salmonella* Stanley and *E. coli* O157:H7, respectively (Niemira and Sites, 2008).

CONCLUSION

From the studies detailed in this chapter, no intervention strategy has been proven capable of inactivating or removing attached bacteria from raw produce surfaces. The reasons for this are complex, but answers may be found in the interaction between bacteria and the plant environment. Pathogens originate from animals, survive well in water and farm wastes, attach strongly to plant tissue, and survive the harsh environment through consumption. The results from the research presented in this chapter provide some answers as to why the public has witnessed an increase in outbreaks arising from produce. The close proximity between animal and vegetable growing operations will continue to present opportunities for outbreaks arising from fresh produce. Our current understanding of the ecology, genetics, and physiology of pathogens attached to plant tissue is rudimentary at best. Researchers will continue to probe the interactions between pathogens and plant surfaces with the goal of finding methods to reduce attachment and promote inactivation. However, the ultimate goal of a pathogen-free produce supply will depend more on the vigilance of individual growers, cooperation from industry groups, and a focused effort from government agencies.

REFERENCES

Agle, M. E. (2003). *Shigella boydii* 18: Characterization and biofilm formation. Ph.D. thesis, University of Illinois, Champaign, IL.

Annous, B. A., Burke, A., and Sites, J. E. (2004). Surface pasteurization of whole fresh cantaloupes inoculated with *Salmonella* Poona or *Escherichia coli*. *Journal of Food Protection* **67,** 1876–1885.

Annous, B. A., Solomon, E. B., Cooke, P. H., and Burke, A. (2005). Biofilm formation by *Salmonella* spp. on cantaloupe melons. *Journal of Food Safety* **25,** 276–287.

Anonymous. (2007). Outbreak alert! Closing the gaps in our federal food-safety net. Center for Science in the Public Interest. Washington, DC. Available at www.cspinet.org

Barak, J. D., Gorski, L., Naraghi-Arani, P., and Charkowski, A. O. (2005). *Salmonella enterica* virulence genes are required for bacterial attachment to plant tissue. *Applied and Environmental Microbiology* **71,** 5685–5691.

Barak, J. D., Jahn, C. E., Gibson, D. L., and Charkowski, A. O. (2007). The role of cellulose and O-antigen capsule in the colonization of plants by *Salmonella enterica*. *Molecular Plant Microbe Interactions* **20,** 1083–1091.

Barak, J. D., Liang, A., and K-E. Narm, K.-E. (2008). Differential attachment to and subsequent contamination of agricultural crops by *Salmonella enterica*. *Applied and Environmental Microbiology* **74,** 5568–5570.

Beattie, G. A. and Lindow, S. E. (1999). Bacterial colonization of leaves: A spectrum of strategies. *Phytopathology* **89,** 353–359.

Beuchat, L. R. (1999). Survival of enterohemorrhagic *Escherichia coli* O157:H7 in bovine feces applied to lettuce and the effectiveness of chlorinated water as a disinfectant. *J. Food Prot.* **62,** 845–849.

Besser, T. E., Richards, D. L., Rice, D. H., and Hancock, D. D. (2001). *Escherichia coli* O157:H7 infection of calves: Infectious dose and direct contact transmission. *Epidemiology and Infection* **127,** 555–560.

Bialka, K. L. and Demirci, A. (2007). Utilization of gaseous ozone for the decontamination of *Escherichia coli* O157:H7 and *Salmonella* on raspberries and strawberries. *Journal of Food Protection* **70,** 1093–1098.

Bowen, A., Fry, A., Richards, G., and Beuchat L. (2006). Infections associated with cantaloupe: A public health concern. *Epidemiology and Infection* **134,** 675–685.

Boyer, R. R., Sumner, S. S., Williams, R. C. et al. (2007). Influence of curli expression by *Escherichia coli* O157:H7 on the cell's overall hydrophobicity, charge, and ability to attach to lettuce. *Journal of Food Protection* **70,** 1339–1345.

Buchanan, R. L., Edelson, S. G., Miller, R. L., and Sapers, G. M. (1999). Contamination of intact apples after immersion in an aqueous environment containing *Escherichia coli* O157:H7. *Journal of Food Protection* **62,** 444–450.

Behrsing, J., Winkler, S., Franz, P., and Premier, R. (2000). Efficacy of chlorine for inactivation of *Escherichia coli* on vegetables. *Postharvest Biology and Technology* **19,** 187–192.

Brandl, M. T. and Mandrell., R. E. (2003). Fitness of *Salmonella enterica* serovar Thompson in the cilantro phyllosphere. *Applied and Environmental Microbiology* **68,** 3614–3621.

Brandl, M. T. (2006). Fitness of human enteric pathogens on plants and implications for food safety. *Annual Reviews of Phytopathology* **44,** 367–392.

Brandl, M. T. (2008). Plant lesions promote the rapid multiplication of *Escherichia coli* O157:H7 on post-harvest lettuce. *Applied and Environmental Microbiology* **74,** 5285–5289.

California Food Emergency Response Team. (2007). Investigation of an *Escherichia coli* O157:H7 outbreak associated with Dole pre-packaged spinach. California Department of Health Services, Sacramento, CA.

CDC. (2007). Multistate outbreaks of *Salmonella* infections associated with raw tomatoes eaten in restaurants-United States, 2005–2006. *Morbidity and Mortality Weekly Report* **56**, 909–911.

Chmielewski, R.A.N. and Frank, J. F. (2003). Biofilm formation and control in food processing facilities. *Comprehensive Reviews in Food Science and Food Safety* **2**(1), 22–32.

Cooley, M. B., Chao, D., Mandrell, R. E. (2006). *Escherichia coli* O157:H7 survival and growth on lettuce is altered by the presence of epiphytic bacteria. *Journal of Food Protection* **69**, 2329–2335.

Cooley, M., Carychao, D., Crawford-Miksza, L. et al. (2007). Incidence and tracking of *Escherichia coli* O157:H7 in a major produce production region in California. *PlOS One* **2**(11), e1159.

Costerton, J. W., Stewart, P. S., and Greenberg, E. P. (1999). Bacterial biofilms: A common cause of persistent infections. *Science* **284**, 1318–1322.

Critzer, F., Kelly-Wintenberg, K., Souther, S. L., and Golden, D. (2007). Atmospheric plasma inactivation of foodborne pathogens on fresh produce surfaces. *Journal of Food Protection* **70**, 2290–2296.

D'Aoust, J.-Y. (1998). *Salmonella* Species. In *Food Microbiology and Frontiers* (Doyle, M. P., Beuchat, L. R., and Montville, T. J., Eds.). ASM Press, Washington, DC.

Danhorn, T. and Fuqua, C. (2007). Biofilm formation by plant associated bacteria. *Annual Review of Microbiology* **61**, 401–422.

Delaquis, P., Stewart, S., Cazaux, S. and Toivonen, P. (2002). Survival and growth of *Listeria monocytogenes* and *Escherichia coli* O157:H7 in ready-to-eat iceberg lettuce washed in warm chlorinated water. *Journal of Food Protection* **65**, 459–464.

Delaquis, P., Bach, S., and Dinu, L-D. (2007). Behavior of *Escherichia coli* O157:H7 in leafy vegetables. *J. Food Prot.* **70**, 1966–1974.

Dong, Y., Iniguez, A. L., Ahmer, B. M., and Triplett, E. W. (2003). Kinetics and strain specificity of rhizosphere and endophytic colonization by enteric bacteria on seedlings of *Medicago sativa* and *Medicago truncatula*. *Applied and Environmental Microbiology* **69**, 1783–1790.

Duffy, G. (2003). Verocytoxigenic *Escherichia coli* in animal faeces, manures and slurries. *Journal of Applied Microbiology*, Supplement: 94–103.

Eblen, S. B., Walderhaug, M. O., Edelson-Mammel, S. et al. (2004). Potential for internalization, growth, and survival of *Salmonella* and *Escherichia coli* O157:H7 in oranges. *Journal of Food Protection* **67**, 1578–1584.

Elder, R. O., Keen, J. E., Siragusa, G. R. et al. (2000). Correlation of enterohemorrhagic *Escherichia coli* O157:H7 prevalence in feces, hides, and carcasses of beef cattle during processing. *Proceedings of the National Academy of Sciences, USA* **97**, 2999–3003.

Erickson, M. C. and Doyle, M. P. (2007). Food as a vehicle for transmission of shiga toxin-producing *Escherichia coli*. *Journal of Food Protection* **70**, 2426–2449.

Franz, E., Visser, A. A., Van Diepeningen, A. D. et al. (2007). Quantification of contamination of lettuce by GFP-expressing *Escherichia coli* O157:H7 and *Salmonella enterica* serovar Typhimurium. *Food Microbiology* **24**, 106–112.

Garrett, E. H., Gorny, J. R., Beuchat, L. R. et al. (2003). Microbiological safety of fresh and fresh-cut produce: Description of the situation and economic impact. *Comprehensive Reviews in Food Science and Food Safety* **2**(Suppl), 13–19.

Greene S. K., Daly, E. R., and Talbot, E. A. (2008). Recurrent multistate outbreak of *Salmonella* Newport associated with tomatoes from contaminated fields, 2005. *Epidemiology and Infection* **136**, 157–165.

Guo, X., Chen, J., Brackett, R. E., and Beuchat, L. R. (2001). Survival of salmonellae on and in tomato plants from the time of inoculation at flowering and early stages of fruit development through fruit ripening. *Applied and Environmental Microbiology* **67**, 4760–4764.

Guo, X., van Iersel, M. W., Chen, J. et al. (2002a). Evidence of association of salmonellae with tomato plants grown hydroponically in inoculated nutrient solution. *Applied and Environmental Microbiology* **68**, 3639–3643.

Guo, X., Chen, J., Brackett, R. E., and Beuchat, L. R. (2002b). Survival of *Salmonella* on tomatoes stored at high relative humidity, in soil, and on tomatoes in contact with soil. *Journal of Food Protection* **65**, 274–279.

Han, Y., Sherman, D. M., Linton R. H. et al. (2000). The effects of washing and chlorine dioxide gas on survival and attachment of *Escherichia coli* O157: H7 to green pepper surfaces. *Food Microbiol.* **17**, 521–533.

Heaton, J. C. and Jones, K. (2007). Microbial contamination of fruit and vegetables and the behaviour of enteropathogens in the phyllosphere: A review. *J. Appl. Microbiol.* **104**, 613–626.

Hellstrom, S., Kervinen, R., Lyly, M. et al. (2006). Efficacy of disinfectants to reduce *Listeria monocytogenes* in precut iceberg lettuce. *Journal of Food Protection* **69**, 1565–1570.

Herman, K, Ayers, T. L., and Lynch, M. (2008). Foodborne disease outbreaks associated with leafy greens, 1973–2006. *International Conference on Emerging Infectious Diseases*, March 16–19, 2008. Atlanta, GA.

Himathongkham, S. Bahari, S., Riemann, H., and Cliver, D. (1999). Survival of *Escherichia coli* O157:H7 and *Salmonella typhimurium* in cow manure and cow manure slurry. *FEMS Microbiology Letters* **178**, 251–257.

Hora, R., Warriner, K., Shelp, B. J., and Griffiths, M. W. (2005). Internalization of *Escherichia coli* O157:H7 following biological and mechanical disruption of growing spinach plants. *Journal of Food Protection* **68**, 2506–2509.

Hutchison, M. L., Walters, L. D., Moore, A., and Avery S. M. (2005). Declines of zoonotic agents in liquid livestock wastes stored in batches on-farm. *Journal of Applied Microbiology* **99**, 58–65.

Ibekwe, M. A., Watt, P. M., Shouse, P. J., and Grieve, C. M. (2004). Fate of *Escherichia coli* O157:H7 in irrigation water on soils and plants as validated by culture method and real-time PCR. *Canadian Journal of Microbiology* **50**, 1007–1014.

Iniguez, A. L., Dong, Y., Carter, H. D. et al. (2005). Regulation of enteric endophytic bacterial colonization by plant defenses. *Molecular Plant Microbe Interactions* **18**, 169–178.

Islam, M., Doyle, M. P., and Phatak, S. C. et al. (2004). Persistence of enterohemorrhagic *Escherichia coli* O157:H7 in soil and on leaf lettuce and parsley grown in fields treated with contaminated manure composts or irrigation water. *Journal of Food Protection* **67**, 1365–1370.

Iturriaga, M. H., Escartin, E. F. et al. (2003). Effect of inoculum size, relative humidity, storage temperature, and ripening stage on the attachment of *Salmonella* Montevideo to tomatoes and tomatillos. *J. Food Prot.* **66**, 1756–1761.

Iturriaga, M. H., Tamplin, M. L., and Escartin, E. F. (2007). Colonization of tomatoes by *Salmonella* Montevideo is affected by relative humidity and storage temperature. *Journal of Food Protection* **70**, 30–34.

Jablasone, J., Warriner, K., and Griffiths, M. (2005). Interactions of *Escherichia coli* O157:H7, *Salmonella typhimurium*, and *Listeria monocytogenes* plants cultivated in a gnotobiotic system. *International Journal Food Microbiology* **99**, 7–18.

Johannessen, G. S., Bengtsson, G. B., Heier, B. T. et al. (2005). Potential uptake of *Escherichia coli* O157:H7 from organic manure intro crisphead lettuce. *Applied and Environmental Microbiology* **71**, 2221–2225.

Klerks, M. M., Franz, E., van Gent-Pelzer, M., et al. (2007a). Differential interaction of *Salmonella enteric* serovars with lettuce cultivars and plant-microbe factors influencing the colonization efficiency. *The ISME Journal* **1**, 620–631.

Klerks, M. M., van Gent-Pelzer, M., Franz, E. et al. (2007b). Physiological and molecular responses of *Lactuca sativa* to colonization by *Salmonella enterica* serovar Dublin. *Applied and Environmental Microbiology* **73**, 4905–4914.

Kolling, G. L. and Matthews, K. R. (2001). Examination of recovery in vitro and in vivo of nonculturable *Escherichia coli* O157:H7. *Applied and Environmental Microbiology* **67**, 3928–3933.

Kudva, I. T., Blanch, K., and Hovde, C. J. (1998). Analysis of *Escherichia coli* O157:H7 survival in ovine or bovine manure and manure slurry. *Appl. Environ. Microbiol* **64**, 3166–3174.

Li, H., Tajkarimi, M., and Osburn, B. I. (2008). Impact of vacuum cooling on *Escherichia coli* O157:H7 infiltration into lettuce tissue. *Applied and Environmental Microbiology* **74**, 3138–3142.

Liao, C-H. and Cooke, P. H. (2001). Response to trisodium phosphate treatment of *Salmonella* Chester attached to fresh-cut green pepper slices. *Canadian Journal of Microbiology* **47**, 25–32.

Marriott, N. (1999). *Principles of Food Sanitation*. Aspen Publishers, Inc., Gaithersburg, MD.

Matthews, L., Low, J. C., Gally, D. L. et al. (2006). Heterogeneous shedding of *Escherichia coli* O157:H7 in cattle and its implications for control. *Proceedings of National Academy of Sciences USA* **103**, 547–552.

Matthysse, A. G., Deora, R., Mishra, M., and Torres, A. G. (2008). Polysaccharides cellulose, poly-B-1,6-N-acetyl-D-glucosamine, and colanic acid are required for optimal binding of *Escherichia coli* O157:H7 strains to alfalfa sprouts and K-12 strains to plastic but not for binding to epithelial cells. *Applied and Environmental Microbiology* **74**, 2384–2390.

Monier, J.-M. and Lindow, S. E. (2005). Aggregates of resident bacteria facilitate survival of immigrant bacteria on leaf surfaces. *Microbial Ecology* **49**, 343–352.

Niemira, B. A. and Sites, J. (2008). Cold plasma inactivates *Salmonella* Stanley and *Escherichia coli* O157:H7 inoculated on golden delicious apples. *Journal of Food Protection* **71**, 1357–1365.

Penteado, A. L., Eblen, B. S., and Miller, A. J. (2004). Evidence of *Salmonella* internalization into fresh mangos during simulated postharvest insect disinfestation procedures. *Journal of Food Protection* **67**, 181–184.

Richards, G. R. and Beuchat, L. R. (2005). Metabiotic associations of molds and *Salmonella* Poona on intact and wounded cantaloupe rind. *International Journal of Food Microbiology* **97**, 329–339.

Seo K. H. and Frank, J. F. (1999). Attachment of *Escherichia coli* O157:H7 to lettuce leaf surface and bacterial viability in response to chlorine treatment as demonstrated using confocal scanning laser microscopy. *Journal of Food Protection* **62**, 3–9.

Sharma, M. and Donnenberg, M. (2008). A novel approach to investigate internalization of *Escherichia coli* O157:H7 in lettuce and spinach. Fresh Express Produce Safety Research Conference. September 11, 2008, Monterey, CA.

Shaw, R. K., Berger, C. N., Feys, B. et al. (2008). Enterohemorrhagic *Escherichia coli* exploits EspA filaments for attachment to salad leaves. *Applied and Environmental Microbiology* **74**, 2980–2914.

Shi, X, Namvar, A., Kostrzynska, M. et al. (2007). Persistence and growth of different *Salmonella* serovars on pre- and postharvest tomatoes. *Journal of Food Protection* **70**, 2725–2731.

Sivapalasingam, S., Friedman, C. R., Cohen, L., and Tauxe, R. V. (2004). Fresh produce: A growing cause of outbreaks of foodborne illness in the United States, 1973 through 1997. *Journal of Food Protection* **67**, 2342–2353.

Solomon, E. B., Yaron, S., and Matthews, K. R. (2002). Transmission of *Escherichia coli* O157:H7 from contaminated manure and irrigation water to lettuce plant tissue and its subsequent internalization. *Applied and Environmental Microbiology* **68**, 397–400.

Solomon, E. B. and Matthews, K. R. (2005). Use of fluorescent microspheres as a tool to investigate bacterial interactions with growing plants. *Journal of Food Protection* **68**, 870–873.

Solomon, E. B., Brandl, M. T., and Mandrell, R. E. (2006). Biology of foodborne pathogens on produce. In *Microbiology of Fresh Produce* (K. R. Matthews, Ed.). ASM Press, Washington, DC.

Stine, S. W., Song, I., Choi, C. Y., and Gerba, C. P. (2005). Effect of relative humidity on preharvest survival of bacterial and viral pathogens on the surface of cantaloupe, lettuce, and bell peppers. *Journal of Food Protection* **68**, 1352–1358.

Sy, K., Murray, M. B., Harrison, M. D., and Beuchat, L. R. (2005). Evaluation of gaseous chlorine dioxide as a sanitizer for killing *Salmonella*, *Escherichia coli* O157:H7, *Listeria monocytogenes*, and yeasts and molds on fresh and fresh-cut produce. *Journal of Food Protection* **68**, 1176–1187.

Takeuchi, K. and Frank, J. F. (2000). Penetration of *Escherichia coli* O157:H7 into lettuce tissues as affected by inoculums size and temperature and the effect of chlorine treatment on cell viability. *Journal of Food Protection* **63**, 434–440.

Takeuchi, K. and Frank, J. F. (2001). Quantitative determination of the role of lettuce leaf structures in protecting *Escherichia coli* O157:H7 from chlorine disinfection. *Journal of Food Protection* **64**, 147–151.

Takeuchi, K. and Frank, J. F. (2001). Direct Microscopic observation of lettuce leaf decontamination with a prototype fruit and vegetable washing solution and 1% NaCl-NaHCO$_3$. *Journal of Food Protection* **64**, 1235–1239.

Takeuchi, K., Matute, C. M., Hassan, A. N., and Frank, J. F. (2000). Comparison of the attachment of *Escherichia coli* O157:H7, *Listeria monocytogenes*, *Salmonella* Typhimurium and *Pseudomonas fluorescens* to lettuce leaves. *Journal of Food Protection* **63**, 1433–1437.

Ukuku, D. O. (2004). Effect of hydrogen peroxide treatment on microbial quality and appearance of whole and fresh-cut melons contaminated with *Salmonella*. *International Journal of Food Microbiology* **95**, 137–146.

Ukuku, D. O. and Sapers, G. M. (2001). Effect of sanitizer treatments on *Salmonella* Stanley attached to the surface of cantaloupes and cell transfer to fresh-cut tissues during cutting practices. *Journal of Food Protection* **64**, 1286–1291.

Ukuku, D. O. and Fett, W. F. (2002). Relationship of cell surface charge and hydrophobicity to strength of attachment of bacteria to cantaloupe rind. *Journal of Food Protection* **65**, 1093–1099.

Ukuku, D. O. and Fett, W. F. (2004). Effect of nisin in combination with EDTA, sodium lactate, and potassium sorbate for reducing *Salmonella* on whole and fresh-cut cantaloupe. *Journal of Food Protection* **67**, 2143–2150.

Ukuku, D. O. and Fett, W. F. (2006). Effects of cell surface charge and hydrophobicity on attachment of 16 *Salmonella* serovars to cantaloupe rind and decontamination with sanitizers. *Journal of Food Protection* **69**, 1835–1843.

Wachtel, M. R., Whitehand, L. C., and Mandrell, R. E. (2002). Association of *Escherichia coli* O157:H7 with preharvest leaf lettuce upon exposure to contaminated irrigation water. *Journal Food Protection* **65**, 18–25.

Wang, G., Zhao, T., and Doyle, M. P. (1996). Fate of enterohemorrhagic *Escherichia coli* O157:H7 in bovine feces. *Applied and Environmental Microbiololgy* **62**, 2567–2570.

Wang, G. and Doyle, M. P. (1998). Survival of enterohemhorrhagic *Escherichia coli* O157:H7 in water. *Journal of Food Protection* **61**, 662–667.

Warner, J. C., Rothwell, S. D., and Keevil, C. W. (2008). Use of episcopic differential interference contrast microscopy to identify bacterial biofilms on salad leaves and track colonization by *Salmonella* Thompson. *Environmental Microbiology* **10**, 918–925.

Weissinger, W. R., Chantarapanont, W., and Beuchat, L. R. (2000). Survival and growth of *Salmonella baildon* in shredded lettuce and diced tomatoes, and effectiveness of chlorinated water as a sanitizer. *International Journal of Food Microbiology* **62**, 123–131.

Yuk, H.-G., Bartz, J. A., and Schneider, K. R. (2006). The effectiveness of sanitizer treatments in inactivation of *Salmonella* spp. from bell pepper, cucumber, and strawberry. *Journal of Food Science* **71**, 95–99.

Yuk, H.-G., Yoo, M-Y., Yoon, J.-W. et al. (2006). Effect of combined ozone and organic acid treatment for control of *Escherichia coli* O157:H7 and *Listeria monocytogenes* on lettuce. *Journal of Food Science* **71**, 83–87.

Zhuang, R-Y, Beuchat, L. R., and Angulo, F. J. (1995). Fate of *Salmonella* Montevideo on and in raw tomatoes as affected by temperature and treatment with chlorine. *Applied and Environmental Microbiololgy* **61**, 2127–2131.

PART 2

Sources of Contamination

CHAPTER 3

Identification of the Source of Contamination

Jeff Farrar, DVM, Ph.D., MPH
Food and Drug Branch, California Department of Public Health

Jack Guzewich, RS, MPH
US Food and Drug Administration, Center for Food Safety and Applied Nutrition

CHAPTER CONTENTS

Introduction	50
Overview: Phases of a Foodborne Outbreak Investigation	50
Surveillance and Detection	50
Epidemiologic	51
Environmental	53
Traceback Investigations	61
Regulatory/Enforcement	65
Prevention/Research	66
Training Needs for Environmental Investigators of Retail, Food-Processing Facilities, Packing Sheds, and Farms	66
Resuming Operations	67
Farm Investigations	68
Packinghouse Investigations	71
Vacuum Cooler/Hydrocooler Investigation	72
Fresh Cut Produce Processor Investigations	72
Intentional Contamination	73
Lessons Learned	73
Recommendations	75

INTRODUCTION

State and local health departments make up the backbone of the system to detect and investigate foodborne outbreaks and illnesses in the United States. Numerous types of expertise are required in these investigations including knowledge and experience in surveillance systems, epidemiological investigations, laboratory methods, bacterial and viral ecology, water engineering, and environmental investigations. Some health departments have staffing, experience, communication capabilities, training, and interest in some of these areas. However, few agencies have standardized, documented, and practiced procedures and performance standards for the entire foodborne outbreak investigation process. Even fewer have dedicated, highly trained, multidisciplinary teams that work together throughout the entire investigative process. Without a multidisciplinary, multiagency approach, using trained and experienced experts throughout, foodborne outbreak investigations will remain unable to consistently and accurately identify the causes of the outbreaks and develop effective prevention measures to reduce the likelihood of recurrence.

This chapter will provide an overview of the entire foodborne outbreak investigative process as a means of identifying the source of contamination but will focus primarily upon the environmental phase of the investigation.

OVERVIEW: PHASES OF A FOODBORNE OUTBREAK INVESTIGATION

Foodborne outbreak investigations can be roughly characterized into five phases: surveillance/detection, epidemiologic, environmental/traceback, regulatory/enforcement, and prevention/research. This characterization assumes that laboratory diagnostics are part of each of these five phases. These separate, yet overlapping and related phases require very different knowledge, methods, expertise, and legal authority. However, investigators involved in all phases must interact and work together throughout the entire process if the complete story of an outbreak is to be revealed and the risk of additional outbreaks is to be minimized.

Surveillance and Detection

The first phase, surveillance and detection, provides a signal to public-health investigators that an unexpected number of similar illnesses occurred during a certain time-period. These signals can be from a variety of active and passive surveillance efforts. Consumers may report illnesses to local health jurisdictions after eating at a restaurant or attending an event. Medical care providers may report diagnoses of foodborne pathogens to local or state public health agencies, local or state laboratories may notice a higher than expected number of similar pathogens during a specific time period, and emergency room care providers may notify local public health agencies of an unusual number of individuals with similar symptoms (e.g., bloody diarrhea). National surveillance systems, such as PulseNet, have proven very effective in identifying small outbreaks composed of a genotypically similar group of pathogens from a number of geographically widespread, sporadic cases that previously were not connected. However, there remains significant variability among states and local jurisdictions regarding the promptness and thoroughness of surveillance and reporting systems for foodborne illnesses and outbreaks.

Epidemiologic

Once the surveillance system has detected an unusual occurrence of disease, an epidemiologic investigation may be initiated. This stage attempts to identify the specific vehicle responsible for the illnesses, and to characterize the ill individuals in terms of age, gender, race, and location. Laboratories attempt to determine the agent in patient specimens concurrent with the epidemiologic investigation. Ill individuals are interviewed, and food histories are collected along with travel history and visits or exposures to known vectors such as animal petting zoos. Epidemiologists use well-described methods such as case control or cohort studies to statistically compare foods eaten by ill individuals to foods consumed by well individuals in order to identify a specific food vehicle. Even without laboratory confirmed vehicles, statistical associations derived from these investigations can provide compelling evidence of what food or foods caused the illnesses. However, epidemiologic investigations may not always be successful in identifying a specific food. Recall of specific food items consumed may dim after a couple of weeks, or multiple, unrelated food items may be identified.

The time required to complete an epidemiologic investigation varies but frequently requires days or weeks of repeated telephone or in-person interviews followed by descriptive and detailed data analysis. However, as in all other phases of a foodborne outbreak investigation, rapid completion is critical, especially when there is a reasonable likelihood of ongoing exposures to contaminated foods. Similar to the demand by public-health agencies for increased industry documentation of best growing, processing, and shipping practices, public-health agencies are beginning to address expectations of written performance standards and standard operating procedures for foodborne outbreak investigations (www.cste.org/pdffiles/fsreportfinal.pdf; www.fda.gov/ora/fed_state/NFSS/Default.htm).

Epidemiological investigations of foodborne outbreaks can result in findings suggestive of two possible broad sources of exposure: point source and multiple or continuing sources (e.g., tomatoes from the eastern shore of Virginia). In point source exposures, ill individuals are found to have a common event or place of exposure, for example eating at a single restaurant or attending the same catered event prior to onset of symptoms. Some point source exposure events may manifest as very low numbers of ill individuals that occur over extended periods of time or that reappear from time to time. These types of foodborne outbreaks warrant special attention to environmental or foodworker sources of contamination within the food facility. These point source exposures must be carefully assessed during the environmental investigation to determine the probability of foodworkers as the most likely source of contamination. Even with cultures, serology, detailed reviews of work logs and daily job responsibilities, preparation of product work flow charts, and one-on-one interviews of foodworkers without managers present, there may not be sufficient information to definitively rule out foodworkers. However, without such comprehensive, standardized efforts, public-health agencies will continue to be criticized for the lack of effort, and further investigation efforts at manufacturers or farms may not be considered practical.

Multiple source exposures, such as illnesses associated with multiple, unrelated restaurants in multiple states, strongly suggest that the source of contamination is a widely distributed product and that contamination likely occurred prior to the retail facility. In these situations, measuring and recording routine practices and procedures at retail facilities may still be useful to identify risk factors that may contribute to or detract from the survival or growth of pathogens (e.g., refrigeration temperatures, pH, water activity, times and temperatures at each stage, and whether the product was washed and how it was washed).

When specific foods are implicated in epidemiological investigations, ill individuals are further interviewed to determine precisely when and where they purchased or consumed the implicated food, to provide as much specific information about the food as possible including brand, type of product, size, grade, color, and whether they have any remaining opened or unopened product. This specific information is critical for tracebacks and environmental investigations. Without this specific information, further investigation may not be possible. Therefore, the highest priority should be assigned to dispatching staff to households of ill individuals to collect any remaining opened or unopened product implicated in the investigation. Asking family members to deliver remaining products only invites delays in initiating valuable laboratory testing efforts. If a specific retail food facility is determined to be associated with illnesses and the epidemiological investigation has not yet identified or is unable to identify a specific food, investigators should collect samples of all foods deemed to be a possible cause of the outbreak. Samples can be refrigerated or frozen until decisions are made on the epidemiological study.

Knowing that lettuce or tomatoes purchased from store A are associated with illness is important information but is of marginal utility in determining the ultimate source and cause of contamination and in preventing additional exposures to contaminated foods. Many stores carry numerous types and sizes of lettuce and tomatoes from multiple suppliers and sources. Standardized commodity-specific questionnaires should be developed in advance for foods implicated in previous outbreaks, and these questionnaires could be posted on Web sites by federal agencies for use by state and local agencies in subsequent multistate/multijurisdiction foodborne outbreak investigations.

Increasingly, epidemiologists attempt to confirm whether consumers may have utilized "club cards" in the purchase of contaminated foods. This objective data can be used to verify specific product information and purchase dates and thus can provide valuable information during investigations. Retailers generally provide club-card data to public-health investigators once permission to share the information is given by the consumer. These contacts, mechanisms, and standardized questionnaires for obtaining permission to use club-card information should be developed in advance by state agencies and shared with local agencies. Having multiple counties or states contacting a retail food chain asking for the same or similar information for different consumers only creates confusion, delays, and duplication of effort, and creates doubt regarding the ability of agencies to coordinate their efforts. Communication should include clarification of roles and responsibilities of participating agencies.

Providing frequent, current, detailed updates to those involved in the environmental phase of the investigation, especially during outbreaks with a reasonable probability of ongoing exposures, will allow regulators to better focus limited resources.

Environmental

When a specific food item is identified in the epidemiologic investigation, the third phase, the environmental investigation, attempts to determine how, where, when, and why the contamination occurred so that further exposures can be reduced or eliminated, and effective prevention measures can be implemented. Before initiating environmental investigations, it is critical that the results of the epidemiologic investigation be carefully reviewed and discussed to understand the inherent strengths and weaknesses. Ideally, outbreak investigators would be able to genetically match laboratory confirmed pathogens from ill individuals with laboratory confirmed pathogens from epidemiologically implicated foods and identify how and where the contamination occurred. However, this ideal investigation is more often the exception rather than the rule. Because of the inherent delays and variability in incubation periods, diagnostic testing, reporting of results, and in completing epidemiologic investigations, many perishable foods with a relatively short shelf-life may no longer be available for laboratory testing. Many fields growing these crops will have been replanted with other crops. Additionally, environmental investigations may take months of work, consuming huge amounts of resources and resulting in tremendous economic damage to firms and industries. Therefore, it is critical that regulatory agencies have trained, experienced epidemiologists and statisticians available to review and discuss epidemiological findings with communicable disease agencies before beginning the environmental investigation.

Although epidemiological studies are extremely powerful tools; analytical mistakes, incomplete investigations, biases in questionnaires/interviews, or purely chance associations with incorrect foods can result in erroneous conclusions with enormous consequences. A previous epidemiological investigation of illnesses in multiple states incorrectly linked illnesses to consumption of domestic strawberries, although imported raspberries were eventually identified as the vehicle responsible for the illnesses. This erroneous conclusion, based on an incomplete epidemiological investigation, resulted in incorrect preventive guidance to consumers, possibly resulting in additional exposures, extraordinary financial losses to the domestic strawberry industry, massive costs to regulators in the investigation of the wrong commodity, significant delays in implementing correct control measures, and a loss of credibility for the public-health community. Although it is critical that foodborne outbreak investigations be quickly completed, it is more important that the findings are correct.

Occasionally, decisions to initiate tracebacks and environmental investigations must be made with only preliminary or limited epidemiological findings. For example, outbreaks with a strong possibility of ongoing exposures to pathogens such as *E. coli* O157:H7 may require immediate actions with less than complete or definitive epidemiological findings.

Once a decision is made to initiate an environmental investigation, detailed written practices and procedures should be implemented. Receiving, storage, and food preparation practices should be carefully measured and documented. Factors such as time, temperature, water activity, and pH within food manufacturing and retail food facilities may contribute to the growth or survival of pathogens. Previous investigations of green onion-associated hepatitis A outbreaks revealed that some retail facilities stored green onions in containers with water without thoroughly cleaning and sanitizing the container between lots of green onions. This practice may have contributed to the spread of contamination to multiple batches. Accurate, thorough descriptions along with objective measurements using recently calibrated equipment, where appropriate, are essential.

Environmental investigations have been erroneously referred to as tracebacks. Tracebacks of contaminated foods attempt to document each stop or location in the farm-to-fork continuum. Tracebacks are one important part of environmental investigations; however, this terminology does not accurately convey the scope and complexity of this phase of the outbreak investigation.

Previously, evaluations of food facilities implicated in foodborne outbreaks were composed of routine inspections of food facilities, using established, entrenched regulatory inspection forms, checklists, and mindsets. Not surprisingly, this approach often yielded little in the way of useful information as to the cause of the outbreak and even less useful information regarding effective prevention measures. Findings of the lack of a suitable shield on fluorescent lights may have been documented as violations of existing law or regulations, but this finding had little to do with how *Salmonella* or *Shigella* was introduced, survived, or grew in or on a specific food, or which prevention measures could be implemented to reduce the risk of recurrence. The term "environmental investigation" includes tracebacks but also includes methodical, scientific reviews of each point in the farm-to-table continuum to assess opportunities for introduction, survival, and growth of pathogens.

Many regulatory agencies have recently adopted a more scientific, methodical, investigative approach to environmental and traceback investigations. Without this approach, foodborne outbreak investigations will continue to yield few clues as to the causes of outbreaks and prevention measures for reducing the probability of recurrence.

In order to better understand where and how contamination occurred and whether pathogens could have been introduced, survived, or could have grown during food preparation or food manufacturing conditions, investigators must have extensive knowledge of specific pathogens, their ecology, and the effect of pH, time, and temperature and other factors on the organism. When the source of the contamination is determined to be prior to the retail food facility, investigators must use the same concepts as in the retail food

CHAPTER 3: Identification of the Source of Contamination

facility to understand and document growing, harvesting, processing, and shipping practices associated with contaminated foods and food ingredients and must complete a scientific assessment of the probability of introduction, survival, and growth of the organism at each point.

Specifics of Environmental Investigations

Environmental investigations must include evaluations of each step in the farm-to-fork continuum to look for opportunities for introduction, survival, or growth of the pathogen. In essence, the investigators attempt to rule out contamination at each step in the process. Flow charts (Figure 3.1), capturing very specific measurements such as time, temperature, pH, amounts of

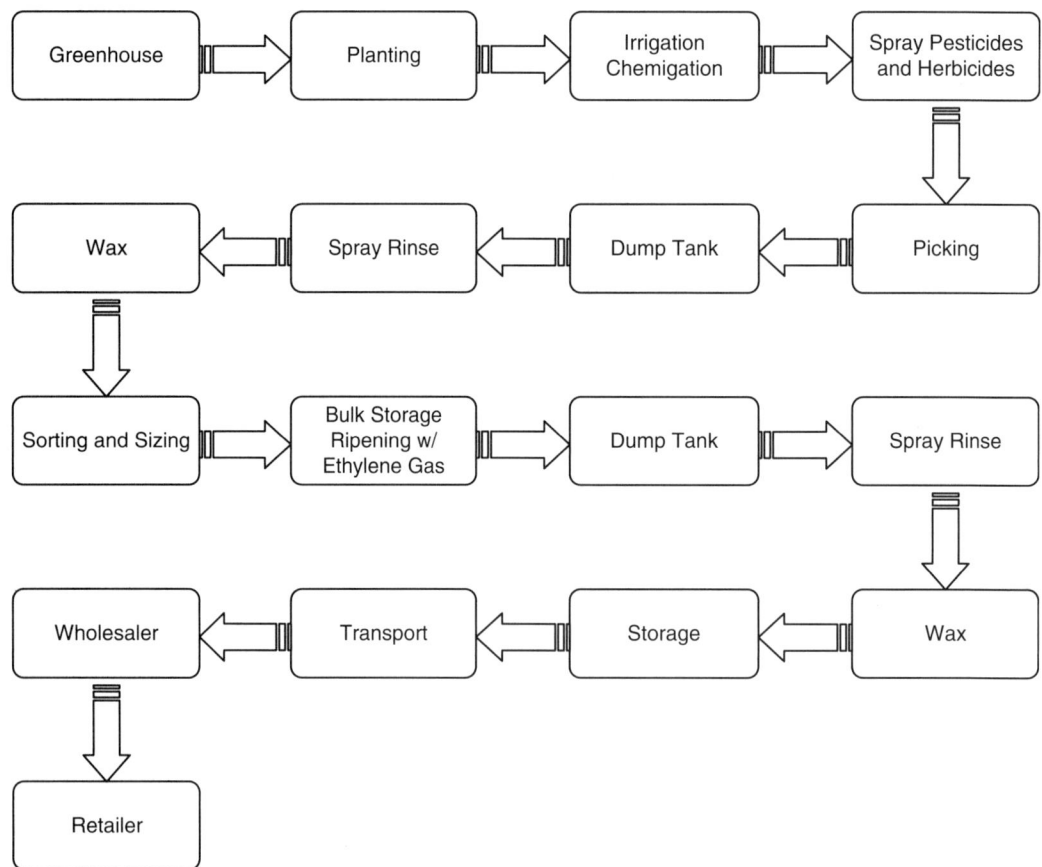

Critical information such as pH, times, temperatures, volume, water activity, and ingredients may need to be included to ensure a thorough understanding of the entire food preparation/food processing event.

FIGURE 3.1 *Generic example of a food process flow chart.*

ingredients, and size of containers are a highly recommended tool in environmental investigations. These charts can provide food scientists with objective data to better determine whether food preparation or processing procedures could have contributed to the outbreak, and more importantly, whether additional preventive steps or barriers should be implemented. Additionally, descriptions and documentation of cleaning and sanitation procedures and chemicals are important.

Environmental investigations of retail, wholesale, and food manufacturing facilities are identical in concept. Thorough assessment and documentation of the entire process at each step is critical. Each type of facility could be segmented into three broad areas: incoming ingredients, food processing/food preparation, and outgoing product. Previous environmental investigations of retail food and food-processing facilities were often conducted by a single individual. However, due to the scope, complexity, and specialized needs of these investigations, a single individual is likely insufficient to complete an environmental investigation promptly and thoroughly, even with single retail food facilities. In large processing facilities, several individuals with targeted training and experience are often necessary, each assigned to lead one of the three segments identified earlier, in order to complete the investigation in a thorough, timely manner.

These investigations cannot be completed thoroughly and accurately without having appropriate equipment. Most environmental investigators have access to obvious equipment such as digital thermometers, stop watches, and digital cameras. However, many have not traditionally utilized specialized equipment such as needle probes for thermocouple thermometers to measure the temperature of hamburgers while cooking, oven probes to monitor the temperature of products throughout the cooking cycle, oxidation-reduction potential meters, water activity meters, and waterproof data loggers that measure the temperature of dishwashers and flume water. Additional areas of expertise and equipment needed may include tracer dyes or compounds to assess cross-connections and flow of water or wastewater, air-stream monitoring and air sampling devices, and water sampling and concentration devices. Digital photographs and videos can often convey in a few seconds conditions or practices that may take pages to explain in writing. However, investigators should ensure that they have legal authority to take photographs prior to initiating the investigation.

Where possible, investigators are strongly encouraged to visually observe "routine" food preparation/food processing procedures and record objective measurements of preparation and processing practices, as well as routine cleaning and sanitation procedures. Often, what is written in procedures manuals is not what actually occurs in the kitchen or in the food-processing facility. Even with the inevitable bias introduced during physical observation by investigators, valuable clues often are obtained by simply observing and documenting a process from start to finish.

Food contamination may have been confirmed by an epidemiologic association between consumption of a food and illness or by laboratory recovery of an agent in a food or by observations of conditions during manufacturing, preparation, or serving that likely lead to contamination. In any of these cases, there may still be additional contaminated food available to consume. Investigators must work quickly with the firm responsible for processing/serving/selling the food to have any remaining food removed from distribution, collect samples where appropriate, determine distribution, and alert the public in all areas where the food may still be available. Many of the produce outbreaks in recent years have taught us that consumers will keep fresh-produce items in their homes much longer than the shelf-life assumed by growers, packers, and retailers. Therefore, the public can still be at risk for exposure long after the shelf-life on the bag has passed. However, in some fresh-produce outbreaks, by the time the epidemiology and food testing steps have been completed and a specific food has been implicated, there truly is no possibility of the food item remaining in distribution, and consequently, no risk of further exposure for the public. In this case, regulatory and public-health agencies must evaluate the benefits of issuing consumer-level advisories. In some situations, previously unknown cases may be identified by public notifications.

Frequently, environmental sampling of retail and food-processing facilities is given only a minimal amount of effort, not surprisingly resulting in negative results. For example, many environmental investigations report collecting five to 10 environmental samples in a restaurant or 15 to 20 samples in a large food processing facility. The number of samples collected in any single facility should be maximized to provide the best opportunity for recovering pathogens, if they are there. Frequently, sampling efforts are determined by figures that laboratories provide regarding the number of samples they can process. Investigators are often instructed not to collect more samples than their laboratory can handle. However, this limitation should not be allowed to drive the environmental investigation. Managers should identify and have in place options to process additional environmental samples once the local or state capacity is exceeded. In some situations, samples from defined areas within a facility can be composited to lessen the total number of samples yet still allow for a realistic opportunity to detect pathogens that may be present in defined areas.

Although there is no precise numerical guideline for the number of samples to be collected in these investigations, 10 to 12 samples from a retail or food-processing facility is likely to be an insufficient number of samples to have a reasonable probability of detecting pathogens if they are present. Investigators should think in terms of a minimum of 50 to 100 samples per retail facility and even larger numbers from food-processing facilities. Portable coolers or freezers, large enough in size to hold the perishable samples, should be provided. Written procedures for adequately packaging samples should include necessary steps to prevent spillage or leaking of samples during transit to laboratories.

Even though many firms may have completed one or more cleaning and sanitation processes prior to environmental sampling, this should not deter investigators from collecting environmental samples. Media exists to help neutralize the effects of cleaning and sanitizing chemicals. Additionally, pathogens can survive and grow in difficult-to-clean places in food facilities. However, where possible, samples should be collected prior to cleaning and sanitation steps.

Within the food facility, investigators should be strongly encouraged to consider the need for disassembling, or have food-facility staff disassemble equipment used to prepare, slice, or process implicated foods and then to take sufficient numbers of samples from the equipment. Do not accept arguments from firms or investigative staff that "taking apart that piece of equipment would take too long." Screwdrivers and wrenches to dissemble equipment should be part of a standard environmental investigation "go kit," and investigators should be encouraged to collect samples from both easy access locations (e.g., floor drains, cleaning rags, mops, brooms, vacuum cleaners, wheels of forklifts/carts) and from hard-to-reach places (e.g., undersides of cutting boards bolted to tables, gears of conveyor belts) as cleaning and sanitation processes may not reach pathogens in these locations. Regulatory actions and possible legal challenges to findings from environmental investigations are more effectively addressed when proper, documented chain-of-custody procedures are followed for all samples (see chain of custody tag, Figure 3.2). Where appropriate, control samples such as unused whirl pack bags and sterile gloves should also be collected. Supervisors should emphasize to investigators that early and frequent communication with the laboratory must occur during the investigation.

Some public-health agencies routinely avoid interviewing foodworkers simply because "we know what they will say" or because "we won't be able to tell if the foodworker got it from the food or introduced it into the facility." Obviously, without effort in this area, potentially useful information might be lost to the investigation. Other agencies are more aggressive with one-on-one interviews, and in collecting stool cultures or serum samples from foodworkers in implicated facilities. Public-health agencies

FIGURE 3.2

Example of a chain of custody form/tag that can be physically attached to samples.

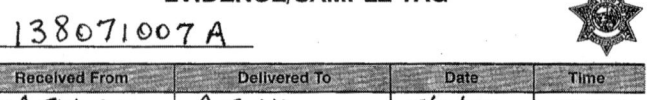

EVIDENCE/SAMPLE TAG

IS# 138071007 A

Received From	Delivered To	Date	Time
A&A Fish Co.	P. Smith	7/10/07	1000 Hrs
P. Smith	J. Jones - Lab	7/10/07	1430 Hrs
J. Jones	Sample Locker	7/10/07	1445 Hrs
J. Jones/Sample Locker	B. Thomas	7/11/07	0800 Hrs
B. Thomas	Sample Locker	7/11/07	1300 Hrs

Comments:

DHS 8067 (7/05)

EVIDENCE/SAMPLE TAG

IS# 138071007A

Received From	Delivered To	Date	Time
Sample Locker/B. Thomas	M. Johnson	7/12/07	1000 Hrs
M. Johnson	Disposal	7/12/07	1300 Hrs

Comments:

DHS 8067 (7/05)

are encouraged to make a concentrated effort to try to rule out foodworker contamination at the point of service. Without these efforts to assess the probability of foodworker contamination in outbreaks with a single point of service, further traceback efforts and environmental investigations may not be initiated. Additionally, certain patterns of illnesses may signal a strong need to test foodworkers and/or conduct more intensive environmental sampling efforts. For example, one or two cases of a specific PFGE matching pathogen may be loosely associated with a specific food facility during a two- to three-month period. Then, three to four months later, one or two additional matching cases may appear, again with some association with the same food facility. This pattern may suggest a need for intensive employee assessment efforts along with intensive environmental sampling.

Traceback Investigations

Traceback investigations are frequently the first step in an environmental investigation. Distribution patterns of foods vary significantly by commodity, season of the year, and type of food. For example, tomatoes destined for dicing at a single processing facility may include product from multiple growers and multiple distributors in multiple states. Following processing, this commingled product may be shipped to dozens of customers who may resell to one or more middlemen until the product reaches the end user (see Example of a Completed Time Line, Table 3.1).

To ensure accuracy in traceback investigations and prevent wasted resources, investigators should carefully select a subsample of those with confirmed illnesses for a traceback investigation. It is not necessary, nor is it logistically feasible, to trace product from every ill case. The subsampling of confirmed ill individuals should give preference to those with the most definitive recall of when and where they purchased or consumed

Table 3.1 Hypothetical Traceback from Joe's Steakhouse to World Distribution, Providing Dates of Receipt at Each Node and Volume of Product Received

Example of a Completed Time Line

DATE	5/06	5/07	5/08	5/09	5/10	5/11	5/12	5/13	5/14	5/15	5/16
At Joe's Steak House (POS) daily inventory	3	0	4	5	1	0	0	0	3	2	0
From XYZ Produce	0	5	0	4	0	4	2	5	0	4	0
From Nations Foods	2	0	2	0	2	0	0	0	0	0	2
At XYZ Produce no inventory											
From Zenith Fresh	0	40	40	30	25	35	50	0	45	35	35
From Best Produce	0	0	10	5	5	10	5	55	10	10	5
From Superior Vegetables	50	10	0	0	0	0	0	0	0	0	0
At Zenith Fresh no inventory											
From New Products	450	500	0	0	0	0	250	300	200	300	0
From Fresh-N-Fast GrowerState F	0	0	550	450	400	500	200	300	300	200	450
At Best Produce no inventory											
From World Dist. Country X	0	0	950	0	0	0	900	0	0	0	950

the contaminated product and to those with singular points of purchase or exposure. Objective data such as club-card records, credit-card bills and receipts, and personal calendars provide additional support to an individual's recall of events that occurred weeks before. This subsample, where appropriate, should also include individuals who purchased or consumed product from different states, retail chains, restaurants, or different brands of product. This provides regulators with a higher degree of confidence in the ability of the traceback to produce the correct answer. This process is referred to as **triangulation**.

Determining the time period of interest is one of the first, and frequently one of the most difficult challenges in traceback investigations. Most frequently, this time period is derived from estimates based upon a sample of the confirmed cases with the most definitive and, ideally, singular recall of exposure to the implicated foods. Unless there are laboratory confirmed samples from implicated products with definitive lot codes or use-by dates, this analysis generally begins with the earliest and latest known exposure/consumption dates from confirmed cases. Then, minimum and maximum incubation periods are included along with the maximum consumable shelf-life under proper refrigeration of the product and minimum and maximum shipment times.

A narrowly defined time period of interest for tracebacks is preferred and allows investigators to more efficiently utilize limited resources. Excessively broad time periods may involve dozens of suppliers, wholesalers, distributors, brokers, and even more farms of origin, making these efforts less likely to be successful or even to be attempted. However, past investigations have demonstrated the need to err on the side of a slightly broader time range to avoid repeated visits to the same facilities as new cases are discovered.

Conducting tracebacks by telephone is tempting because they are quick and are not as resource intensive as in-person tracebacks. However, telephone tracebacks will frequently give you the wrong results very quickly, and thus, are strongly discouraged except in extraordinary circumstances. In-person, on-site tracebacks have repeatedly revealed critical information that would likely not have been obtained by a telephone call. Additionally, these records are necessary for regulators to take enforcement actions when appropriate. Where possible, investigators should personally visit each point in the food-to-table continuum, carefully review and obtain legible copies of all records of incoming and outgoing products during the time period of interest, and review processing practices. Portable, high-speed copiers or scanners can be invaluable when large volumes of records are involved or when business owners do not want to release original copies for photocopying by investigators.

Occasionally, regulators will receive calls from communicable disease staff asking for assistance in conducting an epidemiological traceback. In some epidemiological investigations, a specific ingredient may not be discernable (e.g., tacos). Tracebacks of multiple ingredients in these implicated foods may provide additional insight into which ingredient distribution pattern may best match the geographic distribution of ill individuals. However, all efforts to conduct a thorough and prompt epidemiological investigation should be exhausted first before these requests are made.

Two types of traceback diagrams are currently used to document traceback investigations: a reference traceback flow diagram (Figure 3.3) and a traceback spreadsheet (Table 3.2). These diagrams not only help document findings but also provide investigators with visual depictions of common growers, shippers, processors, or retailers to better discern relationships and commonalities that may exist among the subsample of ill individuals. These traceback diagrams also allow regulators to narrow the focus of the investigation to certain shipment or production dates using purchase dates, exposure dates, incubation periods, daily inventories, and shipping times described earlier. These diagrams should include basic information such as amounts of implicated product shipped/received, lot codes, and dates shipped/received. This information is frequently obtained from invoices, shipping manifest records, and daily inventory and ordering records that are provided when ordering, picking up, and delivering food products.

In addition to collecting and reviewing records, investigators should interview knowledgeable individuals at each food facility to confirm information provided in records. Not infrequently, paper records may not list deviations from usual procedures. For example, retail food facilities may tell suppliers that they only want a specific type, brand, color, or size of food in each shipment (e.g., 3 lb bags of iceberg and romaine blend with carrots from manufacturer X). However, product shortages occur on given days, and suppliers may substitute products from the same or different manufacturers when the originally requested product is not available (e.g., 3 lb bags of iceberg with red leaf lettuce, spring mix, and radicchio from manufacturer Y). Failure to take into account this relatively simple, frequently occurring transaction could result in thousands of hours of investigative time spent pursuing the incorrect commodity or ingredient. Food manufacturers, distributors, and retailers should require suppliers to note any substitutions clearly on shipping manifests. Investigators should collect accurate information on the precise delivery times and preparation times for implicated meals at retail facilities. These times may also assist investigators in narrowing the scope of the traceback and environmental investigation.

Tracebacks of foodborne outbreaks within a single local jurisdiction may be done by local officials. However, these investigations often result in

64 CHAPTER 3: Identification of the Source of Contamination

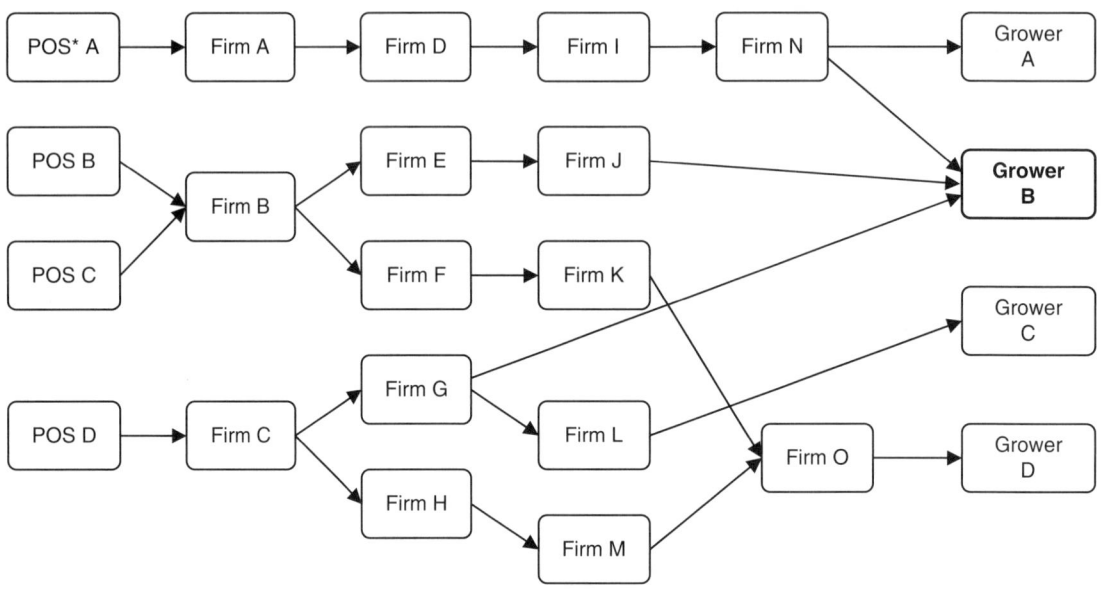

*POS = Point of Service

The boxes and lines that are in bold lettering indicate the implicated product's distribution pathway that links all of the points-of-service (POS) outbreaks to the implicated grower. Tracebacks were initiated at four points of service in response to a multistate outbreak of illnesses that were epidemiologically associated with a specific food product. In the traceback related to POS A, Grower A and B were implicated. In the tracebacks initiated at POS B and POS C, Growers B and D were both implicated. None of these distributors or subdistributors were identified as a common supplier to all four points of service. Therefore, none of these were considered as the most likely source of the contamination. Although a total of four growers were implicated, the only one in common to all of the four tracebacks is Grower B. The only grower assigned for an environmental/farm investigation was Grower B. Distributors/subdistributors G and E were also assigned as a lower priority investigation to document the opportunities for survival and growth during these interim steps. Dates of shipment and receipt, along with volume of shipments can be included in each box to better identify the most likely source.

FIGURE 3.3 *Example of multistate traceback flow diagram.*

findings of suppliers located in other local or state jurisdictions or even in other countries that are outside the jurisdictional authority of the original local agency. Additionally, outbreaks originating in one local jurisdiction often quickly expand to other counties or states. Close coordination of all tracebacks with appropriate state and federal regulators is critical to avoid duplication of effort, to ensure consistency, to ensure that adequate authority exists for collection of records, and to ensure that the highest priorities are addressed first. Multijurisdictional outbreaks are best coordinated by state or federal agencies. State and federal agencies should provide clear direction and protocols, along with frequent feedback to local agencies who wish to assist in the traceback and environmental investigations (see FDA Farm Investigation Guide & Traceback Guide attached).

Table 3.2 Example of a Traceback Spreadsheet

				Timeline November	6	7	8	9	10	11	12	13	14	15	16	17	18	19	20	21	22	23	24	25	26	27	28	29	30
			Retail POS QQ	Case WI_1										Exp						Onset			Cd	Hosp				
			Retail POS RR	Case WI_2												Exp				Onset			Cd					
			Retail POS SS	Case WI_3														Exp			Onset							Cd
			Retail POS XX	Case CA_1													Exp					Onset		Cd				
			Retail POS YY	Case CA_2							Exp				Onset								Cd			Died
			Retail POS ZZ	Case CA_3										Exp						Onset		Cd		Hosp			
Processor A Production Date	Date Ship from Processor A in CO to Distributor A in WI	Distributor A - WI Date Recd (#Cases)	Lots recd by Distributor A - WI																									
10/21/2007	10/25/2007	10/27/07 (18)	7-303-1301	SS(5)	QQ(5)		RR(8)																					
10/31/2007	11/1/2007	11/02/07 (36)	7-303-1302								RR(12)	SS(8)																	
11/1/2007	11/2/2007	11/03/07 (36)	7-310-1303								QQ(5)				RR(15)			SS(10)											
11/13/2007	11/18/2007	11/20/07 (17)	7-317-1306																	RR(10)									
11/21/2007	11/22/2007	11/22/07 (22)	7-318-1404																				QQ(6)	SS(9)					
Processor A Production Date	Date Ship from Processor A to Distributor B in CA	Distributor B - CA Date Recd (# Cases)	Lots recd by Distributor B - CA																									
10/21/2007	10/25/2007	10/27/07 (28)	7-303-1301	XX(10)	ZZ(12)																							
10/31/2007	11/1/2007	11/02/07 (26)	7-303-1302	YY(8)						XX(8)																			
11/1/2007	11/2/2007	11/03/07 (30)	7-310-1303						YY(10)		ZZ(12)		XX(12)																	
11/13/2007	11/18/2007	11/20/07 (17)	7-317-1306																			YY(8)								
11/21/2007	11/25/2007	11/28/07 (28)	7-318-1404																		ZZ(12)		XX(10)							

Regulatory/Enforcement

The regulatory/enforcement phase of the investigation may be done concurrently with, or more preferably after, the initial environmental investigation phase. However, it is critical that the two phases be viewed and implemented separately. The knowledge, skills, and abilities for those involved in the environmental investigation are very different from those in the regulatory/enforcement phase. Attempting to use the same investigator for both purposes at the same time will likely result in an incomplete job in both areas.

Most retail and food manufacturing facilities have limited resources to respond to multiple requests from investigators and can become overwhelmed quickly by the demands associated with outbreak investigations. The highest priority must always be implementation of measures necessary to identify the scope of the distribution of the implicated product and prevent further exposures. Routine inspections of retail and wholesale food facilities should not be placed before information gathering and implementation of measures needed to control the outbreak.

The regulatory/enforcement phase collects evidence of conditions that may have been in violation of statutes, regulations, or guidance documents. Previous inspection findings should also be reviewed. These inspectional findings then need to be discussed to determine if regulatory or enforcement action is warranted.

Prevention/Research

Although investigators involved in the environmental phase of foodborne outbreak investigations have made significant advances in developing standardized methods and in utilizing highly skilled and experienced investigators, the precise mechanism of how pathogens come into contact with fresh produce remains unclear. This lack of understanding has pointed out the need for quick, prioritized research projects so that effective prevention steps can be identified and implemented. Instead of taking three or four years from the request for proposal from a granting agency to a completed scientific paper, federal agencies could consider new approaches to this research including teams of dedicated "high priority/quick turnaround" researchers to address urgent issues identified during the investigation.

Many believe that the contamination first occurs in the field through contaminated water, air, or animal or human feces coming in direct contact with the plants. This contamination, once it is on or within the plant material, is not likely to be eliminated by subsequent washing/sanitizing of the produce and may be further distributed through mixing of small contaminated batches with larger lots. Applied research to confirm or disprove these assumptions and to identify interventions and priorities for their implementation is critically needed. Federal agencies should publish yearly lists of research priorities for commodities involved in recurring outbreaks. These lists can be used by granting agencies and by industry to evaluate and fund produce-related research efforts.

TRAINING NEEDS FOR ENVIRONMENTAL INVESTIGATORS OF RETAIL, FOOD-PROCESSING FACILITIES, PACKING SHEDS, AND FARMS

Training to emphasize an integrated, multidisciplinary team of laboratory, communicable disease, and environmental health staff is important. Training courses such as the NEHA/CDC Epi Ready course emphasize this team approach, building upon the skills, expertise, and authorities each team member brings to the investigation. Core competencies in environmental investigations of retail and food-processing facilities may be established and would include areas such as:

- Basic epidemiology of foodborne outbreak investigations
- Basic microbiology/ecology of common foodborne pathogens
- Aseptic sampling and chain of custody procedures
- Basic traceback and environmental investigation procedures
- Documenting the flow of food and identifying opportunities for contamination, growth, survival, and destruction
- Basic interviewing techniques
- Legal aspects of investigations

Standardized foodworker questionnaires, in appropriate languages, can be developed for interviews with foodworkers and should include questions about specific responsibilities on the dates in question, deviations from these responsibilities, and symptoms consistent with the illness under investigation during the time period of interest. All foodworkers involved in preparing implicated foods or working with implicated ingredients should be interviewed individually where possible. These interviews should be conducted without the manager or supervisor present. If necessary, translators should be provided during these interviews.

RESUMING OPERATIONS

At some point in time following a foodborne outbreak at a retail or food-processing facility, questions will begin to arise regarding what the facility must do or should do to resume operations. Although these requirements may vary by pathogen and by local, state, or federal requirements, the following guidelines may be useful.

- Assess the need to dispose of all foods and ingredients that may have been cross-contaminated.
- Consider the need for one or more "exceptional" cleaning and sanitation cycles using steam and approved cleaning and sanitizer compounds and processes followed by procedures to verify the effectiveness of the cleaning and sanitation.

- Determine whether the firm should be required to implement an ongoing cleaning and sanitation effectiveness monitoring program. Monitoring programs may incorporate ATP devices to give instant feedback on levels of organic matter present after cleaning and sanitizing. Written procedures should also include parameters for frequency of sampling, number of samples, location of sampling, and steps to take if baseline levels are exceeded.

- Discuss the need for requiring the food facility to retrain and provide regular ongoing training to staff, in their native language if necessary, by an approved trainer in basic food safety practices and in specific food preparation and processing procedures.

- Consider whether additional preventive measures such as implementing, modifying, or revalidating HACCP plans for a processing facility; establishing new time or temperature requirements; establishing new guidance for minimizing bare-hand contact and excluding ill foodworkers; providing improved access to handwashing stations; color coding of items such as cutting boards for meat and poultry, equipment, and hardhats for personnel with access to one area of a food-processing facility; and reconfiguring food preparation or processing areas, which may reduce the risk of recurrence.

- Determine whether increased inspections of the food facility for some time period are appropriate, and identify specific areas for more intensive review.

- Provide clear expectations of food facilities along with relevant timelines and expected outcomes (preferably in writing) to the responsible individual to avoid confusion.

FARM INVESTIGATIONS

When environmental investigation findings suggest that the point of contamination may have occurred prior to the retail or food-processing facility, farm investigations may be initiated. Farm investigations require specialized expertise and training in such areas as sanitary surveys of agricultural wells, hydrogeological connections between surface water and underground aquifers, legal requirements and jurisdictions for irrigation water quality, recycled water treatment processes, manure composting processes and requirements, wildlife identification from feces or tracks, wildlife habitat and ranges, wildlife trapping and sampling procedures, environmental (water, air, soil, sediment) sampling procedures, along with crop production, harvesting, and cooling practices for different commodities.

Evidence to date suggests that the point of contamination for most produce-associated outbreaks originates on the farm during growing and harvesting. The agents that have been involved are in two categories: those most commonly associated with animals (e.g., zoonotic including *E. coli* O157:H7 and *Salmonella*) and those associated with humans (e.g., *Cyclospora*, *Shigella* and hepatitis A).

Epidemiologists and microbiologists refer to the place where pathogens are normally found in nature as the reservoir. If an investigator is visiting a farm implicated in an outbreak with an agent that has a human reservoir, the emphasis has to be on identifying how the pathogen moved from humans to the food such as through human feces, or worker hands or water contaminated with human feces. If the investigator is visiting a farm implicated in an outbreak with an animal reservoir, the investigation needs to focus on how the pathogen moved from domestic animals or wildlife to the produce. This would most likely occur through indirect contact with animal feces through water spread or dust spread or direct contact with the animal or its feces.

Farm environmental investigations begin with a thorough inventory of the farm operation including a map of the farm layout; the source and species of seed/seedlings and when they are planted; fertilizers used and how they are applied; pesticides/herbicides used and the water used to dilute them and how they are applied; how and when crops are irrigated and the source of water used; how and when the crop is harvested, and how the crop is transported. Since a farm investigation happens after the crop has been harvested, this information will have to be obtained through an interview with the grower. The field portion of the investigation should include a walk around the field looking for animal tracks and feces and evidence of human feces. The source(s) of water should be evaluated. If it is a surface source, like a stream, pond, or reservoir, conduct a sanitary survey of the watershed looking for opportunities for fecal contamination by humans or animals, depending on the agent of concern. For example, could there have been a discharge of raw sewage into the stream upstream of an irrigation intake due to overload of a municipal system during heavy rain? If the water source is a well, determine if it has been constructed to reduce opportunities for contamination. Is the well properly sealed? Is it down hill from potential sources of contamination? Is it likely to be under the influence of nearby surface water?

Identify nearby domestic animals including cattle. Contamination can spread from the animals to the crops through runoff from fields or feedlots, from waste lagoons and possibly from blowing dust. Document items such as distance of field from possible contamination sources such as feedlots and dairies and prevailing wind direction. Identify wildlife species near or in the fields. This can include birds (and the proximity of fields to flyways or frequent foraging locations), mammals, reptiles, and amphibians. These animals can be attracted to the crops or to water in the fields or near the fields. Determine if manure was used for soil augmentation. When was the manure applied? From what type of animals was the manure? If manure was used, had it been composted? Obtain records that document the composting process.

If the agent is one that may have a human reservoir, the health of the farm workers should be evaluated where possible. Determine if farm workers have been ill with gastroenteritis by speaking with the workers or their supervisors and with local public health officials or clinics that the workers may have visited. How was worker health monitored by supervisors? Were workers ever excluded due to illness? Determine worker practices for harvesting the crop. Did they wash their hands after breaks and after using the toilet? What is the availability of toilet and handwashing facilities in the field? Were gloves worn in the field by workers who would have handled the crop? Since the implicated field may have been harvested weeks before, observations of the same harvest crew in a different location may be helpful where possible.

Examine equipment that could have come in contact with the crop such as bins, ladders, knives, gloves, and boots. Is the equipment clean and cleanable? How was cleaning/sanitizing verified when it was being used? Was water applied to the commodity during or immediately after harvest? What was the source of the water? Document the training fieldworkers have received for sanitation and hygiene, and the language of instruction. Are practices such as harvesting products from the ground (apples dropped from the tree or from the packing shed conveyor belt) or harvesting slightly decayed products in use? Are chemicals such as browning inhibitor solutions recycled to save money?

Collect appropriate environmental samples such as soil, water, crop, and feces in any location that may account for contamination of the crop. It is appropriate to collect a large number of samples in a highly suspect or implicated field. Contamination is likely to be low level and sporadic, so a large number of samples is needed to increase the likelihood of finding the agent. Samples should be tested for the agent of concern and for generic *E. coli* as an indicator of possible fecal contamination.

PACKINGHOUSE INVESTIGATIONS

Past investigations of packinghouses, which are often unenclosed facilities, have documented that the water supply or wildlife could be likely sources of contamination or could result in spreading contamination (Figures 3.4 and 3.5). Water in packinghouses may be used for washing, fluming, or cooling produce. Water sources used in packinghouses should meet standards for potable water or be from a public water supply. This water should be disinfected by a method that maintains a residual of disinfectant throughout the processing period and should include frequent, if not real-time, monitoring of disinfection levels. Examine records for this disinfection, and have operators demonstrate their manual testing while you observe.

Many packing sheds have only a roof and concrete floor, making them attractive locations for birds and other wildlife. Investigators should carefully examine the packinghouse for evidence of animal feces, nests, and tracks. The use of appropriate equipment, such as ultraviolet lights of specific wavelengths and observations of the facility at night, may be helpful in this effort.

Evaluate worker health and hygiene similar to efforts made for field workers. A review of time cards and one-on-one interviews with relevant workers to identify responsibilities during the time period of interest are critical. Evaluate the cleanliness of equipment along with actual cleaning and sanitation procedures, and measures to determine the efficacy of cleaning. Determine the training field workers have received for sanitation and hygiene.

FIGURE 3.4

This photo shows an apple bin drenching apparatus in a packing shed with a pigeon roosting over the equipment and pigeon feces beside the equipment.

FIGURE 3.5

The drenching machine contains an open reservoir of recirculating water in the bottom of the machine used to introduce preservatives/chemicals onto the fruit.

VACUUM COOLER/HYDROCOOLER INVESTIGATION

Various postharvest processes exist to quickly decrease the temperature of produce harvested from the field and thus prolong shelf-life. These include forced chilled air, hydrocoolers, and vacuum coolers. Hydrocoolers flush large volumes of recirculating, chilled water over the produce in wax-coated shipping boxes. This process could serve as the source of contamination or could spread contamination; therefore, adequate water quality, along with monitoring of the disinfection levels of the water and cleaning and sanitation processes, must be maintained at all times to prevent cross-contamination. Similarly, some produce is cooled using vacuum pressurized equipment with or without overhead sprays to partially replenish water removed during the vacuum cooling process. During the investigation, consideration should be given to collecting environmental samples from equipment and thoroughly reviewing and documenting cleaning, sanitation, and disinfection procedures.

FRESH CUT PRODUCE PROCESSOR INVESTIGATIONS

A significant percentage of produce outbreaks have been linked to fresh-cut produce. Investigations of implicated fresh cut processors have not identified obvious sources of contamination at these processors, but investigators have noted that the washing and fluming steps could result in spreading contamination from one or a few contaminated items to a much larger volume or number of items. Similarly, the commingling of cut produce could spread contaminants from one or a few contaminated items to a large number of final packages of produce. Water used in fresh-cut operations should meet potable water standards. Infrequent monitoring of disinfection levels of water may allow a "slug" of contamination to pass through the system. A deliberate, careful assessment of the water quality, disinfection procedures, temperature, and disinfection monitoring system is required. Worker health and hygiene also need to be evaluated.

Environmental investigations at packinghouses, hydrocoolers and fresh-cut processors should include collecting "library" samples from implicated lot codes if available, collecting water samples, documenting the type and frequency of water clarity/disinfection monitoring by processors, and sampling processing equipment surfaces and other locations that could harbor microorganisms.

INTENTIONAL CONTAMINATION

Investigators of foodborne disease outbreaks always need to keep in mind the possibility that the contamination event was intentional. Communication with appropriate local or state law enforcement agencies should occur immediately if intentional contamination is suspected. Some of the features that might tip off investigators to intentional contamination include an unusual agent not previously associated with the vehicle involved, an unusually high attack rate, or an unusually severe disease when compared to past experience. Other information that might indicate intentional contamination include knowledge of existing threats to the food supply; knowledge of previous threats by or issues with employees at the facility or farm; evidence of unauthorized personnel in fields, packing plants, or fresh-cut processors; evidence of unauthorized supplies or equipment in the area where the food was produced or processed; and illness in food or processing plant workers similar to that in outbreak cases.

LESSONS LEARNED

Lessons learned from past produce outbreak investigations include the following.

- Using rotating, inexperienced, untrained investigators on environmental investigations of retail food facilities, farms, packing sheds, and food processors has significantly compromised the effectiveness of past investigations. Investigations need to use a team approach, incorporating persons with expertise and training in epidemiology, microbiology, water quality, animal/wildlife science and regulatory investigations that are used to working with each other. Sending out different investigators on follow-up visits wastes time as these new investigators are not familiar with the setting involved or the past findings, thus requiring them to take additional time to come up to speed before they can provide useful input.

- Produce outbreak investigations usually involve multiple government agencies at the local, state, and federal levels. Coordination of these entities is always a challenge and can result in duplication and gaps in efforts as well as communication problems. One example of the kinds of confusion that can occur is when multiple agencies independently visit the grocery store or restaurant where implicated produce was purchased or served and ask for the same information multiple times. Another example is when regulators have to return to a facility numerous times to collect additional information. Even though it may not be possible to collect all information at the initial visit, careful and methodical planning and daily debriefings can help reduce the number of return visits required. Communication and coordination among investigating agencies throughout the process is critical.

- If the produce implicated in an outbreak was served at a single restaurant, it is necessary to try to determine if the produce became contaminated there. Many times, local health departments conduct routine regulatory inspections of the establishment rather than an epidemiologically based, detailed food preparation review to assess the likelihood that contamination did not occur at the restaurant. These traditional inspection approaches to illness or outbreak investigations are entirely inadequate and should no longer be used as a primary investigative approach. Detailed, well-documented, science-based investigations, as described previously, are imperative. Additionally, making no effort to assess foodworker contamination at the facility may result in erroneous conclusions or an inability to pursue the investigation further.

- Dispatching staff to collect remaining implicated product from ill-consumer households is critical. Advising consumers to "drop off" remaining product is not acceptable. Freezing remaining positive product after testing at state or local labs may prove helpful for further analysis.

- Assessing whether incoming ingredient volume is approximately equal to outgoing finished product volume may be useful in determining if there are gaps in paperwork or processing.

RECOMMENDATIONS

The following guidelines are recommended.

- Local, state, and federal agencies must look for opportunities to improve speed and quality of all phases of the investigative process. For example, population-based comparisons of the number of annual reported foodborne outbreaks may help identify geographic areas (states, counties) that need additional training or infrastructure support to ensure rapid and complete reporting. Additionally, a state-based food complaint system that receives, compiles, and analyzes information from local agencies and then forwards these data to a national database may help identify outbreaks earlier.

- Public-health agencies should develop written performance standards and standard operating procedures and encourage monitoring of these standards during all phases of foodborne outbreak investigations including reporting and surveillance systems.

- Standardized commodity-specific questionnaires should be developed for those commodities previously implicated in more than one outbreak. These questionnaires should include very detailed information needed for tracebacks that may be specific to that commodity. These should be posted on Web sites by federal agencies for use by state and local agencies in foodborne outbreak investigations.

- Contacts, standardized questionnaires, and procedures for obtaining permission from ill individuals to use club-card information should be developed in advance by state agencies.

- Understanding the strengths and weaknesses of epidemiological studies is critical for regulatory agencies, yet many regulatory agencies do not have these resources. Regulatory agencies should consider hiring experienced epidemiologists to quickly review and discuss epidemiological findings before beginning the environmental investigation. These trained, experienced epidemiologists can also help regulatory staff move away from the traditional inspection approach to a more scientific approach.

- Agencies with statutory authority for the specific types of facilities should develop specially trained, multidisciplinary teams to investigate foodborne outbreaks at retail food facilities, food-processing facilities, distributors, brokers, and farms using epidemiological concepts to assess opportunities for introduction, survival, and growth of pathogens. These teams should not be simultaneously tasked with regulatory inspections.

- Federal agencies, in consultation with state and local agencies and industry, should publish yearly lists of research priorities for commodities involved in recurring foodborne outbreaks.

- Food manufacturers, distributors, and retailers should move quickly to improve the traceability of all food products. Although some efforts appear to be underway in the produce area, these efforts may take a decade or longer to result in significant improvements in traceability. Although longer term, industrywide efforts are critical, incremental gains can be made in the short term. For example, buyers can require suppliers to note any substitutions clearly on shipping manifests.

- State and federal agencies should provide clear direction and protocols to local agencies who wish to assist in the traceback and environmental investigations.

- Environmental sampling at retail food facilities, food processors, and farms should be reexamined to require much larger numbers of samples. New methods for on-farm sampling (concentrating large volumes of water, sediment sampling, soil sampling, etc.) are needed.

- Time is critical during fresh produce-related investigations. Outbreaks involving illnesses or exposures in multiple counties in a single state should be aggressively coordinated by appropriate state agencies. Similarly, outbreaks with illnesses or exposures in multiple states should be aggressively coordinated by appropriate federal agencies.

REFERENCES

Anonymous. (2002). National assessment of epidemiological capacity in food safety. Council of State and Territorial Epidemiologists. www.cste.org/pdffiles/fsreportfinal.pdf

FDA. (2001). Guide to traceback of fresh fruits and vegetables implicated in epidemiological investigations. www.fda.gov/ora/Inspect_ref/igs/epigde/epigde.html

FDA. (2005). Guide to produce farm investigations. www.fda.gov/ora/inspect_ref/igs/farminvestigation.html

FDA. (2009). CIFOR outbreak guidelines document. www.fda.gov/ora/fed_state/NFSS/Default.htm

CHAPTER 4

Manure Management

Patricia D. Millner

US Department of Agriculture, Beltsville Agricultural Research Center, Beltsville, MD

CHAPTER CONTENTS

Introduction	79
Manure Use on Crops	82
Survival of Pathogens in Manure	84
Bacteria	84
Protozoan and Helminthic Parasites	87
Viruses	89
Pastures, Lots, and Runoff	90
Manure Treatment Technologies	92
Composting	93
Summary	97

INTRODUCTION

Animal manure is a well-recognized potential source of a wide variety of infectious agents (Table 4.1) that can cause disease in humans, directly or indirectly, particularly through consumption of contaminated water or food (Burger, 1982; Cole et al., 1999; Feachem et al., 1981; Guan and Holley, 2003; Spencer and Guan, 2004; Strauch, 1991; Strauch and Ballarini, 1994). Foodborne illness outbreaks involving fresh fruits and vegetables over the past decade have heightened concerns about contamination of produce from fugitive enteric pathogens at the primary field production level. Possible contamination sources of concern include wildlife and domestic farm animals, insect vectors, runoff from pasture and rangeland grazing or feedlots, contaminated surface water, and manure-based soil amendments. The illness outbreaks and the ensuing industry, consumer, and government responses have increased overall awareness of the potential

Table 4.1 Some Pathogenic Microorganisms of Public-Heath Concern in Manure and Their Animal Sources

Bacteria	Potential Animal Source(s)
Campylobacter coli and C. jejuni	Cattle, sheep, swine, poultry, goats, wildlife
Bacillus anthracis	Cattle, sheep, swine, wildlife
Brucella abortus	Cattle, sheep, goats, wildlife
Escherichia coli patho- & toxigenic strains	Cattle, sheep, swine, wildlife, birds, water fowl
Leptospira spp.	Cattle, swine, horses, dogs, rodents, wildlife
Listeria monocytogenes	Cattle
Mycobacterium bovis	Cattle
Mycobacterium avium paratuberculosis	Cattle
Salmonella spp.	Cattle, sheep, swine, poultry, goats
Yersinia enterocolitica	Swine
Viruses	
Avian – Swine influenza (USDA, 2008)	Poultry, swine, wild birds, water fowl
Hepatitis E	Swine[a]
Parasites	**Disease**
Protozoa	
Balatidium coli	Pigs, swine, guinea pigs, other mammals
Cryptosporidium parvum	Cattle, sheep, swine, amphibians, reptiles, birds, water fowl
Giardia spp.	Cattle, sheep, swine, amphibians, reptiles, birds, water fowl
Toxoplasma spp.	Felines, warm-blooded animals
Helminths	
Ascaris suum	Swine
Taenia spp.	Cattle, swine
Trichuris trichiura	Swine

[a]Herremans, M., Vennema, H., Bakker, J., van der Veer, B., Duizer, E., Benne, C. A. et al. (2007). Swine-like hepatitis E viruses are a cause of unexplained hepatitis in the Netherlands. J. Viral Hepat. **14**, 140–146.

for foodborne pathogen contamination at the field level and the possible linkage to manure pathogens introduced into a complex landscape environment. Characteristics of the different types, virulence, fate, and transport responses of manure pathogens are essential inputs for ultimate use in quantitative microbial risk assessments within a livestock and fresh-produce agroenvironment.

While concern was increasing about manure pathogens inadvertently coming into contact with fresh produce in the field in some production regions, other regions were concerned about the potential for environmental pollution resulting from inadequate handling, storage, stabilization, and land use of animal manure. In response to the last concern, a variety of manure treatment technologies were being developed and evaluated to reduce the potential for environmental overloads of nutrients, pathogens, and air emissions from concentrated animal production facilities and land application of animal manure. Unlike many traditional means of manure management in which pathogen reduction occurs mainly by default rather than intentionally, the new manure management technologies (Burton and Turner, 2003; Williams, 2003) included pathogen destruction as an integral and critical element of the process.

In the United States and many other countries, domestic and municipal sewage sludge is subject to regulated use practices, multicriteria treatment processes designed for pathogen destruction (USEPA, 1994, 2000b), pathogen testing (USEPA, 2003), and storage guidelines (USEPA, 2000a). In the United States, federal regulations establish a standard set of treatment and pathogen limits that correspond to a range of land application circumstances and subsequent landscape, public access, and private agricultural crop and livestock contact situations for biosolids (treated sewage sludge). Individual states may impose additional requirements beyond those established by the US Environmental Protection Agency. In contrast, no federal or state regulations specify pathogen reduction or testing for animal manure prior to land application. However, revised US Clean Water Act regulations emphasize managing land application of manure to reduce input of nutrient, microbial, and other pollutants to surface waters. In such cases, microbial pollution is measured by the traditional indicator bacterial group, fecal coliforms. This is accomplished by limiting application rates or hauling manure from one region to another where soil nutrients are not already overloaded and waterways are not impaired, and therefore, manure application is not severely limited by nutrient management requirements and Total Maximum Daily Load limits. Manure handling and application strategies have been developed to mitigate hauling costs; however, these can lead to ammonia volatilization, which can impair air quality and lead to state regulations on emissions, as has occurred in some jurisdictions.

Although state regulations and guidance, based on Natural Resource Conservation Service (NRCS) guidance (USDA, 1999), are available to assist producers in handling and managing animal manure stockpiles and storage, they emphasize engineering (USDA, 1999) and nutrient management, with some recent attention to air-quality impacts (USDA, 2008) rather than pathogen aspects. Some states also have fact sheets on application-season timing; however, these focus on logistics relative to manure accumulation in confinement facilities, soil conditions, and crop needs, not on pathogen survival and persistence. This focus primarily results from the fact that major use of animal manure is on major grain, forage, and fiber crops, rather than fresh produce.

For confined, rather than grazed or pastured animal production, treatment generally involves the initial collection and removal of manure-urine (slurry) from the animal housing units, with subsequent storage in lagoons followed by spraying onto fields. Alternatively, for solid manure, it may be stacked in high piles, sometimes under roofs and on stabilized surfaces to prevent uptake of precipitation and leaching of nutrients and runoff. It may be further treated by composting or several other means if equipment is available. As newer technologies are implemented, nutrient stabilization, pathogen reduction, and volatile organic emissions reductions will be realized during treatment and with land application.

MANURE USE ON CROPS

Animal manure use as a fertilizer on crop production land has been practiced for millennia worldwide. Although this practice has generally served the farming community's needs and represents an acceptable form of resource conservation on various grain, bean, and cotton crops, it distributes surviving pathogens across large areas (Bicudo and Goyal, 2003). Use of animal manure in primary production of fruit and vegetable crops is a clear hazard and increases the likelihood of contamination by enteric pathogens that survive in the manure-based inputs. Foodborne illness outbreaks with contaminated produce (24%) were calculated to nearly equal those associated with meats (29%) in the United States between 1990 and 1998.

During this period several outbreaks involving produce were reported from small, organic gardens in which raw manure had recently been applied (Cieslak et al., 1993; Guan and Holley, 2003; Nelson, 1997). Organic production relies on animal manures, crop rotation and residues, nitrogen-fixing legumes, composts, and mineral rock powders to maintain soil quality and provide plant nutrients; cultivation, cultural controls, and biocontrols are used to manage insects, weeds, and other pests. Current USDA organic certification regulations require producers to use thermophilic conditions to compost manure, or if raw manure is used, then harvest cannot occur before 90 to 120 days postapplication (USDA, 2000). Neither the USDA National Organics Program (NOP) nor the National Agricultural Statistics Service maintain specific records on the number of certified organic or conventional farms using manure or manure-based products in production of fresh produce. The NOP requires organic farmers and food handlers to meet a uniform organic standard and makes certification mandatory for operations with organic sales exceeding $5000. The NOP implements the regulations through third-party certifiers that it audits. Approximately 50 state and private certification programs in the United States and over 40 foreign programs have been accredited by the NOP.

Many organic growers use manure-based inputs, and many conventional growers also use such inputs, not just for their fertilizer value, but also for their benefit in building and maintaining soil quality. The steady increase in organic food production and distribution worldwide involves adherence to a variety of safety standards (Cooper et al., 2007). Records show that outbreaks with fresh produce have occurred with organically as well as conventionally raised products. Clearly, both methods of production involve similar types of inputs (i.e., seeds, transplants, water, fertilizer, cover crops or previous green crop residue incorporation into soil, employees, etc.). Consequently, the nonpreferential contamination of fresh-produce outbreaks across organic and conventional sources suggests that actual on-site conditions and practices, rather than marketing-based labels (like "organic"), are the critical determinants of the sanitary condition of fresh produce. However, small growers and backyard garden enthusiasts may require continued information regarding good agricultural practices for use of self-prepared composts.

In addition to direct manure application to land, runoff from animal grazing areas or fenced lots to primary fresh market crop production can increase the risk of pathogen contamination to fresh-produce crops. Over the past several decades, the economics and efficiencies of animal husbandry have led to an increase in the animal density per unit area within livestock and poultry production facilities in the United States and abroad. Such intensive production conditions now used for broilers, layers, turkeys, swine, beef, and dairy animals generate major quantities of manure within relatively limited landscape areas. Appropriate use of these manures as fertilizers requires calculation of the nutrient content, particularly nitrogen and phosphorus, relative to crop needs and existing soil test values to avoid use of excessive amounts that may lead to pollution of surface and groundwater by nutrients (nitrate, phosphorous), organic matter, sediments, pathogens, and other materials (Al-Kaisi et al., 1998a, 1998b; Davis et al., 1997). Consequently, several advanced manure management systems have been developed to handle very large volumes of manure from intensive livestock and poultry operations (Vanotti et al., 2003, 2005b; Williams, 2003). Development has focused not only on nutrients, but also on disinfection of pathogens in liquid and solid-phase materials (Vanotti et al., 2005a). Although these developments targeted swine manure, the technologies are applicable to dairy systems using liquid collection schemes.

SURVIVAL OF PATHOGENS IN MANURE

Stressors, such as fluctuations and extremes in temperature, moisture, pH, UV irradiation, and nutrient availability, along with biological pressures from competition, predation, parasitism, toxins, and inhibitory substances, contribute to the natural attenuation of microbial populations in the environment and in manured soil. Microorganisms exhibit a considerable degree of variability relative to their tolerance to these stressors. In addition, viruses and parasites (protozoans and helminths) are dependent on host organisms, sometimes very specific ones, for reproduction, and therefore once shed in manure or other excretions, will not reproduce in the open environment. In contrast, bacteria, such as *E. coli* O157:H7, *Salmonella*, and *Listeria monocytogenes*, are not host dependent, and thus their populations respond dynamically up and down over time to changes in environmental conditions.

Bacteria

The complexity and uncertainties involved in predicting the fate (partitioning, growth, die-off rates) and transport of manure pathogens in the innumerable variety of agricultural situations limit extensive direct application of traditional decay and transfer functions, and likely require alternative and new approaches to modeling such as those suggested in a review of this subject (Pachepsky et al., 2006). Some of the reported variations in fate of manure pathogens relative to environmental and situational conditions are reviewed in this section.

Survival of *E. coli* O157:H7 in stockpiled, raw sheep manure at 21 months contrasts with its elimination after four months in parallel stockpiles that were aerated (oxygen content not reported) by periodic mixing, and survival for only 47 days in bovine manure stockpiles that were periodically mixed (Kudva et al., 1998). However, freezing (−20 °C) and cold storage of bovine manure (4 and 10 °C) prolonged survival up to 100 days, whereas increasing temperatures shorten survival: 70, 56, and 59 days at 5, 22, and 37 °C (Wang et al., 1996). *E. coli* O157:H7 in laboratory-incubated, manured field soil showed a steady population decline to undetectable levels within 165 days at 15 °C and within 231 days at 21 °C, whereas *E. coli* O157:H7 persisted in corresponding samples of this manured soil in which competing microbial factors were absent because the amended soils were autoclaved prior to inoculation (Jiang et al., 2002).

In field studies during late autumn and winter in the United Kingdom, generic *E. coli* in cattle, sheep, and swine manure survived on grassed areas for very long periods (up to six months), and in at least one case, up to 162 days, when initial populations ranged from 4.31 to 5.34 $\log_{10}$CFU/g respectively (Avery et al., 2004). Average D-values for the cattle, sheep, and swine manure for *E. coli* were calculated as 38, 36, and 26 days for the test conditions (Avery et al., 2004). In addition, *E. coli* O157:H7 has been shown to survive up to 28 days in significant numbers on farm structural surfaces that have contacted manure (Williams et al., 2005).

However, plowing and harrowing of soil amended with naturally contaminated pig slurry effectively and rapidly (i.e., immediately) reduced populations of *E. coli* and *Salmonella* DT104 on a clay soil (Boes et al., 2005). In contrast, harrowing only, or surface application to winter wheat stands only, or injection in winter wheat stands only, prolonged survival of *E. coli* to 21 days, and *Salmonella* to seven days.

Interest in the potential contamination of leafy greens and herbs by *E. coli* O157:H7 led to a study of organic iceberg lettuce production using composted bovine manure, solid manure, as well as manure slurry as the fertilizer. Results showed that no bacterial pathogens, not *E. coli* O157:H7, *Salmonella* spp., or *L. monocytogenes* were recovered from the lettuce, even through *E. coli* O157:H7 was present in all the manure amendments applied to the soils (Johannessen et al., 2004). The authors concluded that further research is needed to resolve how contamination of the lettuce was avoided. The absence of *E. coli* and *Enterococcus* spp. from interior or exterior portions of potato skins on tubers harvested 214 days after soil was amended with raw or composted manure containing both of these bacteria, has also been reported (Entry et al., 2005).

Recent reviews of bacterial pathogen survival in animal manures (Bicudo and Goyal, 2003; Guan and Holley, 2003) clearly show that *Salmonella* spp. survive in some situations for up to 60 days in a variety of nonthermophilic manure systems. In contrast, *Salmonella* spp. and *Ascaris suum* ova were destroyed completely after exposure for 24 hours in swine manure biogas digesters operated at 55 °C (Plym-Forshell, 1995). However, salmonellae survived 35 days and 60% of *Ascaris* ova survived up to 56 days in the mesophilic manure pit where the digested manure from the biogas unit was stored. These results show the effectiveness of manure disinfection processes that involve thermophilic temperatures, and the extended survival periods of bacterial and parasitic manure pathogens in mesophilic, facultative storage units even when substantial nutrients have been depleted by the prior digestion.

Ammonia is typically generated by most stored manures; however, its effect on pathogens in the manure had not been specifically evaluated until recently. In disinfection tests with bovine manure, results show that *Salmonella* was destroyed more rapidly and cost-effectively by gaseous ammonia generated from addition of 0.5% aqueous ammonia than from addition of 2% w/w urea to 12% total solids manure slurry (Ottoson et al., 2008). *C. jejuni* was the bacterium most resistant to the anaerobic digestion of cattle slurry (supplemented with pig, hen, and potato waste) at 28 °C, followed by *S. enterica* Typhimurium and *Y. enterocolitica* (Kearney et al., 1993).

Simulated seasonal temperature sequence effects on die-off of *Salmonella* serovars (Agona, Hadar, Heidelberg, Montevideo, Oranienburg, and Typhimurium) were greatest in the first week of the winter-summer sequence (−18, 4, 10, 25 °C) as compared with the spring-summer sequence (4, 10, 25, 30 °C) in a 180-day study with 5 $\log_{10}$CFU/g inoculated directly to moist clay and loamy sand soils, with and without fresh swine manure slurry (Holley et al., 2006). Total die-off (no detection in enrichments) of inocula was more rapid (160 days) in spring-summer temperature sequences regardless of manure, incorporation, or soil moisture content (60–80%) of winter-summer sequence treatments (160 days). By considering the calculated decimal reduction times of 30 days or more for 90% reduction of salmonellae in the application treatments, and common estimated slurry concentrations of 3.0 to 600.0 salmonellae/ml, and application rates of slurry to land, 25 g/kg, the authors concluded that a 30-day delay between field application of manure in spring or fall and use of treated land would minimize risk of environmental contamination and uptake by animals of *Salmonella* (Holley et al., 2006). However, they do not specify the different types of crops for which this 30-day delay would be appropriate, whether the 30-day delay refers to application and planting date or to harvest date, nor do they address other pathogens and their survival within this 30-day period. Such a recommendation also is not consistent with recent leafy green marketing agreement requirements that prohibit use of raw, untreated manures, and the 90- to 120-day required delay between application and harvest for US certified organic producers.

A four-state study of environmental and herd-level risk factors associated with *Salmonella* prevalence in dairy cows, including conventional and organic farms, identified major contributing risk factors as access to surface water, *Salmonella*-positive manure storage, land application of manure slurry or spray irrigation, and cows eating or grazing in fields where manure was surface-applied rather than soil-incorporated within the same growing season (Fossler et al., 2005a, 2005b). In contrast, free-range rearing conditions, sometimes used on organic farms, were found to be slightly beneficial in reducing *Salmonella* spp. but not *Campylobacter* spp. or *L. monocytogenes* contamination in chicken flocks (Esteban et al., 2008).

Animal management practices at organic and conventional farms that focus not only on manure management, but also on measures to control access of wildlife to the housing units and water troughs, has been repeatedly identified as a critical point in maintaining farm hygiene relative to several major zoonotic pathogens, including *E. coli* O157:H7 and other shiga-toxing positive serotypes, *Salmonella* spp., *L. monocytogenes*, *Campylobacter* spp., *Cryptosporidium*, and *Giardia* (Castellan et al., 2004; Meerburg et al., 2006; Murinda et al., 2004). Identification of on-farm pathogen reservoirs and vectors can aid development and use of farm-specific pathogen reduction programs.

Protozoan and Helminthic Parasites

Several eukaryotic parasites in manure are characteristically more resistant to the range of environmental stressors encountered in various agricultural situations and treatment technologies than are viruses or most nonspore-forming bacteria (Bowman, 2009; Fayer and Ungar, 1986; Robertson et al., 1992; Tzipori and Widmer, 2008). Ova of the parasitic helminths, *Ascaris lumbricoides* and *A. suum*, are particularly persistent because the outer shell is resistant to most environmental stressors that adversely effect other groups of microorganisms. Thus, in evaluation of efficacy of treatment technologies, *Ascaris* has been used as a conservative benchmark for microbial destruction.

Environmental stressor effects on the infectivity of *Cryptosporidium* oocysts showed that in water at 4 °C, infectivity is maintained for two to six months (Tzipori, 1983). Oocysts also tolerate a wide variety of common disinfectants such as sodium hydroxide, sodium hypochlorite, and benzylkonium chloride without significant loss in infectivity (Campbell et al., 1982). However, ammonia but not pH (Fayer et al., 1996; Jenkins et al., 1998), desiccation, and very extreme temperatures (e.g., freeze-drying, freezing, or 30 min at 65 °C) completely eliminated viability and infectivity of the oocysts (Tzipori, 1983). Moist heat at 55 °C for 15 to 20 minutes, such as present in the thermophilic phase and core of a composting mass or in a thermophilic digestor, destroyed oocyst infectivity in calf feces and intestinal contents (Anderson, 1985). These temperatures are easily achievable in thermophilic manure treatment technologies that are operated properly. Process management is key to ensuring that pathogens are destroyed, even when a treatment technology with the capacity to meet the pathogen destruction criteria is used.

With controlled-environment laboratory studies, *Cryptosporidium parvum* oocysts were reported to be more resistant to degradation than *Giardia muris* cysts (Olson et al., 1999) in soil, water, and cattle manure. *Giardia* cysts were infective for only one week at 4 and 25 °C, whereas

Cryptosporidium remained infective for eight weeks at 4 °C, and four weeks at 25 °C. At −4 °C, *Giardia* was noninfective within one week, but *Cryptosporidium* remained infective for more than 12 weeks. Field studies in the summer in the United Kingdom have shown that *Cryptosporidium* oocysts in swine manure on grassy fields were reportedly reduced by 1-log during both eight- and 31-day periods, whereas similar 1-log reductions for *Salmonella, E. coli* O157, *L. monocytogenes*, and *C. jejuni* averaged only 1.86, 1.70, 2.80, and 1.86 days, respectively (Hutchison et al., 2005).

In addition, recent advances in molecular characterization of species and genotypes of *Cryptosporidium* and *Giardia* show that such approaches are essential to ensure accurate identifications of organisms in environmental transmission studies (Fayer et al., 2008; Santin et al., 2008). Molecular epidemiological studies strongly indicate that *C. parvum* is the major species pathogenic to both cattle and humans, and that certain genotypes and subtypes predominate in calves worldwide (Xiao et al., 2007). In contrast, molecular data suggest that zoonotic transmission is not as prevalent in the epidemiology of giardiasis (Xiao and Fayer, 2008). In a study of 14 farms from seven eastern US states, *C. parvum*, *C. bovis*, and *C. andersoni* were found on two, six, and eight farms, and infected 0.4, 1.7, and 3.7% of the 541 cows, respectively (Fayer et al., 2007). Low prevalence of *Cryptosporidium* overall, and for each of the previous species individually, in mature cows in this study was very highly significant ($p \leq 0.0001$), compared with young cattle including those previously examined on most of the same farms. The very low level of infection of mature cows with *C. parvum* suggests that field practices be developed and used to manage manure and potential runoff for young cattle likely to shed this pathogen. In the western United States, a study involving more than 5200 fecal samples, from 22 sites in seven states, showed that fresh fecal material from feedlot systems contained about 1.3 to 3.6 *C. parvum* oocysts/g feces, or about 2.8×10^4 to 1.4×10^5 oocysts/animal-day (Atwill et al., 2006).

Clearly, data are needed to evaluate the effectiveness of various field management practices designed to reduce pathogen loading and transport off-site within large-scale animal production areas. Runoff from cattle grazing areas as well as feeding and holding/resting lots may have off-site impacts if appropriate catchment and diversion measures are not available or are incorrectly implemented. Data on field management strategies to reduce prevalence and amounts of pathogens off-site is accumulating. For example, in Canada, a field study with and without vegetative filter strips (VFS), with three slope conditions (1.5, 3.0, and 4.5%) and two 44 minute rainfall intensities (25.4 and 63.5 mm/h) showed that VFS were very effective regardless of slope in reducing *C. parvum* oocysts in surface runoff (Trask et al., 2004). Total recovery of oocysts in runoff from the VFS ranged from 0.6 to 1.7% and 0.8 to 27.2% with low and high rate rainfall, respectively, whereas oocyst recovery from non-VFS sites ranged from 4.4 to 14.5% and 5.3 to 59%, from low- and high-rate rainfall, respectively (Trask et al., 2004).

Viruses

Limited data are available in publicly accessible databases documenting viral bio-burdens in animal manures in the United States. Hepatitis E virus (HEV), a nonenveloped virus that is relatively more environmentally stable than enveloped viruses, was recovered from 15 of 22 swine manure pit and three of eight manure lagoon samples in a multifarm survey, although no HEV was recovered from drinking- or surface-water samples on the 28 farm sites studied (Kasorndorkbua et al., 2005). The presence of exotic Newcastle disease (END) after depopulation and decontamination (D & D) of the infected birds in a California outbreak was assessed with emphasis on manure, compost, and manure conveyors (Kinde et al., 2004). At one ranch, END was recovered up to 16 days postdepopulation, but not thereafter; no END was recovered from a second ranch. Further research on avian influenza and END, both enveloped viruses, resulted in development of a real-time reverse transcriptase-PCR method for rapid quality control checks on composts and D & D in the event of exotic disease outbreaks (Guan et al., 2008).

Although bovine enterovirus (BEV) is not a zoonotic pathogen, it has been suggested as a potential indicator of fecal contamination from animals (cattle/deer) and a molecular epidemiological tool; it is a relatively stable, nonenveloped virus (Ley et al., 2002). It is rapidly inactivated by thermophilic (55 °C) anaerobic digestion of manure as is bovine parvovirus, (BPV; enveloped) (Monteith et al., 1986). Both viruses are also inactivated by composting for 28 days. However, mesophilic (35 °C) anaerobic digestion prolonged BEV and BPV survival to 13 and eight days, respectively, whereas 30 minutes at 70 °C only inactivated BEV. Additional research is needed to determine if either of these bovine viruses would be suitable indicators of manure treatment efficacy. African swine fever virus (ASFV; enveloped) and swine vesicular disease virus (SVDV; extraordinarily resistant to desiccation, freezing, and the fermentation and smoking processes used to preserve food) are rapidly inactivated by thermophilic temperatures (≥ 50 °C) in any of several treatment technologies (Turner et al., 1999). With other exotic viruses of swine, including foot-and-mouth disease virus, Aujeszky's disease virus, and classical swine fever virus (both enveloped), thermophilic temperatures (≥ 60 °C) were found to be essential for rapid viral inactivation (Turner et al., 2000).

A North Carolina state-sponsored study with fresh, swine manures compared virus survival in conventional lagoons to that in five environmentally superior and technologically advanced candidate manure treatments (Costantini et al., 2007; Williams, 2003). Pretreatment manure from all farms had detectable porcine sapoviruses [PoSaVs] and rotavirus A [RV-A], whereas porcine enteric viruses (porcine noroviruses [PoNoVs]) and rotavirus C [RV-C] were present only in some of the farms using the candidate technologies (Costantini et al., 2007). After treatment, only the conventional technology samples contained detectable PoSaV RNA. Candidate farm posttreatment samples with detectable RV-A and RV-C were not infectious by cell culture immunofluorescence assay, nor did they result in clinical signs or seroconversion in inoculated gnotobiotic pigs. Results indicate that the specific environmentally superior manure treatment technologies evaluated would reduce the viral bioburden in treated liquids and solids.

PASTURES, LOTS, AND RUNOFF

Runoff can move significant amounts of pathogens from the original site of manure deposition, be it a pasture, pen, or lot (Thurston-Enriquez et al., 2005). Understanding the transport and survival of zoonotic pathogens potentially present in livestock manure and runoff is critical for development of appropriate and effective practical measures to reduce adverse environmental, food safety, and public-health impacts that nonpoint source releases can potentially have. Results from several complex studies are beginning to inform this issue.

Release of *E. coli* from fecal cowpats during rainfall was reported to occur primarily as individual bacterial cells (Muirhead et al., 2005); bacterial cells preferentially attach to manure colloids and organic matter and small-size silt and clay particles (Guber et al., 2007). However, variation among strains of *E. coli* in dairy manure were shown to have significantly different attachment affinities for various soil textural fractions (Pachepsky et al., 2008).

An agricultural management area scale study with 10 dairies and ranches showed that fecal coliform concentrations were highly variable both within and between animal loading units (Lewis et al., 2005). Fecal coliform concentrations for pastures ranged from 2.3 to 6.36 $\log_{10}$CFU/100 ml and for dairy lots ranged from 3.29 to 8.22 $\log_{10}$CFU/100 ml, with mean concentrations of 5.08 and 6.5 $\log_{10}$CFU/100 ml for pastures and lots, respectively. The investigators (Lewis et al., 2005) noted that the previously cited results are being used by dairy managers to change on-farm practices, by regulatory agency staff, and by sources of technical and financial assistance. Results from a different set of field study tests over 26 months with a cattle feedlot runoff control-vegetative treatment (VT) system showed that the system effectively reduced environmental risk by containing and removing *E. coli* O157:H7 and *Campylobacter* spp. from feedlot runoff (Berry et al., 2007). In this study, *Cryptosporidium* oocysts and *Giardia* cysts were infrequently isolated, and generic *E. coli* populations in the vegetative treatment soil declined with time, although their presence (12 of 30 samples in VT areas vs 1 of 30 samples in nonrunoff impact areas) in freshly cut hay from the VT areas indicated the risk of contamination in that region. Both *E. coli* O157:H7 and *Campylobacter* spp. were absent from the baled hay.

In another agricultural management scale study that included 350 storm runoff samples from dairy lots and other high-cattle-use landscapes (Miller et al., 2007, 2008), a California team found 59 and 41% prevalence of *Cryptosporidium* oocysts and *Giardia duodenalis* cysts, respectively, in runoff associated with areas containing calves less than two months old. In contrast, only 10% of runoff samples associated with cattle older than six months were positive for these protozoans. Percent landscape slope, animal stocking number, and density were not as important factors as animal age, intensity, and cumulative amount of precipitation for *Cryptosporidium* oocyst concentrations. Animal age, stocking number and instantaneous precipitation were major factors in concentrations of *G. duodenalis* cysts in runoff samples. Specific beneficial management practices, notably vegetated buffer strips especially located near calf areas, were associated with reduced runoff loads of these protozoans, whereas cattle exclusion and removal of manure was not. These results support those from other studies cited here that indicate targeted strategies for field management of stock and manure have some potential for reducing manure risk impacts off-site. Additional research is needed to determine if similar field management strategies in land areas adjacent to and surrounding primary fresh produce croplands will reduce fugitive enteric pathogen contamination with the sensitive crop fields.

Another route of off-site transport being examined involves movement of protozoan oocysts through shallow soils via macropores (Harter et al., 2008). A modeling study using packed soil boxes with and without macropores enabled collection of macropore flow data as for that in shallow soils. Macropore flow was shown to be responsible for *C. parvum* transport through the shallow soil to underlying pore spaces when soil bulk density, precipitation, and total shallow subsurface flow rate were taken into account. A risk assessment of oocyst transport was conducted and was determined to be consistent with the reported occurrence of oocysts in springs or groundwater from fractured or karstic rocks protected only by shallow overlying soils (Harter et al., 2008).

MANURE TREATMENT TECHNOLOGIES

Manure is considered a potential source of pathogens, but this does not mean that every sample of manure will contain all the various types of pathogens that have been reported or that they will be present at maximally reported concentrations. However, manure treatment technologies, just like food-processing technologies, are designed to destroy the most resistant types of pathogens likely to be encountered.

Treatment systems for manure from animal housing units have evolved beyond traditional collect, store, land apply approaches and now include processes that aid in protecting soil, water, and air quality (Humenik, 2001; Vanotti et al., 2007, 2008; Westerman and Bicudo, 2005). When manure storage is coupled with managed treatment processes, the result on pathogens essentially acts as a multibarrier system. Some treatment systems address several of these requirements, whereas some are specialized and address only individual factors (USDA, 1999, 2007).

Manure treatment technologies are grouped into two major categories that reflect the primary mechanisms involved in the processes: (1) physico-chemical, or (2) biological. Some systems can be integrated in sequence to meet several treatment criteria in different phases of a system. Some of the physico-chemical approaches to treatment include thermal conversion (combustion, gasification and pyrolysis), solid-liquid separation and filtration, advanced alkaline treatment, and aeration/mixing. Some of the biological approaches to treatment include thermophilic composting (Rynk, 1992), vermicomposting, anaerobic digestion (Bicudo and Goyal, 2003), thermophilic digestion (Ahring et al., 2002; Aitken et al., 2007), autothermal thermophilic aerobic digestion (Layden et al., 2007), sequencing batch reactors (Juteau et al., 2004), and constructed wetlands (Cirelli et al., 2007; Humenik, 2001; Karim et al., 2008).

In general, thermophilic processes, particularly those operated in-vessel, are designed to expose all treated material to extreme lethal temperatures (60–65 °C), while still maintaining sufficient metabolic activity by the nonpathogenic bacteria to sustain the process heat (Juteau et al., 2004).

Thermophilic composting remains one of the most cost-effective treatment technologies for manure solids; it functions well in a variety of environments. Initial capital as well as operations and maintenance costs are minimal compared with other treatment technologies.

COMPOSTING

Use of composting as a cost-effective treatment for manure and the widespread use of compost in large and small primary production systems, even home gardens, warrants special mention of practices that ensure the product will be significantly sanitized of original pathogens. Manure composting as used here refers to the controlled aerobic, thermophilic (self-heating) decomposition of organic matter by microorganisms such that three major objectives are met: (1) nutrient stabilization; (2) pathogen reduction; and (3) odor and vector-attraction reduction. Self heating of stacked manure without attention to the time and temperature needed for all parts of the stockpile to meet pathogen reduction criteria, nutrient stabilization, and vector-attraction reduction does not adequately meet the requirement for a managed process. Such practice is simply stockpiling, with default self-heating in some parts. If the final compost product were intended for corn, wheat, cotton, or other field crops, this type of stockpiling might be adequate relative to pathogen content and the other two major treatment objectives. However, without adherence to a managed process and all that this involves, including regular temperature monitoring-recording at selected places within the pile, along with pathogen testing, the producer would have no basis for asserting that all parts of the pile were subjected to temperatures that would substantially reduce pathogens. Such unmanaged composting is common among backyard garden composts, and hence the strong caution to avoid inclusion of animal feces and diseased plant material in such piles.

To produce compost that is disinfected for use on crops (such as leafy greens, herbs, carrots, radishes, green onions, strawberries, etc.), in which the harvestable portions will directly contact soil, requires quality control and standards at each step in the composting process. A critical control point approach to the composting process could aid compost producers in meeting stringent requirements for use of manure-based products in primary fresh produce cropping systems. Growers, processors, distributors, and buyers need a science-based quality assurance system for compost inputs supported by validated quality assurance test standards for pathogens (such as *E. coli* O157:H7, *Salmonella* spp., *L. monocytogenes, Campylobacter* spp., and parasites). Compost test standards currently used by the US Composting Council Standards of Testing Assurance program for members reflect USEPA biosolids test methods for fecal coliforms, *Salmonella*, and *Ascaris* ova (Thompson et al., 2002; USEPA, 2003). Test methods for other pathogens need to be validated.

During thermophilic composting, the mass of the pile insulates the core and metabolic heat generated by rapid microbial decomposition of the organic matter cannot escape quickly enough to equalize the temperature to ambient. Thus, the core pile temperatures increase, while surface temperatures remain insufficient to destroy pathogens, unless piles are turned so that the outer mass is exposed to thermophilic core temperatures (Shepherd et al., 2007). Turning is best accomplished early in the process, that is, during the first two to three weeks while maximal thermophilic temperatures are generated. Pile temperature usually declines immediately after turning, but rebounds to 55 °C or greater as long as readily decomposable organic matter remains. Microbial respiration rate declines as readily available nutrients are depleted, and maximal temperatures achieved after each turning coincide with the decline in readily available nutrients. With time (e.g., 10–12 weeks or more), the humification begins, and nutrients are further immobilized within the biomass of living and dead microbial cells. Lignin and cellulosic compounds are the slowest to decompose and do not provide readily available carbohydrate for any of the bacteria of concern as foodborne illness or public-health pathogens.

Guidance on composting basics, feedstock mix ratios, technology requirements, and conditions for large and small on-farm composting of manure and other organic feedstocks are available from several sources (Christian et al., 1997; Misra et al., 2003) and from many state agricultural extension programs. Critical factors in composting include aeration, nutrients, C:N ratio, moisture, pile structure, pH, temperature, and time (De Bertoldi et al., 1986; Haug, 1993). Aeration, either mechanically with blowers, by turning, or with passive means, is essential to meet the microbial requirement for oxygen needed for aerobic decomposition. The porosity, air flow characteristics, structure, and physical texture of the biomass mixture also impacts pile aeration. Management of the composting mixtures is needed to ensure that the process achieves the target time-temperature criteria for pathogens, and that turning and other operations are conducted according to good manufacturing practices.

Composting formats comprise a range of process and facility types. Some very mechanized and highly managed composting systems can involve frequent turning of windrows or mechanical aeration and biofiltration of static (stationary) piles that are either free-standing or enclosed within a vessel or containment system. The latter may include plastic polymer silage-type tubing, various types of synthetic material covers, and fully enclosed metal containers equipped with air-flow control devices, temperature sensors, feedback controls, and leachate collection systems. These mechanized approaches are useful for large-scale operations that maximize materials throughput and minimize process times. Other less mechanized compost systems, typical of some on-farm approaches (Christian et al., 1997), utilize passively ventilated static piles. Oxygen transfer rates in these passive formats are less efficient than mechanically aerated formats; hence decomposition proceeds somewhat slowly and temperature maxima are not as great or as sustained as for forced aeration systems. Such formats require a greater footprint than mechanized approaches, but they also have lower capital and operating costs. Because static piles are not turned until after the thermophilic phase ends, constructing the pile on a bed of at least 30 cm of woodchips, old hay, or straw and covering the outer surface of the pile with a layer of similar materials, or unscreened coarse textured compost to a depth of 20 to 30 cm (as is done with the aerated static format; Rynk, 1992) provides an insulated zone sufficient to ensure that all the "new" compostable mixture is within a thermophilic zone. This can also provide an absorptive layer for leachate and moderate creation of anaerobic zones from liquid accumulation pockets. Excessively tall piles run the risk of compaction at the base and diminished heating.

Good manufacturing practices for compost that is intended for use in primary fresh produce cropping systems, such as leafy greens and herbs, can best be met when compost producers conduct their operations within a Hazard Analysis Critical Control Point (HACCP) framework. Though not typical of current composting industry practice in the United States, the main elements for critical control point composting are outlined as follows. In the United Kingdom, national regulations stipulate very high thermophilic temperatures (> 60 °C) for three days in static aerated pile or 14 days in turned windrow composting that includes any catering (i.e., food) wastes (DEFRA, 2004). In the United States, several states and localities use the time-temperatures and operational requirements that apply to biosolids compost (USEPA, 2003). The USEPA requirements of 55 °C for three consecutive days in a static aerated pile format (this includes a base and blanket layer 15–30 cm thick as insulation to ensure that the new material is well within the thermophilic zone) have been applied by states for all other types of composting, including landscape trimmings, wood, food, papermill sludge, and dissolved air flotation waste to mention a few. Carcass composting has special requirements, and the product is not appropriate for use on primary fresh produce crops. For windrow composting, USEPA requirements stipulate 55 °C with five turnings during the two-week period when the thermophilic temperatures are generated. Both formats require testing of final product according to USEPA standards for *Salmonella* and fecal coliforms (USEPA, 2003), plus other microbes as state or industry requirements specify. Material from piles that do not achieve time-temperature or test standards must be recomposted until they meet requirements.

Compost tea is a compost-derived product used by some growers as a foliar spray or soil drench to promote plant growth and protection against phytopathogens (Scheuerell and Mahafee, 2002). Compost tea (CT) production processes that use supplemental nutrients can support regrowth of even a few surviving cells of *E. coli* O157:H7 and *Salmonella* (Ingram and Millner, 2007). Production of CT without supplements avoids growth of these pathogens from initial trace concentrations. Sanitization of CT equipment is essential between batches.

Overall, critical control points suggested for composting operations are associated with each of the major steps in the process as follows.

1. Delivery, Inspection, and Input Preparation. Avoid spread of microorganisms from reception areas and materials to subsequent and especially final product handling areas, notably by delivery and operation vehicles; avoid processing delays that allow microbes to multiply; establish dirty-to-clean areas to avoid cross-contamination; attend to spills in a timely manner; control access of birds and vermin; avoid use of manure delivery trucks for hauling of final product unless equipment is cleaned and sanitized; train employees; monitor adherence to plan.

2. Thermophilic, Decomposition, and Disinfection. Avoid cross-contamination from vehicles, equipment, and containers; maintain the dirty-to-clean sequencing; control access of birds and vermin; periodically calibrate temperature measurement and recording devices; measure and record temperatures according to protocols; troubleshoot temperature failures and segregate material not meeting time-temperature exposure for retreatment; clean and sanitize equipment; train employees; monitor adherence to plan.

3. Curing, Maturation, and Stabilization. Avoid cross-contamination from vehicles and equipment; turn piles as necessary; maintain the dirty-to-clean sequencing; control access of birds and vermin; clean and sanitize equipment; train employees; monitor adherence to plan.

4. Sampling and Analysis. Avoid cross-contamination during sampling via equipment and personnel; submit samples according to protocols for testing at a certified compost testing laboratory; train employees; monitor adherence to plan.

5. End-Product Preparation. Ensure product is stored properly to avoid recontamination by equipment, personnel or vermin; avoid blending with untreated, unstabilized products; transport in clean vehicles that have been sanitized if previously used for hauling manure or potentially contaminated wastes; maintain analyses on batch-lot production and product deliveries for traceback when needed; advise clientele on appropriate storage and use of product; provide copies of temperature and test records to clientele; caution clientele regarding inappropriate usages.

In contrast to thermophilic composting, vermicomposting (worm composting) cannot be conducted at temperatures sufficient to kill pathogens because the epigeic worms used (*Eisenia foetida* and *Lumbricus rubellus*) do not tolerate high temperatures or excessive amounts of ammonia. Thus, manure solids and urine require pretreatment or significant dilution (50%) before being introduced into the vermicompost bins or containers. A two-step process that involves precomposting by a thermophilic method and subsequent vermicomposting or vice versa, was reported to meet pathogen reduction requirements while yielding a stable, consistent, and nutrient-rich product (Ndegwa and Thompson, 2001).

SUMMARY

Advanced treatment technologies that achieve high standards of nutrient stabilization, pathogen reduction, and reduction of odor and vector attraction are available for animal operations that generate large quantities of liquids or solids. In an era when agriculture is increasingly challenged to protect soil, water, and air resources, while maintaining crop and animal production, there is an urgent need to develop, adapt, and use innovative manure management technologies and practices to reduce pathogen loads in manure prior to land application. Storage systems such as traditional lagoons and high stacks cannot provide predictable pathogen control because the environmental factors that impact microbial survival vary widely in such uncontrolled management schemes. Grazing or free-ranging animal systems present different challenges because manure deposition on rangeland can potentially be dispersed across large drainage areas that will impact distant off-site locations.

Vegetated filter strips and treatment areas that intercept run-off from agricultural drainage channels show promise when coupled with grazing and feedlot livestock management practices in reducing off-site pathogen loads. Information about effectiveness of field measures and practices at the field- and watershed-levels is urgently needed. A HACCP framework for composting operations needs to be developed, used, and evaluated for its performance characteristics relative to reducing the number of incompletely disinfected composts and its cost-effectiveness.

REFERENCES

Ahring, B. K., Mladenovska, Z., Iranpour, R., and Westermann, P. (2002). State of the art and future perspectives of thermophilic anaerobic digestion. *Water Sci. Technol.* **45,** 293–298.

Aitken, M. D., Sobsey, M. D., Van Abel, N. A., Blauth, K. E., Singleton, D. R., Crunk, P. L. et al. (2007). Inactivation of *Escherichia coli* O157:H7 during thermophilic anaerobic digestion of manure from dairy cattle. *Water Res.* **41,** 1659–1666.

Al-Kaisi, M. M., Davis, J. G., and Waskom, R. M. (1998a). Fact sheet 1.222. Liquid manure application methods. Colorado State University Extension, Fort Collins, CO.

Al-Kaisi, M. M., Waskom, R. M., and Davis, J. G. (1998b). Fact sheet No. 1.223. Liquid manure application to cropland. Colorado State University Agricultural Extension, Fort Collins, CO.

Anderson, B. C. (1985). Moist heat inactivation of *Cryptosporidium* sp. *Amer. J. Public Health* **75,** 1433–1444.

Atwill, E. R., Pereira, M. D., Alonso, L. H., Elmi, C., Epperson, W. B., Smith, R. et al. (2006). Environmental load of *Cryptosporidium parvum* oocysts from cattle manure in feedlots from the central and western United States. *J. Environ. Qual.* **35,** 200–206.

Avery, S. M., Moore, A., and Hutchison, M. L. (2004). Fate of *Escherichia coli* originating from livestock faeces deposited directly onto pasture. *Lett. Appl. Microbiol.* **38,** 355–359.

Berry, E. D., Woodbury, B. L., Nienaber, J. A., Eigenberg, R. A., Thurston, J. A., and Wells, J. E. (2007). Incidence and persistence of zoonotic bacterial and protozoan pathogens in a beef cattle feedlot runoff control vegetative treatment system. *J. Environ. Qual.* **36,** 1873–1882.

Bicudo, J. R. and Goyal, S. M. (2003). Pathogens and manure management systems: A review. *Environ. Technol.* **24,** 115–130.

Boes, J., Alban, L., Bagger, J., Mogelmose, V., Baggesen, D. L., and Olsen, J. E. (2005). Survival of *Escherichia coli* and *Salmonella* Typhimurium in slurry applied to clay soil on a Danish swine farm. *Prev. Vet. Med.* **69,** 213–228.

Bowman, D. D. (2009). *Manure pathogens.* McGraw-Hill, New York.

Burger, H. J. (1982). Large-scale management systems and parasite populations prevalence and resistance of parasitic agents in animal effluents and their potential hygienic hazard. *Vet. Parasitol.* **11,** 49–60.

Burton, C. H. and Turner, C. (2003). *Manure management—Treatment strategies for sustainable agriculture*. Silsoe Research Institute, Wrest Park. Silsoe, Bedford, UK.

Campbell, I., Tzipori, A. S., Hutchison, G., and Angus, K. W. (1982). Effect of disinfectants on survival of *Cryptosporidium* oocysts. *Vet. Rec.* **111,** 414–415.

Castellan, D. M., Kinde, H., Kass, P. H., Cutler, G., Breitmeyer, R. E., Bell, D. D. et al. (2004). Descriptive study of California egg layer premises and analysis of risk factors for *Salmonella enterica* serotype *enteritidis* as characterized by manure drag swabs. *Avian Dis.* **48,** 550–561.

Christian, A. H., Evanylo, G. K., and Pease, J. W. (1997). On-farm composting—A guide to principles, planning, and operations. Virginia Cooperative Extension, Blacksburg, VA.

Cieslak, P. R., Barrett, T. J., Griffin, P. M., Gensheimer, K. F., Beckett, G., Buffington, J., and Smith, M. G. (1993). *Escherichia coli* O157:H7 infection from a manured garden. *Lancet* **342,** 367.

Cirelli, G. L., Consoli, S., Di Grande, V., Milani, M., and Toscano, A. (2007). Subsurface constructed wetlands for wastewater treatment and reuse in agriculture: Five years of experiences in Sicily, Italy. *Water Sci. Technol.* **56,** 183–191.

Cole, D. J., Hill, V. R., Humenik, F. J., and Sobsey, M. D. (1999). Health, safety, and environmental concerns of farm animal waste. *Occup. Med.* **14,** 423–448.

Cooper, J., Niggli, U., and Leifert, C. (2007). *Handbook of organic food safety and quality*. Woodhead Publishing Ltd, Cambridge, UK.

Costantini, V. P., Azevedo, A. C., Li, X., Williams, M. C., Michel, F. C., Jr., and Saif, L. J. (2007). Effects of different animal waste treatment technologies on detection and viability of porcine enteric viruses. *Appl. Environ. Microbiol.* **73,** 5284–5291.

Davis, J. G., Andrews, J. E., and Al-Kaisi, M. M. (1997). Fact Sheet No. 1.221. Liquid Manure Management. Colorado State University Agricultural Extension, Fort Collins, CO.

De Bertoldi, M., Ferranti, M. P., L'Hermite, P., and Zucconi, F. (1986). *Compost: Production, Quality and Use*. Elsevier Applied Science, London.

DEFRA (2004). Guidance on the treatment in approved composting or biogas plants of animal by-products and catering waste—Animal By-Products Regulation 2003. www.defra.gov.uk/animalh/by-prods/pdf/compost_guidance.pdf. Accessed 8/17/08.

Entry, J. A., Leytem, A. B., and Verwey, S. (2005). Influence of solid dairy manure and compost with and without alum on survival of indicator bacteria in soil and on potato. *Environ. Pollut.* **138,** 212–218.

Esteban, J. I., Oporto, B., Aduriz, G., Juste, R. A., and Hurtado, A. (2008). A survey of food-borne pathogens in free-range poultry farms. *Int. J. Food Microbiol.* **123,** 177–182.

Fayer, R., Graczyk, T. K., Cranfield, M. R., and Trout, J. M. (1996). Gaseous disinfection of *Cryptosporidium parvum* oocysts. *Appl. Environ. Microbiol.* **62,** 3908–3909.

Fayer, R., Santin, M., and Trout, J. M. (2007). Prevalence of *Cryptosporidium* species and genotypes in mature dairy cattle on farms in eastern United States compared with younger cattle from the same locations. *Vet. Parasitol.* **145,** 260–266.

Fayer, R., Santin, M., and Trout, J. M. (2008). *Cryptosporidium ryanae* n. sp. (Apicomplexa: Cryptosporidiidae) in cattle (*Bos taurus*). *Vet. Parasitol.*

Fayer, R. and Ungar, B. L. (1986). *Cryptosporidium* spp. and cryptosporidiosis. *Microbiol. Rev.* **50,** 458–483.

Feachem, R., Bradley, D. J., Garelick, H., and Mara, D. D. (1981). *Appropriate technology for water supply and sanitation: Health aspects of excreta and sewage management-a state-of-the art review.* The World Bank, Washington, DC.

Fossler, C. P., Wells, S. J., Kaneene, J. B., Ruegg, P. L., Warnick, L. D., Bender, J. B. et al. (2005a). Herd-level factors associated with isolation of *Salmonella* in a multi-state study of conventional and organic dairy farms I. *Salmonella* shedding in cows. *Prev. Vet. Med.* **70,** 257–277.

Fossler, C. P., Wells, S. J., Kaneene, J. B., Ruegg, P. L., Warnick, L. D., Eberly, L. E. et al. (2005b). Cattle and environmental sample-level factors associated with the presence of *Salmonella* in a multi-state study of conventional and organic dairy farms. *Prev. Vet. Med.* **67,** 39–53.

Guan, J., Chan, M., Ma, B., Grenier, C., Wilkie, D. C., Pasick, J. et al. (2008). Development of methods for detection and quantification of avian influenza and Newcastle disease viruses in compost by real-time reverse transcription polymerase chain reaction and virus isolation. *Poult Sci.* **87,** 838–843.

Guan, T. Y. and Holley, R. A. (2003). Pathogen survival in swine manure environments and transmission of human enteric illness—A review. *J. Environ. Qual.* **32,** 383–392.

Guber, A. K., Pachepsky, Y. A., Shelton, D. R., and Yu, O. (2007). Effect of bovine manure on fecal coliform attachment to soil and soil particles of different sizes. *Appl. Environ. Microbiol.* **73,** 3363–3370.

Harter, T., Atwill, E. R., Hou, L., Karle, B. M., and Tate, K. W. (2008). Developing risk models of *Cryptosporidium* transport in soils from vegetated, tilted soilbox experiments. *J. Environ. Qual.* **37,** 245–258.

Haug, R. T. (1993). *The Practical Handbook of Compost Engineering.* Lewis Publishers, Boca Raton, FL.

Holley, R. A., Arrus, K. M., Ominski, K. H., Tenuta, M., and Blank, G. (2006). *Salmonella* survival in manure-treated soils during simulated seasonal temperature exposure. *J. Environ. Qual.* **35,** 1170–1180.

Humenik, F. J. (2001). *Lesson 25—Manure treatment options.* MidWest Plan Service, Ames, IA.

Hutchison, M. L., Walters, L. D., Avery, S. M., and Moore, A. (2005). Decline of zoonotic agents in livestock waste and bedding heaps. *J. Appl. Microbiol.* **99,** 354–362.

Ingram, D. T. and Millner, P. D. (2007). Factors affecting compost tea as a potential source of *Escherichia coli* and *Salmonella* on fresh produce. *J. Food Prot.* **70,** 828–834.

Jenkins, M. B., Bowman, D. D., and Ghiorse, W. C. (1998). Inactivation of *Cryptosporidium parvum* oocysts by Ammonia. *Appl. Environ. Microbiol.* **64,** 784–788.

Jiang, X., Morgan, J., and Doyle, M. P. (2002). Fate of *Escherichia coli* O157:H7 in manure-amended soil. *Appl. Environ. Microbiol.* **68,** 2605–2609.

Johannessen, G. S., Frøseth, R. B., Solemdal, L., Jarp, J., Wasteson, Y., and Rørvik, L. M. (2004). Influence of bovine manure as fertilizer on the bacteriological quality of organic Iceberg lettuce. *J. Appl. Microbiol.* **96,** 787–794.

Juteau, P., Tremblay, D., Ould-Moulaye, C. B., Bisaillon, J. G., and Beaudet, R. (2004). Swine waste treatment by self-heating aerobic thermophilic bioreactors. *Water Res.* **38,** 539–546.

Karim, M. R., Glenn, E. P., and Gerba, C. P. (2008). The effect of wetland vegetation on the survival of *Escherichia coli*, *Salmonella typhimurium*, bacteriophage MS-2 and polio virus. *J. Water Health* **6,** 167–175.

Kasorndorkbua, C., Opriessnig, T., Huang, F. F., Guenette, D. K., Thomas, P. J., Meng, X. J., and Halbur, P. G. (2005). Infectious swine hepatitis E virus is present in pig manure storage facilities on United States farms, but evidence of water contamination is lacking. *Appl. Environ. Microbiol.* **71,** 7831–7387.

Kearney, T. E., Larkin, M. J., and Levett, P. N. (1993). The effect of slurry storage and anaerobic digestion on survival of pathogenic bacteria. *J. Appl. Bacteriol.* **74,** 86–93.

Kinde, H., Utterback, W., Takeshita, K., and McFarland, M. (2004). Survival of exotic Newcastle disease virus in commercial poultry environment following removal of infected chickens. *Avian Dis*. **48,** 669–674.

Kudva, I. T., Blanch, K., and Hovde, C. J. (1998). Analysis of *Escherichia coli* O157:H7 survival in ovine or bovine manure and manure slurry. *Appl. Environ. Microbiol.* **64,** 3166–3174.

Layden, N. M., Mavinic, D. S., Kelly, H. G., Moles, R., and Bartlett, J. (2007). Autothermal thermophilic aerobic digestion (ATAD)—Part I: Review of origins, design, and process operation. *J. Environ. Eng. Sci.* **6,** 665–678.

Lewis, D. J., Atwill, E. R., Lennox, M. S., Hou, L., Karle, B., and Tate, K. W. (2005). Linking on-farm dairy management practices to storm-flow fecal coliform loading for California coastal watersheds. *Environ. Monit. Assess.* **107,** 407–425.

Ley, V., Higgins, J., and Fayer, R. (2002). Bovine enteroviruses as indicators of fecal contamination. *Appl. Environ. Microbiol.* **68,** 3455–3461.

Meerburg, B. G., Jacobs-Reitsma, W. F., Wagenaar, J. A., and Kijlstra, A. (2006). Presence of *Salmonella* and *Campylobacter* spp. in wild small mammals on organic farms. *Appl. Environ. Microbiol.* **72,** 960–962.

Miller, W. A., Lewis, D. J., Lennox, M., Pereira, M. G. C., Tate, K. W., Conrad, P. A., and Atwill, E. R. (2007). Climate and on-farm risk factors associated with *Giardia duodenalis* cysts in storm runoff from California coastal dairies. *Appl. Environ. Microbiol.* **73,** 6972–6979.

Miller, W. A., Lewis, D. J., Pereira, M. D. G., Lennox, M., Conrad, P. A., Tate, K. W., and Atwill, E. R. (2008). Farm factors associated with reducing *Cryptosporidium* loading in storm runoff from dairies. *J. Environ. Qual.* **37,** 1875–1882.

Misra, R. V., Roy, R. N., and Hiraoka, H. (2003). *On-farm composting*. Food and Agriculture Organization of the United Nations, Rome.

Monteith, H. D., Shannon, E. E., and Derbyshire, J. B. (1986). The inactivation of a bovine enterovirus and a bovine parvovirus in cattle manure by anaerobic digestion, heat treatment, gamma irradiation, ensilage and composting. *J. Hyg. (Lond.)* **97,** 175–184.

Muirhead, R. W., Collins, R. P., and Bremer, P. J. (2005). Erosion and subsequent transport state of *Escherichia coli* from cowpats. *Appl. Environ. Microbiol.* **71,** 2875–2879.

Murinda, S. E., Nguyen, L. T., Nam, H. M., Almeida, R. A., Headrick, S. J., and Oliver, S. P. (2004). Detection of sorbitol-negative and sorbitol-positive Shiga toxin-producing *Escherichia coli*, *Listeria monocytogenes*, *Campylobacter jejuni*, and *Salmonella* spp. in dairy farm environmental samples. *Foodborne Pathog. Dis.* **1,** 97–104.

Ndegwa, P. M. and Thompson, S. A. (2001). Integrating composting and vermicomposting in the treatment and bioconversion of biosolids. *Bioresour. Technol.* **76,** 107–112.

Nelson, H. (1997). The contamination of organic produce by human pathogens in animal manures. Available on-line at http://eap.mcgill.ca//SFMC_1.htm (accessed 8/17/08). Ecological Agriculture Projects, Faculty of Agricultural and Environmental Sci., McGill Univ. (Macdonald Campus), Ste-Anne-de-Bellevue, Canada.

Olson, M. E., Goh, J., Phillips, M., Guselle, N., and McAllister, T. A. (1999). *Giardia* cyst and *Cryptosporidium* oocyst survival in water, soil, and cattle feces. *J. Environ. Qual.* **28,** 1991–1996.

Ottoson, J., Nordin, A., von Rosen, D., and Vinneras, B. (2008). *Salmonella* reduction in manure by the addition of urea and ammonia. *Bioresour. Technol.* **99,** 1610–1615.

Pachepsky, Y. A., Sadeghi, A. M., Bradford, S. A., Shelton, D. R., Guber, A. K., and Dao, T. (2006). Transport and fate of manure-borne pathogens: Modeling perspective. *Agric. Water Manag.* **86,** 81–92.

Pachepsky, Y. A., Yu, O., Karns, J. S., Shelton, D. R., Guber, A. K., and van Kessel, J. S. (2008). Strain-dependent variations in attachment of *E. coli* to soil particles of different sizes. *Intl. Agrophys.* **22,** 61–66.

Plym-Forshell, L. (1995). Survival of salmonellas and *Ascaris suum* eggs in a thermophilic biogas plant. *Acta Vet. Scand.* **36,** 79–85.

Robertson, L. J., Campbell, A. T., and Smith, H. V. (1992). Survival of *Cryptosporidium parvum* oocysts under various environmental pressures. *Appl. Environ. Microbiol.* **58,** 3494–3500.

Rynk, R. (1992). *On-Farm Composting Handbook, NRAES No. 54*. NRAES (Natural Resource, Agricultural, and Engineering Service) Cooperative Extension, Ithaca, NY.

Santin, M., Trout, J. M., and Fayer, R. (2008). A longitudinal study of cryptosporidiosis in dairy cattle from birth to 2 years of age. *Vet Parasitol* **155,** 15–23.

Scheuerell, S. and Mahafee, W. (2002). Compost tea: Principles and prospects for plant disease control. *Compost Sci. Util.* **10,** 313–338.

Shepherd, M. W., Jr., Liang, P., Jiang, X., Doyle, M. P., and Erickson, M. C. (2007). Fate of *Escherichia coli* O157:H7 during on-farm dairy manure-based composting. *J. Food Prot*. **70**, 2708–2716.

Spencer, J. L. and Guan, J. (2004). Public health implications related to spread of pathogens in manure from livestock and poultry operations. *Methods Mol. Biol*. **268**, 503–515.

Strauch, D. (1991). Survival of pathogenic micro-organisms and parasites in excreta, manure and sewage sludge. *Rev. Sci. Tech*. **10**, 813–846.

Strauch, D. and Ballarini, G. (1994). Hygienic aspects of the production and agricultural use of animal wastes. *J. Vet. Med. Ser. B* **41**, 176–228.

Thompson, W., Legge, P., Millner, P., and Watson, M. (2002). Test methods for the examination of composts and composting, CD-ROM. U.S. Composting Council, http://tmecc.org/tmecc/. Accessed 8/17/08.

Thurston-Enriquez, J. A., Gilley, J. E., and Eghball, B. (2005). Microbial quality of runoff following land application of cattle manure and swine slurry. *J. Water Health* **3**, 157–171.

Trask, J. R., Kalita, P. K., Kuhlenschmidt, M. S., Smith, R. D., and Funk, T. L. (2004). Overland and near-surface transport of *Cryptosporidium parvum* from vegetated and nonvegetated surfaces. *J. Environ. Qual*. **33**, 984–993.

Turner, C., Williams, S. M., Burton, C. H., Cumby, T. R., Wilkinson, P. J., and Farrent, J. W. (1999). Pilot scale thermal treatment of pig slurry for the inactivation of animal virus pathogens. *J. Environ. Sci. Health B* **34**, 989–1007.

Turner, C., Williams, S. M., and Cumby, T. R. (2000). The inactivation of foot and mouth disease, Aujeszky's disease and classical swine fever viruses in pig slurry. *J. Appl. Microbiol*. **89**, 760–767.

Tzipori, S. (1983). Cryptosporidiosis in animals and humans. *Microbiol. Rev*. **47**, 84–96.

Tzipori, S. and Widmer, G. (2008). A hundred-year retrospective on cryptosporidiosis. *Trends Parasitol*. **24**, 184–189.

USDA. (1999). Agricultural waste management field handbook, National Engineering Handbook Part 651. US Department of Agriculture, Natural Resource Conservation Service. Washington, DC. www.wsi.nrcs.usda.gov/products/W2Q/AWM/handbk.html

USDA. (2000). National organic program, 7 CFR Part 205. The Organic Foods Production Act of 1990, as amended (7 U.S.C. 6501–6522) 65 FR 80637, Dec. 21, 2000. US Department of Agriculture, Agricultural Marketing Service. Washington, DC.

USDA. (2007). Animal waste management software and user's guide. U.S. Department of Agriculture, Natural Resource Conservation Service. Washington, DC. [http://www.wsi.nrcs.usda.gov/products/W2Q/AWM/pgrm.html#DOCUMENTS].

USDA. (2008). NRCS air quality and atmospheric change fact sheets—Air quality and animal operations. U.S. Department of Agriculture, Natural Resource Conservation Service. Washington, DC. www.airquality.nrcs.usda.gov/Documents/

USEPA. (1994). A plain English guide to the EPA Part 503 biosolids rule, EPA/832/R-93/003. Office of Wastewater Management, US Environmental Protection Agency, Washington, DC.

USEPA. (2000a). Guide to field storage of biosolids and other organic by-products used in agriculture and for soil resource management, EPA/832/B-00-007. Office of Wastewater Management, US Environmental Protection Agency, Washington, DC.

USEPA. (2000b). Standards for the use or disposal of sewage sludge. Code of Federal Regulations (USEPA CFR), 40CFR 503 Title 40, Parts 425 to 699 [Revised as of July 1, 2000], pp. 754–786, Office of Wastewater Management, US Environmental Protection Agency, Washington, DC.

USEPA. (2003). Environmental regulations and technology control of pathogens and vector attraction in sewage sludge, EPA/625/R-92-013, Office of Research and Development, National Risk Management Research Laboratory. US Environmental Protection Agency, Cincinnati, OH.

Vanotti, M. B., Hunt, P. G., Szogi, A. A., Humenik, F. B., Millner, P. D., and Ellison, A. Q. (2003). Solids separation, nitrification-denitrification, soluble phosphorus removal, solids processing system. North Carolina Animal Waste Management Workshop, pp. 30–35.

Vanotti, M. B., Millner, P. D., Hunt, P. G., and Ellison, A. Q. (2005a). Removal of pathogen and indicator microorganisms from liquid swine manure in multi-step biological and chemical treatment. *Bioresour. Technol.* **96,** 209–214.

Vanotti, M. B., Rice, J. M., Ellison, A. Q., Hunt, P. G., Humenik, F. J., and Baird, C. L. (2005b). Solid-liquid separation of swine manure with polymer treatment and sand filtration. *Trans. Amer. Soc. Agric. Engin* **48,** 1567–1574.

Vanotti, M. B., Szogi, A. A., Hunt, P. G., Millner, P. D., and Humenik, F. J. (2007). Development of environmentally superior treatment system to replace anaerobic swine lagoons in the USA. *Bioresour. Technol.* **98,** 3184–3194.

Vanotti, M. B., Szogi, A. A., and Vives, C. A. (2008). Greenhouse gas emission reduction and environmental quality improvement from implementation of aerobic waste treatment systems in swine farms. *Waste Manag.* **28,** 759–766.

Wang, G., Zhao, T., and Doyle, M. P. (1996). Fate of enterohemorrhagic *Escherichia coli* O157:H7 in bovine feces. *Appl. Environ. Microbiol.* **62,** 2567–2570.

Westerman, P. W. and Bicudo, J. R. (2005). Management considerations for organic waste use in agriculture. *Bioresour. Technol.* **96,** 215–221.

Williams, A. P., Avery, L. M., Killham, K., and Jones, D. L. (2005). Persistence of *Escherichia coli* O157 on farm surfaces under different environmental conditions. *J. Appl. Microbiol.* **98,** 1075–1083.

Williams, C. M. (2003). Development of environmentally superior technologies, year 3, progress report for technology determinations per agreements between the Attorney General of North Carolina and Smithfield Foods and Premium Standard Farms, and Frontline Farmers, reporting period July 26, 2002–July 25, 2003. North Carolina State University Animal and Poultry Waste Management Center, North Carolina State University, Raleigh, NC.

Xiao, L. and Fayer, R. (2008). Molecular characterization of species and genotypes of *Cryptosporidium* and *Giardia* and assessment of zoonotic transmission. *Int. J. Parasitol.* **38,** 1239–1255.

Xiao, L., Zhou, L., Santin, M., Yang, W., and Fayer, R. (2007). Distribution of *Cryptosporidium parvum* subtypes in calves in eastern United States. *Parasitol. Res.* **100,** 701–706.

CHAPTER 5

Water Quality

Charles P. Gerba

Department of Soil, Water and Environmental Science, University of Arizona, Tucson, AZ

Christopher Y. Choi

Department of Agricultural and Biosystems Engineering, University of Arizona, Tucson, AZ

CHAPTER CONTENTS

Introduction	105
Irrigation Water	106
Water Quality Standards for Irrigation Water	108
Occurrence of Pathogens in Irrigation Water	109
Contamination of Produce During Irrigation	111
Survival of Pathogens on Produce in the Field	113
Other Sources	113
Summary and Conclusions	114

INTRODUCTION

Water not only plays an essential role in the growth of produce, but also in cleaning and hygienic uses pre- and postharvest. Ensuring the microbial quality of this water is critical to prevent the contamination of the produce by waterborne enteric pathogens. Water has always played a key role as a vehicle for the transmission of pathogens transmitted by the fecal-oral route. For more than a century now it has been recognized that fecally contaminated water, either used for drinking or irrigation of food crops, can result in the transmission of enteric pathogens. Conventional treatment of drinking water (filtration and disinfection) has eliminated diseases such as cholera and typhoid in the developed world, however waterborne outbreaks still occur because of treatment or distribution system failures. This is because raw water sources are always subject to contamination by animals and sewage discharge. At the same time that modern drinking water treatment began, the use of raw sewage for food crop irrigation was banned or severely restricted. However, in arid regions with limited water resources, guidelines for the use of treated wastewater for food crops traditionally eaten raw have been developed, although in the developing world they are seldom enforced.

Several produce outbreaks have been known or suspected to have arisen from contamination in the field, suggesting contamination by irrigation or during handling (Dentinger et al., 2001; CDC, 2001; Nyard et al., 2008). Perhaps more significant may be the low level transmission of viruses and protozoan parasites by food from contaminated irrigation water. Quantitative microbial risk analysis has suggested that low levels of virus in irrigation water can result in a significant level of risk of infection to consumers (Petterson et al., 2001). Stine et al. (2005) estimated that less than one hepatitis A virus per 10 liters in irrigation water could result in a risk exceeding 1:10,000 per year, considering the efficiency of transfer of the virus to crop and survival until harvest time. The 1:10,000 risk of infection per year is currently the acceptable level used by the United States Environmental Protection Agency for Drinking Water (Regli et al., 1991).

Contamination of produce may also occur through the use of contaminated water to apply pesticides, fertilizers, washing, hydrocooling, handwashing, and icing.

Irrigation Water

The largest user of fresh water in the world is agriculture, with more than 70% being used for irrigation. About 240 million hectares, 17% of the world's cropland, are irrigated, producing one third of the world's food supply (Shanan, 1998). Nearly 70% of this area is in developing countries. Irrigation with sewage or sewage-contaminated waters in developing countries is fairly common and usually not regulated. Although guidelines for wastewater reuse have been developed by the World Health Organization (WHO, 2006), their application in developing countries will remain difficult, due to inadequate institutional capability and general lack of financial resources.

The World Health Organization estimates that 10% of the world's population consumes food that is irrigated with untreated wastewater (WHO, 2006), amounting to about 20 million hectares of crop land (Scott et al., 2004a). Significant irrigation with wastewater of food crops occurs near the major cities in Peru and Bolivia in Latin America (Scott et al., 2004). In Pakistan sewage is used directly for irrigation of vegetables commonly eaten raw (Ensink, 2004).

In the United States, California and Arizona are the nation's major producers of lettuce, carrots, broccoli, and cantaloupe (Arizona Farm Bureau, 2003). All these crops are grown almost entirely by irrigated agriculture. It is thus surprising that we know little about the microbial quality of this water. Most studies have dealt with the occurrence and fate of enteric pathogens in reclaimed water used for irrigation and not the quality of surface waters currently in use. Little data exists on the occurrence of pathogens in irrigation waters, which do not intentionally receive sewage discharges.

There are few published studies on the quality of nonreclaimed wastewater used as an irrigation source (Steele and Odumeru, 2004). Irrigation agriculture requires approximately two acre-feet of water per acre of growing crops. The frequency and volume of application must be carefully programmed to compensate for deficiencies in rainfall distribution and soil moisture content during the growing season. Rivers and streams are tapped by large dams and then diverted into extensive canal systems. Ground water is pumped from wells into canals (which places the water at risk from surface contamination), and catchments are constructed to trap storm-water runoff. Because water availability is often critical, little attention is given to the microbial quality of the irrigation runoff. In water-short areas, available sources are subject to contamination by sewage discharge from small communities (unplumbed housing along canals in developing countries is common), cattle feedlot drainage, animals grazing along canal embankments, storm-water events, and return irrigation water (noninfiltrated water from the field being irrigated is returned to the irrigation channels) (Table 5.1).

Table 5.1 Factors Influencing Water Quality in Man-Made Irrigation Channels

Factor	Remarks
Rainfall	Rain can act to resuspend sediments containing microbes; wash in fecal matter from bank sides and enhance drainage.
Storm-water drainage	In some areas canals may receive storm-water flows from streets and grazing land.
Return flows	Water remaining from flood-field irrigation may be returned to canals and used to irrigate other fields.
Water fowl	Large numbers of water fowl may contribute bacterial pathogens such as *Campylobacter*.
Animal occurrence on bank sides	Animal grazing may occur on bank sides of irrigation channels. In some areas the bank sides are used as urban parks and pet feces may be washed into canals during rainfall events.
Urban areas	Runoff during storms may be diverted to irrigation canals.
Channel size and depth	The impact of storm events or other contamination events are greater in smaller channels because of less water for dilution.
Distance of canal from water source	The greater the distance from the water source, the greater the chance for contamination.
Recreational use of source water	Bathers can contribute significant amounts of pathogens if the lakes or other bodies of water are used for recreational purposes.

Since irrigation channels are frequently small, these occurrences of pollution discharge can result in rapid deterioration of water quality.

WATER QUALITY STANDARDS FOR IRRIGATION WATER

Most of the research on enteric pathogen contamination of vegetables and fruits during production has been done to evaluate the safety of reclaimed wastewater irrigation. Many states in the United States have standards for the treatment of reclaimed water to be used for food crop irrigation (Asano, 1998), and the World Health Organization has also made recommendations (WHO, 2006). The state of California requires advanced physical-chemical treatment and extended disinfection to produce virus-free effluent. A coliform standard less than 2/100 mL must also be met (Asano, 1998). The state of Arizona had a virus standard of one plaque forming unit/40 liters and *Giardia* cysts of one per 40L in addition to a fecal coliform standard of 25/100 mL (Rose and Gerba, 1991). Although standards for the use of reclaimed wastewater exist for food crops eaten raw in the United States, irrigation using reclaimed water for crop irrigation is seldom practiced. In developing countries, raw or partially treated wastewater is often used to irrigate crops, especially in arid regions.

One of the few early studies conducted on irrigation waters documented the wide range in microbial quality of this water (Geldreich and Bordner, 1971). The wide variation was attributed to the discharge of domestic sewage into streams from which the irrigation water was obtained. This study was conducted in the western United States (Wyoming, Utah, and Colorado). Median values of fecal coliform bacteria ranged from 70 to 450,000/100 ml. Based on results obtained with *Salmonella* occurrence in the same waters, they recommend a fecal coliform standard for irrigation waters of 1000/100 mL.

Guidelines for the microbial water quality of surface water tend to be more lenient than those for wastewater because of the belief that enteric viruses and other human pathogens are less likely to be present or less numerous (Steele and Odumeru, 2004). The criteria range from less than 100 to less than 1000 fecal coliforms per 100 mL. Guidelines also include criteria for *Escherichia coli* and fecal streptococcus (Steele and Odumerus, 2004). The United States Environmental Protection Agency guidelines for surface water recommends fewer than 1000 fecal coliforms per 100 mL of surface water, including river water, for irrigation of crops (EPA, 1973). The differences among the guidelines reflect widespread uncertainty about the actual risk of disease transmission by the irrigation water. Obviously, data on the occurrence of pathogens in irrigation waters would aid the development of a risk-based approach to the development of standards.

OCCURRENCE OF PATHOGENS IN IRRIGATION WATER

The microbial quality of irrigation water depends on the source of the water and contamination as it is transmitted through the distribution system. Sources of human enteric pathogens may involve sewage discharges into source water, septic tanks, and recreational bathers, for example (Table 5.1). In the United States, disinfection of wastewater effluents is required before discharge into surface waters, greatly reducing the risks from enteric bacterial and viral enteric pathogens. However, this is not a common practice in much of the world, including Europe. Humans are believed to be the only significant source of enteric viruses in water, although hepatitis E virus may be the exception. Zoonotic pathogens may not only originate from domestic animals, but also from wild animals such as migrating water fowl.

Although groundwater is often considered a microbially safe source for irrigation water, recent studies in the United States have indicated that 8 to 31% of the groundwaters may contain viruses (Abbaszadegan et al., 2003; Borchardt et al., 2003). These viruses may originate from septic discharges; leaking sewer lines; or infiltration from lakes, rivers, or oxidation ponds.

Several outbreaks have been linked to the use of contaminated irrigation water. In a multistate outbreak of *Salmonella* Newport involving tomatoes, the strain involved in the outbreak was detected in the water used to irrigate the fields in which the tomatoes were grown (Greene et al., 2008). Irrigation water was also believed to be involved in another *Salmonella* outbreak involving lettuce grown in Italy (Nygard et al., 2008). In Sweden an outbreak of *E. coli* 0157:H7 associated with locally grown lettuce was traced to the use of river water contaminated by local farm animals.

Several studies have reported the occurrence of enteric pathogens in irrigation water (Table 5.2). *Salmonella* and the enteric protozoan parasites *Giardia* and *Cryptosporidium* have been reported in irrigation waters used for produce (Table 5.2). Human rotaviruses and noroviruses have also been detected in irrigation water. Kayed (2004) studied the occurrence of protozoan parasites, bacterial indicators of fecal contamination, and noroviruses in irrigation waters in central and western Arizona. The irrigation waters in central Arizona are derived from a series of dammed reservoirs. The water is then channeled through a series of canals traveling a distance as great as 40 miles. In western Arizona the water comes from a reservoir on the Colorado River. Noroviruses were detected in 20.7% of the samples

Table 5.2 Occurrence of Pathogens in Irrigation Water

Study	Location	Results	Remarks
Garcia-Villanova Ruiz et al., 1987	Spain	*Salmonella* detected in 54.6% of 181 samples of irrigation water	*Salmonella* also detected on 1.7% of vegetable samples
Robertson and Gjerde, 2001	Norway	*Salmonella* detected in irrigation water used for bean spouts	
Thurston-Enriquez et al., 2002	Arizona and Central America	*Giardia, Cryptosporidium* and microsporidia detected	
Okafo et al., 2003	Nigeria	*Salmonella* detected in 2 to 14% of irrigation water used for produce production	
Gannon et al., 2004a	Western Canada	*E. coli* O157:H7 and *Salmonella* detected	
Chaidez et al., 2005	Western Mexico	*Giardia* detected in 48% of samples and *Cryptosporidium* in 50%	
Duffy et al., 2005	Texas	*Salmonella* detected in irrigation water	
Esponoza-Medina et al., 2006	Brazil	*Salmonella* in 23.5% of irrigation water used to irrigate cantaloupe	*Salmonella* also detected in 9.1% of groundwater samples and 4.8% of chlorinated water samples
Van Zyl et al., 2006	South Africa	Human rotaviruses detected in 14% of irrigation water samples used on raw vegetables	Rotavirus also detected on 1.4% of vegetable samples
Materon et al., 2007.	Southern Texas	*Salmonella* detected in river water used to irrigate cantaloupes	Washing melons with chlorine containing water did not eliminate the *Salmonella*
Izumi et al., 2008	Japan	*E. coli* O157:H7 in irrigation water used to irrigate persimmons	*Salmonella* also isolated in pesticide spray water made from irrigation water

from western Arizona and 18.2% of the samples from central Arizona. Geometric average *E. coli* concentrations were 6.4/100 mL in western Arizona canals and 18/100 mL in central Arizona. *Salmonella* and *Campylobacter* spp. were also detected in the irrigation water, especially after rainfall events. Since polymerase chain reaction was used to detect the norovirus, the infectivity of the viruses could not be determined. Still, these results show that contamination of irrigation water by enteric viruses does occur, even when there is no intentional discharge of sewage into the system.

CONTAMINATION OF PRODUCE DURING IRRIGATION

The likelihood of the edible parts of a crop becoming contaminated depends upon a number of factors including growing location, type of irrigation application, and nature of the produce surface (Table 5.3). If the edible part of the crop grows on or near the soil surface, it is more likely to become contaminated than a fruit growing in the aerial parts of a plant. Some produce surfaces are furrowed or have other sources that may retain water (e.g., tomato vs cantaloupe). There are four distinct methods of irrigation: sprinkler systems, gravity-flow systems (flood irrigation), drip or trickle methods, and subsurface irrigation. In 1998 approximately 50 million acres of farmland were irrigated annually in the United States, 27 million acres with surface irrigation systems, and the remainder by sprinkler systems (USDA, 1998). In a recent survey in the United Kingdom, it was found that overhead irrigation was the predominant method for fruits and vegetables (Tyrrel et al., 2006). Worldwide, less than 1% of the total irrigated area is believed to be irrigated by drip irrigation (Postel et al., 2001). However, drip irrigation is growing rapidly around the world, covering almost 3,000,000 hectares (Postel et al., 2001).

A limited number of studies have looked at the contamination of crops by enteric bacteria in irrigation water, but only a few have evaluated the degree of contamination by viruses. Stine et al. (2005) quantified the transfer of viruses (PRD-1 virus, a coliphage) and bacteria from water used to

Table 5.3 Factors Affecting the Contamination of the Edible Parts of Plants during Irrigation

Growing location of the edible portion of the plant
- distance from the soil or water surface

Frequency of irrigation
- number of days last irrigated before harvest

Surface of the edible portion
- smooth
- webbed
- rough

Type of irrigation method
- furrow or flood
- sprinkler
- drip
  - surface
  - subsurface

prepare pesticide spray to the surface of cantaloupe, iceberg lettuce, and bell peppers. The average transfer of bacteria from the water to the surface of fruit was estimated to range from 0.00021 to 9.4%, and the average viral transfer ranged from 0.055 to 4.2%, depending on the type of produce.

The type of irrigation method can greatly influence the degree of crop contamination. For example, the degree of PRD-1 virus transference to lettuce was found to be 4.4%, 0.02%, and 0.00039% for spray, furrow, and drip irrigation (Stine et al., 2005a, 2005b; Choi et al., 2004). The reduction in contamination by bacteria (*Escherichia coli*) on lettuce by use of subsurface drip versus flood irrigation was 99.9%; however, it was only 99% for virus (PRD-1) (Song et al., 2006). Stine et al. (2005a) compared surface and subsurface irrigation as sources of contamination of cantaloupe, iceberg lettuce, and bell peppers when the water was seeded with coliphage PRD-1 under field-growing conditions in Arizona. Coliphage was detected on both the lettuce and cantaloupe, but not on the bell pepper.

Oron et al. (1995) applied irrigation water containing up to 1000 plaque forming units/mL of a vaccine strain of poliovirus to tomato plants by subsurface drip irrigation in an outdoor setting in Israel. Some virus was detected in the leaves of the plants, but not in the fruits. The authors stated that the high content of the virus in the water might explain the occurrence of the virus in the leaves. No virus was detected in plants irrigated with wastewater containing the same level of virus. The authors suggested that this might be due to the interaction of the virus with particulate or soluble matter present in the wastewater, but absent in the irrigation water, preventing their entrance into the roots.

Alum (2001) studied the effectiveness of drip irrigation in the control of viral contamination of salad crops (lettuce, tomato, cucumber) in a greenhouse in potted plants. The plants were irrigated with secondary effluent using surface drip and subsurface irrigation. Irrigation water was periodically seeded with coliphages MS-2, PRD-1, poliovirus type 1, adenovirus 40, and hepatitis A virus. Surface irrigation always resulted in the surface contamination of both the aboveground and the underground parts of the plants. In lettuce it was observed that only the outer leaves of the plant became contaminated. No contamination of the plants occurred when subsurface drip irrigation was used. No systemic uptake of viruses was observed in any of the crops.

Choi et al. (2004) assessed viral contamination of lettuce by surface and subsurface drip irrigation using coliphage MS-2 and PRD-1. A greater number of coliphages was recovered from the lettuce in the subsurface plots as compared to those in the furrow-irrigated plots. Shallow drip tape installation and preferential water paths through cracks on the soil surface appeared to be the main causes of high viral contamination. In subsurface drip irrigation, penetration of the irrigation water to the soil surface led to the direct contact with the lettuce stems. Thus, drip tape depth can influence the probability of produce contamination. Greater contamination by PRD-1 was observed, which might be due to its longer survival time.

SURVIVAL OF PATHOGENS ON PRODUCE IN THE FIELD

Studies on the survival of viruses on produce postharvest indicate that little inactivation occurs because of the low temperatures of storage (Seymour and Appleton, 2001). Few studies are available on the survival of viruses on growing crops. Tierney et al. (1977) found that poliovirus survived on lettuce for 23 days after flooding of outdoor plots with wastewater. The virus persisted in the soil for two months during the winter and two to three days during the summer months. Sadovski et al. (1978) spiked wastewater and tap water used to irrigated cucumbers with a high titer of poliovirus. They were able to detect the virus on the cucumbers grown with either (1) surface drip irrigation or (2) the soil and drip lines covered with polyethylene sheets, although virus was detected only occasionally on the cucumbers irrigated with plastic covering the soil, right after irrigation. Viruses on the soil-irrigated cucumbers survived for at least eight days after irrigation.

Hepatitis A virus and coliphage PRD-1 survival on growing produce was found to be similar under high and low humidity conditions (Stine et al., 2005). In general the inactivation rates of these viruses were lower than those of *E. coli* 0157:H7, *Shigella sonnei*, and *Salmonella* enteric on cantaloupe, lettuce, and bell peppers. Hepatitis A virus was reduced about 90% after 14 days, indicating that they could survive from an irrigation event to harvest time.

Salmonella and *E. coli* are also capable of growth on produce surfaces preharvest under certain conditions (Stine et al., 2005a). Islam et al. (2004) observed survival of *Salmonella* for weeks on carrots and radishes after irrigation with artificially contaminated irrigation water. Also using artificially contaminated irrigation water containing *Giardia* cysts and *Cryptosporidium* oocysts, Armon et al. (2002) readily contaminated vegetables during flood irrigation. Of all the vegetables studied, the highest prevalence of oocysts occurred on zucchini.

Other Sources

Use of fecally contaminated water for application of pesticides or in wash water may lead to produce contamination. Keratita and Drechsel (2004) reported that wash water used for produce in Ghana was an important source of contamination with enteric bacteria. A hydrocooler was found to be a source of melon rinds with fecal coliforms and fecal enterococci (Gagliardi et al., 2003). The use of ice for hydrocooling after harvesting may be another source of contamination (Cannon et al., 1991). The use of untreated water to make up pesticide spray has been suspected as a source of enteric pathogens, and Izumi et al. (2008) reported the isolation of *Salmonella* in pesticide spray used on persimmons, in which irrigation water was used to dilute the pesticide before application. Finally, it has been observed that use of contaminated hand wash water, resulting from multiple use by different individuals, can result in hand contamination by enteric bacteria (Ogunsola and Adesiji, 2008).

SUMMARY AND CONCLUSIONS

The role water plays in the contamination of produce has been little studied, despite its potential significance. Irrigation water appears to have the greatest potential for crop contamination in developed countries, since it is always subject to contamination by animal feces. Information on the occurrence of bacterial indicators of fecal contamination and pathogens in irrigation water and potential sources of irrigation water contamination is urgently needed. We need to understand the ecology of enteric pathogens and indicator bacteria in terms of transport and survival of enteric pathogens in complex irrigation delivery systems to better define the risks to produce. Meaningful standards for indicator bacteria that better assess the risk of produce contamination and risk of infection to the consumer need to be developed. Irrigation methods and the type of produce affect the degree of contamination. Spray irrigation of produce is common, and this offers the greatest potential for contamination of the edible parts of the produce. Although the percentage of pathogen transfer from contaminated water to produce by some types of irrigation methods (e.g., drip irrigation) may be low, risks can still be considered significant because of the low numbers of some enteric pathogens, such as viruses, necessary to cause infection (Peterson and Ashbolt, 2001). Environmental conditions, such as temperature and humidity, may determine the survival of viruses on produce surfaces. Limited studies suggest that enteric viruses can survive on produce longer than enteric bacteria, and that such viruses introduced on produce surfaces at the time of irrigation can survive through harvesting. Similar data needs to be developed for foodborne protozoan parasites.

REFERENCES

Abbaszadegan, M., LeChevallier, M, and Gerba, C. (2003). Occurrence of viruses in US groundwater. *J. Amer. Water Works Assoc.* **95,** 107–120.

Alum, A. (2001). Control of viral contamination of reclaimed irrigated vegetables by drip irrigation. Ph.D. Dissertation. University of Arizona, Tucson, AZ.

Arizona Farm Bureau. (2003). Arizona Agricultural Statistics. Phoenix, AZ. www.azfb.org/agfacts/

Armon, R., Gold, D., Brodsky, M., and Oron, G. (2002). Surface and subsurface irrigation with effluents of different qualities and presence of *Cryptosporidium* oocysts in soil and on crops. *Water Sci. Technol.* **49,** 115–122.

Asano, T. (1998). *Wastewater and reclamation reuse.* Technomic Publishing, Lancaster, PA.

Borchardt, M. A., Bertz, P. D., Spencer, S. K., and Battigelli, D. A. (2003). Incidence of enteric viruses in groundwater from household wells in Wisconsin. *Appl. Environ. Microbiol.* **69,** 1172–1180.

Cannon, R. O., Hirschhorn, J. R. B., Rodeheaver, D. C. et al. (1991). A multistate outbreak of Norwalk virus gastroenteritis associated with consumption of commercial ice. *J. Infect. Dis.* **164,** 860–863.

Centers for Disease Control and Prevention. (2003). Hepatitis A outbreak associated with green onions at a restaurant—Monaca, Pennsylvania, 2003. *MMWR* **52,** 1155–1157.

Chaidez, C., Soto, M., Gortares, P., and Mena, K. (2006). Occurrence of *Cryptosporidium* and *Giardia* in irrigation water and its impact on fresh produce industry. *Int. J. Environ. Hlth. Res.* **15,** 339–345.

Choi, C., Song, I, Stine, S. et al. (2004). Role of irrigation and wastewater: Comparison of subsurface irrigation and furrow irrigation. *Water Sci. Technol.* **50,** 61–68.

Dentinger C., Bower, W. A., Nainan, O. V. et al. (2001). An outbreak of hepatitis A associated with green onions. *J. Infect. Dis.* **183,** 1273–1276.

Duffy, E. A., Lucia, L. M., Kells, J. M. et al. (2005). Concentrations of *Escherichia coli* and genetic diversity and antibiotic resistance profiling *Salmonell*a isolated from irrigation water, packing shed equipment, and fresh produce in Texas. *J. Food Protect.* **68,** 70–79.

EPA. (1973). *Water quality criteria.* Ecological Research Series, EPA R3-73–033. Environmental Protection Agency, Washington, DC.

Ensink, J., Mahmood, T., van der Hoek, W. et al. (2004). A nation-wide assessment of wastewater use in Pakistan: An obscure activity or a vitally important one? *Water Policy* **6,** 1–10.

Esponoza-Medina, I. E., Rodriguez, F. J., Vargas-Arispuro, I., and Islas-Osuna, M. A. (2006). PCR identification of *Salmonella*: Potential contamination sources from production and postharvest handling of cantaloupes. *J. Food Prot.* **69,** 1422–1425.

Gagliardi, J. V., Millner, P. D., Lester, G., and Ingram, D. (2003). On-farm and postharvest processing sources of bacterial contamination to melon rinds. *J. Food Protect.* **66,** 82–87.

Gannon, V. P. J., Graham, T. A., Read, S. et al. (2004). Bacterial pathogens in rural water supplies in southern Alberta, Canada. *J. Toxicol. Environ. Hlth.* **A 67,** 1643–1653.

Garcia-Villanova, Ruiz, B., Cueto Espinar, A., and Bolanos Carmona, M. J. (1987). A comparative study of strains of *Salmonella* isolated from irrigation waters, vegetables and human infections. *Epidemiol. Infect.* **98,** 271–276.

Geldreich, E. E. and Bordner, R. H. (1971). Fecal contamination of fruits and vegetables during cultivation and processing for market. *J. Milk Food Technol.* **34,** 184–198.

Greene, S. K., Daly, E. R., Talbot, E. A. et al. (2007). Recurrent multistate outbreak of *Salmonella* Newport associated with tomatoes from contaminated fields, 2005. *Epidemiol. Infect.* **136,** 157–165.

Islam, M., Morgan, J. Doyle, M. P. et al. (2004). Fate of *Salmonella enterica* serovar Typhimurium on carrots and radishes grown in fields treated with contaminated manure composts or irrigation water. *Appl. Environ. Microbiol.* **70,** 2497–2502.

Izumi, H., Tsukada, Y., Poubol, J., and Hisa, K. (2008). On-farm sources of microbial contamination of persimmon fruit in Japan. *J. Food Protect.* **71,** 52–59.

Kayed, D. (2004). Microbial quality of irrigation water used in the production of fresh produce in Arizona. Ph.D. Dissertation. University of Arizona, Tucson, AZ.

Keraita, B. N. and Drechsel, P. (2004). Agricultural use of untreated urban wastewater in Ghana. In *Wastewater use in irrigated agriculture* (C. A. Scott, N. I. Faruqui, and L. Raschid-Sally, Eds.), pp. 101–112. Commonwealth Agricultural Bureau International Publishing, Wallingford, UK.

Materon, L. A., Martinez-Garcia, M., and McDonald, V. (2007). Identification of sources of microbial pathogens on cantaloupe rinds from pre-harvest to post-harvest operations. *Wld. J. Microbiol. Biotechnol.* **23,** 1281–1287.

Nyard, K., Lassen, J., Vold, L. et al. (2008). Outbreak of *Salmonella* Thompson infections linked to imported rucola lettuce. *Foodborne Pathog. Dis.* **5,** 165–173.

Ogunsola, F. T. and Adesiji, Y. O. (2008). Comparison of four methods of hand washing in situations of inadequate water supply. *West Afr. J. Med.* **27,** 24–28.

Okafo, C. N., Umoh, V. J., and Galadima, M. (2003). Occurrence of pathogens on vegetables harvested from soils irrigated with contaminated streams. *Sci. Total Environ.* **311,** 49–56.

Oron, G., Goemans, M., Manor, Y., and Feyen, J. (1995). Poliovirus distribution in the soil-plant system under reuse of secondary wastewater. *Water Res.* **29,** 1069–1078.

Petterson, S. R., Teunis, P. F., and Ashbolt, N. J. (2001). Modeling virus inactivation on salad crops using microbial count data. *Risk Anal.* **21,** 1097–1108.

Postel, S., Polak, P., Gonzales, F., and Keller, J. (2001). Drip irrigation for small farmers: a new initiative to alleviate hunger and poverty. *Water Int.* **26,** 3–13.

Regli, S., Rose, J. B., Haas, C. N., and Gerba, C. P. (1991). Modeling the risk from *Giardia* and viruses in drinking water. *J. Am. Water Works Assoc.* **88,** 76–84.

Robertson, L. J. and Gjerde, B. (2001). Occurrence of parasites on fruits and vegetables in Norway. *J. Food Protect.* **64**, 1793–1798.

Rose, J. B. and Gerba, C. P. (1991). Assessing potential health risks from viruses and parasites in reclaimed water in Arizona and Florida, USA. *Water Sci. Technol.* **23**, 2091–2098.

Sadovski, A. Y., Fattal, B., Goldberg, D., Katzenelson, E., and Shuval, H. I. (1978). High levels of microbial contamination of vegetables irrigated with wastewater by the drip method. *Appl. Environ. Microbiol.* **36**, 824–830.

Seymour, I. J. and Appleton, H. (2001). Foodborne viruses and fresh produce. *J Appl. Microbiol.* **91**, 759–773.

Scott, C. A., Faruqui, N. I., and Rachid-Sall, L. (Eds.). (2004). *Wastewater use in irrigated agriculture: Confronting the livelihood and environmental realities.* Commonwealth Agricultural Bureau International Publishing, Wallingford, UK.

Shanan, L. (1998). Irrigation development: Proactive planning and interactive management. In *The arid frontier* (H. Bruins and L. Harvey, Eds.). Kluwer Academic Press, London.

Sivapalasingam, S., Friedman, C. R., Cohen, L., and Tauxe, R. V. (2004). Fresh produce: A growing cause of outbreaks of foodborne illness in the United States, 1973 through 1997. *J. Food Protect.* **67**, 2342–2353.

Soderatorm, A., Osterberg, P., Lindqvist, A. et al. (2008). A large *Escherichia coli* 0157 outbreak in Sweden associated with locally produced lettuce. *Foodborne Pathog. Dis.* **5**, 339–349.

Song, I., Stine, S. W., Choi, C. Y., and Gerba, C. P. (2006). Comparison of crop contamination by microorganisms during subsurface drip and furrow irrigation. *J. Environ. Eng.* **132**, 1243–1248.

Steele, M. and Odumeru, J. (2004). Irrigation water as source of foodborne pathogens on fruits and vegetables. *J. Food Protect.* **67**, 2839–2849.

Stine, S. W., Song, I., Pimentel, J. et al. (2005a). The effect of relative humidity on pre-harvest survival of bacterial and viral pathogens on the surface of cantaloupe, lettuce, and bell pepper were studied. *J. Food Protect.* **68**, 1352–1358.

Stine, S. W., Song, I., Choi, C. Y., and Gerba, C. P. (2005b). Application of microbial risk assessment to the development of standards for enteric pathogens in water used to irrigate fresh produce. *J. Food Protect.* **68**, 1352–1358.

Tierney, J. T., Sullivan, R., and Larkin, E. P. (1977). Persistence of poliovirus 1 in soil and on vegetables grown in soil previously flooded with inoculated sewage sludge or effluent. *Appl. Environ. Microbiol.* **33**, 109–113.

Thurston-Enriquez, J. A., Watt, P., Dowd S. C. et al. (2002). Detection of protozoan parasites and microsporidia in irrigation waters used for crop production. *J. Food Protect.* **65**, 378–382.

Tyrrel, S. F., Knox, J. W., and Weatherhead, E. K. (2006). Microbiological water quality requirements for salad irrigation in the United Kingdom. *J. Food Protect.* **69**, 2029–2035.

USDA. (1998). Farm & Ranch Irrigation Survey, Census of Agriculture. United States Department of Agriculture. www.nass.usda.gov/census/census97/fris/fris.htm.

Van Zyl, W. B., Page, N. A., Grabow, W. O. et al. (2006). Molecular epidemiology of group A rotaviruses in water sources and selected raw vegetables in southern Africa. *Appl. Environ. Microbiol.* **72**, 4554–4560.

World Health Organization (WHO). (2006). *Guidelines for the safe use of wastewater, excreta and greywater, Vol 2. Wastewater use in agriculture.* WHO, Geneva, Switzerland.

CHAPTER 6

Sapro-Zoonotic Risks Posed by Wild Birds in Agricultural Landscapes

Larry Clark

*United States Department of Agriculture, Animal and Plant Health Inspection Service,
National Wildlife Research Center, Fort Collins, CO*

CHAPTER CONTENTS

Introduction	120
Bird Species Commonly Associated with Agriculture	120
Pigeons	121
Gulls	123
Water Fowl	123
Passerines	123
Bacterial Diseases	124
Campylobacter	124
Chlamydia	125
Escherichia coli	125
Listeria	127
Salmonella	127
Fungal Diseases	128
Aspergillus	128
Cryptococcus	128
Histoplasma	129
Parasitic Diseases	130
Cryptosporidia	130
Microsporidia	130
Toxoplasma	130
Mitigation Options	131
Summary	132

INTRODUCTION

There are over 1400 catalogued human pathogens, with approximately 62% classified as zoonotic (Taylor et al., 2001). Most evidence of direct transmission of pathogens to humans involves domestic and companion animals, whereas the reservoir for most zoonoses is wildlife; yet there are relatively few well-documented cases for the direct involvment of transmission from wildlife to humans (Kruse et al., 2004). In part, this absence of evidence reflects the mobility of wildlife, the difficulty in accessing relevant samples, and the smaller number of studies focused on characterizing wildlife pathogens relative to the human and veterinary literature (McDiarmid, 1969; Davis et al., 1971; Hubalek, 2004). Because humans generally do not have direct contact with wild birds, exposure to pathogens is via indirect routes, that is, environmental. This indirect exposure route (sapro-zoonotic) makes identifying the wild bird source of the pathogen all the more difficult. However, with better diagnostic technologies and better understanding of the disease ecology, documenting the risks posed by wildlife to human health become more feasible.

This review focuses on the risks wild birds pose to the contamination of field crops and the risk such contamination poses to human health. For the most part, there are few studies that document the role wild birds have in contaminating field crops and subsequent acquisition of pathogens and onset of disease by humans (Tsiodras et al., 2008). Most of the evidence in the literature focuses on relatively few commensal wildlife species in urbanized environments, or at best, general wildlife surveillance and monitoring efforts. Absent any compelling direct evidence, this review summarizes the circumstantial evidence, relying mostly on the characterization of host range of pathogens, similarities of virulence traits of animal and human pathogens, and habitat use patterns of wild birds in agricultural and urban landscapes. Nonetheless, the material presented here does represent a solid circumstantial case for the potential of wild birds to contaminate the field and to act as agents for the transmission of pathogens to humans. More directed studies will be needed to form a more informed assessment as to what actual human pathogen contamination risks wild birds pose to field crops, and by implication, to human health. Finally, this review briefly covers mitigation efforts that might be undertaken to reduce risks of pathogen transmission by wild birds.

BIRD SPECIES COMMONLY ASSOCIATED WITH AGRICULTURE

Wild birds, and especially migratory species, can become long-distance vectors for a wide range of microorganisms. Moreover, many bird species incorporate agricultural fields into their habitat use patterns. However, for the purpose of this review, focus is limited to only a few groups of birds: gulls (Charadriformes), water fowl (Anseriformes), pigeons and doves (Columbiformes), and selected passerine birds (Passeriformes) such as blackbirds, crows, starlings, and sparrows. These groups of birds tend to have high-use patterns of agricultural habitats, they are abundant, and they have close commensal relationships with human activity. These species also tend to be abundant and gregarious; hence, they provide greater opportunity and capacity to contribute larger fecal loads to the environments they use. In the end, it is the likelihood and magnitude for fecal contamination of soils, substrates, and water that represents the most direct link to risks to human health.

Contamination of produce can occur via many routes; for example, at the field level during the growing season, during harvesting, postharvest handling, processing, shipping, marketing, or in the home (Beuchat and Ryu, 1997). Wild birds are most likely to be involved in contamination while the crops are in the field, and perhaps at field-side processing and storage facilities. Moreover, the likelihood that birds are responsible for contamination of crops with human pathogens will depend largely on the birds' exposure to environmental sources of pathogens (Figure 6.1), their capacity to physically transport the pathogen, and perhaps but not necessarily be limited by their ability to act as a reservoir of the pathogen (Kruse et al., 2004).

Pigeons

Pigeons (*Columba livia*) live in close association with humans (Johnston and Janiga, 1995). They are a gregarious species that feed in flocks, form large roosts, and visit habitats that have a high likelihood of harboring human pathogens (e.g., dairies and feedlots). Pigeons will often flock to agricultural fields to pick up grit to aid in digestion, or to consume spilled grains. As with most commensal bird species considered to be at high risk for transmission of human pathogens, pigeons have a prodigious capacity to produce feces. When occurring in large numbers, the fecal load for contamination of surface water and soils can also be large. Perhaps more importantly, is the propensity of pigeons to use architectural structures as day and night roosting sites. Open crop storage or processing sheds present an ideal condition for attracting pigeons, and in the absence of bird exclusionary mitigation measures, present an opportunity for fecal accumulation. As with most species of birds, humans are not likely to come in direct contact with pigeons; rather humans are likely to come in contact with feces or fecally contaminated substrates. Accumulation of feces presents greater opportunities for direct contact of produce being processed or stored with contaminated soils, feces, dusts, and water sources (Figure 6.1).

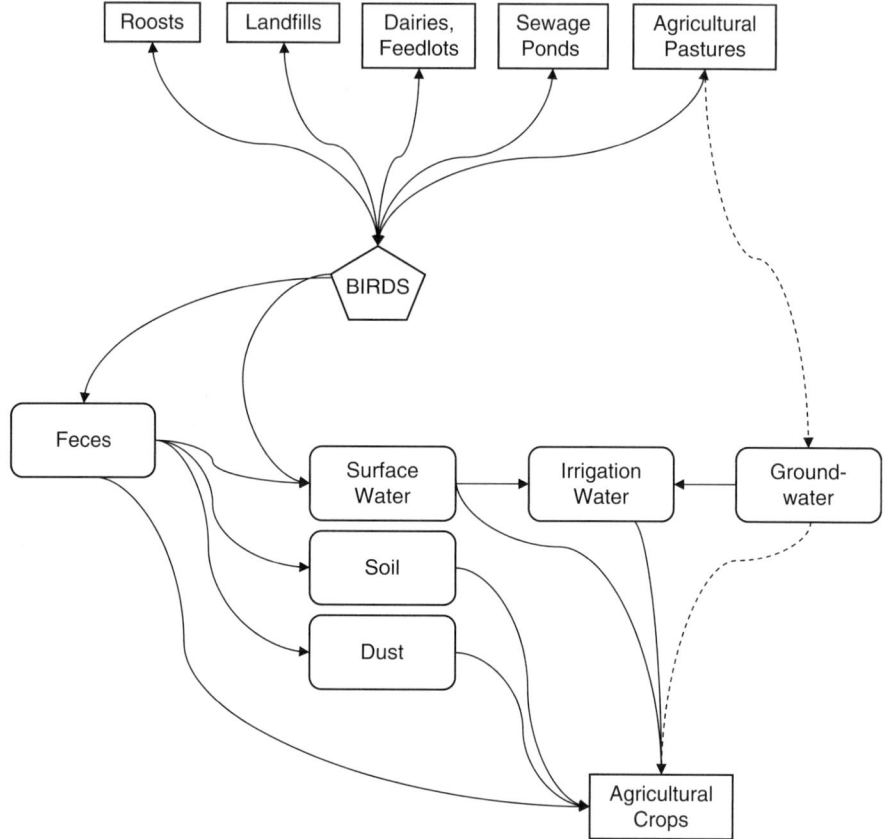

FIGURE 6.1 *Possible routes of contamination of field or orchard crops by wild birds. Rectangles indicate environmental sources/habitats of pathogens. Pentagon represents various species of wild birds likely to visit both the source and agricultural landscapes. Rounded rectangles represent media in which pathogens may reside and be transported.*

Over 60 human pathogens have been isolated from pigeons; however, only five pathogens have been documented to be routinely transmitted to humans (Haag-Wackernagel and Moch, 2004): *Chlamydophila psittaci, Histoplasma capsulatum, Aspergillus* spp., *Candida parapsilosis,* and *Cryptococcus neoformans.* Most cases of disease transmission are related to inhalation of dusts and aerosols (Anon., 2002), emphasizing the risk to human health of accumulated feces at building structures. Only one case of foodborne illness has been linked directly to pigeons, and that was a case of lemon pudding made from the eggs of domestic pigeons (Clarenburg and Dornickx, 1933).

Gulls

A variety of species of gulls (Laridae) have been implicated in the carriage of human pathogens and contamination of watersheds, surface waters, and structures (Wither et al., 2005; Kinzelman et al., 2008). For the most part the number of gulls is increasing, largely owing to the availability of landfills, which provide a source of abundant food. Gulls are attracted to agricultural fields to forage on rodents, insects, and at times the crops themselves (O'Connor, 1992). Gulls can occur in large numbers, produce prodigious quantities of fecal material, and thus act as a source for contamination of soils and substrates. Gulls frequent landfills, dairies, feedlots, sewage ponds, other waste facilities, and agricultural pastures, all sources of human pathogens (Coulson et al., 1983; Fricker, 1984; Yorio and Caille, 2004; Nelson et al., 2008).

Water Fowl

Ducks and geese frequent agricultural fields to feed on crops, spilled grains, acquire grit, or loaf. Water fowl are most likely to be a source of contamination of soils and surface waters, and indirectly, groundwater (Somarelli et al., 2007; Graczyk et al., 2008). Large flocks are likely to frequent agricultural fields beginning in the fall and leaving in the early spring (McKay et al., 2006; Amano et al., 2007). During this period the flocks may move across the landscape on a local, regional, or continental scale, thus increasing the opportunities for transporting pathogens from one site to another. Water fowl are likely to use surface water areas, agricultural pastures, and sewage ponds, all of which can act as environmental sources for human pathogens (Conn et al., 2007).

Passerines

The numbers of passerine species that use agricultural fields is large, but those that travel in large flocks are more limited. Blackbirds (Ictaridae) and starlings (*Sturnus vulgaris*) will frequently use feedlots, dairies, and agricultural pastures, and thus are likely to encounter human pathogens at those sites (Odermatt et al., 1998; Kaleta, 2002; Nielsen et al., 2004). Both species are also likely to use agricultural crop fields as sources of food and grit, thereby providing an opportunity to contaminate soils and crops with their feces. Starlings may also use crop storage and processing facilities as roost sites, providing additional opportunities to contaminate produce. House sparrows (*Passer domesticus*) have a more restricted range. However, this species is commonly associated with poultry houses, feedlots, and dairies, and commonly roosts and nests in crop storage and processing facilities

(Craven et al., 2000; Kirk et al., 2002). Other species of passerines are more likely to use only the agricultural fields; thus their risk of contaminating field crops will be based on their exposure to pathogens in the general environment (Boutin et al., 1999; Laiolo, 2005).

BACTERIAL DISEASES

Campylobacter

Campylobacter is widespread in wild birds (Luechtefeld et al., 1980). High isolation rates have been obtained in gulls (*Larus* spp., 15–50%), crows (*Corvus corone cornix*, *Corvus levaillanti*, and *Corvus corone*, 34–89%), blue magpies (*Cyanopica cyanus*, 20%), gray starlings (*Sturnus cineraceus*, 14%), and pigeons (4–26%) (Ito et al., 1988; Quessy and Messier, 1992; Casanovas et al., 1995). In cases where stomach contents have been analyzed, it is evident that the birds have visited landfills, thus indicating the importance of food habits as a primary factor in the varying prevalence of *C. jejuni* (Kapperud and Rosef, 1983). *Campylobacter* spp. are also widely distributed in aquatic environments and in sewage effluents and agricultural runoff, environments conducive to exposing water fowl to the pathogen (Brown et al., 2004). *Campylobacter* species may survive, and remain potentially pathogenic, for long periods in aquatic environments, but less so on terrestrial substrates (Krampitz and Hollander, 1999).

Campylobacters are emerging as one of the most significant causes of human infections worldwide, and the role that terrestrial birds and water fowl have in the spread of disease is beginning to be elucidated. On a world scale campylobacters are possibly the major cause of human gastrointestinal infections. The possible role of wild birds in causing human disease is still uncertain. For example, starlings shed *Campylobacter* at high rates, suggesting that they may be a source of human and farm-animal infection. However, based on genetic analysis, Colles et al. (2008a) concluded that these bacteria were distinct from poultry or human disease isolates, with the ST-177 and ST-682 clonal complexes possibly representing starling-adapted genotypes. Thus, these authors concluded that there was no evidence that wild starlings represent a major source of *Campylobacter* infections of food animals or humans. Similarly, Colles et al. (2008b) investigated wild geese as a potential source of *Campylobacter* infection for humans and farm animals in waterborne disease outbreaks. The authors found that large numbers of wild geese carry *Campylobacter*; however, there was limited mixing of *Campylobacter* populations among the different sources examined. Thus, they concluded that genotypes of *C. jejuni* isolated from geese are highly host specific, and geese were unlikely to be the source of the human disease outbreaks. In contrast, French et al. (2008) identified members of *Campylobacter* clonal complexes ST-45, ST-682, and ST-177 recovered from starling feces as being indistinguishable from those observed in human cases, and concluded that wild birds could contribute to the burden of campylobacteriosis in preschool children at playgrounds.

Chlamydia

Avian chlamydiosis was originally called "parrot fever." However, recent studies have shown that parrot fever and ornithosis are the same disease manifested in different species and are all caused by the bacterium *Chlamydophila psittaci* (Andersen and Vanrompay, 2000). Chlamydial infections have been identified in over 150 species of wild birds (Brand, 1989). Generally, these wild birds are asymptomatic. Bacteria are shed sporadically in nasal secretions and feces. Although the natural host reservoir systems are unknown, its wide occurrence in wild bird populations and the intermittent infections of farm stock are consistent with exposure to wild birds. Sporadic shedding was seen in experimentally inoculated great-tailed grackles (*Cassidix mexicanus*) and cowbirds (*Molothrus ater*), indicating their potential as host-reservoir systems (Roberts and Grimes, 1978). The most probable risk to farm stock and poultry is when wild birds gain access to feed bins and contaminate the bins with their feces. Infection usually occurs through exposure to contaminated aerosol dusts (Page, 1959). Turkeys can become infected by exposure to starlings, common grackles (*Quiscalus quiscula*), and brown-headed cowbirds (Roberts and Grimes,1978; Grimes et al., 1979). Serovars D and E can result in 50 to 80% morbidity and 5 to 30% mortality in turkeys (Andersen, 1997). In ducks the economic impact is also significant, with morbidity and mortality ranging from 10 to 80% and 0 to 30%, respectively (Andersen et al., 1997). Infections to mammalian farm stock can also be a cause of health and economic concern (Shewen, 1980). Wild avian strains also can infect mammals, including humans, and can cause severe disease or death (Andersen and Vanrompay, 2000).

Escherichia coli

The most important reservoir for verocytotoxin-producing *Escherichia coli* (VTEC) is considered to be ruminants, particularly cattle, though VTEC can be isolated from many mammals and birds (Wallace et al., 1997; Rice et al., 2003). Infection of humans by VTEC may result in combinations of watery diarrhea, bloody diarrhea, and hemolytic uremic syndrome. Severe disease in the form of bloody diarrhea and the hemolytic uremic syndrome is attributable to Shiga toxin (Stx), which exists as two major types, Stx1 and Stx2 (Gyles, 2007). Humans can become infected through contamination of food and water, as well as through direct contact. Given the propensity of certain species of birds to frequent facilities or pastures used by ruminants there remains a distinct possibility that wild birds may play a role in the transmission of VTEC.

Several studies have implicated wild birds in the transmission of VTEC based on the similar characterizations between avian and human isolates (Asakura et al., 2001; Kobayashi et al., 2002). Numerous studies have documented the presence of *Stx*-producing *E. coli* (STEC) in pigeons (Morabito et al., 2001), gulls (Makino et al., 2000), water fowl (Feare et al., 1999; Kullas et al., 2002), and passerines (Nielsen et al., 2004; Kobayashi et al., 2008). Other human virulence factors have also been identified in these species, including *eae*, *cldt*, CNF-1, CNF2, K1, LT, *hly*A, SLT-I, SLT-II, STa, and STb (Morabito et al., 2001; Kullas et al., 2002; Fukuyama et al., 2003; Pedersen et al., 2006). In addition, phylogenic studies have illustrated the similarity of VTEC based on a number of measures (Makino et al., 2000). Though less common, a few studies have shown direct linkages and phylogenetic relatedness between avian VTEC and isolates causing human illness (Sonntage et al., 2005; Eiidokun et al., 2006).

Though VTEC are widely reported in the species groups just discussed, Canada geese have been implicated as the most likely source of nonpoint source pollution of inland waters. Molecular fingerprints of *E. coli* isolated from regional populations showed an unexpectedly high percentage of isolates identified as having a wildlife origin (geese and deer). Geese were the dominant source of *E. coli* (44.7–73.7% of the total sources) in four sub-watersheds, followed by cows (10.5–21.1%), deer (10.5–18.4%), humans (5.3–12.9%), and unidentifiable sources (0.0–11.8%) (Somarelli et al., 2007).

Public pressure is mounting to reduce or eliminate antimicrobials as ingredients of feed for poultry and other agricultural animals, primarily due to the fear of multidrug-resistant bacteria in clinical infections in both animals and humans. Wild birds have been implicated as reservoirs and as vectors for the spread of antibiotic resistant strains of *E. coli*. Gibbs et al. (2007) found drug resistant strains of *E. coli* in feces of yellow-headed blackbirds (*Xanthocephalus xanthocephalus*). Doleiska et al. (2008) reported *E. coli* resistant to 12 antimicrobials in 9% (n = 54) of isolates from house sparrows (*Passer domesticus*). Greater than 95% of *E. coli* isolates from Canada geese in agricultural environments were resistant to penicillin G, ampicillin, cephalothin, erythromycin, lincomycin, sulfathiazole, and vancomycin; no *E. coli* were resistant to bacitracin or ciprofloxacin (Fallacaara et al., 2001; Cole et al., 2005; Middleton and Ambrose, 2005).

Most studies have focused only on the serotype O157:H7/H-; however, there are suggestions that wild birds may be involved in the transmission of other pathogenic serotypes (Kullas et al., 2002). Morabito et al. (2001) described widespread, clonally related isolates of *E. coli* O45, O18ab, and O75 serotypes in several pigeon flocks. The overall prevalence was similar between three flocks (10.8%), with evidence of *Stx*-producing *E. coli* (STEC).

Listeria

Listeriosis in humans is caused by infection by *Listeria monocytogenes*. All serovars of *L. monoctyogenes* are considered human pathogens, and the most potentially virulent are 4b, 1/2b, and 1/2a. *L. monocytogenes* is commonly associated with soils and feces in the environment and may be one of the most ubiquitous microorganisms in the soil. Human exposure is most likely through ingestion of contaminated food, but hand–oral contact or inhalation may also occur. Approximately 1 to 10% of the human population is thought to carry *L. monocytogenes* asymptomatically in the intestines. Healthy people rarely become ill after exposure. Serious cases almost always occur in the elderly, pregnant women, newborns, and those who are debilitated or immunocompromised (Acha and Szyfres, 2003).

Listeria spp. are commonly found in birds (Harken et al., 1986; Fenlon, 1985). Overall, the prevalence detected across species lies within the same range. On the lower end Quessy and Messier (1992) found a prevalence of 9.5% in fecal samples of ring-billed gulls (*Larus delawarensis*). Clark and Sullivan (unpublished data) found prevalence ranging from 8 to 12% from Canada goose (*Branta canadensis*) fecal samples from five states (Colorado, New York, Pennsylvania, Washington, and Wisconsin) that were collected throughout the year. On the higher end, investigators have found prevalences of 43.2% in crows and 36% in gulls (Helstrom et al., 2008). Wild birds may pose a risk to human health. The magnitude of prevalence may be largely due to local environmental conditions that birds may visit, such as landfills, sewage treatment facilities, and livestock facilities. Visitation of such environments may also impact exposure and carriage of virulent strains of *Listeria*. The serovars 1/2a and 4b predominated in the eight serotyped *L. monocytogenes* isolates in the Yoshida et al. (2000) study. Similar geneotypes have been found in wild birds and local fresh food markets as well (Zhang et al., 2007; Mousupye and von Holy, 2000; Hellstrom et al., 2007).

Salmonella

Various *Salmonella* strains have been isolated from a range of wild-bird species. Given the ubiquitous nature of the host range, and the pathogenicity of the organism to humans, caution should be exercised anytime there is exposure to fecally contaminated surfaces or soils. When sapro-zoonotic infections do occur, 99% of the documented cases appear to have resulted from aerosol exposure (Haag-Wackernagel and Moch, 2004).

Gulls appear to pose the greatest risk of infection and carriage. This is perhaps owing to their greater propensity to visit sewage treatment ponds and thus acquire pathogens. *Salmonella* prevalence in gulls ranges between 1 to 55% (Butterfield et al., 1983; Fenlon, 1981, 1983; Sixl et al., 1997; Casanova, 1995). Several genetic and epidemiological studies have linked transport of pathogens from the site of acquisition to distant sites, including food-processing facilities and stockyards (Coulson et al., 1983; Nesse et al., 2005). Persistence in marked free-ranging gulls and experimentally infected sparrows was shown to be limited, approximately 10 days (Snoeyenbos et al., 1967; Palmgren et al., 2006). Other genetic studies have shown that strains carried by gulls are similar to human pathogens.

Pigeons are generally characterized by a low prevalence (3–4 %) of *Salmonella* (Pasmans et al., 2004; Tanaka et al., 2005; González-Acuña et al., 2007). Despite the low prevalence, studies have implicated pigeons and sparrows in the maintenance of the pathogen at feedlots and dairies (Quevedo et al., 1973; Connolly et al., 2006; Pedersen et al., 2006).

FUNGAL DISEASES

Aspergillus

Aspergillus spp. are rapidly growing molds most commonly associated with decaying matter and the feces of water fowl and raptors, although a variety of wild birds and domestic poultry are also known to become infected with *Aspergillus* spp. (Buxton and Sommer, 1980; Friend, 2006). In animals, greater than 90% of infections are caused by *A. fumigatus* (Quinn, 1994). The highest prevalence of *A. fumigatus* in water fowl is in winter. The most susceptible people for the respiratory and allergic complications of infection in populations are those who are immunocompromised or are on extended regimens of medication (Latge, 2001).

Cryptococcus

Cryptococcus neoformans is a fungus typically associated with bird feces (Blaschke-Hellmessen, 2000). *C. neoformans* typically affects only the immunocompromised. Cryptococcal infection may cause a pneumonia-like illness, with shortness of breath, coughing, and fever. Skin lesions may also occur. Another common form of cryptococcosis is central nervous system infection, such as meningoencephalitis. The primary risk for infection is inhaling dusts containing contaminated feces. Pigeons appear to be the primary wild bird involved in transmission to humans, though *C. neoformans* has been detected in a variety of other species (Pollock, 2003; Cafarchia et al., 2006; Rosario et al., 2008). Prevalence in pigeons may range between 9 and 19% (Weber and Shafer, 1991; Soogarun et al., 2006). Unlike *Histoplasma*, *C. neoformans* viability in the environment is limited. Ruiz et al. (1982) showed that viability of *C. neoformans* decreased from 86 to 50% over the course of a year, once pigeons were excluded from a roost. They attributed this decrease in viability to desiccation.

Histoplasma

Histoplasma capsultatum is a zoonotic fungal pathogen, commonly found in soils and bird feces, that affects the respiratory system (Ajello, 1964). In endemic areas of the United States as much as 80 to 90% of the human population is infected (Rubin et al., 1959). Fewer than 10% of those who inhale airborne spores develop a pulmonary infection. However, the pulmonary form can disseminate and is potentially fatal if not treated. Acute pulmonary histoplasmosis is the most dramatic form of the disease and occurs in people who have inhaled massive doses of spores. Chronic infection in humans can result in permanent lung damage. People with HIV are most susceptible to the disseminated form of the illness. People at highest risk are those working in agriculture, particularly poultry operations, or those people coming in contact with bird feces associated with bird roosts (Dodge et al., 1965; Tosh et al., 1966). Such roosts are likely to be associated with dense vegetation (i.e., agricultural wind breaks), trees, or storage or processing sheds. Because transmission is through breathing dust particles containing spores, any disturbance of contaminated soil can cause infection (Storch et al., 1980; Stoberski et al., 1996). Soil studies have shown that the viable spores persist in contaminated soils over many years (9–13+ years; DiSalvo and Johnson, 1979), long after bird activity at a site has since ceased (Gustafson et al., 1981). Moreover, residents downwind from contaminated disturbed soils can become infected (Latham et al., 1980; Chick et al., 1981). Formalin has been used to sterilize soils contaminated with *Histoplasma* and deemed to be at high risk of further infection to local human populations (Smith et al., 1964; Tosh et al., 1967; Bartlett et al., 1984). Although *Histoplasma* has been detected in the feces of many species of birds, pigeons, blackbirds, and starlings are the most likely wild birds to be a source of soil contamination, and the species most likely to be associated with roosts near agriculture production or processing (Schwarz et al., 1957; Pollack, 2003; Cermeno et al., 2006).

PARASITIC DISEASES

Cryptosporidia

Cryptosporidium parvum is an important gastrointestinal parasite of humans and other animals that can be transmitted via contamination of food and water (Mackenzie et al., 1994; Millar et al., 2002). Symptoms may be long-lasting and include diarrhea, loose or watery stool, stomach cramps, upset stomach, and a slight fever (Fayer et al. 1998). Some people have no symptoms. In persons with average immune systems, symptoms usually last about two weeks. Water fowl in general, but Canada geese in particular, have been implicated in the contamination of water (Hatch, 1996; Smith et al., 1993; Graczyk et al., 1997; Fallacara et al., 2004), and oocytes recovered from feces have been demonstrated to be infectious (Graczyk et al., 1998).

Microsporidia

Microsporidians (*Encephalitozoon intestinalis*, *E. hellem*, *E. cuniculi*, *E. bieneusi*) are obligate intracellular parasites that are increasingly involved in opportunistic infections of the immunocompromised and immunocompetent people (Weber et al., 1994). *E. hellem* has been the species most commonly associated with avian hosts (Slodkowicz-Kowalska et al., 2006). Epidemiological evidence strongly supports contaminated water, including water used for crop production, as a significant risk factor for human disease (Dowd et al., 1998; Fournier et al., 2000; Mathis et al., 2005; Thurston-Enriquez et al., 2002); avian contamination of surface water via defecation is an important contributory risk factor for pathogen transmission (Slodkowicz-Kowalska et al., 2006). Although water-borne transmission is the most likely avenue for opportunistic infection of humans, a recent study by Haro et al. (2005) showed that pigeon feces was positive for *E. Bieneusi* (9.7% prevalence), *E. intestinalis* (4% prevalence), and *E. hellem* (1% prevalence), and 4.8% of pigeons were coinfected. Bart et al. (2008) found similar prevalence. The authors concluded that there was no barrier to transmission from pigeons to humans.

Toxoplasma

Toxoplasma gondii is a common single-celled parasite responsible for infection of more than 60 million people in the United States each year. Infection can be acquired via hand-to-mouth contact with feces, contaminated soil, water, or raw meat. In most cases, the hosts' natural immune system clears the disease, and most healthy humans are rarely aware that

Table 6.1 Prevalence of *Toxoplasma gondii* in Selected Species of Birds Commonly Associated with Agricultural Production or Processing

Order	Common Group Name	Prevalence Range (%)	Reference
Charadriiformes	Gulls	6–16	Literak et al., 1992; Burridge et al., 1979
Anseriformes	Ducks, geese	1–28	Literak et al., 1992; Pak, 1976
Galliformes	Pheasants, quail, turkey	2–19	Literak et al., 1992
Columbiformes	Pigeons, doves	5–12	Literak et al., 1992; Gibson and Eyles, 1957; Jacobs et al., 1952; Catar, 1974
Passeriformes	House sparrow (*Passer domesticus*)	1–18	Hejlicek et al., 1981; Ruiz and Frenkel, 1980; Literak et al., 1992; Pak, 1976
	Starling (*Sturnus vulgaris*)	1	Literak et al., 1992; Pak 1976
	Crow (*Corvus brachyrhynchos*)	1	Finlay and Manwell, 1956

they are infected. Symptoms include flu-like symptoms and swollen joints and fatigue. However, people with impaired immune systems, embryos, and neonates are particularly vulnerable to severe consequences of infection (e.g., eye and brain damage). Birds are included in the extensive list of wildlife species implicated as carriers of this parasite (Coutelen et al., 1953; Drobeck et al., 1953; Siim et al., 1963; Dubey, 2002). The prevalence of *T. gondii* in wild birds likely to be associated with agricultural landscapes is moderately high (Table 6.1). This is of some concern because *T. gondii* is readily transmitted through the fecal-water route and represents a risk for contaminating crops or water sources used by humans for consumption or food processing (Bahia-Oliveira et al., 2003). Finally, Human populations can be affected by exposure to feces associated with roosts (Peach et al., 1989).

MITIGATION OPTIONS

Several excellent reviews exist on the general practices of excluding and repelling birds from agricultural landscapes (Hyngstom et al., 1994; Mason, 1995; Conover, 2002; Linz, 2003). Other venues for research and methods for animal damage control can be found in the *Proceedings of the Vertebrate Pest Conference*, *Proceedings of Animal Damage Management Conference*, and the journals of human-wildlife conflict: *Journal of Wildlife Management*, *Wildlife Society Bulletin*, and *Wildlife Research*.

Lethal control is not generally an option for bird control, owing to federal and state statues and prohibitions. Nonlethal methods to scare birds off fields are the only recourses available to growers. However, in reality, it is not practical to exclude all wild birds from large agro-ecosystems. It is feasible using integrated pest management approaches to limit populations in specific fields (Avery, 1989; Mason and Clark, 1992, 1996; Clark, 1998). Methods can include the use of frightening devices, such as visual deterrents, pyrotechnics, propane exploders, or alarm calls, but it is important to avoid presenting cues on a systematic or regular basis so as to avoid habituation. A review of each of these methods and their successes and failures can be found in the handbook *The Prevention and Control of Wildlife Damage* (Hyngstrom et al., 1994). The difficulty with the approaches outlined in the handbook is one of expense and human vigilance, both of which are at the root cause of failure. Stevens et al. (2000) explored the use of demand performance systems employing radar technologies to activate a variety of bird scaring devices. The method was successful at keeping migratory birds off hazardous waste ponds of 180 and 90 acres for over a year. The principal drawback of the system was expense. Most agricultural operations could not afford such protection.

Chemical repellents have been tried, but not all crops are amenable to their use owing to expense and regulatory restrictions (Clark, 1997). Nonetheless, several products have been developed, for example, methyl anthranilate and anthraquinone-based products, and several other investigatory repellents have been evaluated (Avery and Mason, 1997; Cummings et al., 2002; Avery et al., 2005; Werner et al., 2008).

SUMMARY

Wild birds are capable of pathogen carriage, acting as reservoirs, and becoming infected with a wide variety of pathogens, some of which are zoonoses. This review did not attempt to provide an exhaustive list. Rather its intent was to focus on the avian species and pathogens that represented the greatest likelihood to be of concern at the agricultural production and processing levels. Unfortunately, little direct evidence bears on this issue; hence, risk was assessed using information derived from urban human health, water quality, poultry, livestock production, wildlife health, and the veterinary literature. Despite the lack of direct evidence relating to birds and the risks to farm-side production and processing of produce, it appears there is ample evidence to support the notion that birds can pose a human health risk by serving as a source of contamination of produce and crops. Nonetheless, more detailed empirical and risk-modeling studies are needed. Moreover, such studies should be integrated into a larger ecological perspective of the values of birds to agroecosystems balanced against the health risks they pose. Finally, studies and analyses also should incorporate assessments of mitigation management strategies in the context of economic, ecological, and public-health valuations. These approaches are clearly beyond the scope of this review, but should be seriously considered over a simplistic interpretation of disease risk posed by wild birds and measures needed to eliminate them from agro-ecosystems.

REFERENCES

Acha, P. N. and Szyfres, B. (2003). *Zoonoses and communicable diseases common to man and animals. Volume 1. Bacterioses and mycoses. 3rd ed.* Listeriosis, pp. 168–179. Washington, DC: Pan American Health Organization; Scientific and Technical Publication No. 580.

Ajello, L. (1964). Relationship of histoplasma capsulatum to avian habitats. *Public Health Rep.* **79**, 266–270.

Amano, T., Ushiyama, K., Fujita, G., and Higuchi, H. (2007). Predicting grazing damage by white-fronted geese under different regimes of agricultural management and the physiological consequences for the geese. *Journal of Applied Ecology* **44**, 506–515.

Andersen, A. A. (1997). Two new serovars of *Chlamydia psittaci* from North American birds. *J. Vet. Diagn. Invest.* **9**, 159–164.

Andersen, A. A. and Vanrompay, D. (2000). Avian chlamydiosis. *Rev. Sci. Tech.* **19**, 396–404.

Anonymous. (2002). Compendium of measures to control *Chlamydophila psittaci* (formerly *Clamydia psittaci*) infection among humans (Psittacosis and pet birds) In www.avma.org/pubhlth/psittacosis.asp, Vol. 2003. National Association of State Public Health Veterinarians.

Asakura, H., Makino, S., Kobori, H. et al. (2001). Phylogenetic diversity and similarity of active sites of Shiga toxin (stx) in Shiga toxin-producing *Escherichia coli* (STEC) isolates from humans and animals. *Epidemiol Infect.* **127**, 27–36.

Avery, M. L. (1989). Experimental evaluation of partial repellent treatment for reducing bird damage to crops. *Journal of Applied Ecology* **26**, 433–439.

Avery, M. L. and Mason, J. R. (1997). Feeding responses of red-winged blackbirds to multisensory repellents. *Crop Protection* **16**, 159–164.

Avery, M. L., Werner, S. J., Cummings, J. L. et al. (2005). Caffeine for reducing bird damage to newly seeded rice. *Crop Protection* **24**, 651–657.

Bahia-Oliveira, L. M. G., Jones, J. L. Azevedo-Silva, J. et al. (2003). Highly endemic, waterborne toxoplasmosis in North Rio de Janeiro State, Brazil. *Emerging Infectious Diseases* **9**, 55–62.

Bart, A., Wentink-Bonnema, M., Heddema, E. R. et al. (2008). Frequent occurrence of human-associated microsporidia in fecal droppings of urban pigeons in Amsterdam, The Netherlands. *Appl. Environ. Microbiol.* **74**, 7056–7058.

Bartlett, P. C., Weeks, R. J., and Ajello, L. (1982). Decontamination of a *Histoplasma capsulatum*-infested bird roost in Illinois. *Arch. Environ. Health* **37**, 221–223.

Beuchat, L. R. and Ryu, J. H. (1997). Produce handling and processing practices. *Emerg. Infect. Dis.* **3**, 459–465.

Blaschke-Hellmessen, R. (2000). *Cryptococcus* species—Etiological agents of zoonoses or sapronosis? (in German) *Mycoses* **43** Suppl 1, 48–60.

Brand, C. J. (1989). Chlamydial infections in free-living birds. *J. Am. Vet. Med. Assoc.* **195**, 1531–1535.

Brown, P. E., Christensen, O. F., Clough, H. E. et al. (2004). Frequency and spatial distribution of environmental *Campylobacter* spp. *Appl. Environ. Microbiol.* **70**, 6501–6511.

Boutin, C., Freemark, K. E., and Kirk, D. A. (1999). Farmland birds in southern Ontario: Field use, activity patterns, and vulnerability to pesticide use. *Agriculture, Ecosystems and Environment* **72**, 239–254.

Burridge, M. J., Bigler, W. J., Forrester, D. J., and Hennemann, J. M. (1979). Serologic survey for *Toxoplasma gondii* in wild animals in Florida. *J. Am. Vet. Med. Assoc.* **175**, 964–967.

Butterfield, J., Coulson, J. C., Kearsey, S. V. et al. (1983). The herring gull *Larus argentatus* as a carrier of *Salmonella*. *J. Hyg.* (Lond). **91**, 429–436.

Buxton, I. and Sommer, C. V. (1980). Serodiagnosis of *Aspergillus fumigatus* antibody in migratory ducks. *Avian Dis.* **24**, 446–454.

Cafarchia, C., Romito, D., Iatta, R. et al. (2006). Role of birds of prey as carriers and spreaders of *Cryptococcus neoformans* and other zoonotic yeasts. *Med. Mycol.* **44**, 485–492.

Casanovas, L., de Simón, M., Ferrer, M. D. et al. (1995). Intestinal carriage of campylobacters, salmonellas, yersinias and listerias in pigeons in the city of Barcelona. *J. Appl. Bacteriol.* **78**, 11–13.

Catar, G. (1974). Toxoplazmoza v ekologickych podmienkach na Slovensku (in Slovakian). *Biologicke Prace* (Bratislava) **20**, 1–38.

Cermeno, J. R., Hernandez, I., Cabello, I. et al. (2006). *Cryptococcus neoformans* and *Histoplasma capsulatum* in dove's (*Columba livia*) excreta in Bolivar state, Venezuela. *Rev. Latinoam. Microbiol.* **48**, 6–9.

Chick, E. W., Compton, S. B., Pass, T., III et al. (1981). Hitchcock's birds, or the increased rate of exposure to histoplasma from blackbird roost sites. *Chest* **80**, 434–438.

Clarenburg, A. and Dornickx, C. G. J. (1933). Nahrungmittelvergiftung bei Menschen in Zusammendgang mit Tauben paratyphose. *Zscher. Hyg. Inf. Krankh.* **114**, 31–41.

Clark, L. (1997). A review of the bird repellent effects of 117 carbocyclic compounds. In *Repellents in wildlife management* (J. R. Mason, Ed.), pp. 343–352 (August 8-10, 1995, Denver, CO). Colorado State University, Fort Collins, CO.

Clark, L. (1998). Review of bird repellents. In *Eighteenth Vertebrate Pest Conference* (Barker, R. O. and Crabb, A. C., Eds.), pp. 330–337 (March 2-5, 1998, Costa Mesa, CA).

Cole, D., Drum, D. J., Stalknecht, D. E. et al. (2005). Free-living Canada geese and antimicrobial resistance. *Emerg. Infect. Dis.* **11**, 935–938.

Colles, F. M., Dingle, K. E., Cody, A. J., and Maiden, M. C. (2008a). Comparison of *Campylobacter* populations in wild geese with those in starlings and free-range poultry on the same farm. *Appl. Environ. Microbiol.* **J74**, 3583–3590.

Colles, F. M., McCarthy, N. D., Howe, J. C. et al. (2008b). Dynamics of *Campylobacter* colonization of a natural host, *Sturnus vulgaris* (European Starling). *Environ. Microbiol.* **S29**. [Epub ahead of print; www.wiley.com/bw/journal.asp?ref=1462-2912]

Conn, D. B., Weaver, J., Tamang, L., and Graczyk, T. K. (2007). Synanthropic flies as vectors of *Cryptosporidium* and *Giardia* among livestock and wildlife in a multispecies agricultural complex. *Vector-Borne and Zoonotic Diseases* **7**, 643–651.

Connolly, J. H., Alley, M. R., Dutton, G. J., and Rogers, L. E. (2006). Infectivity and persistence of an outbreak strain of *Salmonella enterica* serotype *Typhimurium* DT160 for house sparrows (*Passer domesticus*) in New Zealand. *N. Z. Vet. J.* **54**, 329–332.

Conover, M. R. (2002). *Resolving human-wildlife conflicts: The science of wildlife damage management*. CRC Press, New York.

Coulson, J. C., Butterfield, J., and Thomas, C. (1983). The herring gull *Larus argentatus* as a likely transmitting agent of *Salmonella montevideo* to sheep and cattle. *J. Hyg.* (Lond). **91**, 437–443.

Coutelen, F., Biguet, J., Doby, J. M., and Deblock, S. (1953). Le probleme de toxoplasmoses aviares. Receptivite variable de vuelques oiseaux a une souche humaine de toxoplasmes. *Ann. Parasitol. Hum. Comp.* **28**, 129–156.

Craven, S. E., Stern, N. J., Line, E. et al. (2000). Determination of the incidence of *Salmonella* spp., *Campylobacter jejuni*, and *Clostridium perfringens* in wild birds near broiler chicken houses by sampling intestinal droppings. *Avian Dis.* **44**, 715–720.

Cummings, J. L., Avery, M. L., Mathre, O. et al. (2002). Field evaluation of Flight Control to reduce blackbird damage to newly planted rice. *Wildlife Society Bulletin* **30**(3), 816–820.

Davies, J. W., Anderson, R. C., Karstad, L., and Trainer, D. O. (1971). Infectious and parasitic diseases of wild birds. Iowa State University Press, Ames, IA.

Disalvo, A. F. and Johnson, W. M. (1979). Histoplasmosis in South Carolina: Support for the microfocus concept. *Am. J. Epidemiol.* **109**, 480–492.

Dodge, H. J., Ajello, L., and Engelke, O. K. (1965). The association of a bird-roosting site with infection of school children by *Histoplasma capsulatum*. *Am. J. Public. Health Nations Health* **55**, 1203–1211.

Dolejská, M., Senk, D., Cízek, A. et al. (2008). Antimicrobial resistant *Escherichia coli* isolates in cattle and house sparrows on two Czech dairy farms. *Res. Vet. Sci.* **85**, 491–494.

Dowd, S. E., Gerba, C.P, and Pepper, I. L. (1998). Confirmation of the human pathogenic microsporidia *Enterocytozoon bieneusi*, *Encehphalitozoon intestinalis*, and *Vittaforma corneae* in water. *Appl. Environ. Microbiol.* **64**, 3332–3335.

Drobeck, H. P., Manwell, R. D., Bernstein, E., and Dillon, R. D. (1953). Further studies of toxoplasmosis in birds. *Am. J. Hyg.* **58**, 329–339.

Dubey, J. P. (2002). A review of toxoplasmosis in wild birds. *Vet. Parasitol.* **106,** 121–153.

Ejidokun, O. O., Walsh, A., Barnett, J. et al. (2006). Human Vero cytotoxigenic *Escherichia coli* (VTEC) O157 infection linked to birds. *Epidemiol. Infect.* **134,** 421–423.

Feare, C. J., Sanders, M. F., Blasco, R., and Bishop, J. D. (1999). Canada goose (*Branta canadensis*) droppings as a potential source of pathogenic bacteria. *J. R. Soc. Health* **119,** 146–155.

Fallacara, D. M., Monahan, C. M., Morishita, T. Y., and Wack, R. F. (2001). Fecal shedding and antimicrobial susceptibility of selected bacterial pathogens and a survey of intestinal parasites in free-living waterfowl. *Avian Dis.* **45,** 128–135.

Fallacara, D. M., Monahan, C. M., Morishita, T. Y. et al. (2004). Survey of parasites and bacterial pathogens from free-living waterfowl in zoological settings. *Avian Dis.* **48,** 759–767.

Fayer, R., Graczyk, T.K, Lewis, E. J. et al. (1998). Survival of infectious *Cryptosporidium parvum* oocysts in seawater and Easter oysters (*Crassostrea viginica*) in the Chesapeake Bay. *Appl. Environ. Microbiol.* **64,** 1070–1074.

Fenlon, D. R. (1981). Seagulls (*Larus* spp.) as vectors of salmonellae: An investigation into the range of serotypes and numbers of salmonellae in gull faeces. *J. Hyg.* (Lond). **86,** 195–202.

Fenlon, D. R. (1983). A comparison of *Salmonella* serotypes found in the faeces of gulls feeding at a sewage works with serotypes present in the sewage. *J. Hyg.* (Lond). **91,** 47–52.

Fenlon, D. R. (1985). Wild birds and silage as reservoirs of *Listeria* in the agricultural environment. *J Appl Bacteriol.* **59,** 537–543.

Fournier, S., Liguory, O., Santillana-Hayat, M., et al. (2000). Detection of microsporidia in surface water: A one-year follow-up study. *FEMS Immunol. Med. Microbiol.* **29,** 95–100.

French, N. P., Midwinter, A., Holland, B. et al. (2008). Molecular epidemiology of *Campylobacter jejuni* isolated from wild bird faecal material in children's playgrounds. *Appl. Environ. Microbiol.* [Epub ahead of print; http://aem.asm.org/]

Fricker, C. R. (1984). A note on *Salmonella* excretion in the black headed gull (*Larus ribibundus*) feeding at sewage treatment works. *J. Appl. Bacteriol.* **56,** 499–502.

Friend, M. (2006). Disease emergence and resurgence: The wildlife-human connection. Circular 1285. US Department of Interior and US Geological Survey, Reston, VA.

Fukuyama, M., Furuhata, K., Oonaka, K. et al. (2003). Isolation and serotypes of Vero toxin-producing *Escherichia coli* (VTEC) from pigeons and crows *Kansenshogaku Zasshi.* **77,** 5–9.

Gibbs, P. S., Kasa, R., Newbrey, J. L. et al. (2007). Identification, antimicrobial resistance profiles, and virulence of members from the family Enterobacteriaceae from the feces of yellow-headed blackbirds (*Xanthocephalus xanthocephalus*) in North Dakota. *Avian Dis.* **51,** 649–655.

Gibson, C. L. and Eyles, D. E. (1957). Toxoplasma infections in animals associated with a case of human congenital toxoplasmosis. *Am J. Trop. Med. Hyg.* **6,** 990–1000.

González-Acuña, D., Silva, G. F., Moreno, S. L. et al. (2007). Detection of some zoonotic agents in the domestic pigeon (*Columba livia*) in the city of Chillán, Chile. *Rev Chilena Infectol.* **24**, 199–203.

Graczyk, T. K., Cranfield, M. R., and Fayer, R. (1997). Infectivity of *Cryptosporidium parvum* oocysts is retained upon intestinal passage through a migratory waterfowl species (Canada goose, *Branta canadensis*). *Trop. Med. Int. Health* **2**, 341–347.

Graczyk, T. K., Fayer, R., Trout, J. M. et al. (1998). *Giardia* sp. cysts and infectious *Cryptosporidium parvum* oocysts in the feces of migratory Canada geese (*Brant canadensis*). *Appl. Environ. Microbiol.* **64**, 2736–2738.

Graczyk, T. K., Majewska, A. C., and Schwab, K. J. (2008). The role of birds in dissemination of human waterborne enteropathogens. *Trends in Parasitology* **24**, 55–59.

Grimes, J. E., Owens, K. J., and Singer, J. R. (1979). Experimental transmission of *Chlamydia psittaci* to turkeys from wild birds. *Avian Dis.* **23**, 915–926.

Gustafson, T. L. Kaufman, L., Weeks, R. et al. (1981). Outbreak of acute pulmonary hisotplasmosis in members of a wagon train. *Am. J. Med.* **71**, 759–765.

Gyles, C. L. (2007). Shiga toxin-producing *Escherichia coli*: An overview. *J. Anim. Sci.* **85**(13 Suppl), E45–E62.

Haag-Wackernagel, D. and Moch, H. (2004). Health hazards posed by feral pigeons. *J Infect.* May **48**(4), 307–313.

Harkin, J. M. and Phillips, W. E., Jr. (1986). Isolation of *Listeria monocytogenes* from an eastern wild turkey. *J. Wildl. Dis.* **22**, 110–112.

Haro, M., Izquierdo, F., Henriques-Gil, N. et al. (2005). First detection and genotyping of human-associated nicrosporidia in pigeons from urban parks. *Appl. Environ. Microbiol.* **71**, 3153–3137.

Hatch, J. J. (1996). Threats to public health from gulls (*Laridae*). *Int. J. Environ. Health Res.* **6**, 5–16.

Hejlicek, K., Prosek, F., and Treml, F. (1981). Isolation of *Toxoplasma gondii* in free-living small mammals and birds. *Acta Vet. Brno.* **50**, 233–236.

Hellström, S., Kiviniemi, K., Autio, T., and Korkeala, H. (2008). *Listeria monocytogenes* is common in wild birds in Helsinki region and genotypes are frequently similar with those found along the food chain. *J Appl Microbiol.* Mar **104**(3), 883–888.

Hubálek, Z. (2004). An annotated checklist of pathogenic microorganisms associated with migratory birds. *J. Wildl Dis.* **40**, 639–659.

Hyngstrom, S. E., Timm, R. M., and Larson, G. E. (1994). Prevention and control of wildlife damage. University of Nebraska Cooperative Extension. US Department of Agriculture-Animal and Plant Health Inspection Service—Animal Damage Control. Great Plains Agricultural Council—Wildlife Committee, Lincoln NE.

Ito, K., Kubokura, Y., Kaneko, K. et al. (1988). Occurrence of *Campylobacter jejuni* in free-living wild birds from Japan. *J. Wildl. Dis.* **24**, 467–470.

Jacobs, L. Melton, M. L., and Jones, F. E. (1952). The prevalence of toxoplasmosis in pigeons. *Exp. Parasitol.* **2**, 403–416.

Johnston, R. R. and Janiga, M. (1995). *Feral pigeons*. Oxford Univ. Press, Oxford.

Kaleta, E. F. (2002). Foot-and-mouth disease: Susceptibility of domestic poultry and free-living birds to infection and to disease—A review of the historical and current literature concerning the role of birds in spread of foot-and-mouth disease viruses. *Deutsche Tierarztliche Wochenschrift* **109**, 391–399.

Kapperud, G. and Rosef, O. (1983). Avian wildlife reservoir of *Campylobacter fetus* subsp. *jejuni*, *Yersinia* spp., and *Salmonella* spp. in Norway. *Appl Environ Microbiol.* **45**, 375–380.

Kinzelman, J., McLellan, S. L., Amick, A. et al. (2008). Identification of human enteric pathogens in gull feces at Southwestern Lake Michigan bathing beaches. *Can. J. Microbiol.* **54**, 1006–1015.

Kirk, J. H., Holmberg, C. A., and Jeffrey, J. S. (2002). Prevalence of *Salmonella* spp. in selected birds captured on California dairies. *J. Am. Vet. Med. Assoc.* **220**, 359–362.

Kobayashi, H., Pohjanvirta, T., and Pelkonen, S. (2002). Prevalence and characteristics of intimin- and Shiga toxin-producing *Escherichia coli* from gulls, pigeons and broilers in Finland. *J Vet. Med. Sci.* **64**, 1071–1073.

Kobayashi, H., Kanazaki, M., Hata, E., and Kubo, M. (2008). Prevalence and characteristics of eae- and stx-Positive *Escherichia coli* from wild birds in the immediate environment of Tokyo Bay. *Appl. Environ. Microbiol.* [Epub ahead of print; http://aem.asm.org]

Krampitz, E. S. and Holländer, R. (1999). Longevity of pathogenic bacteria especially *Salmonella* in cistern water. *Zentralbl. Hyg. Umweltmed.* **202**, 389–337.

Kruse, H., Kirkemo, A. M., and Handeland, K. (2004). Wildlife as source of zoonotic infections. *Emerg. Infect. Dis.* **10**, 2067–2072.

Kullas, H., Coles, M., Rhyan, J., and Clark, L. (2002). Prevalence of *Escherichia coli* serogroups and human virulence factors in faeces of urban Canada geese (*Branta canadensis*). *Int. J. Environ. Health. Res.* **12**, 153–162.

Laiolo, P. (2005). Spatial and seasonal patterns of bird communities in Italian agroecosystems. *Conservation Biology* **19**, 1547–1556.

Latgé, J. P. (2001). The pathobiology of *Aspergillus fumigatus*. *Trends Microbiol.* **9**, 382–389.

Latham, R. H., Kaiser, A. B., Dupont, W. D., and Dan B. B. (1980). Chronic pulmonary histoplasmosis following the excavation of a bird roost. *Am. J. Med.* **68**, 504–508.

Luechtefeld, N. A., Blaser, M. J., Reller, L. B., and Wang, W. L. (1980). Isolation of *Campylobacter fetus* subsp. *jejuni* from migratory waterfowl. *J. Clin. Microbiol.* **12**, 406–408.

Linz, G. M., Ed. (2003). Management of North American blackbirds. National Wildlife Research Center, Fort Collins, CO.

Literák, I., Vanko, R., Dolejská, M. et al. (2007). Antibiotic resistant *Escherichia coli* and *Salmonella* in Russian rooks (*Corvus frugilegus*) wintering in the Czech Republic. *Lett. Appl. Microbiol.* **45**, 616–621.

Mackenzie, W. R., Hoxie, N. J., Proctor, M. E. et al. (1944). A massive outbreak in Milwaukee of *Cryptosporidium* infection transmitted through the public water supply. *New England Journal of Medicine* **331**, 161–167.

Makino, S., Kobori, H., Asakura, H. et al. (2000). Detection and characterization of Shiga toxin-producing *Escherichia coli* from seagulls. *Epidemiol. Infect*. **125**, 55–61.

Mason, J. R., Ed. (1997). Repellents in wildlife management: Proceedings of a symposium. Proceedings of the Second DWRC Special Symposium (August 8-10, 1995, Denver, CO). National Wildlife Research Center, Fort Collins CO.

Mason, J. R. and Clark, L. (1992). Nonlethal repellents: The development of cost-effective, practical solutions to agricultural and industrial problems. *Proceedings Vertebrate Pest Conference* **15**, 115–129.

Mason, J. R. and Clark, L. (1996). Grazing repellency of methyl anthranilate to snow geese is enhanced by a visual cue. *Crop Protection* **15**, 97–100.

Mathis, A., Weber, B., and Deplazes, P. (2005). Zoonotic potential of the microsporidia. *Clin. Microbiol. Rev*. **18**, 423–445.

McDiarmid, A. (1969). Diseases in free-living wild animals. Academic Press, London, UK.

McKay, H., Watola, G. V., Langton, S. D., and Langton, S. A. (2006). The use of agricultural fields by re-established greylag geese (*Anser anser*) in England: A risk assessment. *Crop Protection* **25**, 996–1003.

Middleton, J. H. and Ambrose A. (2005). Enumeration and antibiotic resistance patterns of fecal indicator organisms isolated from migratory Canada geese (*Branta canadensis*). *J. Wildl. Dis*. **41**, 334–341.

Millar, B. C., Finn, M., Xiao, L. et al. (2002). *Cryptosporidium* in food-stuffs— An emerging aetiological route of human foodborne illness. *Trends in Food Science & Technology* **13**, 168–187.

Morabito, S., Dell'Omo, G., Agrimi, U. et al. (2001). Detection and characterization of Shiga toxin-producing *Escherichia coli* in feral pigeons. *Vet. Microbiol*. **82**, 275–283.

Mosupye, F. M. and von Holy, A. (2000). Microbiological hazard identification and exposure assessment of street food vending in Johannesburg, South Africa. *Int J Food Microbiol*. **61**(2-3), 137–145.

Nelson, M., Jones, S. H., Edwards, C., and Ellis, J. C. (2008). Characterization of *Escherichia coli* populations from gulls, landfill trash, and wastewater using ribotyping. *Dis. Aquat. Organ*. **81**, 53–63.

Nesse, L. L., Refsum, T., Heir, E. et al. (2005). Molecular epidemiology of *Salmonella* spp. isolates from gulls, fish-meal factories, feed factories, animals and humans in Norway based on pulsed-field gel electrophoresis. *Epidemiol. Infect*. **133**, 53–58.

Nielsen, E. M., Skov, M. N., Madsen, J. J. et al. (2004). Verocytotoxin-producing *Escherichia coli* in wild birds and rodents in close proximity to farms. *Appl. Environ. Microbiol*. **70**, 6944–6947.

O'Connor, K. (1992). *The herring gull*. Dillon Press, Toronto, Canada.

Odermatt, P., Gautsch, S., Rechsteiner, D. et al. (1998). Swarms of starlings in Basel: A natural phenomenon, a nuisance or a health risk? *Gesundheitswesen* **60**, 749–754.

Pak, S. M. (1976). Toxoplasmosis of birds in Kazakhstan (in Russian). *Contributions Nat. Nidality Dis.* **5**, 116–125.

Palmgren, H., Aspán, A., Broman, T. et al. (2006). *Salmonella* in Black-headed gulls (*Larus ridibundus*); prevalence, genotypes and influence on *Salmonella* epidemiology. *Epidemiol. Infect.* **134**, 635–644.

Pasmans, F., Van Immerseel, F., Hermans, K. et al. (2004). Assessment of virulence of pigeon isolates of *Salmonella enterica* subsp. *enterica* serovar *typhimurium* variant *copenhagen* for humans. *J. Clin. Microbiol.* **42**, 2000–2002.

Pedersen, K., Clark, L., Andelt, W. F., and Salman, M. D. (2006). Prevalence of Shiga toxin-producing *Escherichia coli* and *Salmonella enterica* in rock pigeons captured in Fort Collins, CO. *J. Wildl. Dis.* **42**, 46–55.

Pollock, C. (2003). Fungal diseases of columbiformes and anseriformes. *Vet. Clin. North Am. Exot. Anim. Pract.* **6**, 351–361.

Quessy, S. and Messier, S. (1992). Prevalence of *Salmonella* spp., *Campylobacter* spp. and *Listeria* spp. in ring-billed gulls (*Larus delawarensis*). *J Wildl Dis.* **28**, 526–531.

Quevedo, F., Lord, R. D., Dobosch, D. et al. (1973). Isolation of *Salmonella* from sparrows captured in horse corrals. *Am. J. Trop. Med. Hyg.* **22**, 672–674.

Quinn, P. J. (1994). *Clinical veterinary microbiology*. Elsevier Health Sciences, New York.

Rice, D. H., Sheng, H. Q., Wynia, S. A., Hovde, C. J. (2003). Rectoanal mucosal swab culture is more sensitive than fecal culture and distinguishes *Escherichia coli* O157:H7-colonized cattle and those transiently shedding the same organism. *J. Clin. Microbiol.* **41**, 4924–4929.

Roberts, J. P. and Grimes, J. E. (1978). Chlamydia shedding by four species of wild birds. *Avian Dis.* **22**, 698–706.

Rosario, I., Acosta, B., and Colom, M. F. (2008). Pigeons and other birds as a reservoir for *Cryptococcus* spp. *Rev. Iberoam. Micol.* **25**, S13–S18.

Rubin, H., Furcolow, M. L., Yates, J. L., and Brasher, C. A. (1959). The course and prognosis of histoplasmosis. *Amer. J. Med.* **27**, 278–288.

Ruiz, A., Neilson, J. B., and Bulmer, G. S. (1982). A one year study on the viability of *Cryptococcus neoformans* in nature. *Mycopathologia* **77**, 117–122.

Schwarz, J., Baum, G. L., Wang, C. J. et al. (1957). Successful infection of pigeons and chickens with *Histoplasma capsulatum*. *Mycopathol. Mycol. Appl.* **S8**, 189–193.

Siim, J. C., Biering-Sorensen, U., and Moller, T. (1963). Toxoplasmosis in domestic animals. *Adv. Vet. Sci.* **8**, 335–429.

Sixl, W., Karpísková, R., and Hubálek, Z. et al. (1997). *Campylobacter* spp. and *Salmonella* spp. in black-headed gulls (*Larus ridibundus*). *Cent. Eur. J. Public Health* **5**, 24–26.

Slodkowciz-Kowalska, A., Graczyk, T. K., Tamang, L. et al. (2006). Microsporidian species known to infect humans are present in aquatic birds: Implications for transmission via water? *Appl. Environ. Microbiol.* **72**, 4540–4544.

Smith, C. D., Furcolow, M. L., and Tosh, F. E. (1964). Attempts to eliminate *Histoplasma capsulatum* from soil. *Am. J. Hyg.* **79**, 170–180.

Smith, H. V., Brown, J., Soulson, J. C. et al. (1993). Occurrence of oocysts of *Cryptosporidium* sp. in *Larus* spp. gulls. *Epidemiol. Infect.* **110,** 135–143.

Snoeyenbos,. G. H., Morin, E. W., and Wetherbee, D. K. (1967). Naturally occurring *Salmonella* in "blackbirds" and gulls. *Avian Dis.* **11,** 642–646.

Sonntag, A. K., Zenner, E., Karch, H., and Bielaszewska, M. (2005). Pigeons as a possible reservoir of Shiga toxin 2f-producing *Escherichia coli* pathogenic to humans. *Berl. Munch Tierarztl. Wochenschr.* **118,** 464–470.

Somarelli, J. A., Makarewicz, J. C., Sia, R., and Simon, R. (2007). Wildlife identified as major source of *Escherichia coli* in agriculturally dominated watersheds by BOX A1R-derived genetic fingerprints. *J. Environ. Manage.* **82,** 60–65.

Soogarun, S., Wiwanitkit, V., Palasuwan, A. et al. (2006). Detection of *Cryptococcus neoformans* in bird excreta. *Southeast Asian J. Trop. Med. Public Health* **37,** 768–770.

Stevens, G. R., Rogue, J., Weber, R., and Clark, L. (2000). Evaluation of a radar-activated, demand-performance bird hazing system. *International Biodeterioration & Biodegradation* **45,** 129–137.

Stobierski, M. G., Hospedales, C. J., Hall, W. N. et al. (1996). Outbreak of histoplasmosis among employees in a paper factory—Michigan, 1993. *J. Clin. Microbiol.* **34,** 1220–1223.

Storch, G., Burford, J. G., George, R. B. et al. (1980). Acute histoplasmosis. Description of an outbreak in northern Louisiana. *Chest* **77,** 38–42.

Tanaka, C., Miyazawa, T., Watarai, M., and Ishiguro, N. (2005). Bacteriological survey of feces from feral pigeons in Japan. *J. Vet. Med. Sci.* **67,** 951–953.

Taylor, L. H., Latham, S. M., and Woolhouse, M. E. (2001). Risk factors for human disease emeregence. *Philos. Trans. R. Soc. Lond. B. Biol. Sci.* **356,** 983–989.

Thurston-Enriquez, J. A., Watt, P., Dowd, S. E. et al. (2002). Detection of protozoan parasites and microsporidia in irrigation waters used for crop production. *J. Food Prot.* **65,** 378–382.

Tosh, F. E., Weeks, R. J., Pfeiffer, F. R. et al. (1967). The use of formalin to kill *Histoplasma capsulatum* at an epidemic site. *Am. J. Epidemiol.* **85,** 259–265.

Tsiodras, S., Kelesidis, T., Kelesidis, I. et al. (2008). Human infections associated with wild birds. *J. Infect.* **56,** 83–98.

Wallace, J. S., Cheasty, T., and Jones, K. (1997). Isolation of vero cytotoxin-producing *Escherichia coli* O157 from wild birds. *J. Appl. Microbiol.* **82,** 399–404.

Weber, A. and Schäfer, R. (1991). The occurrence of *Cryptococcus neoformans* in fecal samples from birds kept in human living areas. *Berl. Munch. Tierarztl. Wochenschr.* **104,** 419–421.

Weber, R., Bryan, R. T., Schwartz, D. A., and Owen, R. L. (1994). Human microsporidial infections. *Clin. Microbiol. Rev.* **7,** 426–461.

Werner, S. J., Cummings, J. L., Tupper, S. K. et al. (2008). Blackbird repellency of selected registered pesticides. *Journal of Wildlife Management* **72,** 1007–1011.

Wither, A., Rehfisch, M., and Austin, G. (2005). The impact of bird populations on the microbiological quality of bathing waters. *Water Sci. Technol.* **51,** 199–207.

Yorio, P. and Caille, G. (2004). Fish waste as an alternative resource for gulls along the Patagonian coast: Availability, use, and potential consequences. *Mar. Pollut. Bull*. **48,** 778–783.

Yoshida, T., Sugimoto, T., Sato, M., and Hirai, K. (2000). Incidence of *Listeria monocytogenes* in wild animals in Japan. *J. Vet. Med. Sci*. **62,** 673–675.

Zhang, Y., Yeh, E., Hall, G. et al. (2007). Characterization of *Listeria monocytogenes* isolated from retail foods. *Int. J. Food Microbiol*. **113**(1), 47–53.

CHAPTER 7

Produce Contamination by Other Wildlife

Daniel H. Rice

Food Laboratory Division, New York State Department of Agriculture and Markets, Albany, NY

CHAPTER CONTENTS

Introduction	143
Viral Pathogens	146
Bacterial Pathogens	146
Parasitic Pathogens	151
Protozoa	152
Helminths	153
Mitigating Wildlife—Crop Interactions	155
Summary	156

INTRODUCTION

Fresh produce is well established as an important source of foodborne illness in humans, as evidenced by its being a vehicle of transmission in several large, multistate outbreaks of disease involving a variety of pathogens and fresh produce. The 2008 FAO/WHO microbiological hazards in fresh fruits and vegetables meeting report lists leafy green vegetables as level 1 priority commodities for contamination with enterohemorrhagic *E. coli*, *S. enterica*, *Campylobacter* spp., *Shigella* spp., hepatitis A virus, noroviruses, *Cyclospora cayatenensis*, *Cryptosporidium*, *Yersinia pseudotuberculosis*, and *L. monocytogenes*, most of which are zoonotic pathogens (FAO/WHO, 2008). In this report wildlife are singled out as being potential sources of contamination of fresh produce with human pathogens. The report lists a number of questions to be addressed related to what the potential roles are of wildlife in contaminating produce, either directly or by environmental contamination routes. The lack of information on wildlife reservoirs of human diseases that can be transmitted by contaminated fresh produce is evidenced by this report. Some of the gaps regarding this issue will be addressed in this chapter.

The majority of agricultural production takes place in rural areas that also provide habitat to a wide variety of feral domestic and wild animals. Fresh produce is utilized as a food source by a variety of wild animals. Farmers lose an estimated 4.5 billion dollars annually in crop production losses due to wild animals consuming unharvested produce or damaging crops prior to harvest (Conover, 2002). Although there are some measures that can be taken to minimize the opportunity for produce in the field to come in contact with wild animals, the options are limited, and in most instances it is impossible to keep wild animals out of farm land. Very often wildlife control programs to mitigate crop damage are at odds with recreational activities such as hunting. It is for this reason and others that wildlife will continue to be directly associated with most agricultural production practices. Control measures to protect produce postharvest are more readily available and apparent. These are important because, as described in other chapters of this book, there are important considerations in the processing, packaging, and transport sectors that relate to the potential for wildlife to contaminate produce once harvested.

Up to one third of the population of developed countries experiences a foodborne illness each year (Schlundt J, 2004). Wildlife have been implicated as possible sources of contamination in produce-borne illness including several outbreaks of disease (Besser et al., 1993; Beuchat and Ryu, 1997; Brackett, 1999; Burnett and Beuchat, 2001; Cody et al., 1999; Greene et al., 2008; Jay et al., 2007), and although an outbreak of human illness associated with produce directly contaminated by wildlife has never been reported, there are a number of reasons to believe these have occurred. First, each year foodborne pathogens are associated with an estimated 76 million illnesses, 325,000 hospitalizations, and 5200 deaths in the United States (Mead et al., 1999). The CDC and others estimate that only a fraction of foodborne illnesses are actually detected, and of those that are detected, a food vehicle for infection is rarely determined. Second, wildlife are reported to carry many infectious agents that have been associated with foodborne illness in humans. Consequently, the probability that wildlife have been and will continue to be sources of contamination of produce that subsequently cause human illnesses is high; however, the extent to which wildlife contribute to the contamination of produce resulting in human illness is not known. The most compelling evidence for a wildlife role in contaminating produce is that of feral swine being implicated in a 2006 multistate outbreak of *E. coli* O157:H7 linked to packaged fresh baby spinach (Jay et al., 2007). Deer have been implicated as a possible source of *E. coli* O157:H7 contamination of apples used to make both unpasteurized juice and cider that were linked to outbreaks of illness in humans, although the links in these outbreaks were weak at best (Besser et al., 1993; Cody et al., 1999).

For the purpose of discussion in this chapter, the focus will be primarily on the species typically thought of as wildlife, primarily mammals, including feral domestic mammals. Other types of animals will be discussed where relevant. The potential role of birds in contaminating produce is significant and is discussed in another chapter. In addition, this chapter will discuss the potential role that insects, mollusks, and helminths play in transmitting pathogens to produce. This chapter will discuss the potential for occurrence of direct contamination of produce by wildlife. Indirect contamination, for example, via irrigation water, is covered in another chapter. A comprehensive review of the peer reviewed published literature indicates that many zoonotic pathogens have been reported to be carried by one or more species of wildlife and have the potential to be transmitted to humans through fresh produce (Table 7.1). As pointed out by Schlundt J et al. (2004), the top five emerging foodborne diseases are caused by *S. enterica*,

Table 7.1 Zoonotic Pathogens Isolated from both Fresh Produce and Wild Animals

	Zoonotic Pathogen	Known Wildlife Hosts and Relevant Intermediate Hosts
Virus	Hepatitis E	Feral swine, wild boar, rodents, primates, and a wide variety of other mammals
Bacteria	*Campylobacter* spp. *S. enterica*	Birds, mammals, marsupials, reptiles, amphibians
	Pathogenic *E. coli*	Mammals
	L. monocytogenes and *Yersinia enterocolitica*	Mammals, birds
	Arcobacter spp.	Not known
Protozoa	*C. parvum*	Mammals
	Giardia spp.	Mammals
	Toxoplasma gondii	Wild felids
Helminths—nematodes	*Ascaris suum*	Feral swine, wild boar
	Toxocara canis	Wild canids
	Toxocara cati	Wild felids
	Toxascaris leonine	Wild canids
	Lagochilascaris minor	Rodents, felids, and raccoons
	Angiostrongylus cantonensis	Rodents via gastropod intermediate hosts
	Angiostrongylus costaricensis	Rodents via gastropod intermediate hosts
Helminths—trematodes	*Fasciola hepatica* and *gigantica*	Wild ruminants, equids, and lagomorphs via a lymnaeid intermediate host
	Fasciolopsis buski	Feral swine via a lymnaeid intermediate host

Campylobacter spp., enterohaemorrhagic *E. coli*, *Toxoplasma gondii*, and *Cryptosporidium* spp., all of which are zoonotic, and all having the potential of being transmitted to humans through fresh produce.

VIRAL PATHOGENS

It is estimated that 67.2 % of foodborne illnesses are viral (Mead et al., 1999); however, viral pathogens are almost exclusively host specific. Hence the vast majority of these cannot be transmitted to humans by wildlife. The only apparent exception is hepatitis E, which is carried by a variety of wild, domestic, and feral mammals (Goens and Perdue, 2004; Smith, 2001). The primary exposure to humans in industrialized countries is through direct contact with infected animals, primarily swine, or eating undercooked meat and organs from infected animals (Teo, 2006). Seropositive rates for hepatitis E in blood donors from developed countries range from 0.4 to 3.3% positive (Smith, 2001). In nonindustrialized countries, waterborne exposure is the primary route of infection (Seymour and Appleton, 2001), and seropositive rates range from 9.5 to 54.8% (Smith, 2001). Hepatitis E is transmitted mainly via the fecal–oral route, so in theory, wild animals shedding hepatitis E in their feces could contaminate produce that is consumed by humans, resulting in transmission of disease. Viruses are very difficult to detect in food samples, with detection techniques historically based on scanning electron microscopy. More recently, molecular-based technologies have been employed and should provide better information on the incidence of hepatitis E and other viral agents in food, including produce (Seymour and Appleton, 2001). Hepatitis E has not been reported to have been detected in fresh produce.

BACTERIAL PATHOGENS

Bacterial agents cause an estimated 30.2% of all foodborne illnesses (Mead et al., 1999). The zoonotic bacterial organisms most commonly associated with foodborne disease are *C. coli* and *jejuni*, *S. enterica*, *E. coli* O157: H7, and *L. monocytogenes*, all of which have been isolated from fresh produce (Brackett, 1999; Mead et al., 1999; Schlundt J, 2004; Tauxe, 2002). *Yersinia enterocolitica*, also a zoonotic foodborne pathogen, has been isolated from fresh produce although never linked to a produce-borne outbreak of illness in humans (Beuchat, 1995; Brackett, 1999; Burnett and Beuchat, 2001). In addition, *Arcobacter* spp. appear to be involved in a limited number of foodborne illnesses each year, but the data on this potential zoonotic pathogen are very limited (Ho et al., 2006). Additional serotypes of enterohemorrhagic *E. coli* including O26, O103, O111, O118, and O145 in the future may be identified as important zoonotic pathogens associated with fresh produce. These have all been associated with human illness (Schlundt J, 2004), many are associated with livestock (Cobbold et al., 2008; Frank et al., 2008), and at least one (O111) has been implicated in an outbreak of illness associated with unpasteurized apple juice (Cobbold et al., 2008; Frank et al., 2008; Vojdani et al., 2008).

Many of the zoonotic pathogens in humans that have been linked to a produce vehicle of infection, including *S. enterica*, *E. coli* O157:H7, *Camplyobacter* spp., and *L. monocytogenes*, have also been isolated from wildlife (Brown et al., 2004; Jijon et al., 2007; Lyautey et al., 2007; Parish, 1997; Renter et al., 2001, 2006; Rice et al., 2003; Scaife et al., 2006; Wahlstrom et al., 2003). However, wildlife have yet to be demonstrated as being a direct source of produce contamination that resulted in an outbreak of illness in humans. There is compelling evidence that feral swine may have contributed to the contamination of spinach with *E. coli* O157:H7 in a large multistate outbreak in 2006 (Jay et al., 2007). In the course of investigating this outbreak, it was found that a number of environmental origin isolates of *E. coli* O157:H7, including those from feral swine, genetically matched the human outbreak strain. Interestingly, over 33% of cattle feces, 23% of feral swine feces, 4% of water samples, and 8% of soil samples, collected from close proximity to cropland implicated as source fields for contaminated spinach, were *E. coli* O157:H7 positive, and most matched the outbreak strain (Jay et al., 2007). Swine feces were commonly found in fields used to grow spinach. This indicates that a long-term reservoir for *E. coli* O157:H7 must have existed in close proximity to these fields; whether or not this reservoir was maintained in animals or the environment was not determined. Feral swine have previously been identified as carrying *E. coli* O157:H7 in a survey of wildlife in Sweden (Wahlstrom et al., 2003). In this study several species of wild animals were surveyed for zoonotic pathogens. Only one of 68 feral swine samples tested were *E. coli* O157:H7 positive, suggesting that wild pigs were transiently colonized rather than a reservoir of this organism.

Other outbreaks of produce-borne illness in humans with epidemiological links to wildlife include apple juice and cider contaminated with *E. coli* O157:H7, attributed to use of apples that had been picked off the ground (Besser et al., 1993; Cody et al., 1999). The authors hypothesized that these apples may have been contaminated with deer feces positive for *E. coli* O157:H7 since there was evidence that deer frequented the orchard where the apples came from. Enteric illness of unknown etiology subsequent to consuming freshly pressed apple cider that was made using dropped apples, obtained from orchards where presence of deer was evident, has also been reported (Vojdani et al., 2008).

Environmental reservoirs may play an important role in the colonization of wild animals by *E. coli* O157:H7 and other zoonotic pathogens. *E. coli* O157:H7 has been demonstrated to remain in farm environments for extended periods of time, providing an apparent source for farm animal colonization over time (LeJeune et al., 2001; Rice et al., 1999; Van Donkersgoed et al., 2001). LeJeune et al. (2001) demonstrated that *E. coli* O157:H7 can remain viable in cattle water trough sediments for up to 245 days and that contaminated water from troughs with no animal contact for over six months were capable of infecting cattle. Cattle water troughs have been identified as potentially important reservoirs of *E. coli* O157:H7 in additional studies (Hancock et al., 1998; Van Donkersgoed et al., 2001; Wetzel and LeJeune, 2006), and persistence of strains within herds for up to 24 months indicates that environmental reservoirs are important in maintaining this organism in cattle herds (Rice et al., 1999).

An extensive longitudinal survey of range cattle environments by Renter et al. (2003) demonstrated that 0.51% of cattle water tanks, 0.25% of lakes and ponds, and 0.41% of free-flowing rivers and streams were positive for a diverse number of *E. coli* O157:H7 strains. In this study an identical strain of *E. coli* O157:H7 was shared by cattle, water, and wildlife. Since environmental reservoirs have been demonstrated to be important in maintaining *E. coli* O157:H7 in cattle production facilities, by implication, environmental reservoirs may be important in maintaining this and other pathogens in wildlife.

Human illnesses with *C. coli* and *jejuni*, *S. enterica*, *E. coli* O157:H7, and *L. monocytogenes* are often associated with foodborne exposure, and each of these organisms has been isolated from produce and from wild animals. Wildlife have been implicated as sources of foodborne pathogens for both outbreaks and individual cases of illness in humans. Only a limited number of surveys of wild animals for zoonotic pathogens exist. In general, these reports identify species of animals that tested positive for specific pathogens, based on a limited number of samples within a limited geographical area, and most are not capable of providing true prevalence estimates in selected populations of animals. Testing of wild animal feces indicate that deer, moose, rabbits, opossums, and wild boar/feral swine can carry *E. coli* O157:H7 (Jay et al., 2007; Renter et al., 2001, 2003, 2006; Rice et al., 2003; Scaife et al., 2006; Wahlstrom et al., 2003). Renter et al. (2001) demonstrated that *E. coli* O157:H7 prevalence in Nebraska white-tailed deer during the fall hunting season was 0.25%, indicating a generally low prevalence in this population of deer. Other surveys have demonstrated a similarly low prevalence in wildlife. However, a survey of wild rabbits in the UK demonstrated that over 8% of rabbit fecal samples tested during the summer were *E. coli* O157:H7 positive (Scaife et al., 2006).

Surveys of wildlife for *S. enterica* indicate that the majority of animals that test positive are birds; however, this organism has also been found in foxes, opossums, gray squirrels, woodchucks, and toads (Jijon et al., 2007; Parish, 1997; Wahlstrom et al., 2003). Eight of 71 (11%) wild animals tested at a rehabilitation center were positive for four serovars of *S. enterica*, and five (7%) of these were nonavian wildlife (opossum, gray squirrel, and woodchuck), indicating a relatively high prevalence of this organism in this population of nonavian wildlife (Jijon et al., 2007). One survey of Swedish wildlife for *Campylobacter* spp. indicated that hares, moose, and feral swine can carry thermophilic *Campylobacter* with prevalences ranging between 1 and 12%, depending upon species tested (Wahlstrom et al., 2003). Another survey of wild animal feces reported that 11% of nonavian feces tested were *Campylobacter* positive; this study did not report feces by species of animal (Brown et al., 2004).

Given its nearly ubiquitous nature and the fact that it is the most common cause of enteric bacterial illness, it is likely that many if not most wild animals have the potential to carry pathogenic species of *Campylobacter*. Unlike most foodborne bacterial pathogens, *Campylobacter* is very sensitive to a wide variety of environmental stressors and unable to multiply outside of an animal host (Park, 2002). In spite of these limitations, *Campylobacter* is the leading cause of bacterial foodborne illness (Mead et al., 1999) possessing a variety of mechanisms that allow it to persist in the environment and on food once contaminated (Cook and Bolster, 2007; Karenlampi and Hanninen, 2004; Murphy et al., 2006). A study that compared *L. monocytogenes* isolates from livestock, wildlife, and humans in Ontario, Canada demonstrated that this organism could be found in 6% of deer, 5% of moose, and 50% of both otter and raccoon feces, indicating that this organism is readily carried by a variety of wildlife (Lyautey et al., 2007). Identical Pulsed Field Gel Electrophoresis (PFGE) patterns of *L. monocytogenes* were found in deer, moose, and cattle, indicating either exposure to a common source or direct contact-associated serial infections.

None of the reported surveys for zoonotic pathogens in wildlife are comprehensive enough to allow estimates of true geographical prevalence among any species of wildlife or to comprehensively identify which species of animals do and do not carry specific zoonotic pathogens. The value of these reports lies in the fact that a wide range of wild animals have been reported to carry all the major foodborne zoonotic bacterial pathogens.

Yersinia enterocolitica has been isolated from fresh produce, but an outbreak of illness in humans has never been associated with produce (Beuchat, 1995). *Y. enterocolitica* is most commonly associated with domestic pigs but has been isolated from a wide variety of wild animals (Shayegani et al., 1986). This study reported that 10% of wild mammals tested from New York State had detectable *Y. enterocolitica* in their feces. It is not known if *Y. enterocolitica* is responsible for produce-borne illness in humans and whether or not wildlife are sources of produce contamination; however, the potential for both exists.

Arcobacter is a relatively new genus of organisms that has been associated with human illness, is zoonotic, has been isolated from food, and is frequently isolated from the environment (Ho et al., 2006). At this time nothing is known about what role, if any, *Arcobacter* plays in produce-borne illness in humans and whether or not there is a wildlife reservoir.

In addition to wildlife being sources of contamination of fresh produce, insects can be a mechanical vector for many foodborne bacterial pathogens. Fruit flies have been demonstrated under experimental conditions to transport *E. coli* to uncontaminated fruit (Janisiewicz et al., 1999; Sela et al., 2005). Fruit flies exposed to composting apples, inoculated with a genetically distinct strain of *E. coli*, readily transported *E. coli* to uncontaminated apples. Both generic *E. coli* and *E. coli* O157:H7 were shown to grow exponentially in apple wounds, demonstrating that fruit flies are viable vectors for contaminating fresh fruit with pathogens, and that once contaminated, fruit wounds are an excellent site for these pathogens to multiply (Janisiewicz et al., 1999). Fruit flies, exposed to feeding stations and feces artificially contaminated with a genetically unique strain of *E. coli*, were subsequently demonstrated to carry *E. coli* for up to seven days and to introduce *E. coli* to apple wounds (Sela et al., 2005). House flies and filth flies can carry a variety of zoonotic foodborne pathogens including *E. coli* O157:H7, *S. enterica*, and *C. jejuni* (Hancock et al., 1998; Iwasa et al., 1999; Olsen, 1998). These flies are frequently found in crop production areas that are adjacent to livestock facilities or other environments conducive to their reproduction. Additionally, the aquatic midge *Chironomus tentans* has been demonstrated to carry *S. enterica* from the larval stage to the adult fly stage in experimental settings using contaminated aquatic sediments and fresh water as the pathogen source (Moore et al., 2003). The larval stages of this organism inhabit a wide variety of aquatic environments including riparian habitats, irrigation canals, and cattle watering troughs; consequently, this insect may participate in the ecology of *S. enterica* in the environment. It is clear that in order for insects to play a role in contaminating fresh fruit or produce, a nearby source of contamination is needed since these vectors do not generally travel very far, have short life spans, and infections are apparently not passed on from generation to generation. Many orchards and produce farms are located in livestock-intensive areas and areas with abundant wildlife. Consequently, fecal sources of pathogens for contaminating insects exist, and insect vectors are potentially important sources of contamination.

There is an interesting body of literature on the soil nematode *Caenorhabditis elegans*. This nematode can be colonized with zoonotic foodborne pathogens including *S. enterica*, *E. coli* O157:H7, and *L. monocytogenes* when exposed to contaminated soil (Caldwell et al., 2003a, 2003b; Kenney et al., 2005, 2006; Labrousse et al., 2000). These small soil nematodes (1.5 mm long) feed on bacteria, and it has been hypothesized that they are candidates for mechanically contaminating certain crops that come into direct contact with soil where a pathogen source exists. It has been shown experimentally that once *S. enterica* is internalized in *C. elegans*, it is apparently protected from many commercial sanitizers, indicating that produce containing *S. enterica*-infected *C. elegans* may remain a source of human infection in spite of cleaning and sanitizing (Caldwell et al., 2003a).

PARASITIC PATHOGENS

Determining the prevalence of parasitic foodborne diseases in humans is difficult mainly because detecting the infective stages of these organisms in food presents several challenges. Epidemiological evidence indicates that 2.6% of foodborne diseases are parasitic (Mead et al., 1999). The vast majority of parasitic foodborne infections are from protozoa; however, infections with nematodes and trematodes have been documented. The primary foodborne zoonotic protozoa are *Cryptosporidium parvum*, *Giardia duodenalis*, and *Toxoplasma gondii*. All three have been associated with outbreaks of illness in humans, but never with a fresh produce vehicle of infection. Additionally, *Balantidium coli* and *Blastocystis* spp. have been isolated from fresh produce and are known to be carried by animals (Slifko et al., 2000), although infection in humans with these organisms never has been reported to be associated with fresh produce. The only zoonotic helminths linked to fresh produce are the trematodes *Fasciola hepatica* and *gigantica*, *Fasciolopsis buski*, and the nematodes *Ascaris suum*, *Toxocara canis* and *cati*, *Toxascaris leonine*, *Lagochilascaris minor*, and both *Angiostrongylus cantonensis* and *costaricensis* (Polley, 2005; Slifko et al., 2000). Only *Fasciola hepatica* has been responsible for outbreaks of produce-borne infections in humans (Macpherson, 2005; Mas-Coma, 2005; Rondelaud et al., 2005).

PROTOZOA

Most instances of foodborne protozoal disease in humans are linked to poor sanitation by food handlers who are themselves infected or through waterborne exposure. It is estimated that only 1 out of 10 cases of foodborne protozoal infections are reported (Casemore, 1990), and of those reported, an etiological agent is often not identified. Detection of protozoal agents in produce is difficult, and even if foodborne transmission were common, confirmation through detection of the agent in produce would be rare. Foodborne transmission has been documented for *C. parvum*, and *G. duodenalis*, both of which also occur in wild animals (Appelbee et al., 2005; Polley, 2005; Rose and Slifko, 1999). These agents are the most common enteric parasites in humans and are widespread in the environment from a number of sources, including discharges from food animal production facilities and treated and untreated wastewater effluent (Dawson, 2005; Hunter and Thompson, 2005). These organisms are also reported to be widespread in irrigation water used for crop production in the United States and Central America (Thurston-Enriquez et al., 2002). Presumably, environmental contamination is common because these organisms have a wide range of hosts including domestic and wild animals. No outbreaks of foodborne giardiasis have been reported in industrially manufactured or processed food. A few outbreaks of foodborne cryptosporidiosis from fresh produce including green onions (CDC, 1997) and unpasteurized apple juice (Dawson, 2005; Smith et al., 2007) have been reported. A few studies have demonstrated the presence of *C. parvum* on a variety of fresh produce including cilantro, lettuce, blackberries, cabbage, basil, parsley, celery, leeks, green onions, green chilis, mung bean sprouts, and other seed sprouts (Calvo et al., 2004; Ortega et al., 1997; Robertson et al., 2002). Whether any of these were contaminated by wildlife could not be determined.

There have been a number of widespread outbreaks of produce-borne disease caused by *Cyclospora cayetanensis*. This protozoan parasite is apparently host specific and thought to occur only in humans. Thus, it appears that outbreaks of *C. cayetanensis* reported in fresh berries and other produce were not from wild animals (Mansfield and Gajadhar, 2004).

There are many wild and feral animal reservoirs for *Cryptosporidium* spp. and *Giardia* spp., including most mammals, birds, and amphibians; both *C. parvum* and *G. duodenalis* have been reported in a wide variety of mammal species (Appelbee et al., 2005; Polley, 2005; Smith et al., 2007). Molecular-based analyses of *C. parvum* and *G. duodenalis* indicate that many of the variants isolated from wild animals are not apparently associated with human illness, indicating that these organisms may not be ubiquitously pathogenic to humans (Appelbee et al., 2005). Future studies may better define the roles and potential roles of both domestic and wild animals in the epidemiology of both *C. parvum* and *G. duodenalis* infections in humans. Wild and feral cats present the only wildlife sources of *Toxoplasma gondii* with relevance to produce contamination. The feces of wild felids can contain oocysts that are infectious to humans; however, direct transmission of this or any other protozoal agents from wild animals to produce that subsequently infects humans has never been documented. *Balantidium coli* and *Blastocystis* spp. are occasionally associated with human illness but never directly linked to fresh produce. Since these organisms also occur in animals, they also have the potential to contaminate produce although a produce-borne outbreak of illness in humans from these organisms has not been reported.

In addition to the potential for direct contamination by wildlife, mechanical transport of *C. parvum* by house flies and wild filth flies has been demonstrated (Graczyk et al., 1999a, 1999b, 2000). This may play a role in produce contamination from farm to retail distribution. Similar to bacterial pathogens, this mechanism of transport requires a pathogen source in close proximity to produce.

HELMINTHS

Foodborne outbreaks of human illness with zoonotic helminths are rare with the vast majority of infections being from contaminated meat. There are several species of zoonotic helminths that occur in wild animals and can theoretically infect humans through contaminated produce, including the nematodes *Ascaris suum*, *Toxocara canis* and *cati*, *Toxascaris leonine*, *Lagochilascaris minor*, *Angiostrongylus cantonensis* and *costaricensis*, and the trematodes *Fasciola hepatica*, *gigantica*, and *Fasciolopsis buski* (Macpherson, 2005; Polley, 2005; Slifko et al., 2000). The majority of zoonotic nematode infections in humans are attributed to direct contact with companion or farm animals (Polley, 1978); however, produce-borne infection is possible. Only the trematode *Fasciola hepatica* has been linked to human produce-borne infections with a wild animal source (Polley, 2005; Rondelaud et al., 2001). As with bacterial zoonotic foodborne pathogens, there is very limited information on the carriage rates of helminths in wildlife that are infective to humans.

Ascaris suum is carried by feral pigs and has a worldwide distribution. Infective ova from this parasite can contaminate produce through direct contact with the feces of wild pigs. One study in the United States reports the prevalence of *A. suum* in wild pigs from Kansas to be 20% (Gipson et al., 1999). Infections in humans are very rare in industrialized countries, and only limited reports exist of outbreaks of illnesses in humans in developing countries. There are no reports of foodborne *A. suum* infections in humans. Theoretically any fecal–oral route of exposure to infective ova is a possible route of infection in humans.

Toxocara canis and *Toxascaris leonina* are carried by wild and feral canines. *Toxocara cati* is carried by feral and wild felines, and *Lagochilascaris minor* is carried by wild felines, raccoons, and rodents (Polley, 2005; Slifko et al., 2000). The infective stages of these nematodes occur in the feces of these definitive hosts. Consequently, human exposure is fecal–oral, and produce is a potential vehicle for human infection. Human disease, including larval visceral migrans and ocular disease from *Toxocara canis* and *cati* and *Toxascaris leonine*, has been reported although never linked to food (Polley, 1978).

The infective larvae of *Angiostrongylus cantonensis* and *costaricensis* occur in gastropod intermediate hosts, presenting a route of human infection in raw vegetables grown in aquatic environments (Polley, 2005). The definitive hosts for these organisms are wild rodents; however, the host range appears to be expanding to include other mammals (Macpherson, 2005). Although there are no reports of foodborne infections of these nematodes in humans, infective ova could theoretically contaminate produce through direct contact with contaminated feces from wild hosts or their waterborne intermediate hosts. Several of these nematodes are common in domestic animals and are reported to occur in their wild counterparts. However, there are no reported surveys of wild animals for the presence of these nematodes, making it difficult to determine the extent to which wild animals are infected and what the potential is for wild animals to contribute to human illness.

Worldwide, there are an estimated 2.4 million cases of fascioliasis annually (Rim, 1992) resulting in significant morbidity and mortality in certain endemic areas (Garcia et al., 2007). In hyperendemic areas, human prevalence is as high as 72% and is coassociated with proximity to aquatic sources of appropriate lymnaeid intermediate hosts (Mas-Coma et al., 1999, 2005). The primary wildlife reservoirs of *Fasciola hepatica* and *gigantica* are wild ruminants and equids (Polley, 2005; Slifko et al., 2000); *F. hepatica* is also common in wild lagomorphs (Rondelaud et al., 2005). The definitive host range for *F. hepatica* is expanding, and depending on geographic location, can include a wide variety of mammals and marsupials (Mas-Coma, 2005; Mas-Coma et al., 2005).

Human infection with *F. hepatica* in developed countries is almost exclusively due to ingestion of contaminated aquatic vegetation; primarily watercress (Macpherson, 2005; Rondelaud et al., 2005), and a number of aquatic plant-borne infections of *F. hepatica* have been reported (Mas-Coma et al., 2005). The primary routes of exposure to humans in developing countries are contaminated drinking water and food. Wildlife sources of *Fasciolopsis buski* include wild pigs (Slifko et al., 2000), and both human and animal infection with this organism are apparently rare. These trematodes are transmitted to humans through the ingestion of infective metacercariae from a variety of gastropods; consequently, foodborne exposure to humans would result from contact with contaminated water or the gastropod intermediate host.

MITIGATING WILDLIFE—CROP INTERACTIONS

The Leafy Green Marketing Agreement of California published a food safety guidance document for lettuce and leafy greens in 2008 (LGMA, 2008). This document is tied closely to a previous guidance document from the industry, spearheaded by the International Fresh-Cut Produce Association (Gorny, 2006), but goes into greater detail on how to mitigate wild animal contact with lettuce and leafy greens than did the document published in 2006. The LGMA guidance document lists deer and wild pigs as "animals of significant risk" and recommends that lettuce and leafy green crops with evidence of heavy contact with these animals not be harvested. In addition, this document recommends that barriers be used whenever possible to mitigate contact of these crops with animals of significant risk through the removal of habitat and installation of fencing. To their credit the authors of this guidance document caution producers from removing habitat that may be important to beneficial insects, and that local regulations may prevent removal of habitat adjacent to croplands.

Limited information exists on management strategies that have been demonstrated to be successful in minimizing contact of produce with wild animals. Jay and Wiscomb (2008) discuss mitigation strategies for controlling the interaction of feral swine with crop lands. Hunting has been identified as a potentially viable tool for controlling feral swine populations under certain conditions however, the very high reproductive rate of feral swine, combined with a lack of access to private land, makes hunting alone an inadequate tool for controlling feral swine populations. Fencing appears to be the mitigation strategy with the greatest potential for success but requires a significant investment of resources to build and maintain, making it an unattractive option for many producers. Removal of habitat is considered to be of no use in controlling feral swine since these animals are highly mobile and have large home ranges.

Physical barriers such as greenhouses are protective to a limited number of crops; however, the vast majority of fresh produce is grown outdoors with unrestricted access to wild animals. The utility of currently available control interventions to minimize wild animal interactions with the majority of growing crops appears weak at best.

SUMMARY

The list of zoonotic pathogens that have the potential to infect humans through a fresh produce vehicle is extensive. Many of these organisms are rarely associated with human illness; however, several are among the current list of most significant emerging foodborne pathogens. Establishing the role that wildlife and feral domestic animals play in produce-borne human illness is very difficult, as evidenced by the number of outbreak investigations that imply wildlife involvement in spite of a lack of conclusive evidence that this has indeed occurred. Almost certainly, some human illnesses with zoonotic pathogens are directly linked to wild animals. Environmental reservoirs are critical components to the ecology and epidemiology of produce-borne illness in humans, and programs aimed at reducing human illness associated with consuming fresh produce must account for a variety of real and potential environmental reservoirs. Wildlife and feral domestic animals are components of these environmental reservoirs, and control measures aimed at the environmental level could very well reduce the incidence of certain zoonotic pathogens in wildlife.

Mitigating wildlife–produce production interactions is a challenge. In most cases there are no economically feasible mechanisms to prevent wildlife from coming into direct contact with produce while being grown. At best barriers can be installed to prevent some species of wildlife from entering production areas, but these are not all exclusive. Once harvested, the opportunities to keep produce and wildlife separate are greatly expanded, and this stage of the production process presents several critical intervention points that HACCP programs should consider. Historically, the incidence of foodborne illness associated with produce was linked to growing season in the Northern Hemisphere; however, the proportion of fresh produce that is imported versus domestically produced is expanding rapidly. In the last 10 years the amount of fresh fruits and vegetables imported to the United States has nearly doubled and is approaching half the produce that is consumed in this country (USDA:ERS, 2007). The phenomenon of year-round access to fresh produce from many parts of the world, including several developing countries in the Southern Hemisphere, has resulted in increased risks of widespread exposure to zoonotic pathogens on produce and an expanded repertoire of pathogens to consider.

REFERENCES

Appelbee, A. J., Thompson, R. C., and Olson, M. E. (2005). *Giardia* and *Cryptosporidium* in mammalian wildlife—Current status and future needs. *Trends Parasitol.* **21,** 370–376.

Besser, R. E., Lett, S. M., Weber, J. T., Doyle, M. P., Barrett, T. J., Wells, J. G., and Griffin, P. M. (1993). An outbreak of diarrhea and hemolytic uremic syndrome from *Escherichia coli* O157:H7 in fresh-pressed apple cider. *JAMA* **269,** 2217–2220.

Beuchat, L. R. (1995). Pathogenic microorganisms associated with fresh produce. *J. Food Protect.* **59,** 204–216.

Beuchat, L. R. and Ryu, J. H. (1997). Produce handling and processing practices. *Emerg. Infect. Dis.* **3,** 459–465.

Brackett, R. (1999). Incidence, contributing factors, and control of bacterial pathogens in produce. *Postharvest Biology and Technology* **15,** 305–311.

Brown, P. E., Christensen, O. F., Clough, H. E., Diggle, P. J., Hart, C. A., Hazel, S. et al. (2004). Frequency and spatial distribution of environmental *Campylobacter* spp. *Appl. Environ. Microbiol.* **70,** 6501–6511.

Burnett, S. L. and Beuchat, L. R. (2001). Human pathogens associated with raw produce and unpasteurized juices, and difficulties in decontamination. *J. Ind. Microbiol. Biotechnol.* **27,** 104–110.

Caldwell, K. N., Adler, B. B., Anderson, G. L., Williams, P. L., and Beuchat, L. R. (2003a). Ingestion of *Salmonella enterica* serotype Poona by a free-living nematode, *Caenorhabditis elegans,* and protection against inactivation by produce sanitizers. *Appl. Environ. Microbiol.* **69,** 4103–4110.

Caldwell, K. N., Anderson, G. L., Williams, P. L., and Beuchat, L. R. (2003b). Attraction of a free-living nematode, *Caenorhabditis elegans*, to foodborne pathogenic bacteria and its potential as a vector of *Salmonella* poona for preharvest contamination of cantaloupe. *J. Food. Prot.* **66,** 1964–1971.

Calvo, M., Carazo, M., Arias, M. L., Chaves, C., Monge, R., and Chinchilla, M. (2004). Prevalence of *Cyclospora sp., Cryptosporidium sp.,* microsporidia and fecal coliform determination in fresh fruit and vegetables consumed in Costa Rica. *Arch. Latinoam. Nutr.* **54,** 428–432.

Casemore, D. P. (1990). Foodborne protozoal infection. *Lancet* **336,** 1427–1432.

CDC. (1997). Foodborne outbreak of cryptosporidiosis Spokane Washington. *Morb. Mort. Wkly. Rpt.* **73,** 353–355.

Cobbold, R. N., Davis, M. A., Rice, D. H., Szymanski, M., Tarr, P. I., Besser, T. E., and Hancock, D. D. (2008). Associations between bovine, human, and raw milk, and beef isolates of non-O157 Shiga toxigenic *Escherichia coli* within a restricted geographic area of the United States. *J. Food Prot.* **71,** 1023–1027.

Cody, S. H., Glynn, M. K., Farrar, J. A., Cairns, K. L., Griffin, P. M., Kobayashi, J. et al. (1999). An outbreak of *Escherichia coli* O157:H7 infection from unpasteurized commercial apple juice. *Ann. Intern. Med.* **130**, 202–209.

Conover, M. R. (2002). *Resolving human-wildlife conflicts: The science of wildlife damage management*. Lewis Publishers, Boca Raton, FL.

Cook, K. L. and Bolster, C. H. (2007). Survival of *Campylobacter jejuni* and *Escherichia coli* in groundwater during prolonged starvation at low temperatures. *J. Appl. Microbiol.* **103**, 573–583.

Dawson, D. (2005). Foodborne protozoan parasites. *Int. J. Food Microbiol.* **103**, 207–227.

FAO/WHO. (2008). Microbiological hazards in fresh fruits and vegetables. Meeting report, WHO Publications.

Frank, C., Kapfhammer, S., Werber, D., Stark, K., and Held, L. (2008). Cattle density and Shiga toxin-producing *Escherichia coli* infection in Germany: Increased risk for most but not all serogroups. *Vector Borne Zoonotic Dis.* **8**, 635–643.

Garcia, H. H., Moro, P. L., and Schantz, P. M. (2007). Zoonotic helminth infections of humans: echinococcosis, cysticercosis and fascioliasis. *Curr. Opin. Infect. Dis.* **20**, 489–494.

Gipson, P. S., Veatch, J. K., Matlack, R. S., and Jones, D. P. (1999). Health status of a recently discovered population of feral swine in Kansas. *J. Wildl. Dis.* **35**, 624–627.

Goens, S. D. and Perdue, M. L. (2004). Hepatitis E viruses in humans and animals. *Anim. Health Res. Rev.* **5**, 145–156.

Gorny, J. R., Gilcas, H., Gombas, D., and Means, K. (2006). Commodity specific food safety guidelines for the lettuce and leafy greens supply chain (1st ed.). International Fresh-Cut Produce Association, Produce Marketing Association, United Fresh Fruit and Vegetable Association and Western Growers.

Graczyk, T. K., Cranfield, M. R., Fayer, R., and Bixler, H. (1999a). House flies (*Musca domestica*) as transport hosts of *Cryptosporidium parvum*. *Am. J. Trop. Med. Hyg.* **61**, 500–504.

Graczyk, T. K., Fayer, R., Cranfield, M. R., Mhangami-Ruwende, B., Knight, R., Trout, J. M., and Bixler, H. (1999b). Filth flies are transport hosts of *Cryptosporidium parvum*. *Emerg. Infect. Dis.* **5**, 726–727.

Graczyk, T. K., Fayer, R., Knight, R., Mhangami-Ruwende, B., Trout, J. M., Da Silva, A. J., and Pieniazek, N. J. (2000). Mechanical transport and transmission of *Cryptosporidium parvum* oocysts by wild filth flies. *Am. J. Trop. Med. Hyg.* **63**, 178–183.

Greene, S. K., Daly, E. R., Talbot, E. A., Demma, L. J., Holzbauer, S., Patel, N. J. et al. (2008). Recurrent multistate outbreak of *Salmonella* Newport associated with tomatoes from contaminated fields, 2005. *Epidemiol. Infect.* **136**, 157–165.

Hancock, D. D., Besser, T. E., Rice, D. H., Ebel, E. D., Herriott, D. E., and Carpenter, L. V. (1998). Multiple sources of *Escherichia coli* O157 in feedlots and dairy farms in the northwestern USA. *Prev. Vet. Med.* **35**, 11–19.

Ho, H. T., Lipman, L. J., and Gaastra, W. (2006). Arcobacter, what is known and unknown about a potential foodborne zoonotic agent! *Vet. Microbiol.* **115**, 1–13.

Hunter, P. R. and Thompson, R. C. (2005). The zoonotic transmission of *Giardia* and *Cryptosporidium*. *Int. J. Parasitol.* **35,** 1181–1190.

Iwasa, M., Makino, S., Asakura, H., Kobori, H., and Morimoto, Y. (1999). Detection of *Escherichia coli* O157:H7 from *Musca domestica* (Diptera: Muscidae) at a cattle farm in Japan. *J. Med. Entomol.* **36,** 108–112.

Janisiewicz, W. J., Conway, W. S., Brown, M. W., Sapers, G. M., Fratamico, P., and Buchanan, R. L. (1999). Fate of *Escherichia coli* O157:H7 on fresh-cut apple tissue and its potential for transmission by fruit flies. *Appl. Environ. Microbiol.* **65,** 1–5.

Jay, M. T., Cooley, M., Carychao, D., Wiscomb, G. W., Sweitzer, R. A., Crawford-Miksza, L. et al. (2007). *Escherichia coli* O157:H7 in feral swine near spinach fields and cattle, central California coast. *Emerg. Infect. Dis.* **13,** 1908–1911.

Jijon, S., Wetzel, A., and LeJeune, J. (2007). *Salmonella enterica* isolated from wildlife at two Ohio rehabilitation centers. *J. Zoo. Wildl. Med.* **38,** 409–413.

Karenlampi, R. and Hanninen, M. L. (2004). Survival of *Campylobacter jejuni* on various fresh produce. *Int. J. Food Microbiol.* **97,** 187–195.

Kenney, S. J., Anderson, G. L., Williams, P. L., Millner, P. D., and Beuchat, L. R. (2005). Persistence of *Escherichia coli* O157:H7, *Salmonella* Newport, and *Salmonella* Poona in the gut of a free-living nematode, *Caenorhabditis elegans*, and transmission to progeny and uninfected nematodes. *Int. J. Food Microbiol.* **101,** 227–236.

Kenney, S. J., Anderson, G. L., Williams, P. L., Millner, P. D., and Beuchat, L. R. (2006). Migration of Caenorhabditis elegans to manure and manure compost and potential for transport of *Salmonella* newport to fruits and vegetables. *Int. J. Food Microbiol.* **106,** 61–68.

Labrousse, A., Chauvet, S., Couillault, C., Kurz, C. L., and Ewbank, J. J. (2000). *Caenorhabditis elegans* is a model host for *Salmonella* typhimurium. *Curr. Biol.* **10,** 1543–1545.

LeJeune, J. T., Besser, T. E., and Hancock, D. D. (2001). Cattle water troughs as reservoirs of *Escherichia coli* O157. *Appl. Environ. Microbiol.* **67,** 3053–3057.

LGMA. (2008). Commodity specific food safety guidelines for the production and harvest of lettuce and leafy greens; June 13, 2008. www.caleafygreens.ca.gov/members/documents/LGMAAcceptedGAPs06.13.08.pdf

Lyautey, E., Hartmann, A., Pagotto, F., Tyler, K., Lapen, D. R., Wilkes, G. et al. (2007). Characteristics and frequency of detection of fecal *Listeria monocytogenes* shed by livestock, wildlife, and humans. *Can. J. Microbiol.* **53,** 1158–1167.

Macpherson, C. N. (2005). Human behaviour and the epidemiology of parasitic zoonoses. *Int. J. Parasitol.* **35,** 1319–1331.

Mansfield, L. S. and Gajadhar, A. A. (2004). *Cyclospora* cayetanensis, a food- and waterborne coccidian parasite. *Vet. Parasitol.* **126,** 73–90.

Mas-Coma, S. (2005). Epidemiology of fascioliasis in human endemic areas. *J. Helminthol.* **79,** 207–216.

Mas-Coma, S., Angles, R., Esteban, J. G., Bargues, M. D., Buchon, P., Franken, M., and Strauss, W. (1999). The Northern Bolivian Altiplano: A region highly endemic for human fascioliasis. *Trop. Med. Int. Health* **4,** 454–467.

Mas-Coma, S., Bargues, M. D., and Valero, M. A. (2005). Fascioliasis and other plant-borne trematode zoonoses. *Int. J. Parasitol.* **35,** 1255–1278.

Mead, P. S., Slutsker, L., Dietz, V., McCaig, L. F., Bresee, J. S., Shapiro, C. et al. (1999). Food-related illness and death in the United States. *Emerg. Infect. Dis.* **5,** 607–625.

Moore, B. C., Martinez, E., Gay, J. M., and Rice, D. H. (2003). Survival of *Salmonella enterica* in freshwater and sediments and transmission by the aquatic midge *Chironomus tentans* (Chironomidae: Diptera). *Appl. Environ. Microbiol.* **69,** 4556–4560.

Murphy, C., Carroll, C., and Jordan, K. N. (2006). Environmental survival mechanisms of the foodborne pathogen *Campylobacter jejuni*. *J. Appl. Microbiol.* **100,** 623–632.

Olsen, A. R. (1998). Regulatory action criteria for filth and other extraneous materials. III. Review of flies and foodborne enteric disease. *Regul. Toxicol. Pharmacol.* **28,** 199–211.

Ortega, Y. R., Roxas, C. R., Gilman, R. H., Miller, N. J., Cabrera, L., Taquiri, C., and Sterling, C. R. (1997). Isolation of *Cryptosporidium parvum* and *Cyclospora cayetanensis* from vegetables collected in markets of an endemic region in Peru. *Am. J. Trop. Med. Hyg.* **57,** 683–686.

Parish, M. E. (1997). Public health and nonpasteurized fruit juices. *Crit. Rev. Microbiol.* **23,** 109–119.

Park, S. F. (2002). The physiology of *Campylobacter* species and its relevance to their role as foodborne pathogens. *Int. J. Food Microbiol.* **74,** 177–188.

Polley, L. (1978). Visceral larva migrans and alveolar hydatid disease. Dangers real or imagined. *Vet. Clin. North Am.* **8,** 353–378.

Polley, L. (2005). Navigating parasite webs and parasite flow: Emerging and re-emerging parasitic zoonoses of wildlife origin. *Int. J. Parasitol.* **35,** 1279–1294.

Renter, D. G., Gnad, D. P., Sargeant, J. M., and Hygnstrom, S. E. (2006). Prevalence and serovars of *Salmonella* in the feces of free-ranging white-tailed deer *(Odocoileus virginianus)* in Nebraska. *J. Wildl. Dis.* **42,** 699–703.

Renter, D. G., Sargeant, J. M., Hygnstorm, S. E., Hoffman, J. D., and Gillespie, J. R. (2001). *Escherichia coli* O157:H7 in free-ranging deer in Nebraska. *J. Wildl. Dis.* **37,** 755–760.

Renter, D. G., Sargeant, J. M., Oberst, R. D., and Samadpour, M. (2003). Diversity, frequency, and persistence of *Escherichia coli* O157 strains from range cattle environments. *Appl. Environ. Microbiol.* **69,** 542–547.

Rice, D. H., Hancock, D. D., and Besser, T. E. (2003). Faecal culture of wild animals for *Escherichia coli* O157:H7. *Vet. Rec.* **152,** 82–83.

Rice, D. H., McMenamin, K. M., Pritchett, L. C., Hancock, D. D., and Besser, T. E. (1999). Genetic subtyping of *Escherichia coli* O157 isolates from 41 Pacific Northwest USA cattle farms. *Epidemiol. Infect.* **122,** 479–484.

Rim, H.-J., Farag, H. F., Sornmani, S., Cross, J. H. (1992). Food-borne trematodes: Ignored or emerging. *Parasitology Today* **10,** 207–209.

Robertson, L. J., Johannessen, G. S., Gjerde, B. K., and Loncarevic, S. (2002). Microbiological analysis of seed sprouts in Norway. *Int. J. Food Microbiol.* **75,** 119–126.

Rondelaud, D., Hourdin, P., Vignoles, P., and Dreyfuss, G. (2005). The contamination of wild watercress with *Fasciola hepatica* in central France depends on the ability of several lymnaeid snails to migrate upstream towards the beds. *Parasitol. Res.* **95,** 305–309.

Rondelaud, D., Vignoles, P., Abrous, M., and Dreyfuss, G. (2001). The definitive and intermediate hosts of Fasciola hepatica in the natural watercress beds in central France. *Parasitol. Res.* **87,** 475–478.

Rose, J. B. and Slifko, T. R. (1999). *Giardia, Cryptosporidium,* and *Cyclospora* and their impact on foods: A review. *J. Food Prot.* **62,** 1059–1070.

Scaife, H. R., Cowan, D., Finney, J., Kinghorn-Perry, S. F., and Crook, B. (2006). Wild rabbits *(Oryctolagus cuniculus)* as potential carriers of verocytotoxin-producing *Escherichia coli. Vet. Rec.* **159,** 175–178.

Schlundt J, T. H., Jansen, J., Herbst, S. A. (2004). Emerging food-borne zoonoses. *Reviews of Science and Technology Office International Epizootes* **23,** 513–533.

Sela, S., Nestel, D., Pinto, R., Nemny-Lavy, E., and Bar-Joseph, M. (2005). Mediterranean fruit fly as a potential vector of bacterial pathogens. *Appl. Environ. Microbiol.* **71,** 4052–4056.

Seymour, I. J. and Appleton, H. (2001). Foodborne viruses and fresh produce. *J. Appl. Microbiol.* **91,** 759–773.

Shayegani, M., Stone, W. B., DeForge, I., Root, T., Parsons, L. M., and Maupin, P. (1986). Yersinia enterocolitica and related species isolated from wildlife in New York State. *Appl. Environ. Microbiol.* **52,** 420–424.

Slifko, T. R., Smith, H. V., and Rose, J. B. (2000). Emerging parasite zoonoses associated with water and food. *Int. J. Parasitol.* **30,** 1379–1393.

Smith, H. V., Caccio, S. M., Cook, N., Nichols, R. A., and Tait, A. (2007). *Cryptosporidium* and *Giardia* as foodborne zoonoses. *Vet. Parasitol.* **149,** 29–40.

Smith, J. L. (2001). A review of hepatitis E virus. *J. Food Prot.* **64,** 572–586.

Tauxe, R. V. (2002). Emerging foodborne pathogens. *Int J Food Microbiol* **78,** 31–41.

Teo, C. G. (2006). Hepatitis E indigenous to economically developed countries: To what extent a zoonosis? *Curr. Opin. Infect. Dis.* **19,** 460–466.

Thurston-Enriquez, J. A., Watt, P., Dowd, S. E., Enriquez, R., Pepper, I. L., and Gerba, C. P. (2002). Detection of protozoan parasites and microsporidia in irrigation waters used for crop production. *J. Food Prot.* **65,** 378–382.

USDA:ERS. (2007). Increased U.S. imports of fresh fruit and vegetables. *Outlook Reports* **FTS-328-01,** 1–20

Van Donkersgoed, J., Berg, J., Potter, A., Hancock, D., Besser, T., Rice, D. et al. (2001). Environmental sources and transmission of *Escherichia coli* O157 in feedlot cattle. *Can. Vet. J.* **42,** 714–720.

Vojdani, J. D., Beuchat, L. R., and Tauxe, R. V. (2008). Juice-associated outbreaks of human illness in the United States, 1995 through 2005. *J. Food Prot.* **71,** 356–364.

Wahlstrom, H., Tysen, E., Olsson Engvall, E., Brandstrom, B., Eriksson, E., Morner, T., and Vagsholm, I. (2003). Survey of *Campylobacter* species, VTEC O157 and *Salmonella* species in Swedish wildlife. *Vet. Rec.* **153,** 74–80.

Wetzel, A. N. and LeJeune, J. T. (2006). Clonal dissemination of *Escherichia coli* O157:H7 subtypes among dairy farms in northeast Ohio. *Appl. Environ. Microbiol.* **72,** 2621–2626.

PART 3

Commodities Associated with Major Outbreaks and Recalls

CHAPTER 8

Leafy Vegetables

Karl R. Matthews

Department of Food Science, School of Environmental and Biological Sciences, Rutgers, The State University of New Jersey

CHAPTER CONTENTS

Introduction	165
Outbreaks Associated with Leafy Greens	166
Growing Conditions by Geographical Region: Link to Outbreaks	168
Harvesting Practices: Influence on Contamination	169
Processing Practices and Product Contamination	171
Handling Prior to Processing	171
Washing and Sanitizing	173
Packaging	175
Interaction of Microbes with Leafy Greens	177
Plant Leaf Characteristic	177
Native Flora of Leafy Greens	178
Microbe Characteristics	179
Influence of Cutting on Microbial Populations	181
Conclusions	181

INTRODUCTION

Foodborne outbreaks from the consumption of contaminated leafy greens are on the rise in the United States and throughout the world. In the United States, analysis of data from 1973 to 2006 indicates that about 5% of foodborne outbreaks were associated with leafy greens (Smith DeWaal and Bhuiya, 2007). Nearly 60% of those outbreaks were caused by norovirus; *Salmonella* and *Escherichia coli* each accounted for about 10% of the outbreaks. Outbreaks associated with *Salmonella* and *E. coli* O157:H7 tend to receive the most attention due to the severity of the illness and occurrence of deaths. Foodborne outbreaks under the heading "leafy greens" encompass more than just those linked to lettuce and spinach. Leafy green products include romaine lettuce, green leaf lettuce, red leaf lettuce, butter lettuce, baby leaf lettuce, escarole, endive, spring mix, spinach, cabbage, kale, arugula, and chard. However, the majority of produce related outbreaks are associated with lettuce.

OUTBREAKS ASSOCIATED WITH LEAFY GREENS

The major produce outbreaks that have occurred in the United States have been associated with bagged leafy greens. In 2005, bagged salad containing three basic ingredients, romaine lettuce, red cabbage, and carrots, were linked to an outbreak of *E. coli* O157:H7 foodborne illness (Smith DeWaal and Bhuiya, 2007). Spinach contaminated with *E. coli* O157:H7 was at the center of a large outbreak that resulted in 205 confirmed illnesses and three deaths in 2006 (Smith DeWaal and Bhuiya, 2007). This outbreak served as a catalyst for the development of more effective measures to ensure that fresh and fresh-cut leafy greens would be safe for human consumption. Small scale (< 50 cases) outbreaks involving *E. coli* O157:H7-contaminated lettuce continue to occur in the United States despite improved production and handling practices. Voluntary recalls of bagged spinach, lettuce, and other leafy greens products have occurred frequently in recent years, perhaps the result of improved routine testing practices by the industry and government agencies.

In Europe, outbreaks have been attributed to both locally produced and imported leafy greens. In 2004, an outbreak of *Salmonella* Thompson infections was reported in Norway, Sweden, and England. These cases were likely all linked to the consumption of contaminated rucola lettuce imported from Italy (Nygard et al., 2008). In 2005, a nationwide outbreak of *S.* Typhimurium var Copenhagen DT104B occurred in Finland due to contaminated lettuce imported from Spain. Approximately 60 confirmed cases were reported (Takkinen et al., 2005). In Sweden, a total of 135 cases, including 11 cases of hemolytic uremic syndrome (HUS), were linked to the consumption of locally produced lettuce that was contaminated with *E. coli* O157. Water samples from a stream used for irrigation were positive for the outbreak strain, as were cattle at a farm upstream of the irrigation point (Soderstrom et al., 2008).

Foodborne illness outbreaks associated with escarole, endive, kale, arugula, and chard are rare, if they have ever occurred. In Canada and the United States, outbreaks of foodborne illness have been traced back to the consumption of cabbage. The causative agents were *E. coli* and *Listeria*.

In each outbreak the cabbage was used in coleslaw. Improper washing of the cabbage prior to use and the use of raw sheep manure in production fields were indicated as causes (Sewell and Farber, 2001).

Surveys designed to evaluate the microbial safety of fresh fruits and vegetables indicate that few samples test positive for foodborne pathogens (Abadia et al., 2008; Arthur et al., 2007; FDA/CFSAN, 2001, 2003). No *E. coli* O157:H7 was isolated from lettuce purchased in retail establishments in Canada, Spain, and the United States (Abadia et al., 2008; Arthur et al., 2007; FDA/CFSAN, 2003). In each of the studies, *Salmonella* was associated with only one lettuce sample. Two lettuce samples from the study conducted in Spain tested positive for *Listeria monocytogenes*. No samples of parsley and cilantro from Canada yielded *Salmonella*, *E. coli* O157:H7, or *Shigella*. These types of survey studies would suggest that the incidence of contamination with any foodborne pathogen is extremely low, and particularly so for *E. coli* O157:H7. Although not often discussed, parasites may also contaminate produce. A study of leafy vegetables in southwestern Saudi Arabia found that 17% of watercress, 17% of lettuce, and 13% of leek were positive for parasites (Al-Binali et al., 2006).

The widespread global contamination of leafy greens with *E. coli* O157 and subsequent outbreaks of foodborne illness are difficult to explain. No specific genes are attributed to *E. coli* O157:H7 that would explain why the pathogen is so intricately associated with leafy greens outbreaks. Research demonstrates the ability of the pathogen to survive for extended periods in water, soil, and manure. The microbe is also capable of surviving shifts in temperature, exposure to sunlight (ultraviolet), moisture, and nutrients. Cattle are considered the primary reservoir of *E. coli* O157:H7. Fecal shedding of the pathogen by domestic and feral pigs, wild birds, deer, and by other domestic livestock and wildlife has been described (Cooley et al., 2007; Jay et al., 2007). The pathogen is not readily killed by exposure to sanitizing washes and can survive in processed leafy greens for extended periods at refrigeration temperatures (Matthews, 2006). Other foodborne pathogens exhibit similar survival characteristics, but for reasons yet unknown, are not associated with outbreaks of foodborne illness linked to the consumption of leafy greens. Intensive research is being conducted to better understand the interaction of *E. coli* O157:H7 and other pathogens with leafy greens. This research should fill data gaps and ultimately aid in protecting the consumer from human illness linked to the consumption of contaminated leafy greens.

The consumer enjoys having foods that are convenient and require minimal preparation prior to use. The sales of bagged leafy greens have exploded since their introduction over 20 years ago. In recent years a number of large outbreaks associated with bagged leafy greens have made consumers question the safety of such products. When an outbreak occurs associated with leafy greens, most consumers first learn of the outbreak through television broadcasts. The large spinach outbreak that occurred in 2006 in the United States serves as an excellent case study on consumer knowledge and attitudes following a large outbreak. The outbreak involved bagged spinach contaminated with *E. coli* O157:H7. A total of 205 cases in 26 states with 103 hospitalizations, 31 with HUS, and three deaths were reported (California Food Emergency Response Team, 2007).

A survey of public response to the recall was conducted by the Food Policy institute at Rutgers University (Cuite et al., 2007). Prior to the outbreak most consumers perceived bagged leafy greens to be safe to eat. Most consumers (~80%) were aware of the recall, but not certain of the types of spinach involved in the recall. Nearly half of the respondents thought that washing contaminated produce would make it safe to eat. Surprisingly, about 13% of Americans who were aware of the spinach recall continued to eat fresh spinach during the recall. This was despite the efforts of the US FDA during the recall to promote its key message that no fresh spinach was considered safe to eat. Consumption of bagged spinach declined dramatically for months following the outbreak. The impact of the outbreak was felt across the produce industry since retail sales of bagged salad without spinach also declined (Calvin, 2007). This change in purchasing habits of fresh leafy greens was associated in part with the information that consumers received in the weeks and months following the outbreak. Consumers may have found it difficult to make informed decisions concerning the safety of leafy greens given the broad information provided by government agencies and experts to the news agencies (Todd et al., 2007).

GROWING CONDITIONS BY GEOGRAPHICAL REGION: LINK TO OUTBREAKS

In the United States leafy vegetable production moves from California to Arizona and Mexico as the seasons change, ensuring a constant supply of product throughout the year. Spain produces approximately half of all commercially grown lettuce in Europe. Differences in soil, climate, and cultivars independently or collectively impact microbial quality of the crops grown in those regions. Rainfall in these regions averages 5 to 15 cm per year. Irrigation is essential to grow crops under such limited rainfall conditions. Contamination can occur in the field by exposure to contaminated irrigation water or floodwaters. A study designed to track *E. coli* O157:H7 in a major produce production region of California suggests that the pathogen, when found in water, is generally close to a point source (Cooley et al., 2007). The authors do point out that in periods of high water-flow, often associated with flooding, the pathogen may be transported over 30 km.

The handling of water used for irrigation can also result in contamination. Water from wells may be pumped into retaining ditches prior to use. Feral animals can contaminate the water, and under the proper environmental conditions, bacterial populations may increase, placing any crops irrigated with the water at risk of being contaminated. Unregulated release of untreated sewage into rivers and streams can result in the contamination of irrigated crops with a range of microorganisms (Okafo et al., 2003). Growers must be aware of Good Agricultural Practices (GAPs) and be willing to implement GAPs to reduce the risk to human health associated with consumption of contaminated leafy greens (Jackson et al., 2007). Specific food safety guidelines have been developed for the production of lettuce and leafy greens (Western Growers, 2008). An important component of those guidelines is the use of nonsynthetic soil amendments.

Soil amendments are commonly used to add organic and inorganic nutrients to the soil. Human pathogens may potentially be present in animals' manures, and if not composted properly or thermally processed, may provide a source of leafy vegetable contamination. Many large organic and conventional leafy vegetable operations use chicken pellets. Studies addressing the microbial safety of chicken pellets are extremely limited. A recent reported indicated that drying the fresh chicken litter/manure at 250 °C eliminated *Salmonella*, fecal streptococci, and enterobacteria (Lopez-Mosquera et al., 2008). Greater attention must be focused on the microbial safety of chicken pellets and other types of soil amendments. Assumptions cannot be made that these products are safe to use simply because they have been through a thermal process. Standard practices with defined time/temperature parameters must be followed to ensure that human pathogens, if present, will be killed.

HARVESTING PRACTICES: INFLUENCE ON CONTAMINATION

The handling of lettuce during and immediately postharvest can have a dramatic effect on microbial safety of the product. Depending on the market and intended utilization for processing, lettuces may be harvested by hand or mechanically. Head lettuces are usually harvested by hand, cored, trimmed (removal of outer leaves), sprayed with a sanitizing wash, bagged, and boxed in the field. Although moving processing to the field likely has distinct economic advantages, the impact on microbial safety has not been examined adequately. Cross-contamination of lettuce through contact with workers' hands (or gloves), knives, automated equipment (conveyor belt), and wash water may occur. The cut end of the lettuce is laden with nutrients that support bacterial growth (Brandl, 2008). Baby lettuce and young lettuce destined for use in bagged salads are machine harvested, dispensed into bins, placed into a refrigerated truck trailer, and then transported to the packing facility. The lettuce may be used immediately or stored, depending on processing practices. The greater the handling and processing of a product, the more pronounced physiological changes will be in that product. Those changes will shorten the product shelf-life and enhance the growth and survival of microbes associated with the product (Aruscavage et al., 2006). Valentin-Bon et al. (2008) reported that the mean total bacterial count of bagged lettuce and spinach samples was 7.0 log CFU/g. Bacterial counts were similar for conventional and organically grown spinach and lettuce mixes. No *E. coli* count exceeded 10 MPN/g; presently, there are no *E. coli* limits for bagged produce in the United States.

Leafy greens other than lettuce and spinach have received little attention. Specialty crops are more likely to be harvested by hand, but are subjected to steps similar to those used for baby lettuce. The behavior of *E. coli* O157:H7 in association with leafy greens and lettuce has received the greatest attention. The leaf age was shown to be a factor in the growth and subsequent population of *Salmonella* and *E. coli* O157:H7 associated with Romaine lettuce (Brandl and Amundson, 2008). Populations of the pathogens were greater on young inner leaves compared to middle and outer leaves (older). A few studies have investigated the fate of *L. monocytogenes* and *Salmonella* on leafy greens (Brandl and Mandrell, 2002; Jablasone et al., 2005; Lapidot et al., 2006). Research demonstrated that the population of *L. monocytogenes* on the surface of parsley, grown under field conditions, declines rapidly within two days (Dreux et al., 2007). The researchers suggested that the risk is minimal unless contamination of aerial surfaces occurs very shortly before harvest. Regardless of the target microbe, specific measures must be put into place to ensure the microbial safety of leafy vegetables during harvest (FDA/CFSAN, 2004, 2007).

PROCESSING PRACTICES AND PRODUCT CONTAMINATION

Handling Prior to Processing

The improper handling of product immediately after harvest can compromise the safety of leafy greens. This is particularly true for baby greens that are harvested into bins for transport to the processing facility. Placing the bins directly onto the soil could result in contaminants contacting the bottom of the bin. The bins are often stacked one on top of the other, permitting the transfer of contaminants from the bottom of one bin to the contents of the bin below. The bins should be placed into a refrigerated truck trailer as rapidly as possible to cool down the product and limit the ability of microorganisms to grow. A large amount of latex is released from cut stems, providing nutrients for the growth of microbes. *E. coli* O157:H7, artificially inoculated onto cut lettuce stems, increased 11-fold over four hours of incubation at 28 °C (Brandl, 2008). Proper refrigeration is imperative to cool the product, thereby limiting or preventing growth of the pathogen.

Cooling crops to 4 °C or less will slow or prevent the growth of pathogens including *Salmonella*, *E. coli* O157:H7, and *L. monocytogenes*. Leafy greens are generally cooled under forced air, but passive storage under refrigeration is still a widely used method. Vacuum cooling is a common practice in the leafy greens industry. However, research suggests that the process can promote the infiltration of *E. coli* O157:H7 into lettuce (Li et al., 2008). Therefore, the cooling process could have a significant impact on microbial safety of leafy greens. Most studies addressing the influence of storage temperature on growth and survival of pathogens in association with leafy greens focus more on retail product rather than bulk loose greens. The populations of *E. coli* O157:H7 and *L. monocytogenes* on iceberg lettuce increased approximately 1.5 to 2.5 log during 12 days' storage at 8 °C (Francis and O'Beirne, 2001). Reducing the storage temperature to 4 °C limited growth of the pathogens; however, viable populations remained at the end of the storage period. Others have also demonstrated the growth of *E. coli* O157:H7, *Salmonella*, and *L. monocytogenes* on iceberg lettuce held at refrigeration temperatures (Koseki and Isobe, 2005). The population of *E. coli* O157:H7 and *Salmonella* on cilantro, oregano, basil, chive, parsley, and rosemary declined by 1.5 log or less after storage at 4 °C for 19 days (Hsu et al., 2006). These results underscore the ability of both bacteria to persist under common storage conditions for leafy greens.

The ability to track a product during its journey from the field to the table is critical to the prevention of human illness. The traceability of a product to its origin will limit the number of cases of illness during an outbreak and decrease the amount of product that must be recalled. Until recently, the produce industry has not had in place a suitable system for tracking of produce. Difficulties in traceability are associated with the commingling of product from different fields into the same bag. Most systems use barcodes to trace product as it travels from the field to the consumer. There are many disadvantages to the use of barcodes since they are affixed to the outside and are subject to wear and scratching. A direct line-of-sight is required to read the code, and once on the package, the code cannot be changed to reflect changes in product profile or handling. The use of a radio frequency identification device (RFID) would permit the tracking of leafy greens from the field to the retail level. The RFID tags can be continuously written to with information, and read remotely. The RFID system can be monitored via the Internet, and the technology can be used by large and small operations alike.

There exists the possibility that in the future, consumers would be able to access the information about a particular product at the retail level. The Hawaii Farm Bureau Federation and the Hawaii State Department of Agriculture are testing an RFID system for tracking produce grown by some of Hawaii's large produce suppliers (Anon., 2008). Ultimately, alerts can be made throughout the supply chain when and where a food product is not being handled appropriately. A seal will be available to fresh produce buyers, providing assurance that the product purchased is directly traceable to a particular Hawaiian farm or processor. The technology exists so that in the future, temperature and humidity information can be tracked to determine if food products are transported and stored under appropriate conditions. The RFID system can be integrated into Good Manufacturing Practices and Hazard Analysis Critical Control Point systems. In Sweden, a large supplier of pallets to the produce industry uses RFID to track pallets. Proper washing and sanitizing of the pallets can be monitored using RFID, ensuring that pallets potentially contaminated with human pathogens are not used in the harvest, storage, or transport of fresh produce.

The microbial safety of leafy greens is paramount to human health. Microbiological testing of water, soil amendments, equipment (field and processing), and commodity must be conducted to comply with GMPs, HACCP programs, buyer agreements, and other food safety guidelines. In some instances, quantitative microbiological analysis is required for water that is intended for irrigation (Western Growers, 2008). A decision tree was developed whereby crops directly contacted with water exceeding accepted microbiological criteria (most probable number testing for generic

E. coli) would be sampled and tested for *E. coli* O157:H7 and *Salmonella*. Microbiological screening is also being implemented prior to product processing. Many large grower/packers are now practicing test-and-hold programs. In general, palletized harvested product is identified at delivery to the processor. The product is sampled and held in a designated area of the warehouse until microbiological testing can be completed. Typically, product is tested for *E. coli* O157:H7 and *Salmonella*; more recently, processors are testing for *L. monocytogenes*. The testing methods used must be completed within 12 hours; a longer period would require large holding structures and limit production capabilities. PCR-based methods are best suited for this purpose since they are rapid, specific, and cost effective. A fairly comprehensive list of rapid methods for use with water and food samples has been compiled (www.wga.com).

Washing and Sanitizing

Equipment Design

A wealth of literature exists on the efficacy of various wash/sanitizing systems to remove or inactivate microbes associated with leafy greens (reviewed in Doyle, 2005). Chemical sanitizers generally provide a 1- to 2-log reduction in viable bacteria associated with leafy greens. Changes in equipment and processing strategies are required to enhance microbial reduction. Prior to bagging, leafy greens are dumped into water flumes containing sanitizing water. The total exposure time may range from 60 to 120 seconds. Leaves may float on the surface, a phenomenon often referred to as "lily padding," which limits exposure of the entire leaf to the antimicrobial contained in the water. Equipment that effectively submerges all product upon entry into the flume system is now available. Other systems use bends in the flume to redirect product, causing it to become submersed, or air jets that agitate the water and effectively cause leaves floating on the surface to become submerged. Although every processor may have different procedures, leafy greens often go through a triple-wash process. The first wash occurs when product goes into an agitating tank containing weakly chlorinated water. This step is intended to remove gross physical debris (bugs, soil, stones). The major cleaning occurs in a second tank or flume containing chlorinated water. Although bacteria associated with the surface of a product may be killed, the chlorine in the water is intended to control bacterial numbers in the wash water and associated with equipment. The final wash is actually a rinse step that is intended to remove residual chlorine from the product.

Equipment used for conveying product, removing excess water, and for packaging must be user friendly and facilitate ease of cleaning. Cleaning of equipment during shift breaks will minimize the build-up of organic matter on equipment and cross-contamination of product. Microorganisms can rapidly build up on equipment and effectively "inoculate" product during processing. This may result in a shortened shelf-life and increased risk to consumer safety.

The present methods used to wash and sanitize leafy greens will inevitably result in cross-contamination should a few leaves be contaminated initially. Research clearly shows that levels of foodborne pathogens on leafy greens may be only minimally reduced following washing in water containing sanitizing agents such as chlorine. Research demonstrates that during washing, *E. coli* O157:H7 could be transferred between lettuce pieces (Shafaie and Matthews, 2005; Wachtel and Charkowski, 2002). Additional studies showed that lettuce pieces washed in water contaminated with *E. coli* O157:H7 (5.5 log CFU/ml) had populations of *E. coli* O157:H7 of 3.4 log CFU/g (Shafaie and Matthews, 2005). These studies demonstrate the potential for even a few contaminated leaves to disseminate a pathogen to a large mass of leafy greens, exacerbating the magnitude of an outbreak. As discussed, bins of a given commodity (spinach, red lettuce, arugula) are dumped into a hopper and conveyed to a wash tank or flume, ideal conditions for direct contact of leaves and equipment.

Sanitizing Agents

At present chlorine is the primary postharvest sanitizing agent in use by the fresh produce industry. Under the conditions studied, most sanitizers have been able to produce no more than a 1-2 log CFU reduction in pathogen levels (Doyle, 2005). This was the situation more than five years ago (Parish et al., 2003), and remains true today (Matthews, 2006). Indeed, leafy vegetables processors recognize that water disinfectants are used to prevent cross-contamination, and not to surface-sanitize produce. However, such processes do impact microbial populations on the product. Sanitizing/disinfecting agents other than chlorine must be more aggressively investigated.

New technologies being applied to pathogen reduction of leafy greens include electrolytic oxidizing water and ultrasound alone or in combination with a chemical sanitizer. Acid electrolyzed water (AEW) and neutral electrolyzed water (NEW) have been studied as alternative sanitizers. Electrolyzed water is generated through the electrolysis of a dilute NaCl solution. AEW only passes through the anode chamber and has a strong bactericidal effect due to its low pH (2-4), high oxidation-reduction potential (ORP > 1000 mV), and content of hypochlorous acid. NEW is generated by passing the NaCl solution through the anode and cathode chambers of a membrane electrolyzer. NEW is a near-neutral (pH 8 ± 0.5) solution in which the main bactericidal agents are HOCl, ClO⁻, and HO_2 radical. The potential acceptance of NEW by the leafy greens industry is greater since it would be less corrosive to equipment and cause less irritation to hands. Its effectiveness in reducing *E. coli* O157:H7, *Salmonella*, and *Listeria* populations associated with lettuce and other leafy greens was similar to that of chlorine, with reductions of 1–2 log units (Abadias et al., 2008). The advantage was that NEW, containing about 50 ppm free chlorine, was as effective as applying chlorinated water at 120 ppm free chlorine.

Research associated with the application of ultrasound for the microbial decontamination of produce has been revived. The action of ultrasound is associated with the formation of cavitation bubbles. The movement of these bubbles generates the mechanical energy to remove microbes from a surface. The use of ultrasound in combination with chlorinated water reduced levels of *Salmonella* associated with iceberg lettuce by 2.7 log (Seymour et al., 2002). However, the authors concluded that the high capital cost and expensive process of optimization and water treatment would preclude its use by the fresh produce industry. The development of new equipment that can be retrofitted on existing flume and wash tanks may make the method more appealing from a cost standpoint.

A novel control method that has been receiving interest in recent years is the use of bacteriophage. Researchers have investigated the use of bacteriophages to reduce *S. Enteritidis* and *L. monocytogenes* associated with honeydew melons (Leverentz et al., 2001, 2003). The application of bacteriophages for the control of foodborne pathogens is an emerging area of research (Hagen and Loessner, 2007). The technology could be applied to the control of foodborne pathogens associated with leafy greens. Bacteriophages are usually very host-specific. Many can infect only one bacterial genus, and others infect only one serotype within a bacterial species. Therefore, bacteriophages that specifically attack *E. coli* O157:H7 could be applied during postharvest processing. Public acceptance of the application of bacteriophage to fresh leafy greens as a biocontrol agent may be met with resistance. Consumers may perceive bacteriophage as dangerous, potentially capable of causing human illness.

Packaging

A multitude of measures must be employed to ensure the microbial safety of leafy greens. Prolonged exposure of leafy greens to biocidal compounds during transport and retail sale may aid in killing or preventing the growth of foodborne pathogens. The use of novel packaging materials, packaging design, and package inserts may provide an added dimension to strategies for the control of microbes associated with leafy vegetables.

Treatment of inoculated lettuce leaves with ClO_2 gas, generated from dry chemical sachets, for 30 minutes resulted in log reductions of 3.4, 4.3, and 5.0 for *E. coli* O157:H7, *S.* Typhimurium, and *L. monocytogenes*, respectively; treatment for one hour resulted in log reductions of 4.4, 5.3, and 5.2, respectively (Lee et al., 2004). No treatment-induced quality defects were reported. However, the large population reductions reported in this study may reflect the brief time interval (30 minutes) between inoculation and treatment, which might have been insufficient to allow for aggregate formation and internalization of the targeted pathogens, conditions expected to enhance their survival.

Application of ClO_2 gas treatments to green peppers, inoculated with *L. monocytogenes*, resulted in population reductions of 6 logs on uninjured surfaces and 3.5 logs on injured surfaces (Han et al., 2001). Researchers evaluated the efficacy of gaseous chlorine dioxide in killing *E. coli* O157:H7, *Salmonella*, and *L. monocytogenes* on various fresh-cut commodities including cabbage and lettuce. They reported log reductions of 3.13 to 4.42 for cabbage but only 1.53 to 1.58 for lettuce (Sy et al., 2005). It is not clear why population reductions with lettuce were relatively small. The most likely explanations are internalization at cut surfaces and interference from a film of condensate on stomata and other leaf surfaces under the high humidity conditions of the treatment cabinet (ca. 70–80%).

It may be possible to develop packaging that permits extended uniform exposure of leafy greens to ClO_2 gas while in the package. This would mitigate or prevent the development of adverse quality issues such as browning, often associated with high concentration-short period exposure. Treatment of lettuce and cabbage with ClO_2 gas prior to bagging may actually result in high bacteria populations during storage (Gomez-Lopez et. al., 2008). The ClO_2 gas treatment would reduce populations of all bacteria but might allow for certain types of bacteria to rapidly increase without the competition for nutrients. Another potential concern with the use of ClO_2 gas is browning.

Within the past five years new natural, biodegradable, and edible packaging films have been studied that inhibit the growth of foodborne pathogens. An antimicrobial film effective against *E. coli* and *Salmonella* was shown to extend the shelf-life of iceberg lettuce held at 10 °C by five days (Kang et al., 2007). The researchers conducted additional studies that extended by an additional two days the shelf-life of fresh-cut iceberg lettuce packaged in the antimicrobial film under modified atmosphere (MAP) conditions (95% N_2, 2.6% O_2, and 2.4% CO_2; Kang et al., 2008). The atmosphere conditions can independently inhibit spoilage and foodborne pathogenic bacteria, thereby prolonging shelf-life and increasing microbial food safety (Fonseca, 2006). The possibility exists to combine in-package ClO_2 gas treatment with an antimicrobial packaging film to injure or kill foodborne pathogens present on the product and prevent recovery or growth of survivors.

Novel types of packaging and edible films are being tested that will potentially extend product shelf-life, reduce risk of pathogen growth on contaminated food surfaces, and have consumer acceptance. Carvacrol, a major constituent of oregano with antimicrobial activity, has been incorporated into edible tomato and apple films (Du et al., 2008a, 2008b). The antimicrobial films were produced by combining carvacrol with tomato or apple purees. Both films exhibited antibacterial activity against *E. coli* O157:H7. Antimicrobial films made from plants may have wider acceptance than those developed using chemical compounds. Such products could be used in bagged salad mixes where potential flavor notes may be considered acceptable, and add to the appeal of the product.

The antimicrobial agent nisin has been incorporated into nonedible films. Nisin is a broad-spectrum bacteriocin, produced by lactic acid bacteria, that binds to and forms pores in the cell membranes of gram-positive bacteria. Nisin-coated plastic films were shown to reduce levels of *L. monocytogenes* on vacuum-packaged cold smoked salmon (Neetoo et al., 2008). Redesign of packaging to include inserts or bridges increasing the contact area of product with the antimicrobial packaging material would enhance utility of the method.

In the future, these technologies have the potential to be applied in the control of foodborne pathogens associated with produce. Nonconventional technologies must be considered since there exists no single method that will effectively eliminate foodborne pathogens that may be associated with leafy greens.

INTERACTION OF MICROBES WITH LEAFY GREENS

The interaction of human enteric pathogens with plants has been extensively reviewed (Brandl, 2006; Solomon et al., 2006; Aruscavage et al., 2006; Delaquis et al., 2007; Heaton and Jones, 2008). The conclusions that can be drawn from those reviews suggest that researchers have a limited understanding of the behavior of enteric pathogens in the phylosphere. Greater effort must be made to understand plant microbe interactions and the interaction between enteric pathogens and epiphytic microbes (yeast, molds, bacteria).

Plant Leaf Characteristic

The plant leafy surface is not an ideal environment for the survival of enteric pathogens. The leaf surface is exposed to ultraviolet light, shifts in temperature and relative humidity, and presence of available moisture (presence or absence of rain/irrigation). A range of epiphytic microbes are also present, impeding the ability of enteric bacteria to colonize the leaf surface. The availability of carbon and nitrogen sources leached from the plant will also influence survival and growth. The most common carbon sources available on the surface of plants are glucose, sucrose, and fructose (Mercier and Lindow, 2000). Leakage of nutrients from the plant can occur following damage to the cuticle. The leaf environment is not uniform with respect to conditions that would support survival and growth of bacteria. Epiphytic bacteria preferentially colonize the base of trichomes, around the stomata, and along veins in the leaves. Foodborne pathogens were shown to localize near leaf veins. The reason for this may be related to the greater wettability of the area, increasing nutrient leaching and water availability (Brandl and Mandrell, 2002). Leaf age may be an important factor in the growth and survival of epiphytic and opportunistic bacteria. The exudate from young lettuce leaves was reported to be 2.9 and 1.5 times richer in total nitrogen and carbon, respectively, than exudates from middle leaves (older leaves) (Brandl and Amundson, 2008).

Native Flora of Leafy Greens

The community of microbes on the plant leaf surface is complex and includes many species of bacteria, yeasts, and molds. Populations of aerobic bacteria on leafy greens may average 10^5 to 10^6 CFU/g of leaf tissue (Aruscavage et al., 2006). The Gram-negative bacteria are the most predominant group of epiphytic microorganisms found in the plant phylosphere. Microbes on the plant leaf surface form biofilms, whereby cells are encased in an exopolysaccharide matrix that provides protection against adverse environmental conditions. Early research showed that bacteria in biofilms comprise between 10 and 40% of the total bacterial population on leafy greens (parsley and endive) (Morris et al., 1998). Microbial biofilms on leaves may influence the attachment, growth, and survival of enteric pathogens on the leaf surface. Researchers demonstrated that a biofilm-deficient mutant of *S. Typhimurium* persisted on parsley at elevated levels. They suggested that the microbe may have been able to penetrate the preexisting biofilm and therefore persist at levels similar to the parent strain (Lapidot at al., 2006). A more comprehensive discussion of biofilms can be found in Chapter 2.

Interaction of enteric pathogens with the leaf surface may also benefit from the action of phytopathogenic bacteria. Phytopathogens such as *Erwinia* can degrade plant tissue, providing enteric pathogens with a broadened spectrum of nutrients for growth (Agrios, 1997). Whether an enteric pathogen can colonize a leaf surface may be influenced by the native flora found on the leaf and characteristics of the enteric microbe.

Microbe Characteristics

Enteric bacteria have a vast array of cell surface moieties that may influence the ability of the cell to interact with plant tissue. Relatively few reports on the intimate interaction of foodborne pathogens with leafy greens have been published. Specific moieties or factors involved in attachment/interaction include exocellular polysaccharide, cell surface charge, presence/absence of fimbriae, and hydrophobicity. The presence of curli appeared to have no influence on attachment of *E. coli* O157:H7 to lettuce (Boyer et al., 2007). Curli are coiled extracellular structures on the cell surface that bind fibronectin and other proteins facilitating adherence to and invasion of epithelial cells (Chapman et al., 2002). An exhaustive study involving the screening of a mutant library of *L. monocytogenes* and identification of three mutants that had reduced adherence to radish tissue suggested that temperature may affect expression of attachment factors used by *L. monocytogenes* (Gorski et al., 2003).

Solomon and Matthews (2006) reported that gene expression and bacterial processes, such as motility or production of extracellular compounds, were not required for initial attachment of *E. coli* O157:H7 to lettuce. In the study, live and glutaraldehyde-killed *E. coli* O157:H7 and fluorescent polystyrene microspheres were used in experiments to investigate interactions with lettuce. The microspheres are comparable in size to single bacterial cells, yet free of any surface moieties or appendages that have been hypothesized to be involved in attachment. The role of gene expression and cell surface moieties are likely to be important in further colonization and survival on the leaf. The role of exopolysaccharide production by enteric pathogens to facilitate intimate adherence to plants has not been sufficiently explored. Research suggests that the ability to produce biofilm does not play a significant role in initial adhesion of *Salmonella* to lettuce and survival after disinfection (Lapidot et al., 2006). Notwithstanding the role in adherence, exopolysaccharides protect bacteria from the action of antimicrobial agents and desiccation, and may aid in the integration of the enteric bacteria into the epiphytic biofilm.

A comparison of the genomes of the plant pathogen *Erwinia cartovora subsp. atroseptica* and the human pathogen *S. enterica* Typhi suggests the genomes are similar, but that genes associated with plant interactions are absent in *S.* Typhi (Toth et al., 2006). However, gene expression by enteric pathogens associated with plants may influence survival on the plant and enhance virulence within a human host. The plant pathogen *Erwinia chrysanthemi* induces the *sap* operon to provide resistance to plant antimicrobial compounds and enhance its virulence. The *sap* operon may also be induced in *S. enterica* perhaps providing resistance to plant antimicrobials and providing enhanced virulence in a human host (Taylor, 1998). Expression of the type III secretion system (TTSS) by *Salmonella* when in association with plants may actually be detrimental to survival of the microbe in association with the plant (Iniguez et al., 2005). The type III secretion system enables delivery of pathogenicity proteins to host cells; the system is found in Gram-negative bacteria and animal pathogens (Hueck, 1998). Colonization studies of *Arabidopsis thaliana* suggest that the plant recognizes components of the TTSS as part of its defense system (Iniguez et al., 2005).

Internalization

The localization of enteric pathogens at subsurface sites on leafy green plant tissue impedes their removal during washing and the ability of sanitizers to inactivate the microbe. Bacteria may gain access to interior regions of a leaf through stomata, abrasions or cuts, action of plant pathogens, and through the root system. Takeuchi and Frank (2000) reported that *E. coli* O157:H7 localized within stomata of lettuce were protected from sanitation with chlorine. *E. coli* were found within the roots, hypocotyls, and cotyledons of soil-grown plants inoculated with contaminated irrigation water (Wachtel et al. 2002). Solomon et al. (2002) reported the internalization of *E. coli* O157:H7 into edible tissue of lettuce, detected by laser scanning confocal microscopy, through root associated uptake of the pathogen. Internalized cells were detected in plants exposed to 10^8 CFU *E. coli* O157:H7, but not to 10^4 CFU *E. coli* O157:H7. These results are supported by a recent study that showed exposure of lettuce to 10^4 or less CFU *E. coli* O157:H7 in irrigation water, soil, or manure resulted in only a limited number of plants having internalized target bacteria (Mootian et al., 2008). The entry point(s) (roots, stomata, wound) for the pathogen was not determined.

In the environment, the levels of *E. coli* O157:H7 and other enteric pathogens are likely to be extremely low (Cooley et al., 2007). Therefore, mechanistically, enteric pathogens may be internalized via the root system and transported to edible tissue, but the risk of contamination by this route is likely low. The ability of *E. coli* O157:H7 to become internalized in tissues of leafy greens is supported (Franz et al., 2007) and contradicted (Hora et al., 2005; Warriner et al., 2003) in other published reports.

Chapter 2 contains a more in-depth review of internalization of enteric foodborne pathogens in leafy greens and other fruits and vegetables.

INFLUENCE OF CUTTING ON MICROBIAL POPULATIONS

The action of harvesting and processing of leafy greens inherently results in the release of plant exudate along cut edges. *E. coli* O157:H7 were found to attach preferentially to the cut edges of lettuce leaves (Takeuchi and Frank, 2000). *Salmonella* and *Shigella* were reported to grow more rapidly and reach higher populations on chopped leaves of cilantro and parsley, respectively, compared to whole leaves (Campbell et al., 2001; Wu et al., 2000). Brandl et al. (2004) reported that despite the presence of suitable substrates on lettuce and spinach leaves, *Campylobacter jejuni* was unable to grow. Following large foodborne illness outbreaks associated with the consumption of *E. coli* O157:H7 contaminated lettuce and spinach, greater focus has been placed on the growth of the pathogen on intact and damaged leaves. *E. coli* O157:H7 populations increased 4, 4.5, and 11-fold within four hours on romaine lettuce that received mechanical, physiological, and disease-induced lesions, respectively (Brandl 2008). A two-fold increase in *E. coli* O157:H7 populations occurred on leaves that were left intact. The influence of leaf age was also investigated; the *E. coli* O157:H7 population on young leaves was 27-fold greater than on middle-aged leaves. The study underscores the potential for low numbers of a pathogen to rapidly increase to levels that present a significant human health concern.

The growth of native microflora and foodborne pathogens during storage can dramatically impact shelf-life and safety of a product. Microbial populations on leafy greens increased significantly on product held at temperatures considered abusive for chilled foods. Studies investigating the behavior of *E. coli* O157:H7 on lettuce held at 5 to 22 °C showed that at temperatures at or greater than 8 °C, the pathogen population increased during storage (Delaquis et al., 2007). In bagged lettuce held at or less than 5 °C, the population of *E. coli* O157:H7 remained unchanged or decreased. Maintaining the cold chain from the packer through the consumer is essential to prevent the growth of foodborne pathogens that may be present in low numbers.

CONCLUSIONS

Improvement in the microbiological safety of leafy greens is dependent on gaining a greater understanding of the interaction of enteric foodborne pathogens with plant tissue. In broad terms this knowledge will aid in the development of methods for detection and identification of target microbes, facilitate the development of novel strategies for reducing or eliminating target pathogens, and permit integration of Good Agricultural Practices, Good Manufacturing Practices, and Hazard Analysis Critical Control Point programs to ensure that leafy green commodities are handled appropriately from the field to retail establishment to the consumers' home. Great controversy exists concerning routes (uptake by roots, stomata, damage to tissue) by which enteric bacteria may become internalized.

Regardless of the entry point, research clearly shows that enteric bacteria can be found at subsurface locations. New technologies must be sought that will effectively kill internalized bacteria. Research efforts must build upon the information that is presently available and encourage thinking "outside the box." Experiments need to be conducted that better reflect conditions encountered in the real world and that recognize restrictions associated with manufacturing operations. Basic surface sanitizing strategies fail to consistently reduce levels of target pathogens by greater than 2 log. New technologies using bacteriophage, package inserts for delivery of antimicrobial agents, and novel natural antimicrobials must be considered. Greater emphasis should be given to the utilization of existing technologies and the development of new technologies for tracking product from the field to retail establishments. This would facilitate more rapid and precise recalls, potentially reducing the number of human illnesses and limiting monetary loss to an entire industry, the result of blanket recalls of a given commodity.

REFERENCES

Abadias, M., Usall, J., Anguera, M., Solsona, C., and Vinas, I. (2008). Microbiological quality of fresh, minimally-processed fruit and vegetables, and sprouts from retail establishments. *Inter. J. Food Microbiol.* **123,** 121–129.

Abadias, M., Usall, J., Anguera, M., Solsona, C., and Vinas, I. (2008). Efficacy of neutral electrolyzed water (NEW) for reducing microbial contamination on minimally-processed vegetables. *Inter. J. Food Microbiol.* **123,** 151–158.

Agrios, G. N. (1997). *Plant pathology*. Fourth edition. Academic Press. San Diego, CA.

Al-Binali, A. M., Bello, C. S., El-Shewy, K., and Abdulla, S. E. (2006). The prevalence of parasites in commonly used leafy vegetables in South Western, Saudi Arabia. *Saudi. Med. J.* **27,** 613–616.

Allende, A., McEvoy, J. L., Luo, Y., Artes, F., and Wang, C. Y. (2006). Effectiveness of two-sided UV-C treatments in inhibiting natural microflora and extending the shelf life of minimally processed "Red Oak Leaf" lettuce. *Food Microbiol.* **23,** 241–249.

Anon. (2008). USDA approves $50,000 grant for state food safety program. NR07–17. http://hawaii.gov/hdoa/news/2007-news-releases/news-release-nr07-17-august-15-2007/?searchterm=RFID

Arthur, L., Jones, S., Fabri, M., and Odumeru, J. (2007). Microbiological survey of selected Ontario-grown fresh fruits and vegetables. *J. Food Prot.* **70,** 2864–2867.

Aruscavage, D., Lee, K., Miller, S., and LeJune, J. T. (2006). Interactions affecting the proliferation and control of human pathogens on edible plants. *J. Food Sci.* **71,** 89–99.

Bari, M. L., Nazuka, E., Sabina, Y., Todoriki, S., and Isshiki, K. (2003). Chemical and irradiation treatments for killing *Escherichia coli* O157:H7 on alfalfa, radish, and mung bean seeds. *J. Food Prot.* **66,** 767–774.

Beltran, D., Selma, M. V., Mrin, A., and Gil, G. I. (2005). Ozonated water extends the shelf life of fresh-cut lettuce. *J. Agric. Food Chem.* **53,** 5654–5653.

Bernstein, N., Sela, S., Pinto, R., and Ioffe, M. (2007). Evidence for internalization of *Escherichia coli* into the aerial parts of maize via the root system. *J. Food Prot.* **70,** 471–475.

Boyer, R. R., Sumner, S. S., Williams, R. C., Pierson, M. D., Popham, D. L., and Kniel, K. E. (2007). Influence of curli expression by *Escherichia coli* O157:h7 on the cell's overall hydrophobicity, charge, and ability to attach to lettuce. *J. Food Prot.* **70,** 1339–1345.

Brandl, M. T. (2006). Fitness of human enteric pathogens on plants and implication for food safety. *Annu. Rev. Phytopathol.* **44,** 367–392.

Brandl, M. T. (2008). Plant lesions promote the rapid multiplication of *Escherichia coli* O157:H7 on postharvest lettuce. *Appl. Envron. Micrbiol.* **74,** 5285–5289.

Brandl, M. T., Haxo, A. F., Bates, A. H., and Mandrell, R. E. (2004). Comparison of survival of *Campylobacter jejuni* in the phylosphere with that in the rhizosphere of spinach and radish plants. *Appl. Environ. Microbiol.* **70,** 1182–1189.

Brandl, M. T. and Amundson, R. (2008). Leaf age as a risk factor in contamination of lettuce with *Escherichia coli* O157:H7 and *Salmonella enterica*. *Appl. Environ. Microbiol.* **74,** 2298–2306.

Brandl, M. T. and Mandrell, R. E. (2002). Fitness of *Salmonella enterica* serovar Thompson in the cilantro phyllosphere. *Appl. Envron. Microbiol.* **68,** 3614–3621.

California Food Emergency Response Team. (2007). Investigation of an *Escherichia coli* O157:H7 outbreak associated with Dole pre-packaged spinach. Final March 21, 2007. California Dept. of Health Services, Sacramento, CA, and the US Food and Drug Admin., Almeda, CA.

Calvin, L. (2007). Outbreak linked to spinach forces reassessment of food safety practices. Economic Research Service USDA. *Amber Waves* **5,** 24–31. www.ers.usda.gov/amberwaves

Campbell, V. J., Mohle-Boetani, J., Reporter, R., Abbott, S., Farrar, J., Brandl, M. T. et al. (2001). An outbreak of *Salmonella* serotype Thompson associated with fresh cilantro. *J. Infect. Dis.* **183,** 984–987.

Chapman, M. R., Robinson, L. S., Pinkner, J. S., Roth, R., Heuser, J., Hammer, M. et al. (2002). Role of *Escherichia coli* curli operons in directing amyloid fiber formation. *Science.* **295,** 851–855.

Chua, D., Goh, K., Saftner, R. A., and Bhagwat, A. A. (2008). Fresh-cut lettuce in modified atmosphere packages stored at improper temperatures supports enterohemorrhagic *E. coli* isolates to survive gastric acid challenge. *J. Food Sci.* **73**, 148–153.

Cooley, M., Carychao, D., Crawford-Miksza, L., Jay, M. T., Myers, C., Rose, C. et al. (2007). Incidence and tracking of *Escherichia coli* O157:H7 in a major produce production region in California. *Plosone.* **11**, e1159. www.plosone.org

Cuite, C. L., Condry, S. C., Nucci, M. L., and Hallman, W. K. (2007). Public response to the contaminated spinach recall of 2006. Publication number RR-0107-013. www.foodpolicyinstitute.org

Delaquis, P., Bach, S. and Dinu, L-D. (2007). Behavior of *Escherichia coli* O157:H7 in leafy vegetables. *J. Food Prot.* **70**, 1966–1974.

Doyle, M. E. (2005). Food antimicrobials, cleaners, and sanitizers. Food Research Institute, UW-Madison, *FRI Briefings* **9**, 1–15.

Dreuz, N., Albagnac, C., Carlin, F., Morris, C. E., and Nguyen-The, C. (2007). Fate of *Listeria* spp. on parsley leaves grown in laboratory and field cultures. *J. Appl. Microbiol.* **103**, 1821–1827.

Du, W. X., Olsen, C. W., Avena-Bustillos, R. J., McHugh, T. H., Levin, E. E., and Friedman, M. (2008a). Antibacterial activity against *E. coli* O157:H7, physical properties, and storage stability of novel carvacrol-containing edible tomato films. *J. Food Sci.* **73**, 378–383.

Du, W. X., Olsen, C. W., Avena-Bustillos, R. J., McHugh, T. H., Levin, E. E., and Friedman, M. (2008b). Storage stability and antibacterial activity against *Escherichia coli* O157:H7 of carvacrol in edible apple films made by two different casting methods. *J. Agric. Food Chem.* **56**, 3082–3088.

FDA (Food and Drug Administration, USA). (2001). FDA survey of imported fresh produce. www.cfsan.fda.gov/~dms/prodsur6.html

FDA. (2001). FDA survey of imported fresh produce. www.cfsan.fda.gov/~dms/prodsu10.html

FDA/CFSAN. (2004). Produce safety from production to consumption: A proposed action plan to minimize foodborne illness associated with fresh produce consumption. www.foodsafety.gov./~dms/prodplan.html

FDA/CFSAN. (2007). Guide to minimize microbial food safety hazards of fresh-cut fruits and vegetables. www.cfsan.fda.gov/~dms/prodgui3.html

Francis, G. A. and O'Briene, D. (2001). Effects of vegetable type, package atmosphere and storage temperature on growth and survival of *Escherichia coli* O157:H7 and *Listeria monocytogenes*. *J. Ind. Micro. Biotech.* **27**, 111–116.

Franz, E., Visser, A. A., Van Diepeningen, A. D., Klerks, M. M., Termorshuizen, A. J., and van Bruggen, A. H.C. (2007). Quantification of contamination of lettuce by GFP-expressing *Escherichia coli* O157:H7 and *Salmonella enterica* server Typhimurium. *Food Microbiol.* **24**, 106–112.

Fonseca, J. M. (2006). Postharvest handling and processing: Sources of microorganisms and impact of sanitizing procedures. In *Microbiology of fresh produce* (Matthews, K. R., Ed.), pp.85–120. Washington, DC, Amer. Soc. Microbiol.

Gomez-Lopez, V. M., Ragaert, P., Jeyachchandran, V., Debevere, J., and Devlieghere, F. (2008). Shelf-life of minimally processed lettuce and cabbage treated with gaseous chlorine dioxide and cysteine. *Inter. J. Food Microbiol.* **121,** 74–83.

Gorski, L., Palumbo, J. D., and Mandrell, R. E. (2003). Attachment of *Listeria monocytogenes* to radish tissue is dependent upon temperature and flagellar motility. *Appl. Envron. Microbiol.* **69,** 258–266.

Hagens, S. and M. J. Loessner. (2007). Application of bacteriophages for detection and control of foodborne pathogens. *Appl. Microbiol. Biotech.* **76,** 513–519.

Han, Y., Linton, R. H., Nelsen, S. S., and Nelson, P. E. (2001). Reduction of *Listeria monocytogenes* on green peppers (*Capsicum annuum* L) by gaseous and aqueous chlorine dioxide and water washing and its growth at 7°C. *J. Food Prot.* **64,** 1730–1738.

Heaton, J. C. and Jones, K. (2008). Microbial contamination of fruit and vegetables and the behaviour of enteropathogens in the phylosphere: a review. *J. Appl. Micro.* **104,** 613–626.

Hsu, W-Y., Simonne, A., and Jitareerat, P. (2006). Fates of seeded *Escherichia coli* O157:H7 and *Salmonella* on selected fresh culinary herbs during refrigerated storage. *J. Food Prot.* **69,** 1997–2001.

Hueck, C. L. (1998). Type III secretions systems in bacterial pathogens of animals and plants. *Microbiol. Rev.* **62,** 379–433.

Jablasone, J., Warrnier, K., and Griffiths, M. (2005). Interactions of *Escherichia coli* O157:H7, *Salmonella* Typhimurium and *L. monocytogenes* plants cultivated in a gnotobiotic system. *Inter. J. Food Microbiol.* **99,** 7–18.

Jackson, C., Archer, D., Goodrich-Schneider, R., Gravani, R. B., Bihn, E. A., and Schneider, K. R. (2007). Determining the effect of Good Agricultural Practices awareness on implementation: A multi-state survey. *Food Prot. Trends.* **27,** 684–693.

Jay, M. T., Cooley, M., Carychao, D., Wiscomb, G. W., Sweitzer, R. A., Crawford-Miksza, L. et al. (2007). *Escherichia coli* O157:H7 in feral swine near spinach fields and cattle, central California coast. *Emerg. Infect. Dis.* **13,** 1908–1911.

Kang, S-C., Kim, M-J., and Choi, U-K. (2007). Shelf-life extension of fresh-cut iceberg lettuce (*Lactuca sativa* L) by different antimicrobial films. *J. Microbiol. Biotechnol.* **17,** 1284–1290.

Kang, S-C., Kim, M-J., Park, I-S., and Choi, U-K. (2008). Antimicrobial (BN/PE) film combined with modified atmosphere packaging extends the shelf life of minimal processed fresh-cut iceberg lettuce. *J. Microbiol. Biotechnol.* **18,** 568–572.

Koseki, S. and Isobe, S. (2005). Prediction of pathogen growth on iceberg lettuce under real temperature history during distribution from farm to table. *Inter. J. Food Microbiol.* **104,** 239–248.

Lapidot, A., Romling, U., and Yaron, S. (2006). Biofilm formation and the survival of *Salmonella* Typhimurium on parsley. *Inter. J. Food Microbiol.* **109,** 229–233.

Lee, S-Y., Costello, M., and Kang, D-H. (2004). Efficacy of chlorine dioxide gas as a sanitizer of lettuce leaves. *J. Food Prot.* **67,** 1371.

Leverentz, B, Conway, W. S., Camp, M. J., Janisiewicz, W. J., Abuladze, T., Yang, M. et al. (2003). Biocontrol of *Listeria monocytogenes* on fresh-cut produce by treatment with lytic bacteriophages and a bacterocin. *Appl. Environ. Microbiol.* **23,** 4519–4526.

Leverentz, B, Conway, W. S., Alavidze, Z., Janisiewicz, W. J., Fuchs, Y., Camp, M. J. et al. (2001). Examination of bacteriophage as a biocontrol method for *Salmonella* on fresh-cut fruit: A model study. *J. Food Prot.* **64,** 1116–1121.

Li, H., Tajkarimi, M., and Osburn, B. I. (2008). Impact of vacuum cooling on *Escherichia coli* O157:H7 infiltration into lettuce tissue. *Appl. Environ. Microbiol.* **74,** 3138–3142.

Lindow, S. E. and Brandl, M. T. (2003). Microbiology of the phyllosphere. *Appl. Environ. Microbiol.* **69,** 1875–1883.

Lopez-Mosquera, M. E., Cabaleiro, F., Sainz, M. J., Lopez-Fabel, A., and Carral, E. (2008). Fertilizing value of broiler litter: Effects of drying and palletizing. *Biores. Tech.* **99,** 5626–5633.

Matthews, K. R. (Ed.). (2006). *Microbiology of fresh produce: Emerging issues in food safety.* First Edition. American Society for Microbiology Press, Washington, DC.

Mercier, J. and Lindow, S. E. (2000). Role of leaf surface sugars in colonization of plants by bacterial epiphytes. *Appl. Environ. Microbiol.* **66,** 369–374.

Mootian, G., Wu, W-H., Pang, H-J., and Matthews, K. R. (2008). Transfer prevalence of *Escherichia coli* O157:H7 from soil, water, and manure contaminated with low numbers of the pathogen to lettuce plants of varying age. In Program and Abstract book IAFP Annual meeting. p. 67. Columbus, OH

Neetoo, H., Ye, M., Chen, H., Joerger, R. D., Hicks, D. T., and Hoover, D. G. (2008). Use of nisin-coated plastic films to control *Listeria monocytogenes* on vacuum-packaged cold-smoked salmon. *Inter. J. Food Microbiol.* **122,** 8–15.

Nygard, K., Lassen, J., Vold, L., Andersson, Y., Fisher, I., Lofdahl, S. et al. (2008). Outbreak of *Salmonella* Thompson infections linked to imported rucola lettuce. *Foodborne Path. Dis.* **5,** 165–173.

Okafo, C. N., Umoh, V. J., and Galadima, M. (2003). Occurrence of pathogens on vegetables harvested from soils irrigated with contaminated streams. *Sci. Total Environ.* **311,** 49–56.

Parish, M. E., Beuchat, L. R., Suslow, T. V., Harris, L. J., Garret, E. H., Farber, J. N., and Busta, F. F. (2003). Methods to reduce/eliminate pathogens from fresh and fresh-cut produce. *Comp. Rev. Food Sci. Food Safety.* Vol. **2** Suppl.:161–173.

Sewell, A. M. and Farber, J. M. (2001). Foodborne outbreaks in Canada linked to produce. *J. Food Prot.* **64,** 1863–1877.

Seymour, I. J., Burfoot, D., Smith, R. L., Cox, L. A., and Lockwood, A. (2002). Ultrasound decontamination of minimally processed fruits and vegetables. *Inter. J. Food Sci. Tech.* **37,** 547–557.

Shafaie, Y. and Matthews, K. R. (2005). Role of post-harvest processing practices in contamination of lettuce. In Proceedings Inter. Assoc. Food Prot., Baltimore. MD. Poster P3-07, p. 105.

Smith DeWaal, C. and Bhuiya, F. (2007). Outbreak Alert 2007. Center for Science in the Public Interest. Ninth Edition. www.cspinet.org

Soderstrom, A., Osterberg, P., Linquist, A., Johnson, B., Lindberg, A., Blide Ulander, S. et al. (2008). A large *Escherichia coli* O157 outbreak in Sweden associated with locally produced lettuce. *Foodborne Path. Dis.* **5**, 339–349.

Solomon, E. B., Brandl, M. T., and Mandrell, R. E. (2006). Biology of foodborne pathogens on produce, In *Microbiology of Fresh Produce*. First Edition (K.R. Matthews, Ed.). p. 55. American Society for Microbiology Press, Washington, DC.

Solomon, E. B., Yaron, S., and Matthews, K. R. (2002). Transmission and internalization of *Escherichia coli* O157:H7 from contaminated manure and irrigation water into lettuce plant tissue. *Appl. Environ. Microbiol.* **68**, 397–400.

Sy, K. V., Murray, M. B., Harrison, M. D., and Buechat, L. R. (2005). Evaluation of gaseous chlorine dioxide as a sanitizer for killing *Salmonella*, *Escherichia coli* O157:H7, *Listeria monocytogenes*, and yeasts and molds on fresh and fresh-cut produce. *J. Food Prot.* **68**, 1176–1187.

Takeuchi, K. and Frank, J. F. (2000). Penetration of *Escherichia coli* O157:H7 into lettuce tissues as affected by inoculum size and temperature and the effect of chlorine treatment on cell viability. *J. Food Prot.* **63**, 434–440.

Takkinen, J., Nakari, U., Johansson, T., Niskanen, T., Siitonen, A., Kuusi, M. (2005). A nationwide outbreak of multiresistant *Salmonella* Typhimurium var Copenhagen DT104B infection in Finland due to contaminated lettuce from Spain, May 2005. *Eurosur.* **10**, E050630.1

Taylor, C. B. (1998). Defense responses in plants and animals—More of the same. *Plant cell.* **10**, 873–876.

Todd, E.C.D., Harris, C. K., Knight, A. J., and Worosz, M. R. (2007). Spinach and the media: How we learn about a major outbreak. *Food Protect. Trends.* **27**, 314–321.

Toth, I. K., Pritchard, L., and Birch, P.R.J. (2006). Comparative genomics reveals what makes an enterobacterial plant pathogen. *Annu. Rev. Phytopathol.* **44**, 305–306.

Valentin-Bon, I., Jacobson, A., Monday, S. R., and Feng, P.C.H. (2008). Microbiological quality of bagged cut spinach and lettuce mixes. *Appl. Environ. Microbiol.* **74**, 1240–1242.

Wachtel, M. R. and Charkowski, A. O. (2002). Cross-contamination of lettuce with *E. coli* O157:H7. *J. Food Prot.* **65**, 465–470.

Wachtel, M. R., Whitehand, L. C., and Mandrell, R. E. (2002). Association of *Escherichia coli* O157:H7 with preharvest leaf lettuce upon exposure to contaminated irrigation water. *J. Food Prot.* **65**, 18–25.

Warriner, K., Ibrahim, F., Dickinson, M., Wright, C., and Waites, W. M. (2003). Interaction of *Escherichia coli* with growing salad spinach plants. *J. Food Prot.* **66**, 1790–1797.

Western Growers. (2008). Commodity Specific food safety guidelines for the production and harvest of lettuce and leafy greens. www2.wga.com/popups/bestpracticesdraft.html

Wu, F. M., Doyle, M. P., Beuchat, L. R., Wells, J. G., Mintz, E. D., and Swaminathan, B. (2000). Fate of *Shigella sonnei* on parsley and methods of disinfection. *J. Food Prot.* **63**, 568–572.

CHAPTER 9

Melons

Alejandro Castillo
Department of Animal Science, Center for Food Safety, Texas A&M University, College Station, TX

Miguel A. Martínez-Téllez
Dirección de Tecnología de Alimentos de Origen Vegetal, Centro de Investigación en Alimentación y Desarrollo, A.C., Hermosillo, Sonora, México

M. Ofelia Rodríguez-García
Departamento de Farmacobiología, CUCEI, Universidad de Guadalajara, Guadalajara, Jalisco, México

CHAPTER CONTENTS

Introduction	190
Prevalence of Human Pathogens in and on Melons	190
Outbreaks of Foodborne Disease Linked to Melons	191
Characteristics of Outbreaks	191
Contributing Factors	194
Impact of Regulatory Actions	195
Potential Sources and Mechanisms of Contamination and Measures Recommended to Prevent Contamination	196
Preharvest	197
Postharvest	201
Cutting Practices	203
Structural Characteristics of Melons Promoting Microbial Survival and Growth	206
Current Knowledge about Growth and Survival of Pathogens in Melons	206
Cantaloupe Netting	207
Biofilm Formation	208
Microbial Infiltration and Internalization	209
Use of Antimicrobial Treatments to Decontaminate Melons	209
Treatments Tested on Fresh Melons	210

Fresh-Cut Melons	212
Treatment with Antimicrobial Agents	212
Irradiation	213
Conclusions	214

INTRODUCTION

Melons are an important source of phytonutrients. The orange, netted melons (cantaloupes) are among the richest in ascorbic acid and other vitamins and elemental micronutrients (Eitenmiller et al., 1985). Melons are also pleasant to eat due to their water and sugar content (Hakerlerler et al., 1999). Due to the worldwide preference for melons, the growing of these commodities may be a long-term guarantee for profitable agriculture (Karchi, 2000).

However, melons, especially cantaloupes (netted), have become a recurrent source of pathogens causing outbreaks of foodborne disease, especially *Salmonella* infection. Cantaloupes grow at ground level, thus increasing the potential for fruit surface contamination. Factors such as the potential for pathogens to attach to the porous rind of the melon and internalize (Asta, 1999; Fan et al., 2006) and to form biofilms (Annous et al., 2005), may promote the occurrence of melon-linked outbreaks of foodborne disease. Although the contamination may be restricted to the rind, it can be transferred to the flesh during cutting. Cut cantaloupe is considered to be a potentially hazardous food according to the FDA food code because it is capable of supporting the growth of pathogens due to mild acidity (pH 5.2 to 6.7) and high water activity (0.97 to 0.99) (Bhagwat, 2006). In this chapter, information will be discussed with regards to foodborne disease outbreaks linked to melons, contamination sources and mechanisms of melon contamination, and possible mitigating strategies to reduce the risk of illness associated with consumption of melons.

PREVALENCE OF HUMAN PATHOGENS IN AND ON MELONS

Melons have been reported as vehicles of pathogens causing outbreaks. However, relatively little definitive information on sources of human pathogens in melons is available (Ukuku and Sapers, 2006). In a binational study conducted by Texas A&M University and the University of Guadalajara, México, eight cantaloupe farms and packing sheds from the United States and México were sampled to evaluate cantaloupe contamination with *Salmonella* and *E. coli*. Samples collected from external surfaces of cantaloupes and water environments of packing sheds of cantaloupes farms were examined. Of a total of 1735 samples collected, 31 (1.8%) tested positive for *Salmonella*; this pathogen was isolated from 5 (0.5%) of 950 samples in the south of Texas and from 1 (0.3%) of 300 samples in the State of Colima, México. Fifteen *Salmonella* serotypes were isolated from samples collected in the United States, and nine from samples collected in México (Castillo et al., 2004). Cantaloupes may be especially prone to accumulating microorganisms and harboring bacterial pathogens. In a study of the prevalence of *Salmonella* in the growing and processing environment of oranges, parsley, and cantaloupes in South Texas, Duffy et al. (2005) recovered *Salmonella* from only cantaloupes. In another study involving more than 900 field-collected melons produced in different regions of California during 1999 to 2001, no *Salmonella* were ever recovered (Suslow, 2004).

In two independent surveys conducted by the US Food and Drug Administration (FDA) including Mexican cantaloupes, *Salmonella* was isolated from 0.78 and 1.08%, with 8 and 12 serotypes identified in each survey. In another survey *Salmonella* was isolated from 8 (5.3%) and *Shigella* from 3 (2.0%) of 151 cantaloupe samples collected from nine countries exporting to the United States (Bhagwat, 2006). In 1999 the FDA carried out a study to estimate the frequency of *Salmonella*, *Shigella*, and *E. coli* O157:H7 in 1003 samples. Ninety-six percent of the samples tested negative; 35 (3.5%) were positive for *Salmonella*, 9 (0.9%) for *Shigella*, and none were positive for *E. coli* O157:H7. In another study intended to determine *Salmonella* spp. and *Shigella* spp. in domestic and imported melons, these pathogens were isolated from 2.4% and 0.5% of the domestic melons and from 5.3% and 1.2% of the imported fruit, respectively (FDA, 2001, 2006).

As expected, the microbiota of the fresh produce reflects the environment in which they grow. Figueroa-Aguilar (2005) collected cantaloupe and environmental samples at the cantaloupe farms where the melons involved with the *Salmonella* Poona outbreaks of 2000 and 2001 (CDC, 2002) were produced. *Salmonella* spp. were isolated from four samples of river water, two of water from a hydrocooler, four fecal samples from iguanas, and one fecal sample of an unidentified animal. No *Salmonella* was isolated from melons or well water. In this study, the serotypes identified were *S.* Poona, *S.* Infantis, and *S.* Anatum.

OUTBREAKS OF FOODBORNE DISEASE LINKED TO MELONS

Characteristics of Outbreaks

Melons have been associated with foodborne disease outbreaks caused by several serovars of *Salmonella*, *E. coli* O157:H7, *Campylobacter*, and norovirus (Chapman, 2005). In the period of 1973 to 1997, 13 outbreaks were recorded with 341 cases of disease involving cantaloupes (six outbreaks), watermelons (six outbreaks), and one with a combination of musk and honeydew melons. The etiologic agents of melon-associated disease were *Salmonella* (6), *E. coli* O157:H7 (one outbreak), and *Campylobacter jejuni* (one outbreak) (CDC, 1991).

Salmonella is the pathogen most frequently related with cases and outbreaks linked to melon consumption. The first report of *Salmonella* infection associated with melon consumption was in 1955; 17 cases of disease caused by *S.* Miami were linked to sliced watermelon purchased from a local store (Gaylor et al., 1955). Another outbreak linked to precut watermelon was reported in 1979, caused by *S.* Oranienburg. Six people of two families were affected; each family bought a watermelon half from the same vendor on the side of the highway (CDC, 1979).

In 1990 in the United States, a large multistate salmonellosis outbreak was documented by CDC to be associated with melon consumption. It was reported that *Salmonella* Chester caused 245 cases (two fatal), in 30 US states (Ries et al., 1990), although the real number of cases was estimated to exceed 25,000 (CDC, 1991; Tamplin, 1997). Apparently, transmission of pathogens to the interior of the cantaloupe may have occurred while cutting the unwashed melons (Beuchat, 1996). In a survey of melons coming from the same area where the melons involved with the outbreak originated, *Salmonella* was found in 24 of 2200 melons tested. Twelve different serovars were identified; but none of the isolates was *S.* Chester, the serovar that caused the outbreak (Madden, 1992). In another outbreak in 1991, more than 400 cases were reported from 23 US states and 4 Canadian provinces. The causative agent of the illness in this outbreak was *S.* Poona, and the outbreak was linked to the consumption of contaminated cantaloupes produced in Texas (CDC, 1991).

In June of 1991, an outbreak of *S.* Javiana linked to watermelon affected 39 children. The microorganism was isolated from feces of some of the victims as well as from leftovers of the sliced watermelon. Analysis of the plasmid profile and DNA chromosomal restriction led to the conclusion that it was the same strain. Apparently, the causative factor was the cutting of slices without previously washing the fruit and the consumption of the leftover melon that had been stored at room temperature. The contamination probably took place during transportation of the fruit. After these cases the FDA defined cut melons as potentially hazardous to health (Blostein, 1993). After these outbreaks of salmonellosis, linked mainly to imported melons, the melon industry considered as a prudent measure the implementation of the "Melon Safety Plan," which was focused on the chlorination of the water used to wash the melons, as well as on the ice used in cooling or the refrigerated containers used for transporting the melons (Tauxe, 1997).

In 1997, an outbreak of salmonellosis caused by a strain of *S.* Saphra affected 25 California residents. Pulsed Field Gel Electrophoresis (PFGE) was used to identify the outbreak strain, and identical PFGE patterns were found for 24 *S.* Saphra isolates from infected patients. This PFGE pattern was different from that of the 5 *S.* Saphra strains isolated in prior years in California. Most of the patients in this outbreak were young children whose parents recalled feeding them cantaloupe. The epidemiological and traceback study in this outbreak supported the conclusion that the cantaloupes implicated in this outbreak were imported from a small region of México. Only 17% of the patients washed cantaloupes before cutting (Mohle-Boetani et al., 1999). This outbreak is another example of gastrointestinal disease in the United States related to contaminated, imported melons.

An additional outbreak of salmonellosis involving the consumption of contaminated cantaloupes occurred in Ontario in 1998. Twenty-two cases were linked to the consumption of cantaloupes contaminated with *S.* Oranienburg (Deeks et al., 1998). During the time of the outbreak, cantaloupes were imported into Ontario from many sources, including the United States, México, and Central America. In 1999, melons were involved in an outbreak where *S.* Enteritidis caused 82 cases of disease in the United States (CDC, 2003a).

In 2000, 46 cases of salmonellosis were reported during an outbreak that occurred in over six US states. At least 26 of the cases in California were attributed to *S.* Poona and linked to the consumption of contaminated cantaloupe. Three successive multistate outbreaks occurred in the spring of 2000 (47 cases), 2001 (50 cases including 2 deaths), and 2002 (58 cases), each of which was linked to eating cantaloupes from Mexican farms (CDC, 2002). The PFGE patterns of the 2000 and 2002 outbreak strains were indistinguishable, whereas that of 2001 differed (Tauxe et al., 2008). The FDA evaluated the farms in 2000 and 2001, identifying several possible sources of contamination during field operations, such as river water used for irrigation and fecal matter from the iguanas found to thrive in the fields, and during washing and packing, finding that the measures being applied were insufficient to minimize microbial contamination. After the 2002 outbreak, the FDA issued an import alert on all cantaloupes from México (Tauxe et al., 2008).

Between January 18 and March 5, 2008, state health departments identified 50 ill persons in 16 US states who were infected with *S.* Litchfield with the same genetic fingerprint. In addition, nine ill persons with the outbreak strain were reported in Canada. The CDC collaborated with public-health officials in multiple states across the United States, and with the FDA to investigate this multistate outbreak of *S.* Litchfield infections.

Traceback studies pointed to cantaloupes from Honduras as the likely source of infections (CDC, 2008). The Honduran company that produced these melons had to dismiss 1800 employees, and experienced a $13 million loss due to *Salmonella* contamination.

Several outbreaks linked to melons have been caused by pathogens different from *Salmonella*. In 1993, 27 cases of *E. coli* O157:H7 were linked to melon consumption (Del Rosario et al., 1995). In 1997, cantaloupes again were considered to be the vehicle in nine cases of hemorrhagic enteritis caused by *E. coli* O157:H7 (CDC, 2003), and in 2004, a new outbreak of *E. coli* O157:H7 occurred in Montana. It was associated with cantaloupe consumption and involved six cases of HUS and TTP (ISID, 2004). Additionally, 48 cases of *Campylobacter* enteritis were linked to melons in 1993 (CDC, 2003). Melon-associated outbreaks also have been attributed to norovirus. In 1999 three outbreaks involving 23, 61, and 5 cases were reported in the months of May, June, and September, respectively. For these norovirus outbreaks, cantaloupes were identified as the vehicle (CDC, 2003).

Contributing Factors

Melon is a type of produce with unique characteristics. Keeping the cut melons for several hours at ambient temperature was a factor that appeared multiple times during outbreak investigations. Cut melons placed in salad bars, group dinners, and such, can support the growth of *Salmonella* if the temperature of storage is inadequate (Golden et al., 1993). A common theme in these outbreaks was that the melons were cut and then subjected to temperature abuse. In some cases, melons were cross-contaminated through inadvertent contact with raw meat, or a human handler (Iversen et al., 1987). In most instances, contamination was thought to have originated in the field from contact with contaminated soil or dirt (Bhagwat, 2006), and in other cases the contamination was thought to have come from contact of the melon rind with soil during harvesting and packing. The impact of such contamination is subsequently increased by displaying the precut melons at the store, where appropriate refrigeration cannot be guaranteed (Mohle-Boetani et al., 1999). Transfer of bacteria from the rind to edible melon flesh during cutting has been demonstrated in laboratory conditions (Ukuku et al., 2004). Contamination of produce can arise from a variety of sources including soil, water, equipment, and humans (Beuchat, 1996). Cantaloupes can become contaminated during growth, postharvest handling, and packing, transportation, distribution, or during final preparation at food service or in the home. During postharvest handling and packing, melons can become contaminated by equipment used to hold, transport, clean, grade, sort, or pack melons, or from unsanitary washing (immersion of gondolas/trailers in dump tank water) or use of contaminated cooling water or ice (Parnell et al., 2005).

Impact of Regulatory Actions

The Food and Drug Administration (FDA) is responsible for the safety of US produce imports. Generally, the FDA carries out a random sample collection at the border and prevents the entry of products that fail to pass laboratory inspection. The FDA can also detain a product without physical examination, based on past history of a shipping firm or other information indicating that the product might violate standards. When the source of a public-health concern can be identified, only those firms with a problem have their shipments blocked (Calvin, 2003). Integration of imports into previously domestic fruit and vegetable markets also requires that exporters meet the food safety standards of other countries. Over the past several years, US and Mexican authorities have worked to establish a framework that would broaden the allowance for cantaloupe imports from México. In November 2002, the US Food and Drug Administration (FDA) issued an Import Alert on all Mexican cantaloupes, effectively banning all such imports from México. This action followed three successive years (2000, 2001, and 2002) where outbreaks of *Salmonella*, associated with the consumption of contaminated cantaloupe from México, occurred in Canada and the United States. Strong epidemiological evidence linked the outbreaks of 2001 to two deaths in California (Green et al., 2005).

Between 1999 and 2005, Mexican cantaloupe production declined 24% as a result of export restrictions. Between 1999 and 2006, cantaloupe imports from México declined 92% and in 2006 accounted for just 3% of US imports (SIAP-SAGARPA, 2002).

Since the imposition of the countrywide import alert, the FDA has exempted several growers from this ban, and in October 2005, the FDA signed a memorandum of understanding (MOU) with SENASICA (Mexican National Service for Animal and Plant Health, Food Safety and Quality) that allows for the differential treatment of prospective Mexican cantaloupe exporters, based on their past food safety performance (Anon., 2005). Efforts by Mexican growers to respond to the growing demand of consumers for safer food products are likely to further integrate the US and Mexican fruit and vegetable sectors. However, as of October 2008, the Import Alert #22-01 (FDA, 2008, Import Alert 22-01), Detention Without Physical Examination (DWPE) of Cantaloupes from México, is still valid, and there are still differences of opinion about whether the point of contamination has been controlled (Tauxe et al., 2008). Nevertheless, cantaloupe exports are allowed for some Mexican firms.

The MOU between FDA and SENASICA defines three categories of exporting firms: Category 1, firms exempt from DWPE; Category 2, firms that have been directly implicated in an outbreak, or that have shipments from México testing positive for *Salmonella* or other pathogens; and Category 3, firms that have not been implicated in any outbreak, nor have had shipments testing positive for *Salmonella* or other pathogens. Under the terms of this MOU, firms in Categories 2 and 3 must be recognized by SENASICA to comply with stringent requirements that include, but are not limited to, verification of compliance with Good Agricultural Practices. FDA will accompany SENASICA on up to the first 12 inspections, and these inspections must be successful before these firms are accepted as certified and allowed to ship cantaloupes to the United States. However, shipments from Category 2 firms, in addition to being required to be certified, will be subjected to testing for *Salmonella* and will be allowed entrance only after the shipment has tested negative for this pathogen. When a certified firm has had five consecutive shipments testing negative for *Salmonella*, it is classified as Category 1. Currently, only 12 firms (11 producers and 1 processor) can export fresh, frozen, or processed cantaloupes to the United States. Of these, seven are in Category 1, and the other five are certified but still in Categories 2 or 3.

POTENTIAL SOURCES AND MECHANISMS OF CONTAMINATION AND MEASURES RECOMMENDED TO PREVENT CONTAMINATION

In the last decade cantaloupes and other fruits and vegetables have been involved in numerous outbreaks or recalls in the United States and Canada because of *Salmonella* contamination. Investigations conducted by researchers and or government agencies of the involved countries indicate that the main sources of contamination in cantaloupes are (1) water for irrigation or preharvest practices and postharvest management, (2) worker activities, (3) organic fertilizer, (4) animal or human feces, and (5) equipment and installations. The contamination of the products can occur through direct contact, internalization, or cross-contamination. The application of Good Agricultural Practices in the full chain of production is the best tool for preventing biological contamination of the melons.

Harvesting and packaging of cantaloupes are carried out in different ways. In California, melons are harvested and packed in the field, but in Georgia, the harvested melons are transported to a shed for washing and packing (Akins et al., 2008). In México, cantaloupes are handled in a similar way as in Georgia, according to guidelines in the memorandum of understanding of the FDA-Mexican Government for the voluntary implementation of Good Agricultural Practices and Good Management Practices in the production and packing of fruits and vegetables for fresh consumption by humans, released by the Mexican government (SENASICA, 2006).

The US melon industry recognizes that once a melon is contaminated, it is difficult to remove or to kill a pathogen. Therefore, prevention of microbial contamination at all steps from production to distribution is widely favored over treatments to eliminate contamination after it occurred (PMA and UFFVA, 2005).

Preharvest

Water for Irrigation and Production Practices

Water is an essential element used in several production activities, including irrigation, pesticide and liquid fertilizer applications, among others. Some consider water the main contamination vehicle for fruits and fresh vegetables. When water is in contact with fruits or vegetables, there is a high risk of contamination of the products (Martinez-Tellez et al., 1998), which is why it is important to monitor the microbiological quality of water supplies to avoid pathogen spread among agricultural products. Water can be a transmission source of a large number of microorganisms such as the pathogenic species of *Escherichia coli*, *Salmonella*, *Vibrio cholera*, *Shigella*, and also *Cryptosporidium parvum*, *Giardia lamblia*, *Cyclospora cayetanensis*, *Toxoplasma gondii*, and norovirus and hepatitis A viruses (FDA, 1998b).

Before considering the microbiological analysis to evaluate water quality, we must recognize the fact that water sources have a high risk of contamination with human pathogens, and establish necessary measures to avoid such contamination. These measures must apply to both surface and groundwater sources, such as ponds, rivers, streams, or lakes. Water sources close to cattle areas or other potential sources of contamination must be considered for pathogen testing. Additionally, it is necessary to consider water as a potential vehicle for spreading pathogens from one product to another (Zawel, 1999). For example, irrigation water may carry contamination originated from animal feces in the crop field to fruits attached to the irrigated plants. The presence of pathogens in different sources of water has been reported (Falcão et al., 1993). If water, contaminated with a human pathogen, is used for applying pesticides, the presence of the pesticide may favor survival of the pathogen and even support their growth (Ng et al., 2005).

Although the growth of pathogens in water containing pesticides has been reported in laboratory studies, this potential source of pathogen contamination is not generally considered to be important, and the use of contaminated water for pesticide application is a recurrent practice in production of fresh agricultural products. Exposure to contaminated water during production and handling has been identified as a major factor promoting the presence of viable human pathogens on cantaloupe surfaces (CDC, 2002). Once the pathogens adhere to the cantaloupe rind, the rough characteristics of rind surface promote irreversible attachment and subsequent biofilm formation (Annous et al., 2005). Contaminated water should not be used for such purposes.

Worker Activities

During the production of cantaloupes, field workers have minimal contact with the developing fruit; in some cases, they turn the fruit when it is attached to the plant to avoid the formation of soil spots. Due to deficient hygienic practices of workers during fruit handling, human pathogens and parasites may be transmitted from workers to the fruits by direct contact. The use of wool or cotton gloves during fruit handling has been identified as an important contamination source (Martínez-Téllez, 2008, unpublished data). Personnel training and education about biological hazards in agriculture are a priority in order to prevent contamination during primary production. If the field personnel do not know, and hence, do not apply hygienic practices, workers may involuntarily contaminate crops, the water supply, equipment, and other workers (FAO, 2003a). Portable, clean, and sanitary facilities (toilets and washing stations) with enough supplies (clean water, bactericidal soap, toilet paper, and disposable towels) must be installed in the field for the workers' hygiene so that they are no more than a two to three minute' walk from the work location, to favor their utilization.

Biological hazards can also be reduced during the production of melons by establishing preventive health programs for field workers, and avoiding use of those workers who are sick or have symptoms of a certain disease, wounds in their hands, or open sores, in activities involving direct contact with the product.

Organic Fertilizer

Application of organic fertilizers in melon production is a matter of great concern due to the fact that compost, produced under deficient conditions or recontaminated with human pathogens as a consequence of incorrect handling, can present a high risk of biological contamination during production. Materials that are used traditionally for the production of organic fertilizers are animal, crustacean, and vegetable wastes. The first two represent an important source of human pathogens that must be inactivated during the production process of the organic fertilizer. The origin of the materials can define the type of microorganisms present in the compost if it is not produced correctly. Wastes of birds may contribute mainly *Salmonella* contamination, whereas enterotoxigenic *E. coli* can be introduced from waste of pigs or *E. coli* O157:H7 from cattle. Aerobic and anaerobic processes in organic fertilizer production reach adequate temperatures for the elimination of human pathogens, but it is necessary to verify the elimination of human pathogens in the compost by microbiological analysis of each produced batch.

The U.S. Natural Resources Conservation Service (NRCS) has a standard for operation of composting facilities (Code 317) to reduce the pollution potential of organic agricultural wastes to surface and groundwater. The requirements include, but are not limited to, ensuring that an operating temperature of 130 to 170 °F (54–77 °C) be achieved within 7 days and remain at these temperatures up to 14 days to ensure efficient composting (NRCS, 2005). The current regulation in the United States (AMS/USDA, 2008) requires that if raw animal manure is to be used as organic fertilizer, it must be composted unless it meets specific conditions for spontaneous composting, such as allowing at least 120 days between raw manure application and harvesting of a commodity that has been in contact with the soil so fertilized. The farmer must assure the safety of his composts, whether self-produced, or acquired from specialized suppliers. In general, application of fresh manure or waste of marine origin must never be authorized for any reason in the production of fruits and vegetables.

Animal or Human Feces

Fecal matter, whether human or animal, constitutes an important source of human pathogens, such as *E. coli* O157:H7, *Salmonella*, *Campylobacter*, among others. These microorganisms can survive in soil and manure for more than three months, depending on temperature and soil conditions (Craigmill, 2000). For this reason, the presence of domestic animals or pets and their defecation outdoors in or near fields—factors that may favor the spread of contamination to crop areas—must be forbidden. Contamination may be carried involuntarily by workers on their shoe soles or on their work tools, increasing the risk of product contamination. Human waste collected in the sanitary facilities placed in the crop fields must be disposed correctly to avoid contamination.

Wild fauna is also considered a contamination source in the crop fields. The water sources and fresh food available in crop fields are the main attractions for wildlife, and it is necessary to establish barriers to avoid their access to crop areas. Incidents with other commodities, such as the outbreak of *E. coli* O157:H7 that occurred in 2006 linked to spinach, permitted investigators to establish the role of feral pigs and other fauna in spreading contamination (Jay et al., 2007). Maintaining ecological equilibrium and conservation are also priorities in modern agriculture, so it is necessary to consider the establishment of available water sources for wild fauna, away from the production zones, to reduce the access of wild animals to crop areas.

Farmers must know if the neighboring terrains and facilities are used for animal husbandry, and should take the necessary measures to assure that animal fecal matter will not be transported from contaminated land to their crops during a rainy season (FDA, 1998). This can be avoided by the construction of physical barriers (canals) to divert the runoffs to noncultivated zones.

There is a great need for information on mechanisms for transmission of pathogens from feces to the melons, especially whether animal species that may have intimate contact with the melons, such as insects or nematodes, can serve as vectors of the pathogen to the melon surface. Caldwell et al. (2003) reported the ability of *S.* Poona to survive in the digestive tract of *Caenorhabditis elegans*, a free-living microbivorous nematode, which later can shed this pathogen. This study proved the concept that free-living worms can serve as a source of bacterial pathogens in melons and other produce growing on the ground. In a more recent study, Gibbs et al. (2005) demonstrated the ability of *Diploscapter* sp., another free-living nematode commonly found in the rhizosphere of crop soils, to survive in manure, ingest *S.* Poona and other pathogens, and then shed these pathogens 24 hours after exposure to the inoculated manure. These studies indicate that the primary point of contamination of melons may be the field via environmental vectors.

Facilities and Equipment

Buildings for the storage of field packing materials, crop production tools, pesticides, and fertilizers, common facilities in agriculture fields, must be kept in order and waste-free to avoid the nesting of pests and mice. This kind of fauna may contaminate the materials and equipment for field packing that will be used later during harvest. The exterior of these installations should be examined to assure that the surrounding areas (2–3 m) are free of undergrowth. In this zone it is also necessary to install traps for rodents, aimed at preventing their access to the interior of the storage rooms. All the tools used during production must undergo a cleaning and disinfection process each time they are going to be used or when they are going to be utilized for a different activity, for example, in different crop fields, which would favor cross-contamination. This kind of contamination may occur if a plow or a fertilizer spreader is contaminated in one locality and used again in another crop field without prior cleaning and disinfection. This practice has the added advantage of reducing the presence of phytopathogens, since many practices focused on the reduction or elimination of human pathogens also accomplish the goal of eliminating microorganisms that damage the crops.

Postharvest

The melon packing process is the main point where fecal contamination might be introduced postharvest and where most opportunities are provided for spreading and increasing levels of contamination with human pathogens that originated in the field (Castillo et al., 2004; Johnston et al., 2005). Therefore, growers must pay special attention to personnel training and management of sanitary facilities. The main sources of contamination in the postharvest handling of cantaloupes are (1) water for washing and sanitation, (2) worker activities, and (3) equipment and installations. Contamination of the products can occur through direct contact or cross-contamination. The application of sanitation practices, throughout the full chain of packing, is the best tool for preventing contamination of cantaloupe melon (FDA, 2008).

Water for Washing and Sanitizing

In operations where melons are not packed in the field, postharvest procedures include cooling, rinsing, washing, and disinfecting the melons. In such steps water has direct contact with the product, and it is used in great amounts. As mentioned before, water can be a major vehicle of biological contamination of fruits and vegetables (Falcão et al., 1993; Gerba and Choi, 2006), and only potable water should be used in these practices. Once the melons arrive at the packing facilities, the melon rinsing process takes place; this is to eliminate residual soil or organic matter stuck to the melons. Some washing systems use water recirculation in this step, which involves the risk of a massive spread of contamination. That is why recycling of water is not recommended in this part of the process. Washing may be carried out by dipping the melons in a dump tank with chlorinated water, or by spraying chlorinated water, and is intended to eliminate dirt from melons rather than remove microorganisms. The FDA discourages the use of dump tanks due to the potential for cross-contamination and internalization of pathogens (Smith, FDA, 2004, personal communication). Disinfection is the next step after washing. The concentration of antimicrobial chemicals used must be monitored and documented routinely to assure that the disinfectant is kept at adequate levels. A number of investigations have pointed out the effectiveness of various disinfectants (Barak et al., 2003; Kozempel et al., 2002; Materon, 2003). Nevertheless, the different processes and operation conditions of the packinghouse cause variations in the disinfection results; therefore, validation of the disinfection process in each packing unit is recommended. A detailed description of disinfection procedures is provided later in this chapter.

Worker Activities

As a consequence of deficient hygienic practices of the workers, human pathogens and parasites may be transmitted directly to the fruits during packing, by direct contact with contaminated workers. Poor hygiene of the workers has been identified as an important cause of pathogen contamination. Packing facilities with a high degree of mechanization in the processes of selection and classification of the melons reduce the risk of biological contamination by reducing the manipulation of melons by the personnel. The correct application of personal hygienic practices by the workers, and applied to their handling of tools, with constant supervision would help to reduce the incidence of human pathogens in cantaloupes.

It is important to ensure that all the personnel, including those who do not participate directly in the manipulation of the fruits, including visitors, adhere to the hygiene practices established by the management. In this context a training program in Good Hygienic Practices and product handling must be in place and be offered for all the employees including supervisors, temporary, part-time, and full-time personnel, as well as subcontracted personnel (FDA, 1998). Training the personnel in Good Hygienic Practices is not an easy task, given the diversity in cultural backgrounds and languages among workers in this kind of industry. This difficulty is reflected when trying to teach them new ideas due to cultural and language barriers. Nevertheless, a great effort must be made to reduce the risks of contamination, in part through motivating employees by treating them as an important part of the company and showing appreciation of their value and contributions (Hurst and Shuler, 1992).

Equipment and Facilities

Human pathogen sources in the centralized packing, cooling room, and storage room are mainly contributed by animal vectors such as rodents and birds that intrude into the installations. Such installations must be designed to avoid the introduction of these vectors and to counter such contamination with monitoring and documentation programs of hygiene practices and maintenance of installations and equipment. To reduce the possibility of spreading contamination in the sorting and packing areas, it is recommended that a program of insect and rodent control outside and inside the buildings be established and maintained. The design of buildings and selection of washing machinery should favor ease of cleaning and disinfection, and their mechanical parts should be resistant to corrosion to resist detergents and antimicrobial agents. The product conveyor belts must always be kept clean, sanitized, and free of cracks that may damage the melons and make the washing and sanitization process difficult.

The refrigerated containers used to transport the product, either by sea or land, must be inspected before loading the product to assure their cleanliness, absence of odors and visible residues, and that the temperature is adequate to preserve the quality of cantaloupe melon. The loading zone must maintain hygienic conditions so the loading equipment will not introduce contamination to the interior of the transport container. Supervision and documentation of these operations is necessary for the maintenance of the food safety program.

Cutting Practices

Potential Sources of Pathogens

Unsatisfactory postharvest handling and kitchen practices can increase the risk of contamination of the edible portion of cantaloupes. The pathogen is usually located on the rind and is transferred to the flesh during cutting (Beuchat, 1996). In addition, inappropriate kitchen practices such as use of knives or cutting boards for processing different commodities (e.g., meats) without suitable washing and sanitizing, or poor personal hygiene during for the preparation of the product for consumption may result in cross-contamination. An outbreak of cholera in the United States was associated with the consumption of cut cantaloupes. The cantaloupe was sliced and handled by an asymptomatic, infected person, who was suspected to have contaminated the cut melon during preparation and handling (Ackers et al., 1997).

Mechanisms for Contamination

The most likely mechanism for contamination of melon flesh during cutting is the transfer of pathogens from contaminated rind to the flesh via the knife blade. In particular, cantaloupes are prone to harboring large bacterial populations, and the netting material that covers the rind increases the area for bacterial attachment. In a survey of melons collected from grocery stores in Texas, the aerobic plate count and total coliform counts were found to be 1.0 to 1.4 log cycles higher on the surface of cantaloupes than on the surface of honeydew melons (Cabrera-Diaz and Castillo, unpublished data, 2003). Similar information has been published by others (Ukuku, 2004; Ukuku et al., 2005). The heavier bioburden of the cantaloupe rind may increase the probability for pathogen transfer to the flesh during cutting, which when coupled with the ability of pathogens to grow in melon flesh at abusive temperatures and survive for long periods under refrigeration, may increase the risk of infection (Golden et al., 1993; Vadlamudi, 2004).

Although there is scarcity of information about mechanisms for contamination of melons during peeling, slicing, and serving, multiple outbreaks have been linked to cut melons in salad bars. This may indicate that, in addition to a passive moving of pathogens from the rind to the flesh during cutting, other mechanisms may operate in transferring pathogens to cut melons. Cross-contamination from beef to watermelon was the mechanism in an outbreak of *E. coli* O157:H7 disease that was linked to cut watermelon and other salad bar items in the summer of 2000, affecting at least 62 people in Wisconsin (Beers, 2000). During the outbreak investigation, the outbreak strain was also isolated from ground beef that was being made from sirloin chunks in close proximity to where the salads were being prepared, and from a sample of sirloin collected at the meat-packing plant that supplied the restaurant. Besides the cross-contamination mechanism that was likely to have contributed to this outbreak, worker's hygiene may have played a role as well. One of the cooks and another worker were suffering from diarrhea before the outbreak, and it is possible that they were still infectious and contaminated the salad bar items by inappropriate handling. Another possible mechanism was the recycling of leftovers from the salad bar, which were refrigerated overnight and placed in the salad bar on successive days. This may have resulted in continued contamination of the salad bar over several days (Beers, 2000).

Preventing and Minimizing Contamination

The transfer of pathogens to the flesh of cantaloupes during cutting has been well established (Suslow and Cantwell, 2001; Ukuku, 2004; Ukuku and Fett, 2002a; Ukuku and Sapers, 2002). According to Suslow and Cantwell (2001), the particular characteristics of the cantaloupe rind determines that the presence of as few as 150 bacteria per cm^2 of rind may result in contamination during cutting. Although these investigators did not specify whether this number was estimated from studies with native microbiota of cantaloupe or with inoculated pathogens, other studies involving both epiphytic bacteria and inoculated pathogens prove that counts similar or slightly higher than what was indicated by Suslow and Cantwell (2001) have been sufficient to allow transfer of organisms to the fresh-cut pieces (Ukuku and Fett, 2002b; Ukuku, 2004).

Under the premise that reducing the load of pathogens on the surface of melons will ultimately reduce the probability of conveying these pathogens to the melon flesh, a series of studies have been conducted at the USDA's Agricultural Research Service. Ukuku and Fett (2002) studied the effect of sanitizing the surface of cantaloupes inoculated with *L. monocytogenes*, on the presence of this organism on the cubes of fresh-cut cantaloupe. The melons were inoculated with *L. monocytogenes* at a level of 3.5 log CFU/cm^2; allowed to dry; washed with water, chlorine (1000 mg/L), or hydrogen peroxide (H_2O_2, 5%); and then cut. *L. monocytogenes* was consistently found on the cantaloupe cubes obtained from the unwashed controls and from the melons that were subjected to a water wash, but was consistently absent on the cubes obtained from melons subjected to chlorine or H_2O_2 wash. These sanitizers also were tested on cantaloupe inoculated with *S. Stanley*, and the counts of *S. Stanley* on cantaloupe cubes obtained from unwashed or water-washed melons were between 0.20 and 0.22 log CFU/g, whereas the organism was not detectable on cubes obtained from melons washed with chlorine or H_2O_2 unless the melons were stored at 4 or 20 °C for three to five days before cutting (Ukuku and Sapers, 2002). This may indicate that cutting the cantaloupes promptly after surface disinfection can help in minimizing *Salmonella* contamination of the flesh. A delay in cutting after disinfection may allow sufficient growth of surviving pathogens to yet again increase the chances for transfer to the flesh. Therefore, determining the maximum time period between sanitizing and cutting that still prevents microbial growth is paramount in establishing best practices for fresh-cut processing.

Treating inoculated cantaloupe and honeydew melons with a mixture of H_2O_2 (1%), nisin (25 g/ml), sodium lactate (1%), and citric acid (0.5%) reduced *L. monocytogenes* and *E. coli* O157:H7 to undetectable levels on melon surfaces. Except for one instance, these pathogens could not be recovered, even by enrichment, from melon cubes obtained from whole melons treated with this mixture of sanitizers (Ukuku et al., 2005). Submerging cantaloupes inoculated with a mixture of five strains of *Salmonella* for 60 seconds in hot H_2O_2 (70 °C) or water (97 °C), reduced the transfer of this pathogen to the cut cantaloupes to undetectable levels, whereas cubes cut from untreated melons had *Salmonella* counts of 2.9 log CFU/g (Ukuku et al., 2004).

The cutting practices may also play a role in the transfer of *Salmonella* from contaminated rind to the cantaloupe flesh. Vadlamudi (2004) reported that cutting inoculated cantaloupes after first peeling the rind resulted in decreased transfer of *Salmonella* Poona into the tissue in comparison with cutting of melons and removing the rind later. For both cutting methods, this investigator used a new, sterile knife between steps. Handwashing, or handwashing combined with the use of hand sanitizers and gloves have been studied as measures that can reduce the risk of foodborne disease considerably (Paulson, 2000; Taylor, 2000).

STRUCTURAL CHARACTERISTICS OF MELONS PROMOTING MICROBIAL SURVIVAL AND GROWTH

Current Knowledge about Growth and Survival of Pathogens in Melons

Early reports of the ability of melon edible flesh to sustain the survival and growth of pathogens were published in the 1980s. Fredlund et al. (1987) reported an outbreak of shigellosis linked to watermelon in Sweden, which was caused by *Shigella sonnei*. The watermelon had been purchased in Morocco and brought to Sweden, then cut and consumed three days after arrival to Sweden. The hypothesis was that the strain of *S. sonnei* had been internalized into the watermelon, perhaps by deceitful commercial practices including injecting water to increase the weight. To test this hypothesis, the investigators injected the outbreak strain into 17 watermelons, which then were stored at 20 or 30 °C. At intervals during storage, core samples were obtained using a biopsy needle. *S. sonnei* grew rapidly inside the watermelons to reach counts of 8.0 to 9.0 log CFU/g within three days.

In a study by Escartin et al. (1979), different strains of *Salmonella* and *Shigella* were inoculated onto watermelon, papaya, and jicama. Papaya is a tree fruit and jicama is a tuber. Although they are not melons (cucurbits), they are prepared similarly as melons, peeled and cut before consumption. Thus, should the rind be contaminated, pathogen transfer to the edible tissue also would be expected to occur. All the pathogens tested were able to grow on inoculated cubes of the three commodities and in suspensions of papaya and watermelon flesh when stored at 22 °C. Later, Castillo and Escartin (1994) reported the ability of *Campylobacter jejuni* to survive on watermelon and papaya cubes stored at 25 to 29 °C. Golden et al. (1993) found the growth rate of a mixture of *Salmonella* strains on sliced cantaloupe, watermelon, and honeydew melons, to be similar to the growth rate in tryptic soy broth at 23 °C. At 5 °C, *Salmonella* could not grow, but survived with no reduction in counts over 24 hours of storage. Penteado and Leitão (2004a) reported the ability of *Salmonella* Enteritidis to grow in homogenates of cantaloupe, watermelon, and papaya at different temperatures, including 10 °C. In another study, they also observed growth of *L. monocytogenes* at 10 °C (Penteado and Leitão, 2004b). This demonstrates the importance of correctly refrigerating cut melons to prevent growth.

Pathogens such as *Salmonella* and *E. coli* O157:H7 can have a low infectious dose and their growth may not be required to cause illness. However, the higher the concentration of the pathogen, the higher the risk of disease and of the pathogen spreading throughout the entire batch of cut melons, thereby increasing the exposure if the contaminated melons are consumed. If appropriate conditions of temperature and humidity are provided, pathogens may even grow on the rind, as reported for *E. coli* O157:H7 (Del Rosario and Beuchat, 1995). Growth on cantaloupe rind also was observed for *S.* Poona when stored at 37 °C, especially on wounded rinds (Beuchat and Scouten, 2004). Pathogen growth on the rind surface can increase the chances for transfer to the flesh during cutting.

Cantaloupe Netting

The surface of cantaloupe includes a meshwork referred to as the netting (Webster and Craig, 1976). During preliminary work on characteristics of the cantaloupe surface, we inoculated pieces of cantaloupe and honeydew rinds with a suspension of *S.* Poona, and then the inoculated rind was stored at room temperature for 60 minutes. Superficial samples for *Salmonella* counts were collected with a sterile sponge, immediately after inoculating and after the 60 minute storage. The counts on cantaloupe rind decreased by 1 log cycle after 60 minute storage, whereas no count decrease was observed on honeydew rind (Cabrera-Diaz and Castillo, unpublished data, 2003). It seems unlikely that the cantaloupe surface had an antimicrobial activity. Instead, the netting material may have absorbed the inoculum and provided sites for irreversible attachment of *Salmonella*, which resulted in the inability of our sampling device to recover the attached cells. Parnell et al. (2005) observed that *Salmonella* was detached to a greater extent from honeydew than from cantaloupe melons by a simple soaking in plain water for 60 seconds. When testing the effect of antimicrobial agents on the reduction of microorganisms on melons, Ukuku (2004), Ukuku and Fett (2002a), and Ukuku et al. (2005) found the overall magnitude of reduction to be similar for both types of melon; however, the microbial counts were consistently lower on honeydew, which in turn resulted in a reduced transfer of microorganisms to the fresh-cut product.

The formation of the netting in cantaloupes is thought to be a response to cracking of the fruit surface (Meissner, 1952). This raised net tissue gives the surface of the cantaloupe an inherent roughness, and this surface roughness may favor bacterial attachment and hinder microbial detachment. Attachment of S. Poona to cantaloupe rind was reported by Barak et al. (2003). Annous et al. (2005) also reported that *Salmonella* can attach to the rind of cantaloupes, and suggested that the unique characteristics of the cantaloupe surface provide a large number of attachment sites for bacteria and impede contact between bacteria and aqueous sanitizers. Richards and Beuchat (2004) inoculated cantaloupes by dipping in a suspension of S. Poona, at 4 or 30 °C, and then measured the water uptake from the inoculum suspension. The expected greater water uptake when dipping was carried out at 4 °C was observed only for Eastern cantaloupes. Also, the water uptake was significantly greater for Western cantaloupes than for Eastern cantaloupes. These authors attributed these differences to, among other factors, a more dense netting on Western cantaloupes in comparison to the Eastern cantaloupes.

Netting is a manifestation of cuticle disruption (Meissner, 1952). The cuticle is part of the dermal system, which governs the regulation of water loss. The cuticle is composed of surface waxes, cutin embedded in wax, and a layer of mixtures of cutin, wax, and polysaccharides; its thickness and structure vary greatly depending on the level of development of the plant (Kader, 2002). Greater disruption of the cuticle on Western cantaloupes may enable retention of more water than the netting of Eastern cantaloupes allows (Richards and Beuchat, 2004). This water retention is indicative of the potential for microorganisms to internalize within the rind of cantaloupes if they are washed in dump tanks with contaminated water.

Biofilm Formation

This section will be focused on biofilm formation on melons (see Chapter 2 for more information concerning biofilms on produce). The attachment of *Salmonella* onto cantaloupe rind has been documented. If the attached microorganisms can grow, biofilm formation may occur. Annous et al. (2005) observed rapid biofilm formation of *Salmonella* spp. inoculated on cantaloupe rind. The biofilm provides a protective glycocalyx, which will make the organism recalcitrant to the antimicrobial activity of sanitizers (Frank, 2001). In addition, the biofilm structure may enhance the ability of the microorganism to spread to noncontaminated areas of the product, and even to food preparation surfaces during cutting or peeling, which may result in cross-contamination from the melons to other foods.

Microbial Infiltration and Internalization

Any fissures, cuts, and such of the melon will ultimately favor the entry of microorganisms to the flesh of the fruits. Bacterial internalization in melons has not been as studied as for other commodities. Richards and Beuchat (2004) proposed that the adherence or infiltration of microorganisms into the melon may not be entirely promoted by temperature/pressure differentials but by the surface characteristics of the melons as well. Researchers with the FDA tested the potential for fluid infiltration during cantaloupe hydrocooling using brilliant blue as an indicator of water infiltration. They observed dye infiltration in 28% of 170 melons after immersion in iced water containing the dye, although dye buildup was observed in cankers (rind blemishes); intact melons also were infiltrated (Smith, FDA, personal communication, 2004).

Research conducted at the University of California in Davis (Suslow, 2004) showed that *S.* Typhimurium was able to internalize into cantaloupes through the ground spot and, secondarily, through the stem scar. After postharvest processing, the microorganism was found 5 mm under the rind. The ground spot is the area of the melon that is in contact with the ground during melon development. For cantaloupes, ground contact results in an area with a thin and underdeveloped rind, which is also poorly netted and more susceptible to fungal or bacterial growth. These characteristics make the ground spot an area of great potential for microbial internalization during postharvest practices. Soft rot has been reported to promote bacterial internalization. In a study on naturally contaminated fresh market produce, Wells and Butterfield (1997) found a higher incidence of bacteria that were biochemically similar to *Salmonella* on cantaloupes showing soft rots than on healthy cantaloupes.

USE OF ANTIMICROBIAL TREATMENTS TO DECONTAMINATE MELONS

Chapters 16 and 17 cover different aspects of the current technologies for reducing pathogens on produce. Therefore, this chapter will describe efforts to reduce pathogens on fresh and fresh-cut melons. Most studies have focused on cantaloupes and to a lesser extent, on honeydew melons. The effect of surface sanitizers on the transfer of pathogens from contaminated rind to the flesh during peeling or cutting has been discussed already.

Treatments Tested on Fresh Melons
Chemical Disinfectants

Although different researchers indicate that bacteria are recalcitrant to aqueous sanitizers when present in or on fruits and vegetables, mainly due to superficial as well as physico-chemical characteristics of these food products of plant origin (Annous et al., 2005; Ukuku and Fett, 2004; Ukuku et al., 2005), cantaloupe washing is still common practice in the produce industry, except when melons are field-packed. A water wash is applied to remove soil and other dirt from the melon surface, and is usually followed by a wash with a sanitizer. The water wash alone does not have a significant effect at removing bacterial pathogens, and only those that are loosely attached may be removed if not located in areas that are out of the reach of the water. The topography of the melon rind plays a role in the removal of microorganisms. Parnell et al. (2005) reported that soaking cantaloupes in water for 60 seconds resulted in a reduction of 0.7 log cycles in *Salmonella*, whereas soaking honeydew melons in plain water for 60 seconds resulted in a 2.8 log reduction. These authors also studied the rinds of both types of melons by scanning electron microscopy, and observed a large number of crevices and cracks on the cantaloupe rind, whereas honeydew rind was smooth. The natural roughness of the cantaloupe rind is thought to make removal of microorganisms more difficult than the smooth rind of honeydew melons, likely due to a larger number of protective areas on cantaloupe rind, which promotes bacterial attachment and allows the pathogens to evade contact with water.

Applying sanitizers during postharvest operations seems to be more meaningful when the purpose is to reduce microbial populations in the wash water, which can help prevent internalization and cross-contamination among product lots, rather than to sanitize the product. However, regardless of the limitations of produce disinfection procedures using aqueous sanitizers, such procedures can have some antimicrobial effect, reducing pathogen levels to some extent. Cantaloupe sanitizing may be applied as one more hurdle in a holistic approach to food safety, applicable not only to melons but to all fresh and fresh-cut produce, always keeping in mind that these treatments are not sufficient to stand alone as a kill step during processing. Rodgers et al. (2004) inoculated several fruits and vegetables including cantaloupes, with *E. coli* O157:H7 and *L. monocytogenes*, and then dipped the inoculated samples in solutions of chlorinated trisodium phosphate, chlorine dioxide, ozone, and peroxyacetic acid. According to these authors, exposure of cantaloupes for five minutes to these sanitizers resulted in a reduction of both pathogens from about 6.0 log CFU/g to undetectable levels on the rind, by all treatments, and the reduction observed on melons exposed to water alone was of 0.9 log CFU/g. Large reductions (6.7–7.3 log cycles) were also reported for *E. coli* O157:H7 on inoculated cantaloupes by dipping in solutions of lactic acid (1.5%), lactic acid + H_2O_2 (1.5% each), or lactic acid (1.5%) + tergitol (0.3%), and of 4.3 to 5.5 log cycles by using sodium hypochlorite at 200 mg/L free chlorine, all solutions applied at 20 or 30 °C (Materon, 2003). For chlorine, the reduction of this pathogen was significantly smaller for a contact time of one minute than for 10 minutes of contact. For all lactic acid preparations the time of contact did not have an effect.

In contrast to these reports, which show large reductions of bacterial pathogens on melons by treatment with aqueous sanitizers, the majority of studies indicate that cantaloupes are especially difficult to sanitize (Alvarado-Casillas et al., 2007; Parnell et al., 2005; Ukuku and Sapers, 2001; Ukuku and Fett, 2002a; Ukuku, 2004; Ukuku et al., 2005). Alvarado-Casillas et al. (2007) reported that treating cantaloupes with hypochlorite at 200, 600, and 1000 mg/L resulted in reductions of 2.1 to 2.9 log cycles for *S.* Typhimurium and 1.5 to 2.1 log cycles for *E. coli* O157:H7 on the cantaloupe surface. When hot (55 °C), 2% lactic acid solution was sprayed, the reductions obtained were 3.0 log cycles for *S.* Typhimurium and 2.0 log cycles for *E. coli* O157:H7. Parnell et al. (2005) reduced populations of *S.* Typhimurium LT2 (a nonvirulent strain) on cantaloupes by 1.8 log cycles by soaking the melons in a chlorine solution at 200 mg/L free chlorine, and obtained an additional 0.9 log reduction by scrubbing with chlorine solution, in comparison to soaking only. Ukuku (2004) observed a 2.3 to 2.5 log reduction in *Salmonella* populations on cantaloupes by immersion in 2.5% or 5.0% H_2O_2. According to Sapers et al. (2001), washing cantaloupes with 5% H_2O_2 at 50 °C, alone or in combination with a commercial detergent formulation, was more effective than washing with water, surfactant solutions, 1000 ppm Cl_2, trisodium phosphate, or a commercial detergent formulation in reducing the microbial load on cantaloupe rind. The effectiveness of H_2O_2 over other sanitizers at reducing bacterial pathogens on the rind of melons has been extensively documented by a single research group (Ukuku and Sapers, 2001; Ukuku and Fett, 2002a; Ukuku, 2004; Ukuku et al., 2005).

Hot Water Treatment

Hot water treatments are not recommended for all products because of the possibility of product damage; however, when the peeled product is to be further processed, such as in juice preparation, or when the rind is sufficiently strong, surface pasteurization seems to be a very effective alternative for reducing pathogens. The lethal effect of heat can even reach microorganisms located in places that are not reached by chemical sanitizers. The use of hot water dips was proposed by Pao and Davis (1999) for reducing pathogens on oranges that were further used for juice. Later, researchers with the USDA's Agricultural Research Service developed a method for surface pasteurization of melons. Annous et al. (2004) demonstrated that a dip in hot water (76 °C for 6 min) was able to thermally inactivate S. Poona regardless of attachment or biofilm formation, while maintaining the melon firmness. Ukuku et al. (2004) compared the effectiveness of a 60-second dip in hot water (70–90 °C) or H_2O_2 at 70 °C in reducing populations of *Salmonella* (5-serotype cocktail) on inoculated cantaloupes. Both water at 90 °C and H_2O_2 at 70 °C resulted in about a 4 log reduction without affecting the stability of the flesh. They concluded that hot-water pasteurization, or hot H_2O_2 treatment can reduce the risk of enteric disease through the consumption of cantaloupes that have been surface-contaminated with *Salmonella*. This treatment also was found effective at extending the shelf-life of cantaloupes. More recently, Solomon et al. (2006) verified the lethal effect of surface pasteurization against pathogens on cantaloupe surfaces and conducted computer analysis to determine the heat penetration during treatment. They concluded that the edible flesh of the cantaloupe remained cool while the temperature of the outer surface of the rind increased rapidly.

Other Treatments

Kozempel et al. (2002) developed the Vacuum/Steam/Vacuum system, consisting of rapidly applying a vacuum to eliminate air and humidity, which act as heat insulators on the surface to be treated by steam, then applying an antimicrobial steam treatment, and finally applying another vacuum step to cool the surface and prevent heat damage of the product. Using this system, these authors obtained a reduction of *Listeria innocua* (used as a surrogate for *L. monocytogenes*) of 4.0 to 4.7 log cycles.

FRESH-CUT MELONS

Treatment with Antimicrobial Agents

The effects of a water wash, 50 mg/L chlorine, and 10 µg/ml nisin mixed with EDTA on populations of native mesophilic aerobes, *Pseudomonas* spp., lactic acid bacteria, and yeast and molds were tested on whole and on fresh-cut cantaloupes and honeydew melons (Ukuku and Fett, 2002b). Aerobic plate counts (APC) on untreated fresh-cut pieces of cantaloupe were found to be approximately 1 log cycle higher than on fresh-cut pieces of honeydew, which supports the idea that the cantaloupe surface harbors larger populations than other melons, and therefore a larger number of microorganisms may be transferred to the flesh during cutting. According to these authors, treatment with chlorine proved more effective than nisin + EDTA at reducing all microorganisms, and there were no differences in odor, appearance, and overall acceptability ratings for both melons regardless of the treatment applied. Mesophilic aerobes were reduced to undetectable levels on fresh-cut honeydew, and by about 2 log cycles on fresh-cut cantaloupes. All other organisms were reduced to undetectable levels on both types of melon, although these counts increased during refrigerated storage. Even though these authors did not indicate the level of detection in their counting methods, from the APC obtained from the controls, it can be estimated that the reductions obtained were greater than 3 log cycles.

In Spain, Raybaudi-Massilia et al. (2008) developed an edible coating for fresh-cut melon, which includes antimicrobial molecules carried by the alginate-based coating. The antimicrobials used were essential oils, and their active ingredients were added to the edible coating during preparation. In addition, 2.5% malic acid and 2% calcium lactate also were dissolved and mixed with the alginate coating base. Cantaloupe (*piel de sapo* melon) pieces were inoculated with *S.* Enteritidis and covered with the coating linked to the antimicrobials. All antimicrobials were effective at reducing *Salmonella* and other native microbiota on fresh-cut cantaloupes, and also, the use of the edible coatings containing antimicrobials resulted in an increased shelf-life and improved microbiological quality.

Irradiation

A promising technology for destroying pathogens on fresh and fresh-cut produce is ionizing irradiation. Irradiation kills microorganisms by exposing the matrix to be treated to ionizing energy, which may be gamma rays, x-rays, or electron beams. All these types of irradiation follow the same biocidal mechanism. When the product is exposed to irradiation, the gamma or x-rays, or electron beams collide with the microbial DNA, causing multiple breaks in the DNA chain, thus rendering the cells unable to grow. The organization that regulates irradiation internationally is the International Atomic Energy Agency (IAEA), a part of the United Nations Organization. In the United States, food irradiation is approved by the FDA on a case-by-case basis; currently it is not approved for treatment of melons in the United States or any other country with regulations on irradiation of specific commodities. On August 22, 2008, the FDA amended the current regulation to approve irradiation (up to 4 kGy) for controlling foodborne pathogens and extending the shelf-life in fresh iceberg lettuce and fresh spinach (Federal Register, 2008), and it is likely that other commodities will be approved for irradiation in the future.

Studies on the use of electron beam irradiation for reducing pathogens in fresh-cut cantaloupes indicated that doses close to 1 kGy will result in a reduction of *Salmonella* between 2.2 and 3.6 log cycles (Palekar, 2004). Quality studies for irradiated fresh-cut cantaloupes indicated that irradiation at 1.4 kGy did not have any effect on the quality or sensory characteristics of the cut melons (Palekar et al., 2004a). Fan et al. (2006) reported the effect of gamma irradiation on fresh-cut cantaloupes obtained from fresh cantaloupes that were treated by hot water pasteurization. The cantaloupe cubes were treated at doses up to 0.5 kGy, which resulted in a reduction in aerobic plate counts of 0.5 to 1.4 log cycles. This low-dose irradiation treatment, applied to fresh-cut melon cubes, also extended the shelf-life of the product and did not have any adverse effect on sensory characteristics of the fresh-cut melons.

CONCLUSIONS

The relatively frequent occurrence of outbreaks of foodborne disease linked to melons, often imported, indicates the relevance of effective control measures for reducing the risk of contamination with human pathogens. Cantaloupes are particularly impervious to chemical disinfectants, most likely due to the unique composition and structure of their netted rind, which favors bacterial attachment and biofilm formation. Nevertheless, melon disinfection can still reduce pathogens to some extent, and this measure may be linked to food-safety programs that include procedures that prevent contamination during growing, harvesting, and packing this unique commodity. These Good Agricultural Practices, together with postharvest disinfection and introduction of further pathogen reduction strategies during packing, fresh-cut processing, and marketing, can be linked concurrently in a holistic approach to food safety. However, more research is needed to understand the sources and mechanisms for contamination in the field, how this contamination can proliferate and spread over many product units in a shipment, and whether novel technologies can be applied as additional hurdles to reduce pathogen levels and make safe melons available to consumers.

REFERENCES

Ackers, M., Pagaduan, R., Hart, G. et al. (1997). Cholera and sliced fruit: Probable secondary transmission from an asymptomatic carrier in the United States. *Int. J. Inf. Dis.* **1,** 212–214.

Akin, E. D., Harrison, M. A., and Hurst, W. (2008). Washing practices on the microflora on Georgia-grown cantaloupes. *J. Food Prot.* **71,** 46–51.

Alvarado-Casillas, S., Ibarra-Sánchez, S., Rodríguez García, O. et al. (2007). Comparison of rinsing and sanitizing procedures for reducing bacterial pathogens on fresh cantaloupes and bell peppers. *J. Food Prot.* **70,** 655–660.

Annous, B. A., Burke, A., and Sites, J. E. (2004). Surface pasteurization of whole fresh cantaloupes inoculated with *Salmonella* Poona or *Escherichia coli*. *J. Food Prot.* **67,** 1876–1885.

Annous, B. A., Solomon, E. B., Cooke, P. H., and Burke, A. (2005). Biofilm formation by *Salmonella* spp. on cantaloupe melons. *J. Food Saf.* **25,** 276–287.

Anonymous. (1993). Yet another outbreak of hemorrhagic colitis–Corvallis. Center for Disease Prevention and Epidemiology, Oregon Health Division, Portland. *Commun. Dis. Summary* **42,** 1–2.

Anonymous. (2005). Memorandum of Understanding between the Food and Drug Administration Department of Health and Human Services of the United States of America and Servicio Nacional de Sanidad, Inocuidad y Calidad Agroalimentaria of the United Mexican States concerning Entry of Mexican Cantaloupes into the United States of America. Available at www.fda.gov/oia/agreements/-MOU-FDA-SAGARPA.pdf. Accessed 12/10/2008.

Asta, G. (1999). Fresh cut and prepared rockmelon—A discussion on risk. AFS **2,** 8–9.

Barak, J. D., Chue, B., and Mills, D. C. (2003). Recovery of surface bacteria from and surface sanitization of cantaloupes. *J. Food Prot.* **66,** 1805–1810.

Beers, A. (2000). Outbreak at Milwaukee Sizzler provides lessons to industry (*E. coli* prevention). *Food Chem. News* 27-Nov-2000 **42**(41), 4.

Beuchat, L. R. (1996). Pathogenic microorganisms associated with fresh produce. *J. Food Prot.* **59,** 2004–2216.

Beuchat, L. R. and Scouten, A. J. (2004). Factors affecting survival, growth, and retrieval of *Salmonella* Poona on intact and wounded cantaloupe rind and in stem scar tissue. *Int. J. Food Microbiol.* **21,** 683–694.

Bhagwat, A. (2006). Microbiological safety of fresh-cut produce: Where are we now? In *Microbiology of fresh produce* (K. Matthews), Emerging Issues in Food Safety, pp. 121–165. ASM. Press., Washington, DC.

Blostein, J. (1993). An outbreak of *Salmonella* Javiana associated with consumption of watermelon. *J. Environ. Health* **56,** 29–31.

Caldwell, K. N., Anderson, G. L., Williams, P. L., and Beuchat, L. R. (2003). Attraction of a free-living nematode, *Caenorhabditis elegans*, to foodborne pathogenic bacteria and its potential as a vector of *Salmonella* Poona for preharvest contamination of cantaloupe. *J. Food Prot.* **66,** 1964–1971.

Calvin, L. (2003). Produce, food safety, and international trade: Response to U.S. foodborne illness outbreaks associated with imported produce. In *International*

Trade and Food Safety (J. Buzby, Ed.). USDA, ERS, AER Number 828, Nov. 2003. United States Department of Agriculture, pp. 74–96. Available at www.ers.usda.gov/publications/aer828/. Accessed 11/12/2008.

Castillo, A. and Escartin, E. F. (1994). Survival of *Campylobacter jejuni* on sliced watermelon and papaya. *J. Food Prot.* **57,** 166–168.

Castillo, A., Mercado, I., Lucia, L. M. et al. (2004). *Salmonella* contamination during production of cantaloupe: A bi-national study. *J. Food Prot.* **67,** 713–720.

CDC. (1979). *Salmonella oranienburg* gastroenteritis associated with consumption of pre-cut watermelons-Illinois. *MMWR* **28,** 522–523.

CDC. (1991). U.S. epidemiologic notes and reports: Multistate outbreak of *Salmonella* Poona infections, United States and Canada, 1991. *MMWR* **40,** 549–552.

CDC. (2002). U.S. multistate outbreaks of *Salmonella* Poona infections associated with eating cantaloupe from Mexico, United States and Canada, Period 2000–2002, November 2002. *MMWR* **51**(46), 1044–1047.

CDC. (2003). U.S. Foodborne disease outbreaks line listings, 1990–2003. Available at www.cdc.gov/foodborneoutbreaks/outbreak_data.htm. Accessed 11/12/2008.

CDC. (2003a). U.S. Outbreak of *Salmonella* serotype Javiana infections, Orlando, Florida, June 2002. *MMWR* **51,** 683–684.

CDC. (2008). *Salmonella* Litchfield. Investigation update: Outbreak of *Salmonella* Litchfield infection. Available at www.cdc.gov/salmonella/litchfield/. Accessed 11/12/2008.

Agricultural Marketing Service (AMS/USDA). (2008). National Organic Program. Soil fertility and crop nutrient management practice standard. Code of Federal Regulations Title 7 vol. 3, 205.203. Available at http://frwebgate.access.gpo.gov/cgi-bin/get-cfr.cgi. Accessed 12/09/2008.

Chapman, B. J. (2005). An evaluation of an on-farm safety program for Ontario greenhouse vegetable producers; a global blueprint for fruit and vegetable producers. M. Sc. Thesis, The University of Guelph, Master of Science, Appendix 2.1. Produce related outbreaks from 1990–2003, pp. 170–177.

Craigmill A. L. (2000). Good Agricultural Practices are critical to stemming increase in produce outbreaks. Environmental Toxicology Newsletter Vol. 20 No. 2. Available at http://extoxnet.orst.edu/newsletters/ucd2000/nltrapr00.htm. Accessed July 2008.

Deeks, S., Ellis A, Ciebin, B. et al. (1998). *Salmonella* Oranienburg, Ontario. *Can. Commun. Dis. Rep.* **24,** 177–179.

Del Rosario, B. A. and Beuchat, L. R. (1995). Survival and growth of enterohemorrhagic *Escherichia coli* O157:H7 in cantaloupe and watermelon. *J. Food Prot.* **58,** 105–107.

Duffy, E. A., Lucia, L. M., Kells, J. M. et al. (2005). Concentrations of *Escherichia coli* and genetic diversity and antibiotic resistance profiling of *Salmonella* isolated from irrigation water, packing shed equipment, and fresh produce in Texas. *J. Food Prot.* **68,** 70–79.

Falcão, D. P., Valentini, S. R., and Leite, C. Q. F. (1993). Pathogenic or potentially pathogenic bacteria as contaminants of fresh water from different sources in Araraquara, Brazil. *Wat. Res.* **27,** 1737–1741.

Fan, X., Annous, B. A., Sokorai, K. J. B. et al. (2006). Combination of hot-water surface pasteurization of whole fruit and low-dose gamma irradiation of fresh-cut cantaloupe. *J. Food Prot.* **69,** 912–919.

FAO. (2003). Food and Agriculture Organization of the United Nations, Codex Alimentarius. Code of hygienic practice for fresh fruits and vegetables. CAC/RCP 53 – 2003. Available at www.codexalimentarius.net/web/more_info.jsp?id_sta=10200. Accessed 11/12/2008.

Federal Register. (2008). Irradiation in the production, processing and handling of food. Final Rule. 21 CFR 179.26. FR Aug 22, 2008. **164,** 49603.

FDA. (1998). Guidance for Industry. Guide to minimize microbial food safety hazards for fresh fruits and vegetables. Available at www.foodsafety.gov/~dms/prodguid.html. Accessed 11/12/2008.

FDA. (2003). Detention without physical examination of cantaloupes from Mexico, import alert IA2201 attachment, 2003. Available at www.fda.gov/ora/fiars/ora_import_ia2201.html. Accessed 12/10/2008.

FDA. (2008). US Food and Drug Administration Guide to minimize microbial food safety hazards of fresh-cut fruits and vegetables. Available at www.cfsan.fda.gov/~dms/prodgui4.html. Accessed 11/12/2008.

FDA. (1999). Survey of imported fresh produce. Available at www.cfsan.fda.gov/~dms/prodsur66.html. Accessed 11/12/2008.

Eitenmiller, R. R., Johnson, C. D., Bryan, W. D. et al. (1985). Nutrient composition of cantaloupe and honeydew melons. *J. Food Sci.* **50,** 136–138.

Escartin, E. F., Castillo-Ayala, A., and Saldaña-Lozano, J. (1989). Survival and growth of *Salmonella* and *Shigella* on sliced fresh fruit. *J. Food Prot.* **52,** 471–472.

Figueroa-Aguilar, G. A., González-Ramírez M., Molina-García A. J. et al. (2005). Identificación de *Salmonella* spp en agua, melones cantaloupe y heces fecales de iguanas en una huerta melonera. *Med. Int. Mex.* **21**(4), 255–258.

Frank, J. F. (2001). Microbial attachment to food and food contact surfaces. *Adv. Food Nutr. Res.* **43,** 319–370.

Fredlund, H., Back, E., Sjoberg, L., and Tornquist, E. (1987). Watermelon as a vehicle of transmission of shigellosis. *Scand. J. Infect. Dis.* **19,** 219–221.

Gaylor, G. E., MacCready, R. A., Reardon J. P., and McKernan, B. F. (1955). An outbreak of salmonellosis traced to watermelon. *Pub. Health Rep.* **70,** 311–313.

Gerba, C. P. and Choi, C. Y. (2006). Role of irrigation water in crop contamination by viruses. pp. 257–263. In *Viruses in foods* (S. M. Goyal, Ed.). Springer Science, New York.

Gibbs, D. S., Anderson, G. L., Beuchat, L. R. et al. (2005). Potential role of *Diploscapter* sp. strain lkc25, a bacterivorous nematode from soil, as a vector of foodborne pathogenic bacteria to preharvest fruits and vegetables. *Appl. Environ. Microbiol.* **71,** 2433–2437.

Green, T., Hanson, L., Lee, L. et al. (2005). North American approaches to regulatory coordination. In *Agrifood regulatory and policy integration under stress* (Huff, K., Meilke, K. D., Nutson, R. D., Ochoa, R. F., Rude, J., Eds.), pp. 9–47. Proc. 2nd Workshop of the North American Agri-food Market Integration Consortium, San Antonio, Texas, May 5, 2005. Available at http://naamic.tamu.edu/sanantonio.htm. Accessed 12/10/2008.

Golden, D. A., Rhodehamel, E. J., and Kautter, D. A. (1993). Growth of *Salmonella* spp. in cantaloupe, watermelon and honeydew melons. *J. Food Prot.* **56,** 194–1296.

Hakerlerler, H., Okur, B, Irget, E., and Saatç, N. (1999). Carbohydrate fractions and nutrient status of watermelon grown in the alluvial soils of Küçük Menderes Watershed, Turkey. In *Improved crop quality by nutrient management* (D. Anac and P. Martin-Prével, Eds.), pp. 163–165. Kluwer Acad. Pub., Boston.

Hurst, W. C. and Shuler, G. A. (1992). Fresh produce processing—an industry perspective. *J. Food Prot.* **55,** 824–827.

International Society for Infectious Diseases. (2004). *E. coli* O157, cantaloupes—USA (Montana). *Promed Mail*, Available at http://apex.oracle.com/pls/otn/f?p=2400:1001:1720436792652856::::F2400_P1001_BACK_PAGE,F2400_P1001_ARCHIVE_NUMBER,F2400_P1001_USE_ARCHIVE:1001,20040722.1997,Y. Accessed 11/05/2008.

Iversen, A. M., Gill, M., Bartlett, C. L. et al. (1987). Two outbreaks of foodborne gastroenteritis caused by a small round structured virus: Evidence of prolonged infectivity in a food handler. *Lancet* **2,** 556–558.

Jay, M. T., Cooley, M., Carychao, D. E. et al. (2007). *Escherichia coli* O157:H7 in feral swine near spinach fields and cattle, central California coast. *Emerg. Infect. Dis.* **13,** 1908–1911.

Johnston, L. M., Jaykus, L. A., Moll, D. et al. (2005). A field study of the microbiological quality of fresh produce. *J. Food Prot.* **68,** 1840–1847.

Kader, A. A. (2002). Postharvest biology and technology: An overview. In *Postharvest technology of horticultural crops, 3rd edition* (A. A. Kader, Ed.), pp. 39–47. University of California Agriculture and Natural Resources. Publication 3311.

Karchi, Z. (2000). Development of melon culture and breeding in Israel. *Acta Hort.* **510,** 13–17.

Kozempel, M., Radewonuk, E. R., Scullen, O. J., and Goldberg, N. (2002). Application of the vacuum/steam/vacuum surface intervention process to reduce bacteria on the surface of fruits and vegetables. *Innov. Food Sci. Emerg. Technol.* **3,** 63–72.

Madden, J. M. (1992). Microbial pathogens in fresh produce—the regulatory perspective. *J. Food Prot.* **55,** 821–823.

Martínez-Téllez, M. A., Vargas-Arispuro, I., Silva-Bielenberg H. K. et al. (2007). Producción y Manejo Poscosecha de Hortalizas. In *Buenas Prácticas en la Producción de Alimentos* (A. A. Gardea, G. González, I. Higuera, and F. Cuamea, Eds.). Trillas Editorial, México.

Materon, L. A. (2003). Survival of *Escherichia coli* O157:H7 applied to cantaloupes and the effectiveness of chlorinated water and lactic acid as disinfectants. *World J. Microbiol. Biotechnol.* **19,** 867–873.

Meissner, F. (1952). Die korkbildung der fruchte von *Aesculus*-und *Cucumis*-arten. *Osterr. Bot. Ztschr.* **99,** 606–623.

Mohle-Boetani J. C., Reporter R., Werner S. B. et al. (1999). An outbreak of *Salmonella* serogroup Saphra due to cantaloupes from Mexico, *J. Infect. Dis.* **180,** 1361–1364.

Ng, P. J., Fleet, G. H., and Heard, G. M. (2005). Pesticides as a source of microbial contamination of salad vegetables. *Int. J. Food Microbiol.* **101,** 237–250.

NRCS (2005). Conservation Practice Standard. Composting Facility. Code 317. Available at http://efotg.nrcs.usda.gov/references/public/AL/tg317.pdf. Accessed 12/09/2008.

Palekar, M. P. (2004). Attachment of *Salmonella* on cantaloupe and effect of electron beam irradiation on quality and safety of sliced cantaloupe. Ph. D. dissertation, Texas A&M University.

Palekar, M. P., Cabrera-Diaz, E., Kalbasi-Ashtari, A. et al. (2004). Effect of electron beam irradiation on the bacterial load and sensorial quality of sliced cantaloupe. *J. Food Sci*. **69,** M267–M273.

Pao, S. and Davis, C. L. (1999). Enhancing microbiological safety of fresh orange juice by fruit immersion in hot water and chemical sanitizers. *J. Food Prot*. **62,** 756–760.

Parnell, T., Harris, L. J., and Suslow, T. (2005). Reducing *Salmonella* on cantaloupes and honeydew melons using wash practices applicable to postharvest handling, foodservice, and consumer preparation. *Int. J. Food Microbiol*. **99,** 59–70.

Paulson, D. S. (2000). Handwashing, gloving, and disease transmission by the food preparer. *Dairy Food Environ. Sanit*. **20,** 838–845.

Penteado, A. L. and Leitão, M. F. F. (2004a). Growth of *Salmonella* Enteritidis in melon, watermelon and papaya pulps stored at different times and temperatures. *Food Cont*. **15,** 369–373.

Penteado, A. L. and Leitão, M. F. F. (2004b). Growth of *Listeria monocytogenes* in melon, watermelon and papaya pulps. *Int. J. Food Microbiol*. **92,** 89–94.

PMA and UFFVA (Produce Marketing Association and United Fresh Fruit and Vegetable Association). (2005). Commodity specific food safety guidelines for the melon supply chain. 1st edition. Available at www.cfsan.fda.gov/~acrobat/melonsup.pdf. Accessed 12/12/2008.

Raybaudi-Massilia, R. M., Mosqueda-Melgar, J., and Martín-Belloso, O. (2008). Edible alginate-based coating as carrier of antimicrobials to improve shelf-life and safety of fresh-cut melon. *Int. J. Food Microbiol*. **121,** 313–327.

Richards, G. M. and Beuchat, L. R. (2004). Attachment of *Salmonella* Poona to cantaloupe rind and stem scar tissues as affected by temperature of fruit and inoculum. *J. Food Prot*. **67,** 1359–1364.

Ries, A. A., Zaza, S., and Langkop, C. (1990). A multistate outbreak of *Salmonella* Chester linked to imported cantaloupe [Abstract]. In *Programs and abstracts of the 30th Interscience Conference on Antimicrobial Agents and Chemotherapy*. Washington, DC: *American Society for Microbiology*. P238. Abstract. 195.

Rodgers, S. L., Cash, J. N., Siddiq, M., and Ryser, E. T. (2004). A comparison of different chemical sanitizers for inactivating *Escherichia coli* O157:H7 and *Listeria monocytogenes* in solution and on apples, lettuce, strawberries, and cantaloupe. *J. Food Prot*. **67,** 721–731.

Sapers, G. M., Miller, R. L., Pilizota, V., and Mattrazzo, A. M. (2001). Antimicrobial treatments for minimally processed cantaloupe melon. *J. Food Sci*. **66,** 345–349.

SENASICA (Servicio Nacional de Sanidad, Inocuidad y Calidad Agroalimentaria). (2006). Guidelines for the voluntary implementation of good agricultural practices and good management practices in the production and packing processes of fruits and vegetables for fresh consumption by humans. Available at http://senasicaw.senasica.sagarpa.gob.mx/portal/html/inocuidad_agroalimentaria/inocuidad_agricola/SENASICA_guidelines_230206.pdf. Accessed 12/12/2008.

SIAP-SAGARPA (Sistema de Información Estadística Agroalimentaria y pesquera). (2002). SIACON. Mexico. Available at www.siap.sagarpa.gob.mx/. Accessed 11/12/2008.

Solomon, E. B., Huang, L., Sites, J. E., and Annous, B. A. (2006). Thermal inactivation of *Salmonella* on cantaloupes using hot water. *J. Food Sci.* **71**, M25–M30.

Suslow, T. V. (2003). Key points of control and management of microbial food safety for melon producers, handlers and processors. University of California Division of Agriculture and Natural Resources. Publication 8103. Available at http://anrcatalog.ucdavis.edu/pdf/8103.pdf. Accessed 12/12/2008.

Suslow, T. V. (2004). Minimizing the risk of foodborne illness associated with cantaloupe production and handling in California. Available at http://ucce.ucdavis.edu/files/filelibrary/5622/15931.pdf. Accessed 11/8/2008.

Suslow, T. V. and Cantwell, M. (2001). Recent findings on fresh-cut cantaloupe and honeydew melon. *Fresh Cut* April 2001. Available at www.freshcut.com/fc2001.htm#fc20. Accessed 11/05/2008.

Tamplin, M. (1997). *Salmonella* and cantaloupes. *Dairy Food Environ. San.* **17**, 284–286.

Tauxe, R. V. (1997). Emerging foodborne diseases: An evolving public health challenge. Centers for Disease Control and Prevention, Atlanta, GA. *Special Issue* **3**(4).

Tauxe, R., O'Brien, S. J., and Kirk, M. (2008). Outbreak of food-borne diseases related to the International Food Trade. In *Imported foods. Microbiological issues and challenges* (M. Doyle and M. C. Erickson, Eds.), pp. 69–112. ASM Press, Washington, DC.

Taylor, A. K. (2000). Food protection: New developments in handwashing. *Dairy Food Environ. Sanit.* **20**, 114–119.

Ukuku, D. O. (2004). Effect of hydrogen peroxide treatment on microbial quality and appearance of whole and fresh-cut melons contaminated with *Salmonella* spp. *Int. J. Food Microbiol.* **95**, 137–146.

Ukuku, D. O. (2006). Effect of sanitizing treatments on removal of bacteria from cantaloupe surface, and re-contamination with *Salmonella*. *Int. J. Food Microbiol.* **23**, 289–293.

Ukuku, D. O. and Fett, W. (2002a). Behavior of *Listeria monocytogenes* inoculated on cantaloupe surfaces and efficacy of washing treatments to reduce transfer from rind to fresh-cut pieces. *J. Food Prot.* **65**, 924–930.

Ukuku, D. O., and Fett, W. (2002b). Effectiveness of chlorine and nisin-EDTA treatments of whole melons and fresh-cut pieces for reducing native microflora and extending shelf-life. *J. Food Saf.* **22**, 231–253.

Ukuku, D. O. and Fett, W. F. (2004). Method of applying sanitizers and sample preparation affects recovery of native microflora and *Salmonella* on whole cantaloupe surfaces. *J. Food Prot.* **67**, 999–1004.

Ukuku, D. O. and Sapers, G. M. (2001). Effect of sanitizer treatments on *Salmonella* Stanley attached to the surface of cantaloupe and cell transfer to fresh-cut tissues during cutting practices. *J. Food Prot.* **64**, 1286–1291.

Ukuku, D. E., Bari, M. L., Kawamoto, S., and Isshiki, K. (2005). Use of hydrogen peroxide in combination with nisin, sodium lactate and citric acid for reducing transfer of bacterial pathogens from whole melon surfaces to fresh-cut pieces. *Int. J. Food Microbiol.* **104,** 225–233.

Ukuku, D. O., Pilizota, V., and Sapers, G. M. (2004). Effect of hot water and hydrogen peroxide treatments on survival of *Salmonella* and microbial quality of whole and fresh-cut cantaloupe. *J. Food Prot.* **67,** 432–437.

Ukuku, D. O. and Sapers, G. M. (2006). Microbiological safety issues of fresh melons. In *Microbiology of fruits and vegetables* (G. M. Sapers, J. R. Gorny, and A. E. Yousef, Eds.), pp. 231–251. CRC Press/Taylor & Francis Group, Boca Raton, FL.

Vadlamudi, S. (2004). Effect of sanitizer treatments on *Salmonella enterica* Serotype Poona on the surface of cantaloupe and cell transfer to the internal tissue during cutting practices. M. Sc. thesis. Texas A&M University.

Webster, B. D. and Craig, M. E. (1976). Net morphogenesis and characteristics of the surface of muskmelon fruit. *J. Am. Soc. Hort. Sci.* **101,** 412–415.

Wells, J. M. and Butterfield, J. E. (1997). *Salmonella* contamination associated with bacterial soft rot of fresh fruits and vegetables in the marketplace *Plant Dis.* **81,** 867–872.

Zahniser, S. (2006). U.S.–Mexico agricultural trade during the NAFTA era**.** Proceedings of the Doha, NAFTA and California Agriculture Conference. Giannini Foundation, Sacramento, CA, Jan 13, 2006. Available at http://giannini.ucop.edu/US_Mexico_Zahniser_060123.pdf. Accessed 12/10/2008.

CHAPTER 10

Raw Tomatoes and *Salmonella*

Jerry A. Bartz

Postharvest Pathologist, Plant Pathology Department, University of Florida, Gainesville, FL

CHAPTER CONTENTS

Introduction	223
Commercial Tomato Production and Marketing	225
Evidence that Tomatoes were The Source of Outbreaks of Salmonellosis	229
Outbreak from South Carolina Tomatoes, 1990	229
Second Outbreak Traced to South Carolina Tomatoes, 1993	230
Multistate Outbreak of Salmonellosis, 1998	231
Transplant Games Outbreak, 2002	232
Outbreak of *S.* Newport Linked to Virginia Tomatoes, 2002	232
Multiserotype Convenience Store Outbreak, 2004	233
Second Outbreak of Salmonellosis Caused by Serotype Newport Traced Back to Tomatoes Produced on the Delmarva Peninsula, 2005	234
Three-State Outbreak, 2005	235
Recurrence of Serotype Newport in Delmarva Tomatoes, 2006	235
Outbreak Linked to Ohio Tomatoes, 2006	236
Unanswered Questions	236
Recommendations for Commercial Tomato Production and Handling, Farm-To-Fork	243

INTRODUCTION

Up to 15 outbreaks of salmonellosis have been linked to the consumption of fresh tomato fruits in the United States since 1990. Reports on many of these multistate outbreaks conclude that due to a distribution of outbreak cases over several states, sometimes including Canada, the initial contamination of the fruits occurred on the farm. However, there are no reports of isolating

Salmonella directly from field-grown tomatoes. Gorny (2006) reviewed three different surveys of imported and domestic fresh produce for presence of human pathogens including two by the FDA and one by the USDA. A total of 2924 samples of tomatoes were analyzed in these surveys; *Salmonella* was not detected. Moreover, a majority of illnesses linked to fresh produce during the period 1990 to 2001 was linked to handlers, foodservice, or consumers (average of 83%), as compared with 17% linked to farm. Orozco et al. (2008) detected *Salmonella* in 1.8% of tomatoes grown hydroponically in a greenhouse prior to an extreme weather event during which time flood waters entered several of the houses. Immediately after the flood waters had disappeared, the contamination level increased to 9.4%. Wells and Butterfield (1997) found presumptive *Salmonella* in 6 of 11 samples of sound tomatoes and 7 of 11 samples of soft-rotted tomatoes purchased from local supermarkets in New Jersey. Of 24 presumptive strains, 10 were confirmed as *Salmonella* by serological tests. By contrast, Burnett and Beuchat (2000) did not list tomatoes as an example of "raw produce from which bacterial pathogens have been isolated."

As will be discussed here, certain fields of tomatoes cited in outbreak reports were produced by growers using egregious violations of Good Agricultural Practices (GAPs) such as using pond water to mix pesticides. (Note: The use of the surface water to mix pesticides in Guatemala was believed responsible for a Cyclosporosis outbreak in the United States and Canada that was linked to contaminated raspberries imported from that country; Gorney, 2006.) However, neither the outbreak strain nor any other *Salmonella* species was isolated from the tomatoes produced in these fields. A reason for the failure of investigators to detect *Salmonella* in implicated fields may be the relatively short harvest cycle for field tomatoes (up to three weeks) and time required to identify the probable source. Additionally, only a few of the many tomatoes produced in a field could have been contaminated.

The apparent absence of detectable *Salmonella* in most raw tomato fruits should be contrasted with reports of infectious doses. An infectious dose for 50% of a population of healthy individuals has been estimated at 10^9 cells (Todar, 2005), whereas the minimum infectious dose for certain members of the general population has been estimated as 15 to 20 cells (FDA, 2003). The log 9.0 CFU dose should be easily detected in raw tomatoes or raw tomato products, whereas the latter dose would likely require enrichment.

Tomato fruits are not the preferred substrate for *Salmonella*. The non-host-adapted *Salmonella* serovars that have been linked to outbreaks associated with the consumption of fresh tomatoes infect the intestinal tract of a wide range of warm- and cold-blooded animals (Todar, 2005). These serovars do not seem to multiply significantly in surface waters or other natural environments. They do survive for weeks in water and even longer in soil if conditions are favorable. By contrast, *Salmonella* multiplied on fresh-cut tomatoes stored in laboratory tests (Asplund and Nurmi, 1991; Wei et al.,1995, Zhuang et al., 1995). How well these tests model the natural environment is discussed in the section "Unanswered Questions" below.

COMMERCIAL TOMATO PRODUCTION AND MARKETING

A review of commercial tomato production and marketing enables a better understanding of the difficulty in finding outbreak strains of *Salmonella* in tomatoes from implicated fields. Commercial fresh market tomatoes include several different types ranging from rounds to specialty types. The rounds are sometimes separated into standard rounds and the larger beefsteak tomatoes (Costa and Heuvelink, 2005). The size difference between standard rounds and beefsteaks depends on the geographic area of production. In Europe, rounds average 70 to 100 g, whereas beefsteak tomatoes average 180 to 250 g. In the United States, rounds range from 184 to 252 g (Maynard et al., 2000). The "round" shape, which is genetically controlled, is variously termed flattened, oval, or globe-shaped, depending on a fruit's variation from a geometric round object. A specialty tomato, called a Roma or plum tomato, is a derivative of tomatoes grown for processing into sauce or paste. It is oblong or plum-shaped as compared with round and is preferred for certain uses due to its higher solid contents (less water). Cherry and grape tomatoes are also specialty tomatoes. They are small, usually sweet, and are usually eaten without being sliced or diced. The cherry tomato is round, and the grape tomato is oblong. Another specialty tomato is called the tomato-on-the-vine, or TOV. This is a greenhouse-grown type where fruits remain attached to a fruiting truss. The entire truss is marketed as one unit.

In a tomato variety evaluation, the average yield of an acre of field-grown tomatoes in Florida during a fall season ranged from 32,500 to 65,000 pounds with a first harvest of 3450 to 23,300 pounds (Maynard et al., 2000). Two recently interviewed commercial packers noted that an acre of tomatoes cost $7000 or $8000 to produce, and the harvest ranged from 30,000 to 32,500 pounds per acre (VanSickle, 2008). The yield is affected by season, with acreages planted in the spring and harvested in early to mid-summer yielding the most.

Tomato fruits are picked at sizes ranging from 80% to full size, and at ripeness stages ranging from green to table ripe (Saltveit, 2005). Field crews, who may or may not be wearing gloves (DACS, 2008), move down the rows of plants stripping the fruits into buckets. The crews concentrate on fruit size and position on the plant (crown or lower hand/first or second level on the plant, etc.). Each person is paid by the number of buckets they pick; hence, there is an incentive to work rapidly and a disincentive to handle each fruit carefully or to be aware of its potential defects. Crews are now trained not to pick up dropped fruits (DACS, 2008). If noticed by the harvester, decaying fruits are usually tossed to the row middle. The buckets are emptied into field bins or gondolas (modified semitrailer rigs), which are hauled to a packinghouse where the fruits are emptied into vats of water and then flumed to a packingline. At the packingline, fruits are cleaned with a spray wash, then dried, waxed, sorted, sized, and finally boxed. Misshapen, injured, and decaying tomatoes are culled. Breaker (tannish-yellow to pink color noted at the blossom end) to pink fruits (between 30 and 60% of the fruit surface is pink or red) are sorted from the green tomatoes and packed separately. Green tomatoes are sorted mechanically according to size and then placed into cardboard boxes. Mature green fruits are those that will ripen to a satisfactory horticultural quality (Saltveit, 2005). However, fruits ripened from the earliest stages of mature green, defined as locular gel (solid seeds cut when fruit is sliced), are usually not of high quality. Less mature fruits are often part of the harvest of green fruits and have been reported to represent up to 49% of a typical harvest (Chomchalow, 1991).

Sanitation is a critical factor in tomato production and marketing. Historically, sanitation was used to control postharvest decays, specifically bacterial soft rot, sour rot, Rhizopus rot, and gray mold rot (Bartz, 1980, 1991). Sanitation methods (see Chapter 17) also reduce the risk of the contamination of fruit by *Salmonella* and other hazardous microorganisms (Bartz et al., 2002; DACS, 2008). A new addition to the general sanitation practices (*Tomato Best Practices* manual) is the prohibition against applying nonpotable water to plant surfaces, whether in the form of overhead irrigation or spray mixtures (DACS, 2008). Other sanitation practices include regular cleaning of picking buckets and field bins or gondolas. As a cleaning aid, growers are required to use bins or gondolas that are impervious to water (e.g., avoiding wood) (DACS, 2008). An EPA-approved sanitizer must be added to the water system of a packinghouse (Chapter 17). Machinery in the packinghouse must be cleaned and sanitized after each workday, with particular attention paid to food contact surfaces. As a new rule, all tomatoes produced and harvested in Florida that are intended for the commercial market must be treated with a sanitizer that has a demonstrated ability to reduce bacteria on the fruit by 3 or more logs, has an EPA label for treatment of harvested tomato fruit, and is on the list of approved sanitizers maintained by the Florida Department of Agricultural and Consumer Services (DACS, 2008).

Filled boxes of fruit are covered with a lid, and the boxes are stacked onto wooden pallets. A strip of glue is placed on the upper surface of the lid on each box to help with pallet integrity. Straps or meshes of plastic are placed over the boxes to hold them in place. Pallets of fruit are moved into a ripening room where the fruits are exposed to ethylene gas, a natural plant hormone that can be supplied synthetically (Saltveit, 2005). The ethylene treatment is applied to mature green fruit for up to 3 days, at which time most have reached the breaker stage (Maul et al., 2000). In contrast, immature greens may require up to 5 days of treatment before the onset of ripening (breaker stage). The time to table-ripe stage for green fruit varies with storage temperature. At the most desirable temperature for ripening fruits, 20 °C (68 °F), mature green fruit will become table ripe in about 10 days, whereas breaker fruits ripen to table ripe in eight days (Saltveit, 2005).

Prior to being placed into trucks for shipment to markets, all cartons of round tomatoes packed in packinghouses located in the United States are randomly inspected by trained employees (federal-state inspector) of the USDA Agricultural Marketing Service, Fruit and Vegetable Division, Fresh Products Branch (USDA, 1987). The inspectors record whether the package contents meet weight, grade standards, and fruit-size requirements. Tomatoes imported into the United States and specialty tomatoes are not inspected. A separate inspection may occur at the receiving point if requested by the buyer (load condition does not meet buyer expectations). A minimum of 50 tomatoes are examined from each of at least 10 containers and evaluated for size, grade, and ripeness. Fruit characteristics involved with grade standard include softness, smoothness, injury, shape, and decay. Any defects found are ranked as damage, serious damage, very serious damage, and decay. For each grade, there is a limit on the defects allowed. The fill of the boxes is also recorded since underfilling or a loose pack can allow fruits to move during transit and develop surface injury. By contrast, overfilled cartons along with a very tight load stacking arrangement can lead to bruising. Identifying marks on each carton include brand, grower, or shipper name, address, grade, size, pack, weight, and federal or federal-state lot stamp numbers. These are quoted on the inspection form. An inspection form that is filled out and signed by a federal-state inspector is a valid legal document, should load condition be disputed by shipper or receiver in civil court.

Since most final receivers desire tomatoes of uniform ripeness, quality, and sometimes size, the marketing chain for tomatoes may include several levels of handling after the packinghouse including that by a broker, repacker, distributor, and suppliers. Repackers are usually the largest post-packinghouse handlers. They commonly receive truck loads of tomatoes, sometimes from different packinghouses. At the repacker, boxes of tomatoes are emptied onto a wide belt that enables machines or hand labor to sort the fruits by color and defects (substandard color development, decay, wounds, injuries associated with shipment). Those tomatoes that are not at the desired stage of ripeness are placed back into storage for further ripening. The line is usually sanitized with a spray of chlorinated water although individual fruits may be moistened only for a short period of time (Angulo, memorandum). Defective fruits that are salvageable may be placed into boxes for immediate sale at an area of the plant that is open to walk-up buyers (VanSickle, 2008). The boxes enclosing the final product are supposed to be the original ones, with the inspection number, the grower number, and the packinghouse name and address. As such, fruits within a box would be traceable back to the packinghouse and grower. Maintaining fruit contact with the original box can cause inefficiencies in the process such that repackers often commingle fruits from several sources in order to supply the requirements of their customers. The repacker may or may not have facilities for treating fruit with ethylene. Generally, mature green fruits that have been force ripened by treatment with ethylene will be firmer at table ripe and have a longer shelf-life than will field-ripened tomatoes. Consequently, force-ripened tomatoes are often preferred by the food-service industry.

Alternative sources of fresh fruits and vegetables for food service and various small grocers include terminal markets (sometimes called state farmers' markets) (Block, 2008) and field-packed products (Walsh and Rodriquez, 2008). Terminal markets provide a sales outlet for growers of various sizes, although some buyers assert that shippers will dump substandard or distressed loads at these markets, thereby driving down prices and reducing product quality (VanSickle, 2008). Terminal markets located in the southwestern United States may carry items produced in Mexico in addition to those produced by local growers (Walsh and Rodriquez, 2008). Tomatoes may also be field-packed like strawberries and sold directly to food service, terminal markets, or processors. Field packs generally include fruits that have at least started to ripen in the field. Farms shipping field packs may or may not be certified or inspected (Walsh and Rodriquez, 2008). Produce sold in supermarkets as "grown locally" may or may not be produced on inspected farms, depending on the agreement between the supermarket and the grower. However, in states such as Florida, tomatoes sold in this fashion have to be grown, packed, and inspected according to the USDA Tomato Marketing Order for Florida-grown tomatoes. Allures for terminal markets and fresh-pack operations include the ability to purchase odd-lots, to obtain specialty items, to help local agriculture, and to reduce expenses. On the other hand, such sources of fruit often greatly limit the ability of industry or regulators to trace back items to the packer or grower because transactions range from cash purchases to the more usual account and monetary transfers.

A relatively recent development in the marketing of fresh fruits and vegetables including tomatoes is the ready-to-eat product (Cantwell, 1992; Schlimme, 1995). Fresh cut tomatoes may be prepared near the area of production, or at a regional or local site. Additionally, according to outbreak reports, certain restaurants slice or dice tomatoes on-site. However, purchasing precut tomatoes from a processor offers restaurants and other food service institutions several efficiencies, including portion control (prepacks), reduced labor costs, reduction in waste disposal (up to 50% of the raw product can be lost during the preparation of fresh cuts), reduction in refrigeration space, reduction in raw produce inventory, a wider array of menu items, and year-round availability of many items (Schlimme, 1995). In addition, inventory control is simplified in that the period between orders and deliveries by refrigerated truck is typically within a few hours. The source of raw product for processors at the different levels is unclear and may range from contracts with repackers, to walk-up purchases of repackers' seconds, or salvaged fruit, or fruit purchased at a terminal market (Block, 2008).

It is noteworthy that the period of salmonellosis outbreaks traced to tomatoes coincides with a rapid development of the fresh-cut industry. In 2007, linkage of outbreaks of salmonellosis to various forms of cut tomatoes has led the FDA to designate such products as "potentially hazardous food" with a "temperature control for safety" (PHF/TCS) (CFSAN, 2007b).

EVIDENCE THAT TOMATOES WERE THE SOURCE OF OUTBREAKS OF SALMONELLOSIS

Outbreak from South Carolina Tomatoes, 1990

Salmonella enterica, serotype Javiana was recovered from 176 patients with salmonellosis between June 28 and August 2, 1990 in Minnesota, Illinois, Michigan, and Wisconsin (Hedberg et al., 1999). Case/control studies implicated the consumption of raw tomatoes purchased from retail grocers or served in restaurants. The tomatoes were traced back to a single packinghouse in South Carolina, but the marketing chain was not detailed. The packinghouse was inspected, and samples were taken, but neither the outbreak strain nor any other strains of *Salmonella* were isolated. A failure to monitor or maintain free chlorine concentrations in the dump tank of Packer A was cited as a probable cause of the contaminated fruit; however, there was no evidence of a sanitation failure such as excessive postharvest decays.

A few of the illness cases purchased fruit from a local grocer and ate them at home. The tomatoes reportedly were washed and cored before they were eaten. It is unclear how these cleaned tomatoes were served. Most of the bacteria on a clean, dry tomato exist in the stem scar (Samish et al., 1960). If tomatoes were cored, most of the bacteria on a fruit should have been removed, unless the unwashed coring knife was used to slice the tomatoes (Lin and Wei, 1997). Curiously, the age distribution pattern indicated a food that was not often consumed by children; more than half the cases were between the age of 20 and 39 years (Hedberg et al., 1999). Whether children shun sliced fresh tomatoes is unclear.

Second Outbreak Traced to South Carolina Tomatoes, 1993

Salmonellosis caused by the same or similar strains of serotype Montevideo affected 100 people in Illinois, Wisconsin, Minnesota, and Michigan beginning June 29 and ending August 2, 1993 (Hedberg et al.,1999). This is the same multistate area featured in the 1990 outbreak, and once again, a majority of the cases were in the 20- to 39-year-old age group. The illnesses were related to restaurant meals. Traceback again implicated Packer A in South Carolina, and once again linked the packer to the distributors that serviced the restaurants. However, a memorandum from Dr. F. J. Angulo, the EIS officer responsible for the investigation of the outbreak by the CDC, noted that tomatoes from Packer A were shipped to Company A, which is a repacker in the Chicago area and the only one capable of an ethylene gas treatment. On June 21, Packer A shipped one truckload of tomatoes to this repacker and another to Company B in Michigan. These two semis carried 36,000 lbs each, or 5% of the fruits shipped that day. The contaminated fruits were hypothesized to be mixed in with these fruits. The two truck loads, if from the same grower, would represent a single harvest of about four (actually 3.6) acres (assuming a harvest of 20,000 lbs/ac and a modest packout—tomatoes packed/tomatoes harvested $\times$ 100 = packout). Approximately 35 truckloads of fruit were shipped from Packer A to other receivers on the same day without incident. A portion of these truckloads would likely include fruit harvested from the implicated grower since four acres is not representative of the usual field planted by a commercial tomato grower. One report indicated that certain distributers received tomatoes directly from Packer A (Hedberg et al., 1999). The volume of fruit handled by these distributers was not given. Crisp-head lettuce was linked to illness with a slightly higher correlation than tomato fruit. However, the lettuce vehicle was discounted because up to eight distinctly different companies supplied the heads, which were all grown in California.

Multistate Outbreak of Salmonellosis, 1998

Beginning December 6, 1998 and concluding February 2, 1999, 86 patients were sickened by *S. enterica*, serotype Baildon (Cummings et al., 2001). States (and cases) involved include California (44), Virginia (13), Arizona (13), Georgia (8), Illinois (3), Alabama (2), Tennessee (2), and Kansas (1). The percentage of patients 18 years old or older was 93. The outbreak lasted 46 days in California, 19 days in Virginia, and 43 days in Arizona. Traceback implicated two packinghouses in South Florida and one in West Central Florida, which were investigated. Neither the outbreak strain nor any other *Salmonella* were isolated from environmental samples taken from these packinghouses. No significant "sanitation deficiencies" were observed. At least eight farms supplying the packinghouses were visited. Nothing unusual was observed. Environmental samples did not detect *Salmonella*. However, at one of the packinghouses, the investigator noted the lack of records of chlorine concentration, pH, or water temperature. At the time of the investigation, the water temperature was 38.7 °C with a pH of 6.5. The target chlorine concentration was 125 ppm but was not measured.

The demographic of this outbreak in terms of patient age fits the previous two outbreaks. As with the earlier reports, the authors used laboratory models of the ecology of *Salmonella* in tomatoes to support their conclusions and did not address why Florida tomatoes were shipped to California and Arizona despite the relative close proximity of Mexican production areas (Cummings et al., 2001).

The tomato dicer in California was inspected on May 19. Uncored tomatoes were mechanically diced and packaged into five-pound containers, which were sealed and stored at 4.4 °C. Pulp temperature of the product at the time the containers were filled was not given although bath and flume temperatures were 1.1 °C.

Point of service facilities in Virginia processed whole, uncored tomatoes on-site. Of note was the fact that the 13 patients in Virginia were in nursing homes with a median age of 47 and a range of 20 to 86. Nine of the 13 patients were female. This suggests the food was attractive to the nursing staff since six of the cases were equal to or younger than 47. The California cases were linked with the consumption of lettuce, tomatoes, onions, cheese, and sour cream, whereas the Arizona patients were linked with consumption of Mexican cuisine in a specific chain of restaurants. Thus, a raw ingredient other than tomatoes contained in a taco or other Mexican cuisine could have been the vehicle (note discussion of *S.* Saintpaul outbreak, later).

Transplant Games Outbreak, 2002

Up to 141 cases of salmonellosis caused by *S. enterica* serotype Javiana were recorded among visitors to a theme park near Orlando, Florida at what was known as the Transplant Games (Srikantiah et al., 2005). Onset was June 24, and the last case was July 8, 2002. Most (91%) of the ill respondents reported eating food items at specific food courts. Case/control analysis led to diced Roma tomatoes. Plant X supplied diced tomatoes to the theme park and was investigated on August 13. An unopened package of diced Roma tomatoes retrieved from Plant X contained 150 to 1000 CFU fecal coliforms/g but no *Salmonella*. The Materials and Methods section of the report indicated that diced tomatoes from unopened boxes that had been frozen were cultured. No results of that culturing were reported. The "exposure" list of foods included shredded lettuce, shredded cheddar cheese, diced Roma tomatoes, fresh ground beef, presliced beefsteak tomatoes, and frozen ground beef. Only the diced Roma tomatoes were linked to the outbreak. It was not clear if Plant X only serviced the theme park, had other customers. Plant X purchased Roma tomatoes from a wholesaler, who bought most but not all of his fruit from a single grower/packer in Florida. The packinghouse supplying the fruit was investigated, and no sanitation deficiencies were found.

Outbreak of *S.* Newport Linked to Virginia Tomatoes, 2002

Between July 2 and October 30, 2002, *S. enterica* serotype Newport was recovered from 512 patients (CFSAN, 2007a; Green et al., 2008). The infections were recorded in 22 states, but not Florida. Standard round tomatoes were implicated in case-control studies. Traceback from two restaurants and a hospital led to a tomato grower who was farming on the eastern shore of Virginia (Delmarva Peninsula), but whose headquarters were located in South Florida. The packinghouse was visited in December after operations had ceased. Bird feces was observed on processing equipment and throughout the packinghouse. However, environmental samples were negative for *Salmonella*. The packinghouse was visited again after it had begun operations during the following season. No "significant sanitation deficiencies" were found, indicating that the bird feces problem had been corrected. An environmental sample of an irrigation pond on one of the farms was positive for serotype Newport and the PGFE profile matched the outbreak strain. The concentration of serotype Newport in the water was not recorded. Anecdotal reports indicated that the bottom of the pond was stirred prior to sampling the water. It was not clear in the report if this was a farm managed by the South Florida grower. The pond was used for irrigation, but the application method was drip (see the following discussion of possible movement of *Salmonella* in the xylem of tomatoes). Additionally, the pond water may have been used for pesticide applications (see 1995 Newport outbreak).

Multiserotype Convenience Store Outbreak, 2004

During the summer of 2004, *S. enterica* serotype Javiana (429) infections were linked to sliced Roma tomatoes that were consumed in "made-to-order sandwiches" at a chain of convenience stores (CDC, 2005). A total of 471 cases were reported between June 29 and July 27. Other serotypes involved in this outbreak included Anatum (5), Typhimurium (27), Thompson (4), Muenchen (4), and Group D untypable (6). This was the first outbreak report where multiple strains/serotypes of *Salmonella* were recovered. One of the outbreak strains with a characteristic PFGE pattern was isolated from an unopened package of sliced Roma tomatoes on July 13, and all Roma tomatoes were recalled on July 14. Twenty-two patients reported disease onset after July 19, which was beyond the incubation period for Salmonellosis. These were explained as being caused by continued use of contaminated Roma tomatoes, ineffective recall of implicated Roma tomatoes, low infectious dose, food saved, or secondary transmission. The delicatessen chain purchased sliced tomatoes exclusively from a single processor for 302 outlets located in five states, although the outbreak was identified in nine states including Maryland, Michigan, Missouri, North Carolina, New Hampshire, Ohio, Pennsylvania, Virginia, and West Virginia.

A second and third outbreak were reported during the same time-frame as the multiple serotype outbreak discussed earlier. An outbreak caused by serotype Javiana with a distinctive PFGE pattern led to illnesses in 12 patients in Ontario, Canada. The Canadian officials traced the implicated fruits to a grower/packinghouse in South Florida, which was supplied by fruit from one of their farms in the coastal region of South Carolina. It was not clear if the South Carolina tomatoes were packed there or in Florida. The processing plant implicated in the larger multistate outbreak obtained Roma tomatoes from two packinghouses and one field pack operation in Florida. The packinghouses processed fruit from farms in Florida, Georgia, and South Carolina. Neither the environmental samples nor the fruit samples (Roma and standard rounds) from the farms or packinghouses yielded *Salmonella*. Furthermore, no outbreak related cases were reported in Florida.

A GMP (Good Manufacturing Practices) violation noted at the fresh-cut firm was the soaking of stored tomatoes (45 to 65 °F) in cold water (33 to 35 °F) prior to the slicing operation (the period of time that the fruits were soaked was not given, but if prolonged, would have led to a substantial infiltration of fruit tissues by the soak water; see laboratory models, later). Diced tomatoes prepared by the same firm with likely the same preprocess soak step were not associated with illnesses. The cleaning and sanitation schedule for the slicing and dicing parts of the plant was not discussed nor was the postprocessing storage regime. Many environmental samples from areas around the implicated farms in Florida, Georgia, and South Carolina were positive for *Salmonella*, but none produced the outbreak strains (CFSAN, 2007). Six of the eight farms were located near unimproved areas that feature ditches, canals, ponds, woods, and marshes inhabited by various amphibians, water fowl, small mammals, and deer.

The third outbreak, in the summer of 2004, involved 125 cases of salmonellosis attributed to *S. enterica* serotype Braenderup between June 15 and July 21 (CDC, 2005). Restaurant meals were implicated and the managers were asked about specific types of cheese, lettuce, and tomatoes eaten by customers. Roma tomatoes were the only exposure linked to the illnesses. Overall, four packers and five associated farms were visited after the outbreak and traceback had been completed. No clear source of contamination could be identified during the August through November visits.

Second Outbreak of Salmonellosis Caused by Serotype Newport Traced Back to Tomatoes Produced on the Delmarva Peninsula, 2005

Standard round tomatoes were linked with an outbreak totaling 72 culture-confirmed patients in 13 states over the period July 4 through September 24, 2005 (Greene, 2008). The specific strain isolated from the patients was identified as *S. enterica* serotype Newport. This strain had the same PFGE pattern as an outbreak strain isolated in 2002, which also was associated with the consumption of standard round tomatoes. Traceback from an implicated restaurant and convenience store yielded 12 growers/packinghouses located in five states as possible sources. However, two of the grower/packers located on the Delmarva Peninsula, Virginia were common to the two traces. (Note: The outbreak strain of 2002 was linked to a different packinghouse/grower on the Peninsula.) Greene et al. (2008) noted a potential problem with their case-control study was that 30% of the patients did not recall eating raw tomatoes in restaurants (or apparently elsewhere) prior to their illness. It was suggested that these patients were background illnesses and contracted the bacterium from other people or ate foods that were cross-contaminated by the primary vehicle. Since the full menu of foods consumed by the patients was not listed, one cannot determine the validity of these conjectures.

In the farm investigation, serotype Newport matching the outbreak strain was isolated from an irrigation pond in October (Greene et al., 2007). An investigation of the implicated farms in July of the following year found serotypes Newport in one pond (but not outbreak strain) and Javiana. in a second. The second pond was being used by its owner as a source of water for pesticide applications, which is a clear violation of GAPs and a major risk factor for crop contamination (DACS, 2008). The grower agreed to stop this practice.

Three-State Outbreak, 2005

During the period, November 1 through December 31, 2005, 76 culture-confirmed cases with the same PFGE pattern of serotype Braenderup were identified in Ohio, Michigan and Indiana (CFSAN, 2007). Diced tomatoes were linked to the outbreak in case-control comparisons. Twelve cases were linked to 12 different outlets of a national restaurant chain. The restaurant chain began a traceback on the 12, whereas the FDA examined four. Both led to four diced tomato distribution centers, a fresh-cut firm, a repacker, a grower/packer and up to 13 farms. The packinghouse and farms were located in Florida. Only one of the farms could be linked to at least some of the cases. This farm was composed of two fields, one of which was adjacent to a drainage ditch that was flood-prone. Cattle were observed in a neighboring ranch adjacent to the ditch. Feral hogs had broken through an electric fence and had wallowed in the ditch. Samples taken from the packinghouse were negative for *Salmonella*, whereas *Salmonella* was isolated from the ditch and animal feces found around the field, but the strains were different from the outbreak strain. There was no report that the hogs had entered into or defecated onto the field. A violation of GMPs was observed in the fresh-cut processing facility; tomatoes at 50 °F were soaked in 32 to 35 °F water to firm them before they were processed. When advised of this violation, the firm discontinued the practice. The repacker was said to be practicing GMPs and was not commingling fruits from different growers. Also of note: outbreak cases were not reported in Florida, although this state was implicated as the source of the tomatoes.

Recurrence of Serotype Newport in Delmarva Tomatoes, 2006

During the period July through November, 2006, 115 culture-confirmed isolates of *S.* Newport with the same PFGE patterns were isolated from stool samples collected in 19 states (CDC, 2007). This PFGE pattern was identical to that collected from the 2005 serotype Newport outbreak. Consumption of round tomatoes in a restaurant was associated with illness. A specific restaurant or restaurant chain was identified. The source of the fruit was not determined, probably because of the PFGE linkage with earlier outbreaks. This was the third outbreak of Salmonellosis linked to tomatoes grown on the Delmarva peninsula over the past four years (e.g., 2002, 2005, and 2006). Florida was not included in the 19 states in any of the outbreaks, clear evidence that the contaminated product was not shipped to the state, despite certain Florida growers having farming operations on the peninsula during the time period in question.

Outbreak Linked to Ohio Tomatoes, 2006

Standard round tomatoes packed in Ohio were linked to 190 culture-confirmed cases caused by *S. enterica* serotype Typhimurium during the period September 1 through October 15, 2006 (CDC, 2007). These cases were distributed in 21 states and Canada. In a case-control comparison, consumption of "raw large, red, round tomatoes at a restaurant" was correlated with infection. The Ohio packinghouse was supplied by three growers, who produced tomatoes on 25 fields. Investigation of the packinghouse during the subsequent season did not reveal significant sanitation deficiencies. However, some of the bins of fruit were at the same temperature as the dump tank water. Environmental samples from a creek used for irrigation and a drainage ditch at two of the previous year's fields were positive for *Salmonella* but of a different serotype.

UNANSWERED QUESTIONS

Since 1990, statistical correlations strongly support the linkage of between 13 to 15 outbreaks of salmonellosis and the consumption of raw tomatoes. However, a number of questions remain. An assumption in several outbreak reports is that wide, multistate distribution of the illnesses implicates contamination on the farm. However, the outbreak reports do not list the entire marketing chain or menu items considered by cases or the controls. Without this information, one cannot evaluate the "contaminated at farm" assertion. Marketing chains for tomatoes can be quite complex (VanSickle, 2008) and certain parts, particularly those servicing small or nonstandard growers often lack records of transactions (Block, 2008; Schmit, 2008). Additionally, as noted earlier, the yield of a field of fresh market tomatoes is quite large, particularly in relation to the number of cases reported.

General contamination of such a field ought to lead to large clusters of infections located wherever the tomatoes are shipped.

The lack of information about menu items considered during the initial case/control studies reduces confidence in the statistical claim that raw tomatoes were the vehicle for most or all 15 outbreaks. Not everyone consuming fresh tomatoes became ill as many controls (who did not become ill) had consumed such items. Moreover, in one report, 30% of the cases did not remember eating raw tomatoes (Greene et al., 2008). Differences in the susceptibility of individuals may be a factor (Todar, 2005). Items consumed with tomatoes could also be a factor. Cheese or other dairy product consumption promotes infection by *S. enterica* since the fat protects the bacteria against extreme acidity as the food passes through the stomach (FDA, 2003). However, another possibility is that the actual vehicle was a fresh or dried item commonly included with tomatoes in various dishes.

The possible false linkage of tomatoes with outbreaks was illustrated in the summer of 2008, when the US FDA issued an advisory against consuming tomatoes produced in Mexico and South Florida due to an outbreak of *S. enterica*, serotype Saintpaul. This action caused an estimated $200 million loss to growers as tomatoes and other fresh produce items were shunned in the marketplace (Burke, 2008). Despite many attempts, the bacterium was never isolated from tomato fruit. Instead, it was later detected in Serrano and Jalapeno peppers grown in Mexico (CDC, 2008). Chili peppers grown in Mexico have a history of being hazardous (Burke, 2008). Border inspectors had turned back several shipments of chili peppers due to decay or quality problems in the past few years. Apparently, this information was not available to the outbreak investigators. Throughout the investigation, the FDA and CDC emphasized in press conferences that serotype Saintpaul was a rare strain and that this bacterium, with a single PFGE fingerprint, was being isolated from what eventually became 1442 patients located in 43 states, the District of Columbia, and Canada over nearly a four-month period. How a strain with the same PFGE fingerprint could appear in farms located in two widely separated production areas (South Florida and Mexico) is not clear. A single strain of a plant pathogen could infect crops in widely separated production areas, but plant pathogens have evolved to colonize their plant hosts, to multiply to enormous numbers, and to spread among and between fields of plants in a true epidemic. There is no indication or even suggestion that *Salmonella* infects tomato plants in the field, causes a "disease," multiplies to large populations, and spreads within and among fields of tomatoes. As noted earlier, *Salmonella* is an agent of the intestinal tract of many animals (Todar, 2005).

None of the reports of salmonellosis outbreaks associated with raw tomatoes have addressed diversity among the *Salmonella* spp. that cause enteritis in humans. Indeed, only one outbreak report indicated that several serotypes were involved. If a crop were contaminated by *Salmonella* carried by the animal residents in a wetland area adjacent to a tomato farm, such as suggested in at least two outbreak reports, multiple serotypes ought to be involved. A wetland area does not feature confined suscepts (susceptible individuals) as might occur with a feedlot or egg-production farm. With confined animals, a serovar with a single fingerprint could be the dominant strain since it would colonize and multiply. With a free-range environment, strains should be diverse.

Claudon et al. (1971) isolated 12 serotypes from recreational waters at the University Creek-University Bay area of Lake Mendota, a 9730-acre lake on the northern edge of the University of Wisconsin campus in Madison. The authors attributed the continuing contamination to urban and agricultural runoff although that part of the Lake Mendota watershed is host to numerous amphibians, water fowl, avian species that feed on fish and amphibians as well as other birds (Bartz, personal observations, 1960–1968). If just one strain, characterized by a unique PFGE fingerprint, is causing an outbreak, then it is unclear if that would be compatible with a general environmental contamination of a field. A few intrusions into a field by a wild animal could lead to point sources of contamination (individually contaminated fruit) such as occurred in the greenhouse study in Mexico (Orozco, 2007).

Alternatively, perhaps, the causal agent is particularly adept at colonizing tomato fruits. Serotype Montevideo was cited in two separate reports as growing/surviving better in tomatoes than other serotypes, but the differences did not appear striking (Shi et al., 2007; Guo et al., 2002a). These reports both involved hydroponic and not field production methods. The assumption in the outbreak reports that the source was field contamination was based on laboratory models that have inherent problems in that most (or perhaps all) have not been validated under actual farm operations (Doyle and Erickson, 2008).

Laboratory experiments are intended to provide a model of the natural interactions of *Salmonella* and tomato fruit in the field or marketing chain. Model outcomes can sometimes be badly skewed if input parameters are unrealistic or flawed. Unrealistically high *Salmonella* populations such as applied to wounded surfaces in many laboratory protocols favor survival or multiplication. Alternatively, unrealistically warm or anaerobic storage conditions may favor growth, particularly if spoilage by native microorganisms is involved (Wells and Butterfield, 1997). In the initial report of growth of nontyphoidal *Salmonella* on tomatoes, the population applied to diced fruit was low (5 CFU/g), but the pieces were stored in sealed plastic bags at 22 or 30 °C (Asplund and Nurmi, 1991). The size of the pieces was not listed, whereas the pH of the general mixture was reported as 4.0 to 4.4. There was no mention of the cell sap released during processing. With most diced fresh products, the product is washed with water followed by some sort of drying process that removes excess water (Cantwell, 1992). The pH listed by Asplund and Nurmi (1991) for the general mixture is likely low, as the pH of fluid lining the intercellular spaces (apoplast) beneath the wound surface would likely be higher (Sakurai, 1998). Ripe tomatoes were reported to have an apoplastic pH of 4.4 and green tomatoes 6.7 (Almeida and Huber, 1999).

Fresh-cut products would be expected to have an increased rate of respiration as a response to cell injury (Watada et al., 1996). Higher rates would accompany higher tissue temperatures. The atmosphere within the sealed bags in the Asplund and Nurmi (1991) tests was not reported, but at 22 or 30 °C, it may have reached the critical point where an absence of oxygen causes the tomato cells to begin anaerobic respiration (Cameron et al., 1995). A low oxygen environment would favor growth of fermentative microorganisms, and as the plant cells deteriorate, the many different types of bacteria that normally exist within a tomato fruit (Samish et al., 1960, 1963) would begin rapid growth.

At least four other reports (CFSAN, 2007b; Wei et al., 1995; Weissinger et al., 2000; Zhuang et al., 1995) indicate that *Salmonella* grows on cut or diced tomatoes. In the last two, diced tomatoes were incubated in sealed plastic bags for up to 72 hours at 21 or 30 °C, or 20 or 30 °C, respectively. Weissinger et al. (2000) first treated the diced fruits with a solution of $CaCl_2$ and then allowed the product to drain for five minutes before mixing 600 g with 20 ml of inoculum. Zhuang et al. (1995) added a diluted culture of serotype Montevideo grown in trypticase soy broth (TSB) to the pieces and mixed them well. Weissinger et al. (2000) did not comment on the horticultural quality or possible edibility of the diced product at the different sampling times, whereas Zhuang et al. (1995) noted that bacterial growth was observed on the diced fruit stored for 22 hours at 20 or 30 °C and that subjective analysis of the aroma indicated the product was no longer edible. Moreover, the initial population of aerobic mesophylic bacteria in the chopped tomatoes initially exceeded the added *Salmonella* by about 1.5 logs and by 22 hours was nearly 3 logs higher. Diced products are normally stored at temperatures below 5 °C and, even with such handling, have a shelf-life of about seven days (Schlimme, 1995).

The FDA now considers cut tomatoes to be a potentially hazardous food that is subject to temperature control for safety (CFSAN, 2007b). Data accompanying the guidance showed a three- to 7.5-hour lag period between inoculation and the onset of multiplication for serotypes Enteritidis and Newport applied to cut Roma and beefsteak tomatoes at about log 3 CFU/ml and then stored at 22 °C for 24 hours. There was no growth when the inoculated tomatoes were stored at 5 °C. Food service is advised that cut tomatoes at an initial temperature of 5 °C can be stored at room temperature for up to four hours or for up to six hours if the product temperature does not rise above 21 °C. At that point, the food must be cooked, consumed, or discarded.

Survival and growth have been associated with the deposit of aqueous cell suspensions of *Salmonella* on natural openings such as the stem scar in addition to the wounds discussed earlier. Population size and suspending medium affect survival (Wei et al., 1995). When small populations of S. Montevideo, log 2.8 to 3.9 CFU/ml in distilled water were placed on the smooth periderm of tomato fruits, none could be detected after overnight storage. If the concentration of inoculum was increased to log 9.5 CFU/ml the bacterium could be detected up to three days later. Better survival occurred when bacteria were suspended in a buffer as compared with distilled water. Additionally, better survival occurred when the bacterial suspension was placed on wounds or the stem scar or if the bacterium was suspended in TSB. Which of these laboratory simulations best models the contact of fruits with aqueous cell suspensions of *Salmonella* in commercial practice is unclear.

Immersion inoculation of mature green fruits led to survival and growth of serotype Montevideo up to 18 days later depending on the storage temperature (Zhuang et al., 1995). The horticultural quality of the fruit at the end of this storage period was not reported. Tomato storage life has been reported to be two to four weeks under optimum conditions including high humidity and cool temperatures (Saltveit, 2005). The fruit size in the Zhuang et al. (1995) experiment, 110 to 140 g, is small compared to commercial standards (Olson and McAvoy, 2007). Mature green tomatoes were supplied for this research, but their mature green status did not appear to have been independently confirmed by the authors. A second concern with this report was that the inoculated tomatoes were stored in plastic bags. Although the bags were kept open to enable air-exchange, contact between the plastic and the fruit surface could have created an artificial environment that would favor survival as certain types of plastic film greatly restrict oxygen exchange while trapping moisture (Cantwell, 1992).

Certain laboratory models have attempted to explore internalization of *Salmonella* by tomato roots and then movement up the xylem of the plant as would occur with the wilt pathogen *Ralstonia solanacearum* (Guo et al., 2001).

High populations (log 7.5 CFU) applied to tomato stems by stab inoculation led to an approximately 40% detection of the *Salmonella* in the fruit, but enrichment techniques were used. Possible movement of the bacterium, either actively or passively, in the intercellular spaces of the stem was not discussed. Evidence for vascular movement of *Salmonella* was presented when tomato plants were grown hydroponically in a contaminated Hoagland's nutrient solution log 4.6 CFU/ml (Guo et al., 2002b). The plants were started in sand culture. When the cotyledons emerged, the plantlets were transferred to a hydroponic system based on Hoagland's nutrient solution. After seven days, the plants were transferred to Hoagland's nutrient solution containing log 4.6 CFU *Salmonella*/ml. *Salmonella* was isolated from hypocotyls, cotyledons, stems, and leaves of plants growing in the contaminated nutrient solution. However, the integrity of the root system was not discussed. It is unknown if the roots had completely healed after the transfer from sand to hydroponic culture or if the roots were disturbed by the transfer from the initial nutrient solution to the contaminated one. By contrast, Jablasone et al. (2004) applied 350 ml of an aqueous cell suspension containing log 5.0 CFU/ml as irrigation water every other day to patio tomato plants grown in the laboratory and did not detect (no colonies on any of six plates) *Salmonella* in the stems or fruits during a period of five weeks. Populations in the soil ranged from log 2.3 to 3.7 CFU/g soil. Van der Schloot (1989) concluded that the xylem fluid in tomato stems must pass through pit membranes before moving from a stem into a petiole. Pit membranes have pores of about 0.3 µm in diameter (Goodman, 1967). These allow water to move freely but would filter out particles larger than the pore diameter, which would include bacteria. Bacteria that can digest the membranes, such as the wilt pathogens, could breach this barrier although certain tomato cultivars possess a type of resistance that prevents *R. solanacearum* from entering the xylem (Prior et al., 1990).

Laboratory models have also been developed to explain the infiltration of natural openings on tomato fruits by water (Bartz, 2006). Particulate matter suspended in the water would be internalized if smaller than the surface openings. Soft rot bacteria suspended in water internalize in tomatoes that are cooled while submerged or subjected to hydrostatic forces such as a deep submergence (Bartz, 1982) or a pressurized stream of water. These bacteria cause internal bacterial soft rot lesions. Bacteria that are embedded in tomato tissues are no longer vulnerable to surface sanitizing treatments such as chlorinated water because the sanitizer cannot penetrate into these areas (Watada, 1996). The infiltration/internalization model is frequently cited as how tomatoes could have become contaminated at the packinghouse. With a 20 °C differential (water cooler than fruit), at least a 10-minute immersion is required for fruit weight increases (Bartz, 1982). If chlorine is present in the water, weight increases due to temperature differential may not be followed by bacterial soft rot (Clement, 2000). In the absence of chlorine or with rapid penetration by chlorinated water, decay usually follows infiltration (Bartz, 1988). Thus, it is not clear how *Salmonella* could internalize via this route, particularly given that tomato deterioration from decay or general spoilage were not mentioned in the laboratory or outbreak reports. Moreover, the stem scar is not an area of tomato fruit normally associated with bacterial multiplication since cells in it are not succulent, and it has a large population of natural microflora, which ought to biologically buffer the area against the development of a *Salmonella* population. By contrast, stem scars on freshly harvested fruits seem analogous to the cut stem of a flower. If water contacts the ruptured vascular tissues before an air bubble forms, then the water will move into those tissues as transpiration (water loss from the fruit surface) occurs (Bartz, unpublished). Additionally, water congestion of stem scar tissues (free water infused into the tissues) creates a channel that enables rapid internalization of bacteria and other particulate matter (Bartz, 2006).

How an infectious dose of *S. enterica* can develop on what was once a clean washed tomato is unclear. Most commercial tomatoes are washed with chlorinated water. Water chlorination has been practiced for years by tomato packers as a decay control measure (Bartz and Eckert, 1987). If a packer failed to maintain chlorine concentrations during a warm season harvest period, postharvest decays should have become an issue since in many growing areas, this is a time for rainfall and warm temperatures, which are ideal conditions for bacterial soft rot and sour rot, two of the more damaging decays (Bartz, 1991). Additionally, most commercial tomatoes are inspected and certified to be free of damaging defects including postharvest decays (USDA, 1987).

The tomato industry is currently under suspicion by governmental agencies as well as a large portion of the consuming public (FDA, 2007). The Florida Tomato Industry is attempting to recover the public's trust that fresh tomatoes are a wholesome product by instituting a mandatory inspection and certification program (DACS, 2008). This is a grower-driven program that is being enforced by the state. However, tomatoes grown in the field will always have a more diverse microbial ecosystem than those grown in protected cultures such as greenhouses. There does not appear to be a single process that can make field tomatoes completely safe. On the other hand, a recent report has revealed that greenhouse-grown fruits are not always free of *Salmonella* (Orozco, 2007). But then, tomatoes have been grown in home gardens (unregulated production fields) and eaten fresh by gardeners and neighbors for centuries without apparent incident. Costa and Heuvelink (2005) noted that small plots (unregulated production fields) of tomatoes are grown throughout the tropics and subtropics, providing an important food source for the local population. The layout and management of these small plantings is likely as diverse as the microorganisms in them. Pet dogs and cats, local birds, foxes, rabbits, rodents, snails and slugs, toads, turtles, snakes, and even deer and wild hogs could be everyday visitors to these unregulated plantings. Fertilization ranges from mineral to well-composted animal manures to not so well-composted manures. One difference, of course, is that small plot growers (home gardeners) don't usually dice or slice tomatoes, put them in a sealed plastic container, put the container in the refrigerator, and then store them for a week or more before consuming them.

RECOMMENDATIONS FOR COMMERCIAL TOMATO PRODUCTION AND HANDLING, FARM-TO-FORK

Assuming a 1:38 ratio of reported and clinically established cases versus unreported cases of Salmonellosis (Greene et al., 2007), an outbreak of 500 cases would actually mean 19,000 illnesses. However, an acre of tomatoes yields roughly twice that number of half-pound meals of tomatoes. If an acre of tomatoes were uniformly contaminated, the magnitude of the resulting outbreak would be historic as well as tragic. If fruits are, in fact, contaminated in the field, then it is likely to be sporadic and difficult to pinpoint except where bird or animal intrusion is evident. The prudent control measure then becomes an attempt to prevent contact between the crops and environmental sources of *Salmonella*. Commercial production must include a requirement that all producers use GAPS and all postharvest handlers use GMPS (DACS, 2008).

All types of tomato fruits at a packinghouse or other point of production must be inspected and given an appropriate grade designation. A sanitation step should be required for all tomatoes, which could include a sanitizer wash followed by drying procedure or a gas treatment with ozone or chlorine dioxide (see Chapter 17). The packed tomatoes must be evaluated for cleanliness in addition to the other product requirements for a safe dry pack. Free moisture or soiled containers must not be allowed, since free water enables bacteria to internalize particularly through water congested wounds and natural openings (Bartz, 2006). Soiled or previously used containers invite contamination in addition to restricting the markets' ability to traceback problem lots. All packages of fresh tomatoes must have positive lot identification with grower, date of harvest, packer, shipper, and such clearly listed on each package. Fresh-cut processors should be governed by HACCP where microbial population and other product parameters are measured at critical control points (Gorny, 2006). The processor should keep records of raw product purchased, product dwell time at the plant, as well as product sold. The temperatures of raw material, processing step, and packaged product must be continuously recorded. Fruits salvaged from defective shipments or culled from regular shipments should not be used. Additionally, the plant should not accept fruit from noncertified growers or packers or from uncontrolled markets. The final product package must have a clearly marked date and time of production, a clearly marked "use-by date," and embedded sensors that would indicate if temperature abuse had occurred or that package contents had become anoxic.

Periodic restaurant inspections that have become commonplace should be expanded to include food-service suppliers and distributors. During these periodic and unannounced inspections, the firm's fresh-cut product inventory should be examined including records of product received, product stored, continuous product temperatures, product sold or consumed, and a check of the use-by dates and temperatures of representative packages. Finally, market basket inspections by the FDA or USDA need to focus more on the microbiology of fresh products and less on chemical contamination.

REFERENCES

Asplund, K. and Nurmi, E. (1991). The growth of *Salmonella* in tomatoes. *Internat. J. Food Microbiol.* **13,** 177–182.

Almeida, D. P. F. and Huber, D. J. (1999). Apoplastic pH and inorganic ion levels in tomato fruit: A potential means for regulation of cell wall metabolism during ripening. *Physiol. Plant.* **105,** 506–512.

Bartz, J. A. (2006). Internalization and infiltration. In *Microbiology of fruits and vegetables* (G. M. Sapers, J. R. Gorny, and A. E. Yousef, Eds.), pp. 75–94. CRC Press, Taylor and Francis, Boca Raton, FL.

Bartz, J. A. (1991). Postharvest diseases and disorders in tomato fruit. In *Compendium of tomato diseases* (J.B. Jones, J. P. Jones, R. I. Stall, and T. A. Zitter, Eds.), pp. 44–49. Am. Phytopath. Soc. St. Paul, MN.

Bartz, J. A. (1988). Potential for postharvest disease in tomato fruit infiltrated with chlorinated water. *Plant Dis.* **72,** 9–13.

Bartz, J. A. (1982). Infiltration of tomatoes immersed at different temperatures to different depths in suspensions of *Erwinia carotovora* subsp. *carotovora*. *Plant Dis.* **66,** 302–306.

Bartz, J. A. (1980). Causes of postharvest losses in Florida tomato shipments. *Plant Dis.* **64,** 934–937.

Bartz, J. A. and Eckert, J. W. (1987). Bacterial diseases of vegetable crops after harvest. In *Postharvest physiology of vegetables* (J. Weichmann, Ed.), pp. 351–376. Marcel Dekker, NY.

Bartz, J. A., Schneider, K., Sargent, S. et al. (2002). Addressing microbial hazards in tomato fruit after harvest. In Proc. 2002 Florida Tomato Institute, *Citrus and Veg. Mag.*, pp. 21–23. U. Florida. IFAS, Extension Service, Pro 519, Gainesville.

Block, J. (2008). Blog: Food safety. AgWeb Blogs, Sept. 12, 2008.

Burke, G. (2008). Mexican peppers posed problem long before outbreak. Assoc. Press. Yahoo News, 08/18/2008.

Burnett, S. L. and Beuchat, L. R. (2000). Human pathogens associated with raw produce and unpasteurized juices, and difficulties in decontamination. *J. Indust. Microbiology and Biotech.* **25,** 281–287.

Cameron, A. C., Talasila, P. C., and Joles, D. W. (1995). Predicting film permeability needs for modified atmosphere packaging of lightly processed fruits and vegetables. *Hort. Sci.* **30,** 25–34.

Cantwell, M. (1992). Postharvest handling systems: Minimally process fruits and vegetables. In *Postharvest technology of horticultural crops. 2nd edition*. (A. A. Kader, Ed.)., pp. 277–281. UC Davis, Div. of Agric. and Nat. Resources. Publ. 3311.

CDC. (2002). Outbreak of *Salmonella* serotype Javiana infections—Orlando, Florida, June 2002. *MMWR* **51,** 683–684.

CDC. (2005). Outbreaks of *Salmonella* infections associated with eating Roma tomatoes—United States and Canada, 2004. *MMWR.* **54,** 325–328.

CDC. (2007). Multistate outbreaks of *Salmonella* infections associated with raw tomatoes eaten in restaurants—United States, 2005–2006. *MMWR* **56,** 909–911.

CDC. (2008). Outbreak of *Salmonella* serotype Saintpaul infections associated with multiple raw produce items—United States, 2008. *MMWR* **57,** 929–934.

CFSAN. (2007a). CFSAN outbreak abstracts, epidemiology, traceback, farm and packinghouse investigations, tomatoes 1998–2007. Center for Food Safety and Applied Nutrition, US Food and Drug Administration.

CFSAN. (2007b). Program information manual, retail food protection, storage and handling of tomatoes. www.cfsan.fda.gov/~ear/pimtomat.html

Chomchalow, S. (1991). Storage conditions and timing of ethylene treatment affect uniformity and marketability of tomato fruit. Univ. Florida, MSc. thesis.

Clément, V., Bartz, J. A., and Sargent, S. A. (2000). Postharvest decay risk associated with hydrocooling tomatoes. *Plant Dis.* **84,** 1314–1318.

Costa, J. M. and Heuvelink, E. (2005). Introduction: The tomato crop and industry. In *Tomatoes* (E. Heuvelink, Ed.), pp. 1–19. CABI Publishing. Cambridge, MA.

Cummings, K., Barrett, E., Mohle-Boetani, J. C. et al. (2001). A multistate outbreak of *Salmonella enterica* serotype Baildon associated with domestic raw tomatoes. *CDC Emerging Infectious Diseases*. Vol. 7: Nov–Dec 2001. www.cdc.gov/ncidod/eid/vol7no6/cummings.htm

DACS. (2008). Tomato best practices manual—A guide to T-GAP and T-BMP. The Florida Department of Agriculture and Consumer Services, Tallahassee, FL.

Doyle, M. P. and Erickson, M. C. (2008). Summer meeting 2007—The problems with fresh produce: An overview. *J. Appl. Microb.* **105,** 317–330.

FDA. (2003). *The bad bug book*. Foodborne Pathogenic Microorganisms and Natural Toxins Handbook. U.S. Department of Health and Human Services, Food and Drug Administration, Center for Food Safety and Applied Nutrition. Updated: 01-30-2003. www.cfsan.fda.gov/~mow/intro.html

FDA. (2007). FDA News. FDA Implementing initiative to reduce tomato-related foodborne illnesses. US Department of Health and Human Services, Food and Drug Administration. June 12, 2007. www.fda.gov/bbs/topics/NEWS/2007/NEW01651.html

Goodman, R. N., Király, Z., and Zaitlin, M. (1967). *The biochemistry and physiology of infectious plant disease*. Van Nostrand, Princeton, NJ.

Gorny, J. R. (2006). Microbial contamination of fresh fruits and vegetables. In *Microbiology of fruits and vegetables* (G. M. Sapers, J. R. Gorny, and A. E. Yousef, Eds.), pp. 3–32. CRC Press, Taylor and Francis, Boca Raton, FL.

Greene. S. K. Daly, E. R., Talbot, E. A. et al. (2008). Recurrent multistate outbreak of *Salmonella* Newport associated with tomatoes from contaminated fields, 2005. *Epidemiol. Infect.* **136,** 157–165.

Guo, X., Chen, J., Brackett, R. E., and Beuchat, L. R. (2001). Survival of *Salmonella* on tomato plants from the time of inoculation at flowering and early stages of fruit development through fruit ripening. *Appl. Environ. Microbiol.* **67,** 4760–4764.

Guo, X., Chen, J., Brackett, R. E., and Beuchat, L. R. (2002a). Survival of *Salmonella* on tomatoes stored at high relative humidity, in soil, and on tomatoes in contact with soil. *J. Food Prot.* **65,** 274–279.

Guo, X., Chen, J., van Iersel, M. W., Chen, J. et al. (2002b). Evidence of association of salmonellae with tomato plants grown hydroponically in inoculated nutrient solution. *Appl. Environ. Microbiol.* **68,** 3639–3643.

Hedberg, C. W., Angulo, F. J., White et al. (1999). Outbreaks of salmonellosis associated with eating uncooked tomatoes: implications for public health. *Epidemiol. Infect.* **122,** 385–393.

Jabalasone, J., Brovko, L. Y, and Griffiths, M. W. (2004). A research note: The potential for transfer of *Salmonella* from irrigation water to tomatoes. *J. Sci., Food and Agric.* **84,** 287–289.

Lin, C. M. and Wei, C. I. (1997). Transfer of *Salmonella* Montevideo onto the interior surfaces of tomatoes by cutting. *J. Food Prot.* **60,** 858–863.

Maul, E., Sargent, S. A., Sims, C. A. et al. (2000). Tomato flavor and aroma quality as affected by storage temperature. *J. Food Sci.* **65,** 1228–1237.

Maynard, D. N., Scott, J. W., and Dunlap, A. M. (2000). Tomato variety evaluation Fall 1999. U.F. IFAS. GCREC Res. Rept. BRA 2000–2001.

Olson, S. M. and McAvoy, E. (2007). Tomato varieties for Florida. *2007 Tomato Proc.* pp. 30–34. U.F.l, IFAS, Fl. Coop. Extension Serv., *Citrus and Veg. Mag.*

Orozco, L., Rico-Romero, L. and Escartín, E. F. (2008). Microbiological profile of greenhouses in a farm producing hydroponic tomatoes. *J. Food Prot.* **71,** 60–65.

Prior, P. H., Beramis, M., Chillet, M., and Schmit, J. (1990). Preliminary studies for tomato bacterial wilt (*Pseudomonas solanacearum* E.F.Sm.) resistance mechanisms. *Symbiosis* **9,** 393–400.

Todar, K. (2005). *Salmonella* and salmonellosis. In Todar's online textbook of bacteriology. www.textbookofbacteriology.net/salmonella.html.

Sakurai, N. (1998). Dynamic function and regulation of apoplast in the plant body. *J. Plant Res.* **111,** 133–148.

Saltveit, M. E. (2005). Fruit ripening and fruit quality. In *Tomatoes* (E. Heuvelink, Ed.), pp. 145–170. CABI Publishing. Cambridge, MA.

Samish, Z., Etinger-Tulczynska, R., and Bick, M. (1960). Microflora within healthy tomatoes. *Appl. Microb.* **9,** 20–24.

Samish, Z. Etinger-Tulczynska, R., and Bick, M. (1963). The microflora within the tissue of fruits and vegetables. *J. Food Sci.* **28,** 259–266.

Schlimme, D. V. (1995). Marketing lightly processed fruits and vegetables. *Hort. Sci.* **30,** 15–17.

Schmit, J. (2008). TEXAS: tracing tainted fruit isn't easy. *USA Today*, Aug 14, 2008.

Srikantiah, P. Bodager, D., Toth, B. et al. (2005). Web-based investigation of multi-state Salmonellosis outbreak. *CDC Emerging Infectious Diseases.* Vol. 11. www.cdc.gov/ncidod/eid/vol11no4/04-0997.htm

USDA. (1987). Fresh Tomatoes. Shipping Point Inspection Instructions. USDA, ARS, Fresh Products Branch. Washington DC.

Van der Schoot, C. and Van Bel, A. J. E. (1989). Architecture of the intermodal Xylem of tomato (Solanum lycopersicum) with reference to longitudinal and lateral transfer. *Amer. J. Bot.* **76,** 487–503.

VanSickle, J. J. (2008). Field study for analyzing the opportunities to change the box size for mature green tomatoes shipped from Florida to a 10 kg carton. In Tomato Research Report for 2007–2008, pp. 40–60. U. of Florida, IFAS, Gainesville.

Walsh, M. and Rodriguez, O. R. (2008). Few safeguards for Mexican produce heading north. Sept. 13, 2008. The Assoc. Press. ap.google.com/article/ALeqM5hc-caaOQsaUlBEKLLR0Ax-GSH8-RwD935S1MG4

Watada, A. E., Ko, N. P., and Minott, D. A. (1996). Factors affecting quality of fresh-cut horticultural products. *Postharvest Biol. Tech.* **9,** 115–125.

Wei, C. I., Huang, J. M., Lin, W. F., Tamplin, M. L., and Bartz, J. A. (1995). Growth and survival of *Salmonella* Montevideo on tomatoes and disinfection with chlorinated water. *J. Food Prot.* **8,** 829–836.

Wells, J. M. and Butterfield, J. E. (1997). *Salmonella* contamination associated with bacterial soft rot of fresh fruits and vegetables in the marketplace. *Plant Dis.* **81,** 867–872.

Weissinger, W. R., Chantarapanont, W., and Beuchat, L. R. (2000). Survival and growth of *Salmonella* Baildon in shredded lettuce and diced tomatoes, and effectiveness of chlorinated water as a sanitizer. *Int. J. Food Microbiol.* **62,** 123–131.

Zhuang, R. Y., Beuchat, L. R., and Angulo, F. J. (1995). Fate of *Salmonella* Montevideo on and in raw tomatoes as affected by temperature and treatment with chlorine. *Appl. Environ. Microbiol.* **61,** 2127–2131.

CHAPTER 11

Tree Fruits and Nuts: Outbreaks, Contamination Sources, Prevention, and Remediation

Susanne E. Keller

FDA/CFSAN, National Center for Food Safety and Technology, Summit-Argo, IL

CHAPTER CONTENTS

Introduction	249
Organisms of Concern	250
Outbreaks Associated with Tree Fruits	252
Outbreaks Associated with Tree Nuts	254
Routes of Contamination	255
Prevention	258
Remediation	259
Conclusions	260

INTRODUCTION

According to the USDA, the typical American consumes over 280 pounds of fruits and nuts each year, a large percentage of which are tree fruits and nuts (US Department of Agriculture, 2004). These are consumed both as fresh and processed products. Consumption of produce, particularly fresh produce, continues to increase worldwide in part due to an increased recognition of health benefits and an increase in its year-round availability. Historically, tree fruits and nuts have not been associated with a high risk for causing foodborne disease. However, recent increases in foodborne illnesses associated with fresh produce in general have led to an increasing concern regarding the safety of all fresh fruits and vegetables that are not processed to eliminate any microbial hazard. As a result, the FDA promulgated rules and guidelines concerning not only the handling of fresh produce, but its growth and subsequent processing, including the processing and production of fresh juices in 2001. In this chapter, the risks of foodborne illness, particularly as it affects the consumption of fresh or minimally processed tree fruits and nuts, will be discussed.

ORGANISMS OF CONCERN

The overall incidence of foodborne illness for fruits and nuts remains low, particularly as compared to other food types such as meats, eggs, or dairy products (Centers for Disease Control, 2003). None of the usual human pathogens causing foodborne illness such as enterohemorrhagic *E. coli* (EHEC), *Salmonella* and *Shigella* species, *Cryptosporidium*, or *Listeria monocytogenes* are considered endogenous microflora of fruits and nuts, but all may occur as contaminants. In general, fruits and nuts provide an environment hostile to the growth and survival of these pathogens (Aruscavage et al., 2006; Brandl, 2006; Winfield and Groisman, 2003). However, despite their low incidence, their presence on these foods can be particularly problematic since fruits and nuts are frequently consumed raw and are considered to be part of a healthy lifestyle. Consequently, any outbreaks related to these products could be viewed more negatively by the general public than an outbreak associated with a food that is typically cooked or processed.

In circumstances where pathogenic microorganisms are unlikely to grow or multiply such as may be found on most fruits and nuts, those pathogens with lowest infectious dose and the greatest propensity for survival are likely to be of greatest concern. Of those pathogens with low infectious dose and considerable ability to withstand harsh environments, EHEC stands out, particularly with its association with hemorrhagic colitis, hemolytic uremic syndrome (HUS), and thrombotic thrombocytopenic purpura (TTP). HUS occurs primarily in children under 10 years of age and has a mortality of 3 to 5% (Buchanan and Doyle, 1997). Several serotypes of EHEC are known. The most common serotype, particularly in the United States, Canada, Great Britain, and parts of Europe, is *E. coli* O157:H7 (Buchanan and Doyle, 1997). In the years from 1998 to 2000, the Centers for Disease Control (CDC) recorded 86 outbreaks attributed to *E. coli*. Of these, 68 were identified as outbreaks caused by *E. coli* O157:H7 (Centers for Disease Control, 2003). The great majority of these outbreaks was either from meat products or had an unknown source.

A second group of pathogenic microorganisms with a low infectious dose is the *Shigella* species. As with *E. coli*, some strains produce enterotoxin and Shiga toxin (US Food and Drug Administration, 2006). Epidemics with fatalities as high as 5 to 15% have occurred in Africa and Central America (Centers for Disease Control, 2005c). Outbreaks related to tree fruits and nuts products caused by *Shigella* species are rare and appear to be more associated with poor hygiene (Castillo et al., 2006).

Although infectious doses are typically not as low for *Shigella* species, *Salmonella* species represents another group of foodborne pathogens that is also of concern with respect to tree fruits and nuts. *Salmonella* species includes over 2000 serotypes that cause human disease, half of which are serotypes Enteritidis or Typhimurium (Centers for Disease Control, 2002). There are an estimated 1.4 million cases of Salmonellosis annually, with an estimated 500 fatalities. Again, as with EHEC, these infections are more commonly associated with animal derived foods, such as meat, seafood, dairy, and egg products, rather than produce. Their occasional association with fruits and nuts is facilitated by their tolerance to some extreme conditions. *Salmonella* species is resistant to desiccation, which aids in its survival on the surface of fruits and nuts. Its survival on tree nuts and particularly almonds is now well documented (Beuchat and Heaton, 1975; Danyluk et al., 2007; Uesugi et al., 2006, 2007; Uesugi and Harris, 2006). *Salmonella* is also resistant to acids, a resistance it shares with both EHEC and *Shigella* species (Bagamboula et al., 2002; Lin et al., 1995).

In both *Salmonella* and *E. coli*, acid tolerance is inducible and increases when cells have been adapted either to acid conditions or are in stationary phase (Benjamin and Datta, 1995; Buchanan and Edelson, 1996; Foster and Hall, 1990; Lin et al., 1995). In *E. coli*, tolerance to high acid levels involves three distinct inducible mechanisms and is enhanced in stationary cells (Benjamin and Datta, 1995; Buchanan and Edelson, 1996; Lin et al., 1995, 1996). For *S.* Typhimurium, two major acid tolerance systems have been identified, one associated with log phase and one associated with stationary phase (Bang et al., 2000). Not surprisingly, survival in acidic fruit juices for extended periods has been observed for both *E. coli* and *Salmonella* species (Centers for Disease Control, 1996; Goverd et al., 1979; Parish, 1997). Like both *Salmonella* species and *E. coli*, *Shigella* is also resistant to acids, however, to a somewhat lesser extent. It can survive at a pH as low as 2 to 2.5 and has some of the same acid tolerance mechanisms as does *E. coli*. Its less frequent appearance on tree fruits and nuts may be due to its lesser ability to survive harsh environments.

Listeria monocytogenes is yet another foodborne pathogen of concern with fruits and nuts, particularly fresh-cut products, that is also acid tolerant. *L. monocytogenes* is ubiquitous within the environment and frequently found on fruits and vegetables (Beuchat, 1995; Beuchat and Ryu, 1997; Cox et al., 1989; Fenlon et al., 1996; Gombas et al., 2003; Johnston et al., 2005). The minimum pH for growth of *L. monocytogenes* is dependant on the acidulant. For malic acid, the primary acid found in apple cider/juice, the lowest pH value for growth for some strains of *L. monocytogenes* is from 4.4 to 4.6 (Beuchat, 1995; Beuchat and Ryu, 1997; Cox et al., 1989; Fenlon et al., 1996; Gombas et al., 2003; Johnston et al., 2005; Sorrells et al., 1989). *L. monocytogenes* will survive at lower pH similar to *E. coli* O157:H7 and *Salmonella* (Beuchat and Brackett, 1991; Sorrells et al., 1989). Although no foodborne outbreaks of listeriosis have been attributed specifically to tree fruits or nuts, *L. monocytogenes* has been isolated from unpasteurized apple juice (Sado et al., 1998). Its presence in fresh cut produce in general has resulted in numerous product recalls with significant economic losses (Anon., 2001, 2003a, 2003b, 2006).

In addition to the procaryotic pathogenic bacteria mentioned, fresh fruits and nuts are also susceptible to contamination by protozoa, particularly by *Cryptosporidium parvum*, a highly infectious protozoan parasite causing persistent diarrhea (Guerrant, 1997). Although *Cryptpsoridium* cannot replicate in the environment, the thick-walled oocysts are resistant to acids and chlorine and persist in the environment. Infection does not always manifest itself in severe symptoms but can be dangerous for the immuno-compromised population. Historically, the largest US outbreak of cryptosporidiosis occurred in Milwaukee, Wisconsin in 1993 and affected an estimated 403,000 people (Guerrant, 1997).

The inability to grow and multiply in the environment clearly does not limit the ability of pathogens such as *Cryptosporidium parvum* to cause significant foodborne outbreaks such as those in apple cider in 1993, 1996, 2003, and 2004 (Blackburn et al., 2006; Centers for Disease Control, 1997; Millard et al., 1994). However, *Cryptosporidium* is not the only microorganism to produce foodborne illnesses despite an inability to grow and multiply in the food environment. Viruses have also resulted in a significant number of outbreaks related to fresh produce.

Viruses transmitted by food or water fall into three groups: hepatovirus, enterovirus, and norovirus. Of these, the hepatovirus and norovirus appear to be of greatest concern with tree fruits and nuts. Viral outbreaks are frequently the result of poor sanitation or poor worker hygiene (Einstein et al., 1963; Herwaldt et al., 1994).

OUTBREAKS ASSOCIATED WITH TREE FRUITS

Despite the general perception that foodborne outbreaks related to tree fruits is a recent concern, incidents of foodborne illness associated with these products have been recorded as far back as the early to mid 1900s (Duncan et al., 1946; Parish, 1997). In more recent history, the CDC reported over 100 outbreaks of foodborne illness associated with fruits from 1998 through 2005. Of these outbreaks, one is particular noteworthy. In October 1996, an outbreak of *E. coli* O157:H7 was traced to unpasteurized apple juice produced by Odwalla Inc. The outbreak involved three states and British Columbia, with more than 60 people sick and one death (Centers for Disease Control, 1996). Odwalla Inc. eventually pled guilty to violating Federal Food Safety laws and was fined $1.5 million for selling tainted apple juice (Belluck, 1998). As a consequence of this and other juice-related foodborne outbreaks, the FDA promulgated HACCP regulation for fresh juice in 2001 (US Food and Drug Administration, 2001). All juice in the United States is now required to undergo processing that will achieve a 5-log reduction in the most pertinent pathogen. The 5-log reduction standard was established based on recommendations by the National Advisory Committee on Microbiological Criteria for Foods (NACMCF) (Anon., 1999). NACMCF considered worst-case scenarios, such as might occur if apples were contaminated directly with bovine feces, and included a 100-fold safety factor. Regulatory precedence was also considered when the 5-log pathogen reduction performance standard was established (US Food and Drug Administration, 2001).

The legally required 5-log reduction process must treat the whole juice, with the exception of citrus juice where the 5-log reduction can be achieved through treatments to the surface of the fruit before extraction of the juice. The reasoning behind the exemption for citrus fruit was based on the propensity of microorganisms to become internalized within the fruit. With apples, internalization has been well documented in the literature (Buchanan et al., 1999; Burnett and Beuchat, 2001; Burnett et al., 2000; Fatemi et al., 2006; Fleischman et al., 2001). However with citrus fruit, NACMCF determined that although internalization could occur, it was unlikely that such an event would occur, particularly with sound, tree-picked fruits (Anon., 1999). Consequently, to be eligible for this exemption, citrus fruit must be tree-picked, not wind-fall, or ground harvested, and it must be clean and free of blemishes. In addition, if a citrus juice processor obtains a 5-log reduction based on surface treatments and not through treatment of the whole juice, an end-product testing program must be in place.

Since the implementation of the Juice HACCP rules, outbreaks in the United States related to juice have decreased. However, some outbreaks, particularly related to tree fruits, continue to occur. These outbreaks appear related to unpasteurized (not thermally processed) product, or product given an alternative disinfection treatment but not heat-pasteurized.

In 2003, an outbreak of cyptosporidiosis was linked to ozonated apple cider in Ohio (Blackburn et al., 2006). In this case the processor attempted to destroy pathogens in juice through the use of ozone. The FDA has since issued guidance disallowing the use of ozone to achieve a 5-log reduction of pathogens in juice until such time that data indicating efficacy of treatment can be provided (US Food and Drug Administration, 2004). A second similar case occurred in New York State involving both *E. coli* O157:H7 and *Cryptosporidium* in apple cider in 2004 (Centers for Disease Control, 2004a). In 2005, there was yet another incident concerning *S.* Typhimurium in unpasteurized orange juice (Centers for Disease Control, 2005a, 2005b). In all these cases, problems occurred when novel disinfection treatments were used in lieu of a heat pasteurization treatment without appropriate validation of treatment efficacy. Following these outbreaks, the FDA issued a letter to state agencies and firms that produce treated, but not pasteurized juice, restating the need for processors to comply with current juice HACCP regulations (US food and Drug Administration, 2005). This requires that alternative treatments be validated through a processing authority.

Fresh juice, due to the nature of its production, is intuitively a more hazardous product than the original fruit from which it is produced. With juice, a single contaminated fruit can result in a widespread outbreak since the contamination will be spread through production and be consumed by multiple individuals. When consumed whole, a single contaminated fruit will likely result in illness of only a single consumer. Consequently, detection or characterization of outbreaks due to individual contaminated fruit may be difficult, resulting in greater underreporting than that caused by fruit juices. Occasional outbreaks, however, have been attributed to fresh products other than juices made from fresh tree fruits, most notably fruit salads, where once again multiple pieces of fruit and nuts may be commingled. Although it is not always possible to pinpoint the exact ingredient in a fruit salad responsible for the initial contamination, the association with fruit salads illustrates the vulnerability of fruits, particularly during preparation and processing. One outbreak of this type involved a variant strain of Norwalk virus (Herwaldt et al., 1994). Tree fruits can clearly become contaminated at any point during production and/or processing and result in foodborne illness.

OUTBREAKS ASSOCIATED WITH TREE NUTS

As with fruits, tree nuts were not typically associated with the types of acute foodborne illnesses caused by bacteria. However, recent outbreaks of salmonellosis in raw almonds have shown that even low moisture levels will not completely protect this product from contamination with foodborne pathogens. The first recorded outbreak of this type occurred from the late fall of 2000 to the spring of 2001 (Isaacs et al., 2005). Traceback investigations found *S.* Enteritidis PT30 on equipment surfaces used to process the suspect almonds and on the almond orchard floors. A second outbreak caused by *S.* Enteritidis occurred from September 2003 until April 2004 (Centers for Disease Control, 2004b). In this outbreak there were 29 confirmed cases in 12 states with seven hospitalizations. The strain causing this outbreak was distinguished from the first outbreak strain by its pulsed-field electrophoresis (PFGE) pattern. Again, *Salmonella* was isolated from environmental samples collected at the packaging facilities and from huller-shellers that supplied the almonds.

These outbreaks resulted in considerable concern on the part of the industry and regulators, and in 2006, new proposed rules were published by the USDA outlining a mandatory program to reduce the potential for *Salmonella* in almonds (US Department of Agriculture, 2006). This regulation requires that handlers subject their almonds to a process that achieves a minimum 4-log reduction in *Salmonella* prior to shipment. Handlers are defined in 7 CFR 981.13 to exclude roadside sale of almonds, but includes anyone who receives almonds from a grower for later sale. Exemptions were provided to handlers who ship untreated almonds to manufacturers within the United States, Canada, or Mexico who also agree to treat almonds to meet a 4-log pathogen reduction (under a directly verifiable program). These rules became effective March 31, 2007 and mandatory compliance began September 1, 2007.

Although outbreaks related to acute foodborne illnesses are rare, mycotoxins, and particularly aflatoxins, on tree nuts have historically been a major concern and their presence on nuts cannot be completely eliminated. Since some are teratogenic, mutagenic, or carcinogenic in susceptible animal species and may not produce readily identifiable acute symptoms, an outbreak as such would be difficult if not impossible to identify. Nonetheless, the ingestion of these mycotoxins would have a significant and deleterious effect. Consequently, aflatoxin levels are strictly controlled (US Food and Drug Administration, 2000b). Although some controversy may exist over health risks at low levels, current standards are set at less then 4 ppb aflatoxin by the European Union (Bayman et al., 2002).

ROUTES OF CONTAMINATION

Foodborne pathogens are not typically considered to be part of the normal epiphytic populations on fresh tree fruits or nuts. The surfaces of fresh produce, such as tree fruits, have traditionally been viewed as hostile environments for human pathogens. However, favorable conditions may exist in microsites on plant surfaces where growth and survival may occur (Brandl, 2006). Abundant evidence exists for the survival and even growth of foodborne pathogens on the surface of fruits and nuts as well as all produce. Such growth and survival may occur as part of a biofilm firmly attached to the surface. In some fruits, such as apples, growth/survival may occur internally, particularly when the fruit is damaged (Dingman, 2000; Fatemi et al., 2006). Consequently, once introduced to the fruit surface, the pathogens are not easily removed.

The reservoir for *E. coli* O157:H7 is generally considered to be domestic livestock (Keller and Miller, 2006). The same reservoir exists for *Cryptosporidium*. From this reservoir, these pathogens can be transferred to wild animals or can contaminate water that may be used in irrigation (Cole et al., 1999; Ingham et al., 2004; Steele and Odumeru, 2004). They may also become airborne in dust and be transferred by wind. Only *L. monocytogenes* is considered ubiquitous in nature.

For many foodborne outbreaks related to tree fruits, the actual mechanism through which the contamination occurred remains unknown. In part, this can be attributed to the low incidence of occurrence of foodborne pathogens in general and on tree fruit in particular. A study examining the microbial quality of Golden Delicious apples in Spain did not find virulent *E. coli* strains or any *Salmonella* on any apples regardless of where they were selected in the production stream (Abadias et al., 2006). However *E. coli* was identified on three out of 36 field samples. These results were similar to a study of US orchards to identify potential sources of *E. coli* O157:H7 (Riordan et al., 2000). In this study 14 different orchards were surveyed during autumn of 1999 to determine the incidence and prevalence of *E. coli* O157:H7. *E. coli* was found in the soil and on 6% of fruit samples, but no *E. coli* O157:H7 was found.

In the United States, Good Agricultural Practices (GAPs) have been published jointly by the USDA and FDA to help ensure the safe production of fresh produce (US Food and Drug Administration, 1998). Adherence to GAPs should ensure an overall low incidence of foodborne pathogens on fresh tree fruit and nuts. However, since even a strict adherence to GAPs cannot reduce the incidence of foodborne pathogens completely, and the risk of an outbreak will be increased in a juice product, juice producers are also required to follow FDA's Juice HACCP regulations. Recent outbreaks can be attributed to failure of producers to follow those regulations.

Several studies have examined potential sources of microbial contamination during the manufacture of apple cider (Garcia et al., 2006; Keller et al., 2002, 2004). Risk factors cited included the use of ground harvested apples, improper cleaning, and contaminated wash water. Poor sanitation and hygiene practices can significantly increase risk and the level of foodborne pathogens present on fresh tree fruit and in juice. In a survey by Castillo et al. (2006), freshly squeezed orange juice and fresh oranges from public markets and street vendors in Guadalajara, Mexico were examined for the incidence of *Salmonella* and *Shigella*. *Salmonella* was isolated from 9% of orange juice samples collected and from 10% of orange surfaces tested. *Shigella* was isolated from 5% of juice and 8% of orange surfaces tested. Contamination of the juice was attributed to poor hygienic practices. Poor hygienic practices have been cited as the cause of foodborne illness in numerous outbreaks involving tree fruits. These include an outbreak of infectious hepatitis through contaminated orange juice, an outbreak of *Shigella* in orange juice, and an outbreak of Norwalk virus gastroenteritis from fruit salad on a cruise ship (Einstein et al., 1963; Herwaldt et al., 1994; Thurston et al., 1998).

Poor hygiene practices and sanitation are not the only risk factors contributing to outbreaks in fresh tree fruits. In December of 1999, an outbreak of *Salmonella* serotype Newport occurred in imported mangoes. There were 78 reported illnesses in 13 states with 12 hospitalized and two deaths. Traceback of the contaminated fruit pinpointed a farm in Brazil where hot water treatment was used to control Mediterranean Medfly (Sivapalasingam et al., 2003). The same mangoes were also exported to Europe where no outbreak occurred. The single difference in the fruit was the use of a USDA/APHIS-mandated hot water treatment (46.7 °C for 75 to 90 minutes), implemented to replace the use of dibromide fumigation. The hot water was not routinely chlorinated but was followed by a chlorinated cold water rinse. Subsequent studies with a simulated process indicated that 80% of green mangoes internalized *Salmonella* when first dipped in hot water followed by a cool water rinse. Internalization facilitated by a temperature gradient has also been demonstrated in apples, oranges, and tomatoes (Buchanan et al., 1999; Merker et al., 1999; Zhuang et al., 1995). USDA/APHIS now recommends appropriate filtration and chlorination of hot water dips and a 20-minute period preceding any cold rinse. This single outbreak demonstrates the importance of a thorough investigation and evaluation of technologies used for food processing.

A thorough investigation and evaluation should not be limited to new technologies. Changes in normal procedures can also lead to unsafe practices. One example of this is in the use of browning inhibitors for fresh-cut apples. To conserve costs, processors may be tempted to extend the solution life beyond a single day or shift. Unfortunately, during extended use such solutions can become contaminated with suspended solids and microorganisms. Once contaminated, such solutions will now contain nutrients and provide an environment far more hospitable for survival and growth of pathogens. The survival of *Listeria innocua* has been demonstrated in calcium ascorbate solutions, particularly at higher pH (Karaibrahimoglu et al., 2004). Although no specific source of *L. monocytogenes* found in recalled apple slices was reported (Food and Drug Administration, 2001), contamination through reuse of antibrowning solutions does represent one possible route.

For nuts, as well as for fruits, harvesting or processing methods may play a significant role in their subsequent contamination. In both almond outbreaks of 2001 and 2004, the outbreak strains were isolated from orchard soil samples and equipment (Centers for Disease Control, 2004b; Isaacs et al., 2005). Nonetheless, a source for the contamination was not identified. Persistence of *Salmonella* in the orchard environment was documented by Uesugi et al. (2007), who also found that the rate of isolation increased during months where harvesting occurred. In addition, the highest rate of isolation was found following a rain event while almonds were collected for harvest in windrows on the orchard floor. Previous work by Uesugi and Harris (2006) has also suggested that rainfall during harvest when almonds are collected in windrows may play a role in their subsequent contamination. Higher rates of isolation during harvest were also explained by the presence of harvesting equipment and the generation of dust during harvest. The data suggests that *Salmonella* can survive for long periods of time on orchard floors, and almonds may become contaminated during harvesting operations.

PREVENTION

To reduce the risk of contamination, each step in the production of fresh fruits and nuts must be examined, beginning with field conditions. Ideally, even seemingly unimportant aspects such as drainage or type of soil may play a role in risk reduction. In 1998, the FDA, jointly with the USDA, issued the *Guide to Minimize Microbial Food Safety Hazards for Fresh Fruits and Vegetables* (US Food and Drug Administration, 1998). This document outlined common best practices that may be employed to avoid contamination of fresh produce based on the available science. Fundamental issues addressed included the use of water, both for irrigation and subsequent processing. Since the plan is by necessity general in its outline, each grower/processor must tailor the plan more specifically to meet the requirements of each specific product.

For tree fruits meant for the fresh market, it is critical that GAP guidelines be followed. Domestic animals should not be allowed access to orchards, and an effort should be made to discourage foraging by wildlife (Cole et al., 1999). Irrigation water should be free of dangerous pathogens to reduce the risk of contamination on the field (Steele and Odumeru, 2004). Likewise any water used for the make-up of pesticide should be free of pathogens, and such pesticide solutions should be made fresh for each use. Contaminated pesticide was cited as the probable cause of contamination of Mandarin oranges (Poubol et al., 2006). In addition, pathogens have been demonstrated to grow in some pesticide solutions (Ng et al., 2005).

Fruits destined for the fresh market should be tree-picked, not ground harvested, and blemish free. Several studies have linked ground harvested fruit and damaged fruit with a greater risk of contamination (Dingman, 2000; Fatemi et al., 2006; Keller et al., 2004; Wells and Butterfield, 1997). E. coli O157:H7 has been shown to survive and grow in damaged apple tissue (Dingman, 2000; Fatemi et al., 2006). The extent of growth depended upon the apple variety (Dingman, 2000). Damaged and ground-harvested fruits should be directed to those products that will receive processing designed to destroy any pathogens present.

REMEDIATION

Once tree fruits and nuts have become contaminated with foodborne pathogens, they are extremely difficult to decontaminate while still retaining their "fresh" character. Numerous studies have examined the efficacy of various surface treatments for the removal of pathogens on produce of all types (Alvarado-Casillas et al., 2007; Annous et al., 2001; Beuchat et al., 1998; Bialka and Demirci, 2007; Das et al., 2006; Deza et al., 2003; Fatemi and Knabel, 2006; Kenney and Beuchat, 2002; Kim et al., 2006; Kondo et al., 2006; Kozempel et al., 2001; Kumar et al., 2006; Lang et al., 2004; McWatters et al., 2002; Pao and Davis, 2001; Pao et al., 1999, 2000, 2007; Parnell et al., 2005; Pierre and Ryser, 2006; Pirovani et al., 2000; Raiden et al., 2003; Venkitanarayanan et al., 2002). Typical reductions range from 1 to 4 log depending on the produce type, method of inoculation, level of inoculation, and method of pathogen recovery.

None of the surface methods listed would remove any internally occurring pathogens. Pathogens may also exist in a biofilm or other protected state rendering surface decontamination methods ineffective (Burnett and Beuchat, 2001). Only treatment methods with greater penetration can be expected to destroy these hard to reach pathogens. Currently, the only method able to achieve such penetration other than thermal processing to "cook" produce is irradiation. Several studies have investigated the effects of radiation, either alone or in combination with other treatments, on the decontamination of fresh produce (Fan et al., 2006; Nthenge et al., 2007; Saroj et al., 2006; Schmidt et al., 2006; Young Lee et al., 2006). Although irradiation may be effective on some types of produce, it is not currently approved as a food additive for use on fresh produce (US Food and Drug Administration, 2000a). Irradiation at up to 1 kiloGray (kGy) may be used for growth and maturation inhibition of fresh foods and for deinfestation of arthropod pests; however, these levels would likely be insufficient for reduction or elimination of some microbial pathogens.

Surface decontamination, although unable to destroy pathogens borne internally, can still be viewed as efficacious for tree fruits and nuts for which internalization is unlikely to occur. Such fruits would include citrus fruits, for which the FDA has provided an exemption in the HACCP regulation, allowing a 5-log reduction in the pertinent pathogen to be applied to the surface of the fruit as opposed to the whole fruit. As with any processing step aimed at pathogen reduction, efficacy of the process is predicated on assumed initial microbial loads. Therefore, processes applied to the surface must be applied to clean fruit without visible blemishes or damage. Blemishes or damage to the protective peel could allow pathogens to become internalized, rendering processing treatments ineffective. Consequently, culling or quality sorting can be viewed as a critical step prior to the application of surface treatments (Keller, 2006).

In the production of fruit juices, again a variety of processing methods are available to achieve an appropriate 5-log reduction in pertinent pathogens. The most commonly used method remains thermal pasteurization; however, UV light treatment and ozone treatment have also been explored (Duffy et al., 2000; Harrington and Hills, 1968; Koutchma et al., 2004; Williams et al., 2004; Wright et al., 2000). Regardless of the method used, the juice processor must validate that the process used will achieve the prescribed 5-log reduction with the juice processed in the specific equipment used. Failure to do so has resulted in recent outbreaks. In addition, it should be noted that such treatments will result in a processed, not fresh product, and must be labeled as such.

CONCLUSIONS

Although foodborne illnesses associated with tree fruits and nuts are not common, their occurrence is particularly troublesome since they are associated with a healthful diet. Once tree fruits or nuts become contaminated with foodborne pathogens, their removal is extremely difficult while still maintaining their "fresh" state. Consequently, a fundamental key in the prevention of foodborne illness in fresh produce is to remove the vectors that may transmit the pathogen from the animal source to the produce. Unfortunately, when the vectors may be wild animals, wind, or irrigation water, risk can be reduced, but cannot be eliminated. An adherence to the FDA/USDA *Guide to Minimize Microbial Food Safety Hazards for Fresh Fruits and Vegetables* is critical to ensuring the safety of these foods. In addition, although adherence to GAPs is important, the role of postharvest contamination cannot be overlooked. A principal cause of foodborne outbreaks related to fresh produce of all types remains poor worker and facility hygiene. Attention to the entire production chain, from the farm, throughout processing and transport, to the table of the consumer is required to ensure that tree fruits and nuts do not become contaminated with foodborne pathogens. Such attention should result in the continued safety of tree fruits and nuts.

REFERENCES

Abadias, M., Canamas, T. P., Asensio, A., and Anguera, V. I. (2006). Microbial quality of commercial 'Golden Delicious' apples throughout production and shelf-life in Lleida (Catalonia, Spain). *International Journal of Food Microbiology* **108,** 404–409.

Alvarado-Casillas, S., Ibarra-Sanchez, S., Rodriguez-Garcia, O., Martinez-Gonzales, N., and Castillo, A. (2007). Comparison of rinsing and sanitizing procedures for reducing bacterial pathogens on fresh cantaloupes and bell peppers. *Journal of Food Protection* **70,** 655–660.

Annous, B. A., Sapers, G. M., Mattrazzo, A. M., and Riordan, D. C. R. (2001). Efficacy of washing with a commercial flatbed brush washer, using conventional and experimental washing agents, in reducing populations of *Escherichia coli* on artificially inoculated apples. *J. Food Protect.* **64,** 159–163.

Anonymous. (1999). National Advisory Committee on Microbiological Criteria for Foods: Meeting on Fresh Citrus Juice: Transcript of Proceedings. Accessed 11-4-03, at www.cfsan.fda.gov/~comm/tr991208.html.

Anonymous. (2001). Duck Delivery Produce, Inc. recalls apple snack trays because of possible health risk. Accessed 27 Nov 2007, at www.fda.gov/oc/po/firmrecalls/duck3_01.html.

Anonymous. (2003a). Duck Delivery Produce recalls cut honeydew and cut cantaloupe melon for possible health risk. Accessed 27 Nov 2007, at www.fda.gov/oc/po/firmrecalls/duck07_03.html.

Anonymous. (2003b). Georgia Ag. Department finds contaminated mushrooms. Accessed 27 Nov 2007, at www.fda.gov/oc/po/firmrecalls/southmill08_03.html.

Anonymous. (2006). Monterey Mushrooms recalls fresh sliced white and baby bella mushrooms in PA, MD, NC, NJ, NY, OH, and VA because of possible health risk. Accessed 27 Nov 2007, at www.fda.gov/oc/po/firmrecalls/monterey09_08.html.

Aruscavage, D., Lee, K., Miller, S., and LeJeune, J. T. (2006). Interactions affecting the proliferation and control of human pathogens on edible plants. *Journal of Food Science* **71**, R89–R99.

Bagamboula, C. F., Uyttendaele, M., and Debevere, J. (2002). Acid tolerance of *Shigella sonnei* and *Shigella flexneri*. *Journal of Applied Microbiology* **93**, 479–486.

Bang, I. S., Kim, B. H., Foster, J. W., and Park, Y. K. (2000). OmpR regulates the stationary-phase acid tolerance response of *Salmonella enterica* servovar Typhimurium. *Journal of Bacteriology* **182**, 2245–2252.

Bayman, P., Baker, J. L., and Mahoney, N. E. (2002). *Aspergillus* on tree nuts: Incidence and associations. *Mycopathologia* **155**, 161–169.

Belluck, P. (1998). Juice-poisoning case brings guilty plea and a huge fine. In *The New York Times*. Accessed 3 Dec 2007, at http://query.nytimes.com/gst/fullpage.html?res=9C06EED61239F937A15754C0A96E958260.

Benjamin, M. M. and Datta, A. R. (1995). Acid tolerance of enterohemorrhagic *Escherichia coli*. *Applied and Environmental Microbiology* **61**, 1669–1672.

Beuchat, L. R. (1995). Pathogenic microorganisms associated with fresh produce. *Journal of Food Protection* **59**, 204–216.

Beuchat, L. R. and Brackett, R. E. (1991). Behavior of *Listeria monocytogenes* inoculated into raw tomatoes and processed tomato products. *Applied and Environmental Microbiology* **57**, 1367–1371.

Beuchat, L. R. and Heaton, E. K. (1975). *Salmonella* survival on pecans as influenced by processing and storage conditions. *Applied Microbiology* **29**, 795–801.

Beuchat, L. R., Nail, B. V., Adler, B. B., and Clavero, M. R. S. (1998). Efficacy of spray application of chlorinated water in killing pathogenic bacteria on raw apples, tomatoes, and lettuce. *Journal of Food Protection* **61**, 1305–1311.

Beuchat, L. R. and Ryu, J.-H. (1997). Produce handling and processing practices. *Emerging Infectious Diseases* **3**, 459–465.

Bialka, K. L. and Demirci, A. (2007). Utilization of gaseous ozone for the decontamination of *Escherichia coli* O157:H7 and *Salmonella* on raspberries and strawberries. *Journal of Food Protection* **7**, 1093–1098.

Blackburn, B. G., Mazurek, J. M., Hlavsa, M., Park, J., Tillapaw, M., Parrish, M. et al. (2006). Cryptosporidiosis associated with ozonated apple cider. *Emerging Infectious Diseases* **12**, 684–686.

Brandl, M. T. (2006). Fitness of human enteric pathogens on plants and implications for food safety. *Annual Review of Phytopathology* **44**, 367–392.

Buchanan, R. L. and Doyle, M. E. (1997). Foodborne disease significance of *Escherichia coli* O157:H7 and other enterohemorrhagic *E. coli*. *Food Technology* **51**, 69–76.

Buchanan, R. L. and Edelson, S. G. (1996). Culturing enterohemorrhagic *Escherichia coli* in the presence and absence of glucose as a simple means of evaluating the acid tolerance of stationary-phase cells. *Applied and Environmental Microbiology* **62**, 4009–4013.

Buchanan, R. L., Edelson, S. G., Miller, R. L., and Sapers, G. M. (1999). Contamination of intact apples after immersion in an aqueous environment containing *Escherichia coli* O157:H7. *Journal of Food Protection* **62,** 444–450.

Burnett, S. L. and Beuchat, L. R. (2001). Human pathogens associated with raw produce and unpasteurized juices, and difficulties in decontamination. *Journal of Industrial Microbiology and Biotechnology* **27,** 104–110.

Burnett, S. L., Chen, J., and Beuchat, L. R. (2000). Attachment of *Escherichia coli* O157:H7 to the surfaces and internal structures of apples as detected by confocal scanning laser microscopy. *Applied and Environmental Microbiology* **66,** 4679–4687.

Castillo, A., Villarruel-Lopez, A., Navarro-Hidalgo, V., Martinez-Gonzalez, N. E., and Torres-Vitela, M. R. (2006). *Salmonella* and *Shigella* in freshly squeezed orange juice, fresh oranges, and wiping cloths collected from public markets and street booths in Guadalajara, Mexico: Incidence and comparison of analytical routes. *Journal of Food Protection* **69,** 2595–2599.

Centers for Disease Control. (1996). Outbreaks of *Escherichia coli* O157:H7 infections associated with drinking unpasteurized commercial apple juice—October, 1996. *Morbidity and Mortality Weekly Report* **45,** 975.

Centers for Disease Control. (1997). Outbreaks of *Escherichia coli* O157:H7 infection and cryptosporidiosis associated with drinking unpasteurized apple cider—Connecticut and New York, October 1996. *Morbidity and Mortality Weekly Report* **46,** 4–8.

Centers for Disease Control. (2002). Salmonellosis, Vol. 2003. Accessed Oct 23, at www.cdc.gov/ncidod/dbmd/diseaseinfo/salmonellosis_t.htm.

Centers for Disease Control. (2003). U.S. Foodborne disease outbreaks, annual listing 1990–2000., Vol. 2003. Accessed 17 Oct 2007, at www.cdc.gov/foodborneoutbreaks/report_pub.htm.

Centers for Disease Control. (2004a). 2004 Summary Statistics, Vol. 2007. Accessed 27 Nov 2007, at www.cdc.gov/foodborneoutbreaks/us_outb/fbo2004/Outbreak_Linelist_Final_2004.pdf.

Centers for Disease Control. (2004b). Outbreak of *Salmonella* Serotype Enteritidis Infections Associated with Raw Almonds—United States and Canada, 2003–2004. *Morbid. Mortal. Weekly Rep.* **53,** 484–487.

Centers for Disease Control. (2005a). 2005 Summary Statistics, Vol. 2007. Accessed 6 Dec 2007, at www.cdc.gov/foodborneoutbreaks/us_outb/fbo2005/2005_Linelist.pdf.

Centers for Disease Control. (2005b). Preventing health risks associated with drinking unpasteurized or untreated juice, Vol. 2007. Accessed 6 Dec 2007, at www.cdc.gov/foodborne/juice_spotlight.htm.

Centers for Disease Control. (2005c). Shigellosis. Accessed 30 Nov 2007, at www.cdc.gov/ncidod/dbmd/diseaseinfo/shigellosis_t.htm.

Cole, D. J., Hill, V. R., Humenik, F. J., and Sobsey, M. D. (1999). Health, safety and environmental concerns of farm animal waste. *Occupational Medicine* **14,** 423–428.

Cox, L. J., Kleiss, T., Cordier, J. L., Cordellana, C., Konel, P., Pedrazzini, C. et al. (1989). *Listeria* spp. in food processing, non-food and domestic environments. *Food Microbiology* **6,** 49–61.

Danyluk, M. D., Jones, T. M., Abd, S. J., Schlitt-Dittrich, F., Jacobs, M., and Harris, L. J. (2007). Prevalence and amounts of *Salmonella* found on raw California almonds. *Journal of Food Protection* **70,** 820–827.

Das, E., Gurakan, G. C., and Bayindirli, A. (2006). Effect of controlled atmosphere storage, modified atmosphere packaging and gaseous ozone treatment on the survival of *Salmonella* Enteritidis on cherry tomatoes. *Food Microbiology* **23,** 430–438.

Deza, M. A., Araujo, M., and Garrido, M. J. (2003). Inactivation of *Escherichia coli* O157:H7, *Salmonella* Enteritidis and *Listeria monocytogenes* on the surface of tomatoes by neutral electrolyzed water. *Letters in Applied Microbiology* **37,** 482–487.

Dingman, D. W. (2000). Growth of *Escherichia coli* O157:H7 in bruised apple (Malus domestica) tissue as influenced by cultivar, date of harvest, and source. *Applied and Environmental Microbiology* **66,** 1077–1083.

Duffy, S., Churey, J., Worobo, R., and Schaffner, D. W. (2000). Analysis and modeling of the variability associated with UV inactivation of *Escherichia coli* in apple cider. *Journal of Food Protection* **63,** 1587–1590.

Duncan, T. G., Doull, J. A., Miller, E. R., and Bancroft, H. (1946). Outbreak of typhoid fever with orange juice as the vehicle, illustrating the value of immunization. *American Journal of Public Health* **36,** 34–36.

Einstein, A. B., Jacobsohn, W., and Goldman, A. (1963). An epidemic of infectious hepatitis in a general hospital: Probable transmission by contaminated orange juice. *JAMA* **185,** 171–174.

Fan, X., Annous, B. A., Sokorai, K. J. B., Burke, A., and Mattheis, J. P. (2006). Combination of hot-water surface pasteurization of whole fruit and low-dose gamma irradiation of fresh-cut cantaloupe. *Journal of Food Protection* **69,** 912–919.

Fatemi, P. and Knabel, S. J. (2006). Evaluation of sanitizer penetration and its effect on destruction of *Escherichia coli* O157:H7 in golden delicious apples. *Journal of Food Protection* **69,** 548–555.

Fatemi, P., LaBorde, L. F., Patton, J., Sapers, G. M., Annous, B. A., and Knabel, S. J. (2006). Influence of punctures, cuts, and surface morphologies of golden delicious apples on penetration and growth of *Escherichia coli* O157:H7. *Journal of Food Protection* **69,** 267–275.

Fenlon, D. R., Wilson, J., and Donachie, W. (1996). The incidence and level of *Listeria monocytogenes* contamination of food sources at primary production and initial processing. *Journal of Applied Bacteriology* **81,** 641–650.

Fleischman, G. J., Bator, C., Merker, R., and Keller, S. E. (2001). Hot water immersion to eliminate *Escherichia coli* O157:H7 on the surface of whole apples: Thermal effects and efficacy. *Journal of Food Protection* **64,** 451–455.

Food and Drug Administration. (2001). Enforcement report. Recalls and Field corrections: Foods—class I. Recall number F-535-1. Accessed March 27, 2008, at www.fda.gov/bbs/topics/enforce/2001/enf00708.html.

Foster, J. W. and Hall, H. K. (1990). Adaptive acidification tolerance response of *Salmonella* Typhimurium. *Journal of Bacteriology* **172,** 771–778.

Garcia, L., Henderson, J., Fabri, M., and Moustapha, O. (2006). Potential sources of microbial contamination in unpasteurized apple cider. *Journal of Food Protection* **69,** 137–144.

Gombas, D. E., Chen, Y., Clavero, R. S., and Scott, V. N. (2003). Survey of *Listeria monocytogenes* in ready-to-eat foods. *Journal of Food Protection* **66**, 559–569.

Goverd, K. A., Beech, F. W., Hobbs, R. P., and Shannon, R. (1979). The occurrence and survival of colifoms and salmonellas in apple juice and cider. *Journal of Applied Bacteriology* **46**, 521–530.

Guerrant, R. L. (1997). Cryptosporidiosis: An emerging highly infectious threat. *Emerging Infectious Diseases* **3**, 51–57.

Harrington, W. O. and Hills, C. H. (1968). Reduction of the microbial population of apple cider by ultraviolet irradiation. *Food Technology* **22**, 117–120.

Herwaldt, B. L., Lew, J. F., Moe, C. L., Lewis, D. C., Humphrey, C. D., Monroe, S. S. et al. (1994). Characterization of a variant strain of Norwalk virus from a foodborne outbreak of gastroenteritis on a cruise ship in Hawaii. *Journal of Clinical Microbiology* **32**, 861–866.

Ingham, S. C., Losinski, J. A., Andrews, M. P., Breuer, J. E., Breuer, J. R., Wood, T. M., and Wright, T. H. (2004). *Escherichia coli* contamination of vegetables grown in soils fertilized with noncomposted bovine manure: Garden-scale studies. *Applied and Environmental Microbiology* **70**, 6420 – 6427.

Isaacs, S., Aramini, J., Ciebin, B., Farrar, J. A., Ahmed, R., Middleton, D. et al. (2005). An international outbreak of salmonellosis associated with raw almonds contaminated with a rare phage type of *Salmonella* enteritidis. *Journal of Food Protection* **68**, 191–198.

Johnston, L. M., Jaykus, L., Moll, D., Martinez, M. C., Anciso, J., Mora, B., and Moe, C. L. (2005). A field study of the microbiological quality of fresh produce. *Journal of Food Protection* **68**, 1840–1847.

Karaibrahimoglu, Y., Fan, X., Sapers, G. M., and Sokorai, K. (2004). Effect of pH on the survival of *Listeria innocua* in calcium ascorbate solutions and on quality of fresh-cut apples *Journal of Food Protection* **67**, 751–757.

Keller, S. E. (2006). Chapter 16: Effect of quality sorting and culling on the microbiological quality of fresh produce. In *"Microbiology of Fruits and Vegetables"* (G. M. Sapers, J. R. Gorny, and A. E. Yousef, Eds.), pp. 365–373. CRC Press, Boca Raton, FL.

Keller, S. E., Chirtel, S. J., Merker, R. I., Taylor, K. T., Tan, H. L., and Miller, A. J. (2004). Influence of fruit variety, harvest technique, quality sorting, and storage on the native microflora of unpasteurized apple cider. *Journal of Food Protection* **67**, 2240–2247.

Keller, S. E., Merker, R. I., Taylor, K. T., Tan, H. L., Melvin, C. D., Chirtel, S. J., and Miller, A. J. (2002). Efficacy of sanitation and cleaning methods in a small apple cider mill. *Journal of Food Protection* **65**, 911–917.

Keller, S. E., and Miller, A. J. (2006). Chapter 9: Microbiological safety of fresh citrus and apple juices. In *"Microbiology of Fruits and Vegetables"* (G. M. Sapers, J. R. Gorny, and A. E. Yousef, Eds.), pp. 211–230. CRC Press, Boca Raton, FL.

Kenney, S. J. and Beuchat, L. R. (2002). Comparison of aqueous commercial cleaners for effectiveness in removing *Escherichia coli* O157:H7 and *Salmonella* muenchen from the surface of apples. *International Journal of Food Microbiology* **74**, 47–55.

Kim, H., Ryu, J.-H., and Beuchat, L. R. (2006). Survival of *Enterobacter sakazakii* on fresh produce as affected by temperature, and effectiveness of sanitizers for its elimination. *International Journal of Food Microbiology* **111**, 134–143.

Kondo, N., Murata, M., and Isshiki, K. (2006). Efficiency of sodium hypochlorite, fumaric acid, and mild heat in killing native microflora and *Escherichia coli* O157:H7, *Salmonella* Typhimurium DT104, and *Staphylococcus aureus* attached to fresh-cut lettuce. *Journal of Food Protection* **69**, 323–329.

Koutchma, T., Keller, S. E., Parisi, B., and Chirtel, S. J. (2004). Ultraviolet disinfection of juice products in laminar and turbulent flow reactors. *Innovative Food Science and Emerging Technologies* **5**, 179–189.

Kozempel, M. F., Marshall, D. L., Radewonuk, E. R., Scullen, O. J., Goldberg, N., and Bal'a, M. F. A. (2001). A rapid surface intervention process to kill *Listeria innocua* on catfish using cycles of vacuum and steam. *J. Food Sci.* **66**, 1012–1016.

Kumar, M., Hora, R., Kostrzynska, M., Waites, W. M., and Warriner, K. (2006). Inactivation of *Escherichia coli* O157:H7 and *Salmonella* on mung beans, alfalfa, and other seed types destined for sprout production by using an oxychloro-based sanitizer. *Journal of Food Protection* **69**, 1571–1578.

Lang, M. M., Harris, L. J., and Beuchat, L. R. (2004). Survival and recovery of *Escherichia coli* O157:H7, *Salmonella*, and *Listeria monocytogenes* on lettuce and parsley as affected by method of inoculation, time between inoculation and analysis, and treatment with chlorinated water. *Journal of Food Protection* **67**, 1092–1103.

Lin, J., Lee, I. S., Frey, J., Slonczewski, J. L., and Foster, J. W. (1995). Comparative analysis of extreme acid survival in *Salmonella typhimurium*, *Shigella flexneri*, and *Escherichia coli*. *Journal of Bacteriology* **177**, 4097–4104.

Lin, J., Smith, M. P., Chapin, K. C., Baik, H. S., Bennett, G. N., and Foster, J. W. (1996). Mechanisms of acid resistance in Enterohemorrhagic *Escherichia coli*. *Applied and Environmental Microbiology* **62**, 3094–3100.

McWatters, K. H., Doyle, M. P., Walker, S. L., Rimal, A. P., and Venkitanarayanan, K. (2002). Consumer acceptance of raw apples treated with an antibacterial solution designed for home use. *Journal of Food Protection* **65**, 106–110.

Merker, R., Edelson-Mammel, S., Davis, V., and Buchanan, R. L. (1999). Preliminary experiments on the effect of temperature differences on dye uptake by oranges and grapefruit, Vol. 2007. Accessed 6 Oct 2007, at http://vm.cfsan.fda.gov/~comm/juicexp.html.

Millard, P. S., Gensheimer, K. F., Addiss, D. G., Sosin, D. M., Beckett, G. A., Houck-Jankoski, A., and Hudson, A. (1994). An outbreak of Cryptosporidiosis from fresh-pressed apple cider. *JAMA* **272**, 1592–1596.

Ng, P. J., Fleet, G. H., and Heard, G. M. (2005). Pesticides as a source of microbial contamination of salad vegetables. *International Journal of Food Microbiology* **101**, 237–250.

Nthenge, A. K., Weese, J. S., Carter, M., Wei, C., and Huang, T. (2007). Efficacy of gamma radiation and aqueous chlorine on *Escherichia coli* O157:H7 in hydroponically grown lettuce plants. *Journal of Food Protection* **70**, 748–752.

Pao, S. and Davis, C. L. (2001). Maximizing microbiological quality of fresh orange juice by processing sanitation and fruit surface treatments. *Dairy, Food and Environmental Sanitation* **21**, 287–291.

Pao, S., Davis, C. L., and Kelsey, D. F. (2000). Efficacy of alkaline washing for the decontamination of orange fruit surfaces inoculated with *Escherichia coli*. *Journal of Food Protection* **63**, 961–964.

Pao, S., Davis, C. L., Kelsey, D. F., and Petracek, P. D. (1999). Sanitizing effects of fruit waxes at high pH and temperature on orange surfaces inoculated with *Escherichia coli*. *Journal of Food Science* **64**, 359–362.

Pao, S., Kelsey, D. F., Khalid, M. F., and Ettinger, M. R. (2007). Using aqueous chlorine dioxide to prevent contamination of tomatoes with *Salmonella enterica* and *Erwinia carotovora* during fruit washing. *Journal of Food Protection* **70**, 629–634.

Parish, M. E. (1997). Public health and nonpasteurized fruit juices. *Critical Reviews in Microbiology* **23**, 109–119.

Parnell, T. L., Harris, L. J., and Suslow, T. V. (2005). Reducing *Salmonella* on cantaloupes and honeydew melons using wash practices applicable to postharvest handling, foodservice, and consumer preparation. *International Journal of Food Microbiology* **99**, 59–70.

Pierre, P. M. and Ryser, E. T. (2006). Inactivation of *Escherichia coli* O157:H7, *Salmonella* Typhimurium DT104, and *Listeria monocytogenes* on inoculated alfalfa seeds with a fatty acid–based sanitizer. *Journal of Food Protection* **69**, 582–590.

Pirovani, M. E., Guemes, D. R., Di Pentima, J. H., and Tessi, M. A. (2000). Survival of *Salmonella* hadar after washing disinfection of minimally processed spinach. *Letters in Applied Microbiology* **31**, 143–148.

Poubol, J., Tsukada, Y., Sera, K., and Izumi, H. (2006). On-the-farm contamination of Satsuma mandarin fruit at different orchards in Japan. *Acta Hort.* **712**, 551–560.

Raiden, R. M., Sumner, S. S., Eifert, J. D., and Pierson, M. D. (2003). Efficacy of detergents in removing *Salmonella* and *Shigella* spp. from the surface of fresh produce. *Journal of Food Protection* **66**, 2210–2215.

Riordan, D. C. R., Sapers, G. M., and Annous, B. A. (2000). The survival of *Escherichia coli* O157:H7 in the presence of *Penicillium expansium* and *Glomerella cingulata* in wounds on apple surfaces. *Journal of Food Protection* **63**, 1637–1642.

Sado, P. N., Jinneman, K. C., Husby, G. J., Sorg, S. M., and Omiecinski, C. J. (1998). Identification of *Listeria monocytogenes* from unpasteurized apple juice using rapid test kits. *Journal of Food Protection* **61**, 1199–1202.

Saroj, S. D., Shashidhar, R., Pandey, M., Dhokane, V., Hajare, S., Sharma, A., and Bandekar, J. R. (2006). Effectiveness of radiation processing in elimination of *Salmonella* Typhimurium and *Listeria monocytogenes* from sprouts. *Journal of Food Protection* **69**, 1858–1864.

Schmidt, H. M., Palekar, M. P., Maxim, J. E., and Castillo, A. (2006). Improving the microbiological quality and safety of fresh-cut tomatoes by low-dose electron beam irradiation. *Journal of Food Protection* **60**, 575–581.

Sivapalasingam, S., Barrett, E., Kimura, A., Van Duyne, S., De Witt, W., Ying, M. et al. (2003). A multistate outbreak of *Salmonella enterica* Serotype Newport infection linked to mango consumption: Impact of water-dip disinfestation technology. *Clinical Infectious Diseases* **37**, 1585–1590.

Sorrells, K. M., Enigl, D. C., and Hatfield, J. R. (1989). Effect of pH, acidulant, time, and temperature on the growth and survival of *Listeria monocytogenes*. *Journal of Food Protection* **52**, 571–573.

Steele, M. and Odumeru, J. (2004). Irrigation water as a source of foodborne pathogens on fruit and vegetables. *Journal of Food Protection* **67**, 2839–2849.

Thurston, H., Stuart, J., McDonnell, B., Nicholas, S., and Cheasty, T. (1998). Fresh orange juice implicated in an outbreak of *Shigella flexneri* among visitors to a South African game reserve. *Journal of Infection* **36**, 350.

U.S. Department of Agriculture. (2004). Fruit and tree nuts: Background. In "ERS/USDA Briefing room," Vol. 2007. Accessed 17 April, 2007, at www.ers.usda.gov/Briefing/FruitandTreeNuts/background.htm.

U.S. Department of Agriculture. (2006). 7 CFR Part 981: Almonds grown in California; Outgoing quality control requirements and request for approval of new information collection. *Fed. Reg.* **71**, 70683–70692.

U.S. Food and Drug Administration. (1998). Guide to minimize microbial food safety hazards for fresh fruits and vegetables. Vol. 2007. Accessed 6 Dec 2007, at http://vm.cfsan.fda.gov/~dms/prodguid.html.

U.S. Food and Drug Administration. (2000a). 21 CFR Part 179. Irradiation in the production, processing and handling of food. *Federal Register* **65**, 71056–71058.

U.S. Food and Drug Administration. (2000b). Action levels for poisonous or deleterious substances in human food and animal feed, Vol. 2007. Accessed 27 Nov 2007, at www.cfsan.fda.gov/~lrd/fdaact.html.

U.S. Food and Drug Administration. (2001). 21 CFR Part 120. Hazard analysis and critical control point (HACCP); Procedures for the safe and sanitary processing and importing of juice. Final rule. *Federal Register* **66**, 6137–6202.

U.S. Food and Drug Administration. (2004). Recommendations to processors of apple juice or cider on the use of ozone for pathogen reduction purposes. Vol. 2007. Accessed 7 Aug 2007, at www.cfsan.fda.gov/~dms/juicgu13.html.

U.S. Food and Drug Administration. (2005). Letter to state regulatory agencies and firms that produce treated (but not pasteurized) and untreated juice and cider, Vol. 2007. Accessed 7 Aug 2007, at www.cfsan.fda.gov/~dms/juicgu14.html.

U.S. Food and Drug Administration. (2006). *The bad bug book*. Accessed 30 Nov 2007, at www.cfsan.fda.gov/~mow/intro.html.

Uesugi, A. R., Danyluk, M. D., and Harris, L. J. (2006). Survival of *Salmonella* Enteritidis phage type 30 on inoculated almonds stored at −20, 4, 23, and 35 °C. *Journal of Food Protection* **69**, 1851–1857.

Uesugi, A. R., Danyluk, M. D., Mandrell, R. E., and Harris, L. J. (2007). Isolation of *Salmonella* Enteritidis Phage Type 30 from a single almond orchard over a 5-year period. *Journal of Food Protection* **70**, 1784–1789.

Uesugi, A. R. and Harris, L. J. (2006). Growth of *Salmonella* Enteritidis phage type 30 in almond hull and shell slurries and survival in drying almond hulls. *Journal of Food Protection* **69**, 712–718.

Venkitanarayanan, K. S., Lin, C., Bailey, H., and Doyle, M. E. (2002). Inactivation of *Escherichia coli* O157:H7, *Salmonella* Enteritidis and *Listeria monocytogenes* on apples, oranges, and tomatoes by lactic acid and hydrogen peroxide. *Journal of Food Protection* **65**, 100–105.

Wells, J. M. and Butterfield, J. E. (1997). *Salmonella* contamination associated with bacterial soft rot of fresh fruits and vegetables in the marketplace. *Plant Disease* **81,** 867–872.

Williams, R. C., Sumner, S. S., and Golden, D. A. (2004). Survival of *Escherichia coli* O157:H7 and *Salmonella* in apple cider and orange juice as affected by ozone and treatment temperature. *Journal of Food Protection* **67,** 2381–2386.

Winfield, M. D. and Groisman, E. A. (2003). Role of nonhost environments in the lifestyles of *Salmonella* and *Escherichia coli*. *Applied and Environmental Microbiology* **69,** 3687–3694.

Wright, J. R., Sumner, S. S., Hackney, C. R., Pierson, M. D., and Zoecklein, B. W. (2000). Efficacy of ultraviolet light for reducing *Escherichia coli* O157:H7 in unpasteurized apple cider. *Journal of Food Protection* **63,** 563–567.

Young Lee, N., Jo, C., Hwa Shin, D., Geun Kim, W., and Woo Byun, M. (2006). Effect of g-irradiation on pathogens inoculated into ready-to-use vegetables. *Food Microbiology* **23,** 649–656.

Zhuang, R. Y., Beuchat, L. R., and Angulo, F. J. (1995). Fate of *Salmonella* Montevideo on and in raw tomatoes as affected by temperature and treatment with chlorine. *Applied and Environmental Microbiology* **61,** 2127–2131.

CHAPTER 12

Berry Contamination: Outbreaks and Contamination Issues

Kalmia E. Kniel and Adrienne E.H. Shearer

Department of Animal and Food Sciences, University of Delaware, Newark, DE

CHAPTER CONTENTS

Introduction	271
The Impact of Major Outbreaks	273
History of Viral Contamination of Berries	274
Hepatitis A Outbreaks with Raspberries and Strawberries	275
Norovirus Associated Outbreaks with Raspberries	277
The Role of *Cyclospora cayetanensis* in Berry-Associated Outbreaks	279
Transmission of *Cyclospora* Oocysts and the Role of Foods	280
Bacterial Contamination of Berries	283
Contamination Reduction Strategies	285
In Summary	297

INTRODUCTION

Fruits commonly referred to as berries are popular for many reasons including taste, nutrition, and convenience. Berries include strawberries, brambles (raspberries, blackberries, and associated hybrids), blueberries, cranberries, currants, grapes, gooseberries, and elderberries. However, although these are all technically considered berries, only blueberries and grapes are true berries, as the fruits are multiseeded and derived from a single ovary (Bowling, 2000). This diverse group of fruits has been a source of sustenance throughout history beginning with the earliest hunting and gathering people (Bowling, 2000) and remains an important crop today. Recently the nutritional importance of berries has been suggested, including their high levels of antioxidants and anticancer compounds (Liu, 2007).

Berry production and consumption have increased steadily over the past decade. In the United States, berry production has increased dramatically; whereas only 5.3 million pounds of berries (fresh weight equivalent) were produced in 1970, more than 14.4 million pounds were produced in 2006 (ERS, 2008). Raspberries have seen the greatest increase; production values have increased seven-fold from 1991 to 2006. During that same time period strawberry production doubled. Berry consumption is also growing in the United States (ERS, 2008). Over the past six years the consumption of fresh blueberries has more than doubled, raspberry consumption has increased four-fold, and strawberry consumption has steadily increased over that time. These numbers don't account for the imported berries that are shipped into the United States during much of the year. In 2005 more than 6.3 million tons of berries were produced worldwide (Strik, 2007).

Berry crops are produced and harvested through three marketing channels, including customer-harvested through U-pick farms, fresh sales via local stores or distant domestic and international markets, and processed as frozen fruit, puree, dried fruit, or juice (here processed berries may be sold directly to retail). From a food-safety standpoint, the berries picked fresh and sold internationally or at local markets are of the greatest concern; however, as discussed later, further processing steps including freezing do not kill pathogenic microorganisms, and cases of foodborne illness have been associated with frozen berries.

In terms of impact on food safety, the most significant berries to date have been blackberries, raspberries, blueberries, and strawberries. As with all other produce commodities, the facts that berries are often consumed raw and unwashed may influence the removal of potential contaminants and pathogenic microorganisms. It is necessary to prevent the initial contamination by utilizing Good Agricultural Practices, including using treated or composted manures, using high quality irrigation water at the production level and potable water for making ice after harvesting, using clean and sanitized equipment and transportation vehicles, providing proper sanitation systems for workers, assuring worker health, and maintaining a proper cold chain through delivery to the final customer.

The 22 documented outbreaks of foodborne illness associated with berries are discussed within this chapter. Additionally, it is likely that numerous other outbreaks have gone undetected or unconfirmed. There may be several reasons why berries have been involved in a myriad of outbreaks; contamination by farm and packing-plant workers, use of unsafe agricultural practices, and global sourcing to provide yearlong availability. In general, berry crops are best watered using trickle irrigation, as this allows the grower to apply water at the critical period of fruit development and avoids wetting the fruit, which could foster the development of disease and rot (Bowling, 2000). However, this does not preclude the possibility that nonpotable or even potentially contaminated water might be used for spray irrigation or for pesticide or fertilizer applications. This point is mentioned later in the discussion on contamination of raspberry plants with the protozoan parasite *Cyclospora cayetanensis*. For the most part berries are highly perishable and require minimal handling. In order to do this, sorting and packing into the shipping containers are often performed in the field by the pickers (Ryall and Pentzer, 1982).

THE IMPACT OF MAJOR OUTBREAKS

Over the past two decades, several outbreaks associated with contaminated berries stand out from the rest. Interestingly, these are not associated with bacteria. The protozoan parasite *Cyclospora* caused illness throughout North America associated with raspberries (Herwaldt, 2000), and hepatitis A virus caused numerous illnesses associated with strawberries (Niu et al., 1992). At the time, these outbreaks initiated discussion on the role of "emerging" foodborne organisms in produce contamination, issues of detection, and the need for improved diagnostic methods (Tauxe et al., 1997). These are still important issues today as we seek better diagnostic and prevention methods. We have a greater understanding of the process our food undergoes along the farm-to-fork continuum, but we are still missing some basic information concerning the growth, survival, and transmission of these pathogenic organisms along this route. Historically, these berry-associated outbreaks showed the necessity for reacting to strong epidemiologic data without laboratory confirmation in order to have better control over the course of an epidemic (Tauxe et al., 1997). Although epidemiological data is always important in an outbreak investigation, detection of viruses and protozoa is traditionally difficult and forces investigators to rely on epidemiological data for traceback. Additionally, two factors, relating to these organisms in particular, increase our need for greater understanding, and these are the extent of the *Cyclospora* outbreaks (1996–2001) and the long incubation period of hepatitis A virus (10–50 days, average 30 days).

The economic effects of these outbreaks are long-lasting and often linger more with the perception that specific foods (i.e., imported Guatemalan raspberries and imported Mexican strawberries, contaminated either in Mexico or in the United States) are suspect (Calvin, 2004). The outbreaks discussed as follows provide multiple examples of the importance of good epidemiology. The epidemiological investigations were able to implicate the berry-containing food item and then led to proper recalls and environmental investigations. This is not always the case. In the spring of 1996 with the first reports of illness attributed to *Cyclospora*, the Texas Department of Health erroneously issued a report identifying the source of the problem as strawberries from California (Herwaldt and Beach, 1999). This was disastrous for the California strawberry industry that was in peak production at this time. However, when the CDC issued a statement that the source was Guatemalan raspberries (Herwaldt and Beach, 1999), the Guatemalan spring export season had just concluded, and the growers suffered few effects. Although this led to a huge economic loss in California, the California strawberry growers were able to develop an enhanced food safety system after this problem. This system was tested in 1997 when there was a problem associated with hepatitis A contamination of Mexican strawberries, and consumers questioned California-grown strawberries for the second year in a row (Calvin, 2004). The California produce industry was able to survive these issues; however, after more than two years of repeated outbreaks involving Guatemalan raspberries in outbreaks in North America, the industry there never recovered.

HISTORY OF VIRAL CONTAMINATION OF BERRIES

Both raspberries and strawberries (raw and frozen) have been associated with outbreaks of hepatitis A virus and norovirus. Hepatitis A, a virus spread by human feces, is thought to have contaminated the berries by contact with infected farm workers during harvest or contaminated irrigation water. Frozen and fresh raspberries have also been associated with illness due to norovirus, also spread through contact with human feces and infected food handlers. Processing berries, including freezing and mild cooking may be an important issue in the case of virally contaminated berries. These processing steps do not necessarily clear berries from viral contamination. The stems of strawberries destined for freezing are removed in the field, either using a metal device or a thumbnail. The berries are then transported at ambient temperature to a processing facility where they are washed with water, sliced if applicable, and often mixed with up to 30% sucrose before freezing. The extra human handling during harvesting and comingling in the processing facility is believed to place these berries at greater risk for viral contamination (CFSAN, 2001).

An increased awareness that berries play a role in the transmission of viruses led to epidemiological surveys that in turn increased awareness of contaminated berries associated with illness (Butot et al., 2007). For example, approximately 15 outbreaks of viral illness involving berries were recently identified in Finland between 1998 and 2001 (Ponka et al., 1999). Similarly, frozen berries imported from Poland were found to be responsible for more than 1100 illnesses in Europe in six different outbreaks (Cotterelle et al., 2005; Falkenhorst et al., 2005; Korsager et al., 2005).

HEPATITIS A OUTBREAKS WITH RASPBERRIES AND STRAWBERRIES

One of the earlier recorded berry outbreaks associated with viral contamination was an outbreak of hepatitis A in Scotland linked to consumption of raspberry mousse prepared from frozen raspberries (Reid and Robinson, 1987). Raspberries previously had been noted in epidemiological investigations as potential carriers of virus (Noah, 1981). The raspberry mousse was prepared specifically for a banquet held for 10 people at a large hotel in Aberdeen. The mousse was prepared from two 3 lb tubes of frozen raspberries, gelatin, sugar, and pasteurized cream. Some of the leftover mousse was sent home with the staff or was served on the "sweet trolley" in the dining room the next day. Twenty-four individuals were diagnosed with jaundice, deranged liver functions, fever, malaise, nausea, and flu-like symptoms approximately 24 to 28 days after consuming the mousse. The raspberries were blast-frozen at a distribution center. The raspberries had been obtained from several farms, including small holdings and large private gardens. Three of these growers were implicated indirectly in a previous outbreak (Noah, 1981). Contamination of raspberries apparently occurred at the time of picking or packing, probably by a food handler who was unknowingly shedding hepatitis A virus. A local physician reported that one of the pickers had a hepatitis A infection at the time of picking (Reid and Robinson, 1987). This restates the importance and impact of good personal hygiene and sanitation practices, along with the need for good education in food safety of food handlers at each stage from the farm to the consumer.

A multistate outbreak of hepatitis A was traced to frozen strawberries processed in a single plant in California in 1990 (Niu et al., 1992). Nine hundred students, teachers, and staff in Georgia and Montana developed hepatitis A infection from eating strawberry shortcake and other desserts. Epidemiological data indicated that contamination did not occur from an infected worker within the processing plant but most likely from an infected picker, perhaps when the stems were being removed by hand rather than with a metal tool. Strawberries still are often destemmed prior to being brought into the processing facility.

In the months of February and March of 1997, in Michigan and Maine, there was a similar outbreak, at first linked to frozen raspberries and strawberries (Hutin et al., 1998). More conclusive epidemiological evidence from case-control studies determined that the illness was associated only with frozen strawberries, and the cases involved school children and employees. In Michigan, as in the outbreak in 1990, the frozen strawberries were consumed in strawberry shortcake desserts served in the school cafeterias. A total of 287 cases of hepatitis A were reported from 23 schools in Michigan and 13 schools in Maine. Traceback analysis implicated strawberries grown in Mexico, and processed and distributed through a California processing facility. There was no indication of specific lots that were contaminated as these records were not maintained by the schools at this time. A thorough investigation of the California processing facility did not identify any problems, showed good sanitation and manufacturing practices, limited hand contact from the employees with the berries, and no record of employees with illnesses at the time the strawberries were processed. The FDA also conducted an investigation in the strawberry fields in Mexico. These fields were drip irrigated rather than spray irrigated, which eliminated the likelihood that berries were contaminated by contaminated water.

This investigation revealed several potential problems, including limited slit latrines for the workers and limited access to hand washing facilities that were on trucks circulating through the fields. Although no records were maintained on worker illnesses, the workers did not wear gloves and removed the stems from the strawberries with their fingernails in the fields. Direct hand contact with the berries combined with poor hygienic practices was a possible source of contamination. Other strawberries from the same distributor were placed on hold, and of the 13 other states that received the frozen strawberries, only two cases of hepatitis A were reported in Tennessee, nine cases in Arizona, five cases in Wisconsin, and four cases in Louisiana. All these cases with the exception of those from Louisiana were associated with state school-lunch programs. The Louisiana cases were traced back to consumption of a commercially prepared smoothie drink. For these clusters of cases no epidemiological studies were conducted. The viruses isolated from the majority of cases of hepatitis A described in this multistate outbreak showed high genetic similarity. Due to the relatively low number of cases compared to the large quantity of frozen strawberries that were consumed, it is likely that contamination was not uniform and perhaps at low levels. The findings of this investigation played a role in many of the food safety initiatives designed by Congress and the Clinton Administration within the United States (Lindsay, 1997).

NOROVIRUS ASSOCIATED OUTBREAKS WITH RASPBERRIES

Over the past seven years, several outbreaks were associated with the consumption of raspberries contaminated with norovirus. Perhaps as virus detection methods improve, more outbreaks associated with these viruses will be detected. The eight outbreaks discussed here occurred in Europe. In November 2001, an outbreak of norovirus in 30 individuals involved baked raspberry cakes (Le Guyader et al., 2004). At first there was an apparent association with both pear and raspberry cakes, but epidemiological evidence indicated raspberry cakes with a stronger association with illness. The pink cakes were made with a cream topping containing whole frozen raspberries. Multiple norovirus strains were detected in the raspberries after implementing a complex series of extractions coupled with polymerase chain reaction and genetic sequencing methods.

In France in March 2005, 75 students and teachers reported symptoms of nausea, vomiting, and diarrhea lasting for one to two days (Cotterelle et al., 2005). Epidemiology showed that the illness was strongly associated with consumption of raspberries blended with *fromage blanc*, a fresh cheese similar to cottage cheese, that was served with lunch in the school cafeteria. Stool samples were positive for norovirus, Musgrove strain, but the virus was not successfully isolated from raspberry samples. As in the previous cases, the raspberries in this outbreak were deep frozen and were blended with the cheese while frozen. The blended desserts were topped with individual frozen berries placed by hand; however, the workers were not ill prior to or at the time of the outbreak.

In May 2005, nearly 200 patients and employees at two hospitals in Denmark fell ill with symptoms of norovirus (Korsager et al., 2005). Again epidemiology linked illness to consumption of a *fromage blanc* cheese dessert made with raspberries. Again fecal samples were found to be positive for norovirus. When these illnesses occurred, the Regional Food Inspectorate called for withdrawal of the frozen raspberries. Unfortunately the recall did not happen quickly enough, and just shortly after the recall in early June, nearly 300 cases of norovirus were associated with the same dessert served to approximately 1100 people in a "meals on wheels" system. As earlier, many of the fecal samples were positive for norovirus. The three outbreaks described earlier were not believed to be linked to each other since the raspberries came from a different producer in the outbreak in France compared to those in Denmark. The exact cause of the outbreak was not determined, but it is clear that contamination was spotty due to the relative low number of illnesses compared to the numbers that consumed the frozen raspberries.

During the summer months of 2006, 43 individuals became ill with norovirus associated with the consumption of contaminated raspberries in four outbreaks (Hjertqvist et al., 2006). A homemade cake containing raspberries and cream was the cause of one outbreak. Another was associated with cheesecake and raspberries. Norovirus was detected only in the fecal samples of these patients. The raspberries were of the same brand and imported from China. In a third outbreak at a school, drinks made from the same brand of imported raspberries caused 30 illnesses. In the fourth outbreak a homemade raspberry parfait, made from the same brand of imported raspberries, was served to nine participants of a meeting who all became ill with norovirus.

It is not clear at this time why there appears to have been a sudden increase in outbreaks in Europe associated with fresh or frozen raspberries. This may be a real increase due to contaminated irrigation water, farm workers, or food handlers. Alternatively, this may be an artificial increase due to an increase in reporting or detection in connection with the Foodborne Viruses in Europe network (FBVE).

Viruses are able to survive a variety of environmental pressures (Pirtle and Beran, 1991; Koopmans and Duizer, 2004), including those related to the processing of foods as evident from the outbreaks of hepatitis A and norovirus just described. As discussed previously, several berry-associated outbreaks linked to viruses involved frozen berries. Simple removal or inactivation by washing and freezing varies depending on berry and virus type. Rinsing berries or other soft fruits has been shown to remove bacteria; however, poliovirus was not removed from strawberries after rinsing with warm water (Lukasik et al., 2003), whereas a cool water rinse removed about 2 logs of the norovirus surrogate, feline calicivirus (Gulati et al., 2001). Butot et al. (2008) showed limited effectiveness of washing in removing enteric viruses altogether from berries with either cold or warm water. It is important to note that berries are not washed in the field nor in the packaging plant as this could induce tissue decay, making this an important step for the consumer. As stated earlier, frozen berries have been responsible for numerous cases of illness. Enteric viruses, with the exception of feline calicivirus, were reduced by less than 1 log during freezing on strawberries and raspberries (Butot et al., 2008). Feline calicivirus was reduced by 2 logs in this study; however, it has been noted previously that as a respiratory virus, feline calicivirus is not likely to be the ideal surrogate for norovirus (Cannon et al., 2006).

THE ROLE OF *CYCLOSPORA CAYETANENSIS* IN BERRY-ASSOCIATED OUTBREAKS

Raw raspberries and blackberries imported from Guatemala have been associated with several large outbreaks of gastrointestinal illness attributed to *Cyclospora cayetanensis*; a food and waterborne protozoan parasite that infects the upper small intestine of humans and can cause severe diarrhea, stomach cramps, and nausea, which may be accompanied by fever (Dawson, 2005; Shields et al., 2003). Cyclosporiasis is treatable with trimethoprim-sulamethoxazole (Hoge et al., 1995). *Cyclospora* oocysts first were observed in stool samples in Papua, New Guinea 30 years ago (Ashford, 1979), but interestingly, it is still referred to as an emerging pathogen due to the many unknowns regarding its transmission (Chacin-Bonilla, 2008). Cyclosporiasis is not thought to be associated primarily with immunocompromised individuals like other human protozoan pathogens. It was identified as a new coccidian species in 1993 by Ortega et al., when they successfully induced oocyst sporulation and excystation of the sporozoites *in vitro* (Ortega et al., 1993).

Cyclospora cayetanensis oocysts are quite large at 7.5 to 10 µm in diameter. These oocysts have a strong outer membrane composed of complex carbohydrates and lipids that make the oocysts acid fast. The oocyst membrane protects two oblong sporocysts that surround the infective life stages, with four sporozoites in each sporocyst. The oocyst and sporocyst membranes are strong structures that provide great stability to environmental pressures and ensure that the sporozoites remain viable along their journey to the small intestine. Like many protozoa, *Cyclospora* oocysts are shed unsporulated and sporulate outside the host within 7 to 10 days under favorable environmental conditions (Ortega et al., 1994). In comparison *Cryptosporidium* oocysts are shed already sporulated and infectious, whereas *Toxoplasma gondii* oocysts sporulate within 48 to 72 hours of being in the environment. The infection process begins when the oocysts are ingested by the host. Coccidian oocyst outer membranes respond to the acidic pH of the stomach. When the sporocysts reach the intestinal tract of their hosts, the sporocyst wall breaks down and the sporozoites are released to invade host epithelial cells and undergo multiple cycles of asexual multiplication followed by sexual development for the formation of the unsporulated oocysts that are shed in the host feces.

In total, the 10 events that have involved *C. cayetanensis* and contaminated raspberries accounted for 2864 illnesses. Subsequently, eight traceback investigations were conducted, including five farm investigations, four of these in Guatemala and one in Chile (Timbo et al., 2007). The first reported outbreaks of cyclosporiasis associated with raspberries were in New York and Florida in 1995. These outbreaks did not involve traceback investigations of any kind, and approximately 71 individuals were involved. In New York, drinking water from portable coolers at a country club and raspberries that were served during the outbreak period were both suspected (Carter, 1996). In Florida, raspberries were suspected, but were a component of a fruit cup and desserts served at various social events (Koumans et al., 1998). During 1996, over 1660 individuals in the United States became ill from raspberries contaminated with *Cyclospora*. Together these events involved 20 states and the District of Columbia (Herwaldt et al., 1997, 1999; Careres et al., 1998). There were traceback and farm investigations associated with these three large outbreaks. Raspberries were traced back to Guatemala for many of these events, and berries that were implicated were harvested from between three and 30 farms. In the large multistate outbreak in 1996, a majority of the raspberries were traced to one exporter; however, nothing concrete came of the farm investigations, as exporters included raspberries from different farms in a single shipment (Timbo et al., 2007; Herwaldt et al., 1999). Another large multistate outbreak occurred the following year, again associated with Guatemalan raspberries (Careres et al., 1998). Several farms were identified during the investigation by the FDA (Timbo et al., 2007). In 1998, an outbreak of cyclosporiasis occurred in Massachusetts, but no farm investigation was pursued as it was not possible to determine whether the raspberries originated in Chile or Guatemala (Catherine et al., 1998).

In 2000, raspberries associated with a cake were involved in an outbreak, and the berries were traced to three possible sources (a Guatemalan farm, a Chilean farm, and an unknown US farm) (Ho et al., 2000). The farm in Guatemala was later implicated in an outbreak in Pennsylvania associated with a cake served with cream and raspberries where over 50 people became ill (Ho et al., 2000), and also found to be associated with an outbreak in the state of Georgia the same year when raspberries were served with other fruit over ice cream at a bridesmaids' luncheon (Marrow et al., 2002; Timbo et al., 2007). Raspberries from Chile also were suspected in this latter outbreak in 2000 (Murrow et al., 2002). In 2002, raspberries from Chile again were suspected in an outbreak that involved 22 individuals (CFSAN, 2003).

TRANSMISSION OF *CYCLOSPORA* OOCYSTS AND THE ROLE OF FOODS

The first reported outbreak in the United States of cyclosporiasis involved contaminated water in Chicago, Ilinois, in 1990 (Timbo et al., 2007). Cyclosporiasis has been associated with fresh fruits, vegetables, and herbs, likely contaminated by water, soil, or handlers. In particular, raspberries, basil, parsley, snow peas, and leafy greens have been implicated as probable transmission vehicles in 19 outbreaks of cyclosporiasis in the United States (Timbo et al., 2007; Dawson, 2005; CDC 2004; Shields et al., 2003; Lopez et al., 2001). As with hepatitis A virus, *C. cayetanensis* oocysts are shed by humans, and contamination can occur at both preharvest points (soil, feces, irrigation water, dust, insects, or animals) and postharvest points (human handling, equipment, or transport containers) (Beuchat, 2002).

The role of water has been questioned in the transmission of oocysts to berries and other foods, including basil. The water used to mix pesticides was previously identified as a possible source of contamination in the outbreaks of cyclosporiasis associated with contaminated raspberries (Herwaldt, 2000; Herwaldt and Ackers, 1997). Water was found to be a main vehicle of transmission in a study in Egypt assessing irrigation canals, groundwater, and finished water (El-Karamany et al., 2005). This study also named contact with soil and poultry litter as potential risks for transmission. Water and soil are of concern in many parts of the world where cyclosporiasis is endemic, and individuals' shedding of oocysts may be asymptomatic (Chacin-Bonilla, 2008; El-Karamany et al., 2005; Katz and Taylor, 2001; Sturbaum et al., 1998; Hoge et al., 1993, 1995; Eberhard et al., 1999; Bern et al., 1999; Chacin-Bonilla et al., 2003). Interestingly, several recent studies focused on risks associated with soil. In Peru, contact with the soil was observed to be an important risk factor among children (Mansfield and Gajadhar, 2004). Contact with soil among healthcare and farm workers in Guatemala was a risk factor for cyclosporiasis infection (Bern et al., 1999). In similar studies in communities in Haiti and Venezuela, statistical models showed that contact with soil was an important mode of transmission for *Cyclospora* oocysts (Chacin-Bonilla, 2008; Lopez et al., 2001). In these same studies, poverty was identified as a risk factor for the prevalence of infection in endemic areas (Chacin-Bonilla, 2008).

In several documented outbreaks, raspberries were the likely vehicle of contamination, but they were associated with other foods, including wedding cake (Herwaldt and Ackers, 1997) and lemon tart (Herwaldt, 2000). Although studies attempting to determine the infectious dose have not been successful (Alfano-Sobsey et al., 2004), epidemiological investigations have suggested that the infective dose is low or that in the berry-associated outbreaks, the number of oocysts per berry was high (Herwaldt, 2000). Compared to other coccidians, *Cyclospora* oocysts require a large amount of time to sporulate in the environment. However, the rate of sporulation does not appear to be fixed, as sporulation increased with increasing temperatures from 4 to 22 °C (Smith et al., 1997). *Cyclospora cayetanensis* is difficult to study in the laboratory, as humans are its only known host, making access to oocysts and methods to evaluate viability difficult and limited. Experimental work using the related poultry coccidian parasite *Eimeria acervulina* as a surrogate for *Cyclospora* has shown possible interactions between the oocyst and the raspberry (Kniel et al., 2007). Phylogenetic analysis supports the conclusion that *Eimeria* and *Cyclospora* could belong to the same family (Relman et al., 1996), and one author even suggested that *Cyclospora* should be considered a member of the genus *Eimeria*, based on small subunit ribosomal RNA gene alignment (Pieniazek and Herwaldt, 1997).

Regarding oocyst breakdown, slight differences were observed in oocysts recovered from raspberries; however, this difference was not observed during excystation, but rather in the number of sporocysts present before bead-beating, which is used with *Eimeria* in the laboratory to mechanically disrupt the oocyst membrane. This phenomenon was obvious by microscopy as there was no clumping of sporocysts. There was a visual increase in the number of sporocysts recovered from raspberries, compared to oocysts in suspension or those recovered from basil; where more than 2.2 times the sporocysts were observed from oocysts that had contact with intact raspberries compared to those that did not. The increased release of sporocysts observed after interactions with raspberries may be a combination of factors including pH (pH 3.4 ± 0.2), flavonoids, and other plant phenols. The greater release of sporocysts in the presence of raspberries as compared to other matrices may influence the apparent infectivity of oocysts. An increase in sporulation coupled with an increase in the breakdown of the oocyst membranes on acidic berries like raspberries (Kniel et al., 2007) could lead to a higher infection rate. It is important to note that sporulation can be inactivated by exposing oocysts to extreme temperatures that would be used at home in food preparation or by the food industry (Sathyanarayanan and Ortega, 2006).

The berry surface topography certainly plays a large role, as seen in the interaction of *Toxoplasma gondii* oocysts with the hair-like projections on the raspberry surface, as compared to the relatively smooth surface of a blueberry (Figure 12.1) (Kniel et al., 2002). *Toxoplasma* is a protozoan parasite related to both *Eimeria* and *Cyclospora* and similar in size and shape to both. Due to the facts that cyclosporiasis is often associated with imported produce, and unknown reservoirs or routes of contamination exist, alternative treatment methods for fresh berries should be examined. These are discussed further in this chapter. There are many questions that still need to be addressed, including the potential seasonality of prevalence of infections and outbreaks (Herwaldt, 2000).

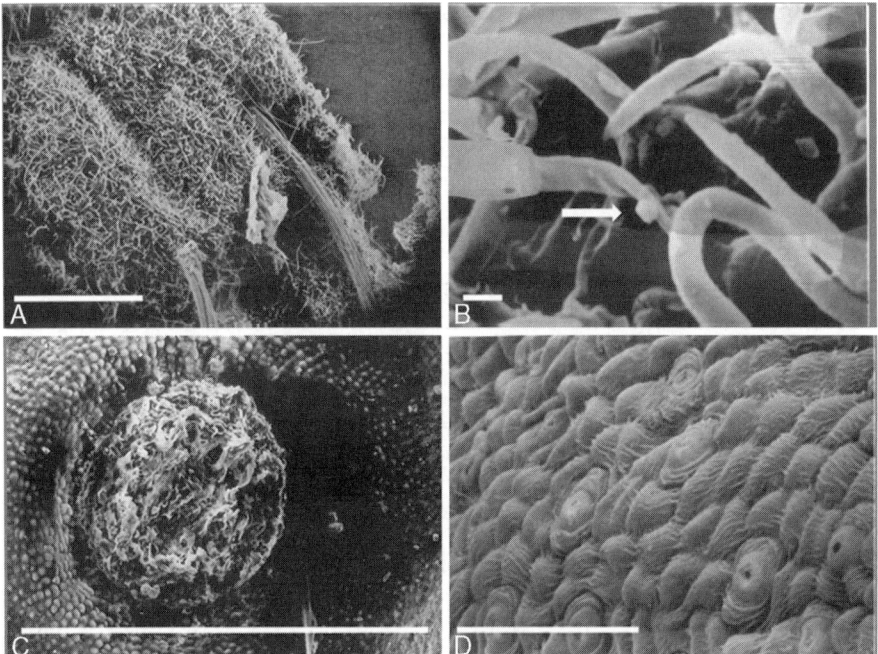

FIGURE 12.1 *Scanning electron micrographs of raspberry and blueberry surfaces inoculated with 2.0×10^4 Toxoplasma oocysts. Hair-like structures on the raspberry likely aid in retention of oocyst contamination (A, bar = 1 mm). One oocyst is visibly attached to one of these structures in B (bar = 10 μm). The blueberry surface lacks these structures (C, bar = 1 mm and D, bar = 0.1 mm). (Adapted from Kniel et al., 2002. Copyright permission from Allen Press and Copyright Clearance Services.)*

BACTERIAL CONTAMINATION OF BERRIES

Whereas other types of produce have been more frequently implicated as vehicles for foodborne outbreaks of bacterial origin, five outbreaks of bacterial etiology have been attributed to berries in the United States since 1973 (Sivapalasingam et al., 2004; CDC, 2008). These include strawberries contaminated with *Staphylococcus aureus* (1985), *Salmonella* Group B (2003), and strawberries/blueberries contaminated with enterohemorragic *E. coli* O26 (2006). Red grapes (2000) and green grapes (2001) were identified as the vehicles for outbreaks of enterohemorrhagic illnesses and *Salmonella* Senftenberg, respectively. Collectively, these outbreaks involved 86 known cases of illness. The outbreaks were sourced to berries from a variety of locations including a grocery store, home, and restaurant/daycare/school/delicatessen. A number of other outbreaks have been attributed to fruit salad, but did not single out berries as the original source (CDC, 2008). Two additional outbreaks were attributed to berries, but the etiology was not determined (Sivapalasingam, 2004). The limited number of traced outbreaks does not mean that additional bacterial illnesses did not result from consumption of contaminated berry products. The food vehicle in outbreaks cannot always be determined, and it is assumed that foodborne illnesses are generally underreported.

This outbreak history does not reveal a pattern in type or source of bacterial contamination for berry products. The sporadic outbreaks of bacterial illness that were not necessarily traced to farm-level contamination raises the possibility that the berries were cross-contaminated postharvest by other foods or surfaces. This also highlights challenges for outbreak investigations with perishable products that are not easily traced to source and receive substantial handling prior to receipt by the end customer.

Surveys of produce have provided some information on the incidence of bacterial contamination of berries. In 1999 and 2000, the FDA initiated surveys of imported and domestic produce, respectively, to determine the incidence of contamination and the research and education needs to reduce foodborne illnesses resulting from consumption of contaminated produce (FDA, 2001a, 2003). Strawberries were among the produce samples analyzed for *Salmonella* and *E. coli* O157:H7. For the survey of imported produce, 143 strawberry samples from five countries (Argentina, Belgium, Canada, Mexico, and New Zealand) were analyzed, and only one sample (0.7%) was found positive for *Salmonella*. None tested positive for *E. coli* O157:H7 (FDA, 2001a). For domestic strawberries, 136 samples from two states (California and Florida) were tested, and all were negative for *Salmonella* and *E. coli* O157:H7 (FDA, 2003). Another published survey (Mukherjee et al., 2006) of bacterial contamination of berries involved sampling of strawberries, blueberries, and raspberries from farms in the Upper Midwest region of the United States. Samples were collected preharvest over a two-year period (2003 to 2004) from farms that used conventional, organic, and semiorganic (organic practices used, but not certified organic) growing practices. Berries were tested for coliforms, *E. coli*, *Salmonella*, and *E. coli* O157:H7. Coliform counts on berries were approximately 1 to 2 $\log_{10}$ MPN/g, and 2 of 194 (1%) berries tested positive for *E. coli*. None of the berries tested positive for *Salmonella* or *E. coli* O157:H7 (Mukherjee et al., 2006). Results from this very limited number of publicly available studies would suggest that the incidence of contamination of berries with pathogenic bacteria is low.

Berry extracts are reported to have antibacterial properties (Ryan et al., 2001; Puupponen-Pimia et al., 2005) owing to their acidity and phenolic compounds (Puupponen-Pimia et al., 2005). However, several studies with artificially contaminated berries suggest these antibacterial properties should not be relied upon for safety. Research studies have been conducted to determine, in the event of contamination, the fate of microbes after exposure to various intrinsic stresses as well as intervention strategies to remove or inactivate pathogens.

CONTAMINATION REDUCTION STRATEGIES

The common processing strategy of thermal pasteurization as a means to inactivate viruses, parasites, and bacteria on berry products has received limited attention. Typical conditions for hot-filled, shelf-stable, single-strength white grape juice (pH 3.9) were calculated to achieve at least a 5-log reduction of *E. coli* O157:H7, *S. enterica*, and *L. monocytogenes* (Mazzotta, 2001). However, mild heat treatment (75°C for 15s) of raspberry puree yielded less than a 3-log reduction in murine norovirus 1 and less than a 4-log reduction of *E. coli* and *B. fragilis* HSP40 infecting phage B40-8 (Baert et al., 2008). Additionally, heat processing renders berries as different products from their fresh counterparts. This may be acceptable for berries used for juices, cereals, and purees; however, alternative processes have been explored as potential means of rendering berries microbiologically safe while maintaining their fresh-like characteristics. Technologies evaluated for application to berries include washing with and without disinfectants, frozen storage, high hydrostatic pressure, exposure to ultraviolet (UV) light, and irradiation. Details of these studies are summarized in Table 12.1.

Washing produce in plain water is generally recommended to consumers as a means of removing visible soil from produce and reducing microbial populations. However, washing is not a means of rendering fresh produce completely free of risk from potential pathogen contamination. A study with strawberries, raspberries, and blueberries, artificially contaminated with various viruses, demonstrated that washing these berries with either cold or warm water yielded viral population reductions of less than 1.5 $\log_{10}$ units (Butot et al., 2008). Likewise, washing inoculated strawberries with water at 22 or 43°C, either with or without scrubbing, yielded less than 1-log reductions in populations of *E. coli* O157:H7, *S.* Montevideo, poliovirus 1, and three different bacteriophages (Lukasik et al., 2003). In another study with protozoa, raspberries contaminated with *Eimeria acervulina* as a surrogate for *C. cayetanensis* were washed with cold tap water and fed to chickens. Washing was not a fail-safe method for removing protozoa as some chickens were symptomatic of infection (Lee and Lee, 2001).

Table 12.1 Efficacy of Intervention Strategies to Reduce or Eliminate Microbial Contamination of Berries

	Target Microbe	Treatment	Conditions	Effectiveness	Reference
Raspberry	*Eimeria acervulina* (*Cyclospora cayetanensis* surrogate)	Wash	Flowing, cold tap water, 5 min	Incomplete removal, duodenal lesions detected in natural host	Lee and Lee, 2001
Raspberry	*Eimeria acervulina* (*Cyclospora cayetanensis* surrogate)	Freezing	$-18\,°C$	No duodenal lesions detected in natural host	Lee and Lee, 2001
Raspberry	*Eimeria acervulina* (*Cyclospora cayetanensis* surrogate)	Heat	Water bath, minimum internal temperature of berry of 80 °C maintained for 1 h	No duodenal lesions detected in natural host	Lee and Lee, 2001
Raspberry	*Eimeria acervulina* (*Cyclospora cayetanensis* surrogate), 10^4 and 10^6 initial inocula levels	HP	550 MPa, 2 min, 40 °C	No symptoms of infectivity in natural host	Kniel et al., 2007
Raspberry	*Eimeria acervulina* (*Cyclospora cayetanensis* surrogate), 10^4 and 10^6 initial inocula levels	UV light	80, 160, or 261 mW/cm^2	80 mW/cm^2 for 10^6 inoculum: reduced severity of intestinal lesions in natural host (chicken); 160 mW/cm^2 for 10^4 inoculum: asymptomatic but shed oocysts	Kniel et al., 2007
Raspberry	*E. coli* O157:H7 (5 strains), *Salmonella enterica* (5 serotypes), 10^5 CFU/g initial inoculum	Pulsed UV light	72 J/cm^2	Approximately 3.5 and 4.5 log reductions in *Salmonella* and *E. coli* O157:H7, respectively	Bialka et al., 2008
Raspberry	Fungi	Pulsed UV light			Lagunas-Solar et al., 2006
Raspberry	*E. coli* O157:H7, *Salmonella*	Ozone – aqueous	21 mg/L, 64 min, 4 °C	Log reductions (CFU/g): 5.6 for *E. coli* O157:H7; 4.5 for *Salmonella*	Bialka and Demirci, 2007a

Raspberry	E. coli O157:H7, Salmonella enterica, (initial inoculum 10^5 CFU/g)	Ozone–gaseous	5% wt/wt, 64 min continuous treatment followed by 64 min pressurized treatment (83 kPa)	Log reductions (CFU/g): 3.55 for Salmonella; 3.75 for E. coli O157:H7	Bialka and Demirci, 2007b
Raspberry	Eimeria acervulina (Cyclospora cayetanensis surrogate)	Gamma irradiation	0.5 to 4.0 kGy dosages	Partial inactivation achieved with 0.5 kGy; no duodenal lesions detected in natural host-fed oocysts treated with 1.0 kGy	Lee and Lee, 2001
Raspberry puree	Murine norovirus 1, E. coli, B. fragilis HSP40 infecting phage B40-8	Mild heat	75 °C for 15 s		Baert et al., 2008
Strawberry	Salmonella, E. coli O157:H7 (5- or 6-strain cocktails of each)	Refrigerated, frozen storage	−20, 5, 24 °C storage, 20% sucrose, cut or intact strawberry surfaces	At 5 °C for 7 d, no reduction of pathogens on cut strawberries, declined < 2 logs on intact surfaces; −20 °C for 30d E. coli reduction of 1- to 2 logs; survival but not growth at 24 °C, sucrose protective to bacteria at −20 °C	Knudsen et al., 2001
Strawberry	L. monocytogenes (5-strain cocktail, initial inocula 10^6 or 10^8 CFU/3 berries)	Refrigerated, frozen storage	−20, 4, 24 °C storage, 20% sucrose	3-log reduction over 7 d at 4 °C on intact surfaces, no reduction on cut surfaces; 1.2 log or less reduction after 28 d at −20 °C, sucrose protective	Flessa et al., 2005

Continued

Table 12.1 Efficacy of Intervention Strategies to Reduce or Eliminate Microbial Contamination of Berries *(continued)*

	Target Microbe	Treatment	Conditions	Effectiveness	Reference
Strawberry	*E. coli* O157:H7, *Salmonella enterica*, (initial inoculum 10^5 CFU/g)	Ozone–gaseous	5% wt/wt, 64 min continuous treatment followed by 64 min pressurized treatment (83 kPa)	Log reductions (CFU/g): 2.60 for *Salmonella*; 2.96 for *E. coli* O157:H7	Bialka and Demirci, 2007b
Strawberry	*E. coli* O157:H7 (5 strains), *Salmonella* (5 serotypes), initial inoculum ca. 10^5 CFU/ml	Ozone–aqueous	8.9 mg/L, −20 °C, 64 min	Log reductions (CFU/g): 2.9 for *E. coli* O157:H7; 3.3 for *Salmonella*	Bialka and Demirci, 2007a
Strawberry	Natural microflora (mesophiles, psychrotrophs, yeasts, and molds)	Ozone–aqueous	1 to 10 mg/L, 5 or 10 min, 4 or 23 °C	Ca. 1 log unit reductions at 10 mg/L	Wei et al., 2007
Strawberry	*E. coli* O157:H7, *L. monocytogenes* (3 strain cocktails), 10^5 to 10^6 initial population	Electrolyzed oxidizing water using NaCl or NaOCl	1, 5, 10 min ambient temp, storage 5, 10, 15 d	2 to 3 log reductions, greater with increased storage time, incomplete inactivation	Udompijitkul et al., 2007
Strawberry	Poliovirus 1, bacteriophages, *Salmonella* Montevideo, *E. coli* O157:H7	Physical and disinfectant washes	NaOCl (50 to 300 ppm), stabilized ClO_2 (200 ppm), peroxyacetic acid (100 or 200 ppm), detylpyridiniumchloride (0.1%), trisodium phosphate, acidified sodium chlorite, Fit, Healthy Harvest, NaCl (2%), vinegar (10%), detergent	Population reductions of less than 1 log to ca. 2 logs	Lukasik et al., 2003
Strawberry	*E. coli* O157:H7, *L. monocytogenes* (10^6 CFU/g)	Various sanitizers	Submerged 5 min in ozone (3 ppm), ClO_2 (3 or 5 ppm), chlorinated trisodium phosphate (CTP, 100 or 200 ppm chlorine), or peroxyacetic acid (80 ppm)	Log reductions: ~5.6 with ozone and ClO_2 (5 ppm); ~4.9 with ClO_2 or CTP; ~4.4 with peroxyacetic acid	Rodgers et al., 2004

Commodity	Microorganism	Disinfectant	Treatment	Observations	Reference
Strawberry	Feline calicivirus	Various disinfectants	NaOCl (5.25%) Peroxyacetic acid (15%) H_2O_2 (11%) QAC 10 min contact various concentrations	None effective at manufacturer recommended concentration. At 4x recommended concentration: 1 log reduction with NaOCl and QAC, 3 log reduction with H_2O_2 and peroxyacetic acid	Gulati et al., 2001
Strawberry	Hepatitis A virus, coliphage MS2 (4 log_{10} recovered at initial time)	Chlorine	20 ppm free chlorine initial, 5 to 10 min	Recovered ca. 2 log from strawberry after 10 min (ca. 2 log reduction from initial recovery)	Casteel et al., 2008
Strawberry	Natural microflora (mesophiles, psychrotrophs, yeasts, molds)	Chlorine	200 mg/L, 5 min	Ca. 2 log unit reductions	Wei et al., 2007
Strawberry	*Salmonella* spp. (7 log_{10} CFU/produce inoculum)	Chlorine dioxide	1 h gas treatment	Log reductions: > 4.7 on smooth surface, ~ 4 log on stem scar, ~ 1 log in puncture wound	Yuk et al., 2006
Strawberry, blueberry, red raspberry	*Salmonella enterica*	Chlorine dioxide	Gaseous ClO_2 at 4.1, 6.2, 8.0 mg/L air, treatment times of 30, 60, 120 min, 23 °C	Reductions on: strawberry (~ 4 log), blueberry (~ 3 log), raspberry (~ 1.5 log) depending on location and treatment time/concentration Greater lethality generally on smooth surface	Sy et al., 2005

Continued

Table 12.1 Efficacy of Intervention Strategies to Reduce or Eliminate Microbial Contamination of Berries (continued)

	Target Microbe	Treatment	Conditions	Effectiveness	Reference
Strawberry puree	Hepatitis A virus initial inoculum >10^6 PFU	HPP	375 MPa, 5 min, 21 °C	4.32 log PFU reduction	Kingsley et al., 2005
Strawberry	Hepatitis A virus, aichi virus, feline calicivirus	UV light	240 mW s/cm^2	1.9 to 2.6 log TCID$_{50}$/ml reductions.	Fino and Kniel, 2008
Strawberry	E. coli O157:H7 (5 strains), Salmonella enterica (5 serotypes), 10^5 CFU/g initial inoculum	Pulsed UV light	64.4 J/cm^2	< 2.5 log reductions	Bialka et al., 2008
Strawberry	Hepatitis A virus, 6.68 log$_{10}$ HAV titre (PFU/ml) initial inoculum	Gamma irradiation	^{60}Co source, total doses of 1 to 10 kGy ambient temperature	3.83 log$_{10}$ HAV titre (PFU/ml) recovered after 10 kGy treatment	Bidawid et al., 2000
Blueberry	Listeria monocytogenes, Salmonella spp., Escherichia coli O157:H7 (3 strains each at ~ 10^6 CFU/g inoculum) and spoilage fungi	Chlorine dioxide gas	4 mg/liter, 0.16 mg/g for 12 h in sealed 20-l container at 22C	Reductions (log CFU/g): 3.94, 3.62, and 4.25 for L. monocytogenes, Salmonella, and E. coli O157:H7, respectively	Popa et al., 2007
Blueberry	L. monocytogenes, P. aeruginosa, S. Typhimurium, S. aureus, Y. enterocolitica (2 strains each, initial inocula ca. 5 log CFU/g)	Chlorine dioxide (aqueous)	1, 3, 5, 10, and 15 ppm for 0.17, 1, 5, 10, 20, 30, 60, and 120 min	Reductions (log CFU/g) at 15 ppm for 2 h: 4.88, 4, 48, 2.28, 4.33, 3.54, and 2.82 for L. monocytogenes, P. aeruginosa, S. Typhimurium, S. aureus, Y. enterocolitica, and natural fungi, respectively	Wu and Kim, 2007
Blueberry	Pseudomonas fluorescens, Enterobacter agglomerans	Chemical oxidants			Crowe et al., 2007

Food	Microorganism	Technology	Treatment Conditions	Results	References
Strawberries, Blueberries, Grapes	*Saccharomyces cerevisiae, Listeria innocua* (4-strain cocktail) as pathogen surrogate	High pressure	300 MPa, 21 °C, 1.5 min for *S. cerevisiae* 450 MPa, 21 °C, 1.5 or 3 min hold time for *L. innocua*	*S. cerevisiae* reduced to 6 CFU/ml on grapes and strawberries; resistance higher in blueberries. *L. innocua* reduced to 7 CFU/ml on strawberries and blueberries with 1.5 min treatment; grapes required 3 min for reduction to < 1 CFU/ml	Chauvin et al., 2005
Strawberries, Raspberries, Blueberries	Norovirus, hepatitis A virus, rotavirus, feline calicivirus	Chlorine, chlorine dioxide, cold or warm water	Chlorine (200 ppm), ambient temperature, 1 or 10 min ClO$_2$ (5, 10, 25 ppm) Cold water wash, warm water wash	HAV most resistant to chlorine, least reduction on raspberries < 1 log reduction NV and HAV < 1.5 log10 unit reduction with just water	Butot et al., 2008
Cranberry Juice Concentrate	*E. coli* O157:H7, *Salmonella*, *L. monocytogenes* (5 strain cocktails of each, 6 to 7-log initial inoculum level)	Frozen storage of inoculated concentrates	14 to 46 °Brix (pH 2.2 to 2.5), 0 °C storage for 6 to 96 h	5-log reduction within 24 h at 0 °C for 18 to 46 °Brix; 14 °Brix required 96 h at 0 °C for 5-log reduction of *E. coli* O157:H7	Enache and Chen, 2007

The addition of sanitizers or disinfectants to water washes is one of the most commonly studied strategies to remove or inactivate pathogens on berries. Those agents evaluated have included chlorine, chlorine dioxide, ozone, peroxyacetic acid, quaternary ammonium compounds, trisodium phosphate, and hydrogen peroxide, among others. Chlorine is a common sanitizer used in food-processing facilities with broad efficacy against many foodborne pathogens and spoilage organisms. It is approved for use in the United States as a wash water additive for the produce industry (Seymour and Appleton, 2001). Chlorine washes have been applied at various concentrations (20 to 300 ppm) to strawberries, blueberries, and raspberries inoculated with several different types of viruses and bacteria (Butot et al., 2008; Casteel et al., 2008; Gulati et al., 2001; Lukasik et al., 2003; Rodgers et al., 2004; Udompijitkul et al., 2007; and Wei et al., 2007). Chlorine washes of berries have generally yielded 1- to 2-log unit reductions in bacteria and viruses. The degree of inactivation observed was incomplete for the inoculation levels studied and would not provide a 5-log reduction that serves as the standard for processing technologies utilized to render juices made from fruits safe from pertinent pathogens (typically *Salmonella* or *E. coli* O157:H7) (FDA, 2001).

It has been reported that upper levels of free available chlorine that are at least partially effective for disinfection (200 ppm) can also cause bleaching and off-flavors in produce (Hurst and Schuler 1992); however, these effects may be product specific and need to be evaluated for each type of produce (Wei et al., 2007). At levels normally used in food processing, chlorine is typically not as effective on protozoan parasites (King and Monis, 2007), which have been implicated in outbreaks associated with consumption of contaminated berries. Chlorine efficacy is affected by the presence of organic material and pH. Additionally, some by-products resulting from chlorine reactions with organic material are considered potentially mutagenic or carcinogenic (Sapers, 2001; Wu and Kim, 2007), and alternatives have been sought.

Chlorine dioxide has been evaluated as an alternative to chlorine as it appears to be less affected by pH changes and the presence of organic material (Seymour and Appleton, 2001). Chlorine dioxide can be applied in both aqueous and gaseous form. Aqueous systems are easier to administer whereas gaseous systems offer good penetration without residual surface moisture to support subsequent growth of spoilage organisms (Wu and Kim, 2007). The reported efficacy of chlorine dioxide for berry decontamination varies widely from approximately 1-log unit reduction (Butot et al., 2008) to 5-log unit reductions (Yuk et al., 2006) depending on the method of delivery and produce type. Generally, delivery in gaseous systems has been reported to have greater efficacy.

Ozone can also be delivered in gaseous or aqueous phase. Ozone has had GRAS status since 2001 and is approved as an antimicrobial treatment (Wei et al., 2007). Ozone safely decomposes to oxygen and water (Wei et al., 2007), although safety precautions during use are needed. Ozone applied to raspberries inoculated with *S. enterica* or *E. coli* O157:H7 reduced the pathogen populations by approximately 3.5 to 5.5 $\log_{10}$ CFU/g; less dramatic reductions were obtained on strawberries (Bialka and Demirci, 2007a, 2007b). Shorter treatment times with ozone on strawberries only yielded approximately 1-log reductions in the natural microbial flora (Wei et al., 2007). Further analysis of the inactivation kinetics of *E. coli* O157:H7 and *S. enterica* on raspberries and strawberries by ozone indicated that the Weibull model was more suitable for estimation of microbial inactivation than first-order kinetics (Bialka et al., 2008). Ozone applied to grapes has also been shown to increase shelf-life by reducing fungal decay (Sarig et al., 1996).

Several other disinfectants have been tested against pathogens on berries (Table 12.1), but with fewer studies than have been conducted for chlorine, chlorine dioxide, or ozone. Results are variable, with 1- to 4.4-$\log_{10}$ CFU/g reductions reported (Gulati et al., 2001; Lukasik et al., 2003; Rodgers et al., 2004; Udompijitkul et al., 2007; Table 12.1). Quality attributes were not the focus of these studies, but it was noted that hydrogen peroxide caused slight discoloration of strawberries (Lukasik et al., 2003).

The efficacy of any disinfectant wash or external treatment would be impacted by microbial inaccessibility. Pathogens that are irreversibly attached to surfaces or protected in aggregations or biofilms may be more resistant to inactivation. Additionally, pathogens may be less accessible if they reside in surface crevices, which are abundant on raspberries and strawberries, or are internalized through surface damage (Bassett and McClure, 2008; Sapers, 2001). A microbial risk assessment of various produce types suggested that the irregular topography of certain berries, and consequent difficulty in washing, makes them high-risk products despite their low pH values, which are believed to be below the growth boundaries of pathogenic bacteria (Bassett and McClure, 2008). The enhanced survival of microbes to disinfection as well as freezing, when located in scars or puncture wounds, has been demonstrated (Flessa et al., 2005; Knudsen et al., 2001; Yuk et al., 2006). Internalization of pathogens by other means has been reported for other types of produce in some laboratory-simulated growth conditions (Guo et al., 2002; Solomon et al., 2002) or processing conditions, such as due to temperature differentials between the fruits and wash water (Bassett and McClure, 2008; Sapers, 2001). It remains to be determined whether or not these latter causes of internalization occur with fresh berries. Despite the known and theoretical limitations of washes for inactivation of pathogens on berry surfaces, disinfectant washes can be effective in the processing environment and thereby serve an important role in minimizing cross-contamination (Sapers, 2001).

The effects of refrigeration, freezing, and frozen storage on the survival of pathogenic microbes on berries or in berry juice have been studied. The survival of *Salmonella* spp., *E. coli* O157:H7, and *L. monocytogenes* on surfaces of unwashed and intact or cut strawberries was evaluated at ambient, refrigerated, and frozen storage. Bacteria were allowed to dry on the strawberry surfaces, and this drying step resulted in a slight population reduction. Populations of *L. monocytogenes* declined by 1 to 3 logs during storage at 24 °C for 48 h (dependent on initial inocula of either 10^6 to 10^8 per sample; Flessa et al., 2005), but no changes were observed in *E. coli* or *Salmonella* (Knudsen et al., 2001). During refrigerated storage for seven days, all pathogen cocktail populations declined (1 to 3 logs) on intact strawberries, but no population reduction was observed on cut surfaces. Frozen storage for one month resulted in population reductions of approximately 2 logs or less, and this reduction was incomplete, even including the reductions observed during drying. These studies indicate that these bacterial pathogens are capable of survival through the normal shelf-life of strawberries at 24 and 4 °C and at least one month in frozen storage. Survival on cut surfaces was greater than on intact strawberry surfaces. In another study, survival of stationary phase and acid-adapted *Salmonella*, *E. coli*, and *L. monocytogenes* was evaluated in concentrated cranberry juice at 0 °C (Enache and Chen, 2007). The population reduction of the three bacterial pathogen cocktails was dependent on growth phase, °Brix, and pathogen type. Stationary-phase cells were more resistant than acid-adapted cells for all three pathogens. A °Brix of 18 to 46 (pH 2.2 to 2.5) resulted in population reductions of 5 logs within 24 hours, but at 14 °Brix (pH 2.5), a 5-log reduction took longer (96 hours) for *E. coli* O157:H7 (Enache and Chen, 2007).

As mentioned previously, the effects of freezing and frozen storage on persistence of viruses are similarly troublesome. Unwashed strawberries, raspberries, and blueberries inoculated with 10^4 to 10^6 TCID$_{50}$ or PCRU (RT-PCR units) with norovirus (NV), hepatitis A virus (HAV), rotavirus (RV), or feline calicivirus (FCV) generally showed less than 1 log unit reduction after frozen storage at −20 °C for two days (Butot et al., 2008). Somewhat greater reductions were observed with FCV on strawberries and raspberries and with RV on blueberries. Even prolonged frozen storage of three months' duration had minimal effect on HAV and RV survival in all berries types.

The survival of a protozoan parasite on frozen raspberries was evaluated in one study (Lee and Lee, 2001). Raspberries were inoculated with the poultry parasite *Eimeria acervulina* as a surrogate for the human parasite *C. cayetenensis*. Berries were frozen at −18 °C and then fed to chickens to determine infectivity. Lesions were not found in the duodenal loop of the intestinal tract of chickens fed previously frozen berries containing *E. acervulina*, indicating that freezing may damage oocysts to some degree.

High hydrostatic pressure processing (HPP) has been studied at length for the inactivation of microbial pathogens on various food products, and often, with minimal to no detriment to fresh characteristics. HP has also been applied to various berry products, and the conditions required for inactivation of microbes depend on berry type and microbial target, as described later. Strawberry puree inoculated with hepatitis A virus in excess of 10^6 PFU was pressure-treated at 375 MPa for 5 minutes, and a 4.3 log PFU reduction was obtained (Kingsley et al., 2005). *L. innocua* (approximately 10^6 CFU/sample inoculum), as a surrogate for *L. monocytogenes*, could be inactivated with 450 MPa on strawberries and blueberries for 1.5 minutes and on grapes with 3 minutes treatment (Chauvin et al., 2005). *Eimeria acervulina* (10^6 oocysts/sample inoculum) as a surrogate for *C. cayetanensis* was inactivated within 2 minutes of treatment at 550 MPa and 40 °C (Kniel et al., 2007). The results of these studies suggest that pressure processing is promising for enhancing the microbial safety of berries. Although of greater significance for quality than safety, HP is also effective against *Saccharomyces cerevisiae* on grapes, strawberries, and blueberries (Chauvin et al., 2005) at pressures less than those required for the previously mentioned human pathogens and their surrogates. Organoleptic properties of pressure-treated berries were not reported in the aforementioned studies, although other studies have demonstrated that pressure levels affect the pigment stability of red raspberries with greatest color retention at low (200 MPa) and high (800 MPa) pressures (Suthanthangjai et al., 2005), and pressure treatment of berry purees better preserves volatiles as compared to heat processing (Dalmadi et al., 2007).

Because ultraviolet light (UV-C) in sufficient doses can cause irreparable damage to genetic material, it has broad efficacy against microorganisms. UV light also offers advantages of being relatively low in cost and without irritating or toxic by-products (Fino and Kniel, 2008). Its potential application for food safety, however, depends in part on whether the UV light adequately reaches microorganisms that could be protected by shadows or turbidity. Thus, for whole fruits, UV treatment bears similarity with external washes in that its application would be limited to accessible surface microorganisms. UV light treatment has been studied for the inactivation of bacteria, viruses, a protozoan, and fungi on berries. The degree of inactivation of *S. enterica* and *E. coli* O157:H7 populations depends on berry type as greater reductions were observed on raspberries (3.5 to 4.5 logs) than on strawberries (< 2.5 logs) with 72 and 64.4 J/cm^2 dosages, respectively (Bialka et al., 2008). These researchers further noted that population reductions were better characterized by the Weibull model than log-linear estimations (Bialka et al., 2008).

Similar reductions in viruses on strawberries were achieved with UV light (240 mW s/cm^2) with 1.9, 2.3, and 2.6 log TCID$_{50}$/ml reductions in aichi virus, feline calicivirus (as a surrogate for norovirus), and hepatitis A virus, respectively (Fino and Kniel, 2008). Research from the same laboratory on raspberries inoculated with *E. acervulina* (as a surrogate for *C. cayetanensis*) demonstrated that UV treatment (261 mW/cm^2) can inactivate the protozoa, but inoculum level was an important factor, with populations of 10^4 to 10^6 oocysts partially inactivated (Kniel et al., 2007). In each of these studies, the researchers noted that accessibility was a likely factor in efficacy. UV light has been evaluated for inactivation of fungal microorganisms of significance to the quality and shelf-life of berries. UV light has yielded partial inactivation of fungal populations in strawberry nectar (Keyser et al., 2008), and variable results have been reported for UV effects on extension of berry shelf-life (0 to 2 days) (Fino and Kniel, 2008; Marquenie et al., 2003).

Irradiation is approved for use in the United States for several food commodities for the control of insects and microorganisms and to delay maturation of fresh commodities (FDA, 2008). The effects of irradiation on the quality of berry products have been studied; however, fewer studies have been conducted to fully characterize irradiation needs for enhanced microbial safety of berries. Bidawid and colleagues (Bidawid et al., 2000) determined that gamma irradiation at a dose of 10 kilograys (kGy) reduced titers of hepatitis A virus inoculated onto strawberries by approximately 3 log$_{10}$ (PFU/ml). An irradiation dosage of 1 kGy is permitted for fresh foods for inhibition of growth and maturation (FDA, 2008); at this dose, there was virtually no reduction in HAV titers on strawberries (Bidawid et al., 2000). Irradiation has been reported to extend the shelf-life of berries, but effects on quality depend on the berry type and dosage. Electron-beam irradiation of strawberries at doses of 1 and 2 kGy provided shelf-life extension of two and four days, respectively, although a decrease in firmness and increase in off-flavors were reported during storage of strawberries (Yu et al., 1995). Studies on the qualities of various types of blueberries receiving doses above 1 kGy reported decreases in ascorbic acid content (Moreno et al., 2008) and decreases in berry firmness (Moreno et al., 2007; Miller et al., 1994), but the berries were not necessarily deemed unacceptable. Higher doses up to 3.2 kGy are reported to affect flavor and color of blueberries (Morena et al., 2007). Gamma irradiation (2 kGy) of grapes reduced fungal spoilage with no apparent affects on berry firmness or soluble solids, but with reduction in titratable acids and ascorbic acid (Thomas et al., 1995). Low levels (ng/g) of furan were detected in gamma-irradiated (5 kGy) grapes (Fan and Sokorai, 2008).

Use of bacteriophage as a means to control bacterial contamination of foods has been suggested as a possible intervention strategy at the farm level to protect crops from plant diseases and to minimize contamination of animal products. Approvals for various bacteriophage products have been granted by the EPA or the FDA for growing or processing tomatoes and poultry products since 2006 (Hagens and Offerhaus, 2008). To our knowledge, no studies have been published to date on application of bacteriophage technology for berry production.

A number of other approaches have been tested for the improvement of microbiological quality of berries, including hot water for blueberries (Fan et al., 2008), modified atmosphere packaging in conjunction with natural antimicrobials for grapes (Guillén et al., 2007), and carbonate or bicarbonate salts for grapes (Gabler and Smilanick, 2001). These studies did not address microbiological safety for human pathogens.

IN SUMMARY

Prevention of produce contamination is a preferred practice, compared to decontamination especially with products such as berries, for which post-harvest intervention strategies are limited, provide incomplete protection, or compromise quality attributes. Since the increase in foodborne illness outbreaks associated with produce, additional measures have been taken to determine and implement good practices aimed at reducing the incidence of contamination on the farm and in handling. In 1998, the *Guide to Minimize Microbial Food Safety Hazards for Fresh Fruits and Vegetables* was released. This guide was developed by the Food and Drug Administration (FDA) and US Department of Agriculture (USDA) in conjunction with the produce industry. Subsequently, the Produce Safety Action Plan of 2004 was implemented. These documents serve as guides rather than regulatory requirements and have broad application to all types of produce. Specific guidance documents have been written to expand and customize insight for certain types of produce including leafy greens, tomatoes, and fresh-cut produce. Such specific guidelines have not been prepared for berries at this writing. Given the diversity of growing methods for berries from ground level to bush to vine, specific guidelines that suit all berry products would be particularly challenging to devise.

Good Agricultural, Manufacturing and Management Practices address major potential sources of contamination throughout the production and processing of produce including water, crop treatment, handling, facilities, and transportation. The reader is referred to other chapters of this book for a detailed discussion of recommendations to minimize microbial hazards in produce production, some of which may apply to berries.

REFERENCES

Alfano-Sobsey, E. A., Eberhard, M. L., Seed, J. R., Weber, D. J., Won, K. Y., Nace, E. K., and Moe, C. L. (2004). Human challenge pilot study with *Cyclospora cayetanensis*. *Emerg. Infect. Dis.* **10**, 726–728.

Ashford, R. W. (1979). Occurrence of an undescribed coccidian in man in Papua New Guinea. *Ann. Trop. Med. Parasitol.* **73**, 479–500.

Baert, L., Uyttendaele, M., Van Coillie, E., and Debevere, J. (2008). The reduction of murine norovirus 1, *B. fragilis* HSP40 infecting phage B40-8 and *E. coli* after a mild thermal pasteurization process of raspberry puree. *Food Microbiol.* **25**, 871–874.

Bassett, J. and McClure, P. (2008). A risk assessment approach for fresh fruits. *J. Appl. Microbiol.* **104**, 925–943.

Bern, C., Hernandez, B., Lopez, M. B., and Arroword, M. (1999). Epidemiological studies of *Cyclospora cayetanensis* in Guatemala. *Emerg. Infect. Dis.* **5**, 766–774.

Beuchat, L. R. (2002). Ecological factors influencing survival and growth of human pathogens on raw fruits and vegetables. *Microbes Infect.* **4**, 413–423.

Bialka, K. L. and Demirci, A. (2007a). Efficacy of aqueous ozone for the decontamination of *Escherichia coli* O157:H7 and *Salmonella* on raspberries and strawberries. *J. Food Prot.* **70**, 1088–1092.

Bialka, K. L. and Demirci, A. (2007b). Utilization of gaseous ozone for the decontamination of *Escherichia coli* O157:H7 and *Salmonella* on raspberries and strawberries. *J. Food Prot.* **70**, 1093–1098.

Bialka, K. L., Demirci, A., and Puri, V. M. (2008). Modeling the inactivation of *Escherichia coli* O157:H7 and *Salmonella enterica* on raspberries and strawberries resulting from exposure to ozone or pulsed UV-light. *J. Food Engineer.* **85**, 444–449.

Bidawid, S., Farber, J. M., and Sattar, S. A. (2000). Inactivation of hepatitis A virus (HAV) in fruits and vegetables by gamma irradiation. *Int. J. Food Microbiol.* **57**, 91–97.

Butot, S., Putallaz, T., and Sanchez, G. (2008). Effects of sanitation, freezing and frozen storage on enteric viruses in berries. *Int. J. Food Microbiol.* **126**, 30–35.

Butz, P., García, A. F., Lindauer, R., Dieterich, S., Bognár, A., and Tauscher, B. (2003). Influence of ultra high pressure processing on fruit and vegetable products. *J. Food Engineer.* **56**, 233–236.

Caceres, V. M., Ball, R. T., Somerfeldt, S. A., Mackey, R. L., Nichols, S. E., MacKenzie, W. R., and Herwaldt, B. L. (1998). A foodborne outbreak of Cyclosporiasis caused by imported raspberries. *J. Family Practice* **47**, 231–234.

Calvin, L. (2004). Response to US foodborne illness outbreaks associated with imported produce. Economic Research Service. Issues in Diet, Safety, and Health/Agriculture Information Bulletin Number 789–795.

Cannon, J. L., Papafragkou, E., Park, G. W., Osborne, J., Jaykus, L. A., and Vinje, J. (2006). Surrogates for the study of norovirus stability and inactivation in the environment: A comparison of murine norovirus and feline calicivirus. *J. Food Protection* **69,** 2761–2765.

Carter, R., Guido, F., Jacquette, G., Rapoport, M. (1996). Outbreak of cyclosporiasis associated with drinking water. In Program of the 30th Interscience Conference on Antimicrobial Agents and Chemotherapy, New Orleans, September 15–18, 1996. Washington, DC, American Society for Microbiology, 259.

Casteel, M. J., Schmidt, C. E., and Sobsey, M. D. (2008). Chlorine disinfection of produce to inactivate hepatitis A virus and coliphage MS2. *Int. J. Food Microbiol.* **125,** 267–273.

Center for Food Safety and Applied Nutrition, FDA. (2001). Analysis and evaluation of preventive control measures for the control and reduction/elimination of microbial hazards on fresh and fresh-cut produce. Chapter IV, Outbreaks associated with fresh and fresh-cut produce.

Center for Food Safety and Applied Nutrition, FDA. (2003). Food Emergency Response (FER) outbreak folders, raspberry-associated outbreak, Vermont.

Centers for Disease Control and Prevention. (2004). Outbreak of cyclosporiasis associated with snow peas—Pennsylvania. *Morb. Mortal. Wkly. Rep.* **53,** 876–878.

Centers for Disease Control and Prevention, CDC. (2008). Available at www.cdc.gov/foodborneoutbreaks/ Accessed 22 July 2008.

Chacin-Bonilla, L. (2008). Transmission of *Cyclospora cayetanensis* infection: A review focusing on soil-borne cyclosporiasis. *Trans. Trop. Med. Hyg.* **102,** 215–216.

Chauvin, M. A., Lee, S. Y., Chang, S., Gray, P. M., Kang, D. H., and Swanson, B. G. (2005). Ultra high pressure inactivation of *Saccharomyces cerevisiae* and *Listeria innocua* on apples and blueberries. *Journal of Food Processing and Preservation* **29,** 424–435.

Cotterelle, B., Drougard, C., Rolland, J., Becamel, M., Boudon, M., Pinede, S. et al. (2005). Outbreak of norovirus infection associated with the consumption of frozen raspberries, France, March 2005. *Euro. Surveill.* 10:E050428.1. www.eurosurveillance.org/ew/2005/050428.asp#1

Crowe, K. M., Bushway, A. A., Bushway, R. J., Davis-Dentici, K., and Hazen, R. A. (2007). A comparison of single oxidants versus advanced oxidation processes as chlorine-alternatives for wild blueberry processing (*Vaccinium angustifolium*). *Int. J. Food Microbiol.* **116,** 25–31.

Dalmadi, I., Polyák-Fehér, K., and Farkas, J. (2007). Effects of pressure- and thermal-pasteurization on volatiles of some berry fruits. *High Pressure Research* **27**(1), 169–172.

Dawson, D. (2005). Foodborne protozoan parasites. *Int. J. Food Microbio.* **103,** 207–227.

Diels, A. M. J., Callewaert, L., Wuytack, E. Y., Masschalck, B., and Michiels, C. W. (2005). Inactivation of *Escherichia coli* by high-pressure homogenization is influenced by fluid viscosity but not by water activity and product composition. *Int. J. Food Microbiology* **101,** 281–291.

Eberhard, M. L., Nace, E. K., Freeman, A. R., Streit, T. G., and Lammie, P. J. (1999). *Cyclospora cayetananesis* infections in Haiti: A common occurrence in the absence of watery diarrhea. *Am. J. Trop. Med. Hyg.* **60,** 584–586.

Falkenhorst, G., Krusell, L., Lisby, M., Madsen, S. B., Bottiger, B., and Molbak, K. (2005). Imported frozen raspberries cause a series of norovirus outbreaks in Denmark, 2005. *Euro. Surveill.* 10:E050922.2. www.eurosurveillance.org/ew/2005/050922.asp#2

Fan, L., Forney, C. F., Song, J., Doucette, C., Jordan, M. A., McRae, K. B., and Walker, B. A. (2008). Effect of hot water treatments on quality of highbush blueberries. *J. Food Sci.* **73,** M292–M297.

Fan, X. and Sokorai, K. L. B. (2008). Effect of ionizing radiation on furan formation in fresh-cut fruits and vegetables. *J. Food Sci.* **73,** C79–C83.

Fino, V. R. and Kniel, K. (2008). UV light inactivation of hepatitis A virus, aichi virus, and feline calicivirus on strawberries, green onions, and lettuce. *J. Food Prot.* **71,** 908–913.

Fleming, C. A., Caron, D., Gunn, J. E., and Barry, M. A. (1998). A foodborne outbreak of *Cyclospora cayetanensis* at a wedding; clinical features and risk factors for illness. *Arch. Int. Med.* **158,** 1121–1125.

Flessa, S., Lusk, D. M., and Harris, L. J. (2005). Survival of *Listeria monocytogenes* on fresh and frozen strawberries. *Int. J. Food Microbiol.* **101,** 255–262.

Food and Drug Administration (FDA). (2001a). FDA Survey of Imported Fresh Produce. Available at www.cfsan.fda.gov/~dms/prodsur6.html. Accessed 21 July 2008.

Food and Drug Administration (FDA). (2001b). The Juice HACCP Regulation, Questions and Answers. Available at www.cfsan.fda.gov/~comm/juiceqa.html#F. Accessed 3 September 2008.

Food and Drug Administration (FDA). (2003). FDA Survey of Domestic Fresh Produce. Available at www.cfsan.fda.gov/~dms/prodsu10.html. Accessed 21 July 2008.

Food and Drug Administration (FDA). (2008). Foods permitted to be irradiated under FDA's regulations (21 CFR 179.26). Available at www.cfsan.fda.gov/~dms/irrafood.html. Accessed 11 September 2008.

Garcia-Palazon, A., Suthanthangjai, W., Kajda, P., and Zabetakis, I. (2004). The effects of high hydrostatic pressure on β-glucosidase, peroxidase and polyphenoloxidase in red raspberry (*Rubus idaeus*) and strawberry (*Frafaria* x *ananassa*). *Food Chem.* **88,** 7–10.

Guillén, F., Zapata, P. J., Martínez-Romero, D., Castillo, S., Serrano, M., and Valero, D. (2007). Improvement of the overall quality of table grapes stored under modified atmosphere packaging in combination with natural antimicrobial compounds. *J. Food Science* **72,** S185–S190.

Gulati, B. R., Allwood, P. B., Hedberg, C. W., and Goyal, S. M. (2001). Efficacy of commonly used disinfectants for the inactivation of Calicivirus on strawberry, lettuce, and a food-contact surface. *J. Food Prot.* **64,** 1430–1434.

Guo, X., van Iersel, M. W., Chen, J., Brackett, R. E., and Beuchat, L. R. (2002). Evidence of association of Salmonellae with tomato plants grown hydroponically in inoculated nutrient solution. *Appl. Environ. Microbiol.* **68,** 3639–3643.

Hagens, S. and Offerhaus, M. L. (2008). Bacteriophages—New weapons for food safety. *Food Technology* **62**(4), 46–54.

Herwaldt, B. L., Ackers, M. L., and the *Cyclospora* Working Group. (1997). An outbreak in 1996 of cyclosporiasis associated with imported raspberries. *New Eng. J. Med.* **329,** 1504–1505.

Herwaldt, B. L., Beach, M. J., and the Cyclospora Working Group. (1999). The Return of *Cyclospora* in 1997: Another outbreak of cyclosporiasis in North America associated with imported raspberries. *Annal. Intern. Med.* **130,** 210–220.

Herwaldt, B. L. (2000). *Cyclospora cayetanensis*. A review, focusing on the outbreaks of cyclosporiasis in the 1990s. *Clin. Infect. Dis.* **5,** 1040–1057.

Hjertqvist, M., Johansson, A., Svensson, N., Abom, P. E., Magnusson, C., Olsson, M. et al. (2006). Four outbreaks of norovirus gastroenteritis after consuming raspberries, Sweden, June–August 2006. *Euro. Surveill.* **11,** 3038.

Ho, A. Y., Lopez, A. S., Eberhart, M. G., Levenson, R., Finkel, B. S., da Silva, A. J. (2000). Outbreaks of cyclopsporiasis associated with imported raspberries, Philadelphia, Pennsylvania, 2000. *Emerg. Infect Dis*. **8,** 783–788.

Hoge, C. W., Shlim, D. R., Rajan, R., Triplett, J., Shear, M., and Rabold, J. G. (1993). Epidemiology of diarrheal illness associated with coccidian like organisms among travelers and foreign residents in Nepal. *Lancet* **341,** 1175–1179.

Hoge, C. W., Echeverria, P., Rajan, R., Jacobs, J., Malt, S., and Chapman, E. (1995). Prevalence of *Cyclospora* spears and the enteric pathogens among children less than 5 years of age in Nepal. *J. Clin. Microbiol.* **33,** 3958–3960.

Hurst, W. C. and Schuler, G. A. (1992). Fresh produce processing—An industry perspective. *J. Food Prot.* **55,** 824–827.

Hutin, Y. J. F., Pool, V., Cramer, E. H., Nainan, O. V., Weth, J., Williams, I. et al. (1998). A multistate foodborne outbreak of hepatitis A. *New Eng. J. Med.* **340,** 595–602.

Katz, D. E. and Taylor, D. N. (2001). Parasitic infection of the gastro-intestinal tract. *Gastroenterol. Clin*. **30,** 635–653.

Keyser, M., Müller, I. A., Cilliers, F. P., Nel, W., and Gouws, P. A. (2008). Ultraviolet radiation as a non-thermal treatment for the inactivation of microorganisms in fruit juice. *Innovat. Food Sci. Emerg. Technol.* **9,** 348–354.

King, B. J. and Monis, P. T. (2007). Critical processes affecting *Cryptosporidium* oocyst survival in the environment. *Parasitol.* **134**(3), 309–323.

Kingsley, D. H., D. Guan, and D. G. Hoover. (2005). Pressure inactivation of hepatitis A virus in strawberry puree and sliced green onions. *Journal of Food Protection* **68**(8), 1748–1751.

Kniel, K. E., Lindsay, D., Sumner, S., Hackney, C. R., Pierson, M., and Dubey, J. P. (2002). Examination of attachment and survival of Toxoplasma gondii oocysts on raspberries and blueberries. *J. Parasitol.* **88,** 790–793.

Kniel, K. E., Shearer, A. E. H., Cascarino, J. L., Wilkins, G. C., and Jenkins, M. C. (2007). High hydrostatic pressure and UV light treatment of produce contaminated with *Eimeria acervulina* as a *Cyclospora cayetanensis* surrogate. *J. Food Prot.* **70**, 2837–2842.

Knudsen, D. M., Yamamoto, S. A., and Harris, L. J. (2001). Survival of *Salmonella* spp. and *Escherichia coli* O157:H7 on fresh and frozen strawberries. *J. Food Protection* **4**, 1483–1488.

Koopmans, M. and Duizer, E. (2004). Foodborne viruses: An emerging problem. *Int. J. Food Microbiol.* **90**, 23–41.

Korsager, B., Hede, S., Boggild, H., Bottiger, B., and Molbak, K. (2005). Two outbreaks of norovirus infections associated with the consumption of imported frozen raspberries, Denmark, May–June 2005. *Euro. Surveill.* **10**, E050623.1. www.eurosurveillance.org/ew/2005/050623.asp#1

Koumans, E. H., Katz, D. J., Malecki, J. M., Kumar, S., Wahlquist, S. P., Arrowood, M. J. et al. (1998). An outbreak of cyclosporiasis in Florida in 1995: A harbinger of multistate outbreaks in 1996 and 1997. *Am J Trop Hyg.* **59**, 235–242.

Lagunas-Solar, M. C., Piña, C., MacDonald, J. D., and Bolkan, L. (2006). Development of pulsed UV light processes for surface fungal disinfection of fresh fruits. *J. Food Prot.* **69**, 376–384.

Le Guyader, F. S., Mittelholzer, C., Haugarreau, L., Hedlund, K., Alsterlund, R., Pommepuy, M., and Svensson, L. (2004). Detection of norovirus in raspberries associated with a gastroenteritis outbreak. *Int. J. Food. Micro.* **97**, 179–186.

Lee, M. B. and Lee, E. H. (2001). Coccidial contamination of raspberries: Mock contamination with *Eimeria acervulina* as a model for decontamination treatment studies. *J. Food Protect.* **64**, 1854–1857.

Lindsay, R. (1997). Hepatitis A and Mexican strawberries. American University, The School of International Service, Washington, DC. www.american.edu/TED/strwberr.htm. Accessed August 2008.

Liu, R. H. (2007). The potential health benefits of phytochemicals in berries for protecting against cancer and coronary heart disease. In *Berry fruit value added products for health promotion* (Y. Zhao, Ed.), pp. 187–204. CRC Press, Boca Raton, FL.

Lopez, A. S., Dodson, D. R., Arrowood, M. J., Orlandi, P. A., da Silva, A. J., Bier, J. W. et al. (2001). Outbreak of cyclosporiasis associated with basil in Missouri in 1999. *Clin. Infect. Dis*. **32**, 1010–1017.

Lukasik, J., Bradley, M. L., Scott, T. M., Dea, M., Koo, A., Hsu, W. Y. et al. (2003). Reduction of Poliovirus 1, bacteriophages, *Salmonella* Montevideo, and *Escherichia coli* O157:H7 on strawberries by physical and disinfectant washes. *J. Food Protect.* **66**, 188–193.

Lukasik, J., Bradley, M. L., Scott, T. M., Hsu, W. Y., Farrah, S. R., and Tamplin, S. M. (2001). Elution, detection, and quantification of Polio 1, bacteriophages, *Salmonella* Montevideo, and *Escherichia coli* O157:H7 from seeded strawberries and tomatoes. *J. Food Protect.* **64**, 292–297.

Marrow, L. B. (2002). Outbreak of cyclosporiasis in Fulton County, Georgia. *Georgia Epidemiology Report* **18**, 1–2.

Mazzotta, A. S. (2001). Thermal inactivation of stationary-phase and acid-adapted *Escherichia coli* O157:H7, *Salmonella*, and *Listeria monocytogenes* in fruit juices. *J. Food Protect.* **64**, 315–320.

Miller, W. R., McDonald, R. E., McCollum, T. G., and Smittle, B. J. (1994). Quality of climax blueberries after low-dosage electron-beam irradiation. *J. Food Quality* **17,** 71–79.

Mlikota Gabler, F. and Smilanick, J. L. (2001). Postharvest control of table grape gray mold on detached berries with carbonate and bicarbonate salts and disinfectants. *American J. Enol. Viticulture* **52,** 12–20.

Moreno, M. A., Castell-Perez, M. E., Gomes, C., Da Silva, P. F., and Moreira, R. G. (2007). Quality of electron beam irradiation of blueberries (Vaccinium corymbosum L.) at medium dose levels (1.0–3.2 kGy). *LWT-Food Science and Technology* **40,** 1123–1132.

Moreno, M. A., Castell-Perez, M. E., Gomes, C., Da Silva, P. F., and Moreira, R. G. (2008). Treatment of cultivated highbush blueberries (Vaccinium corymbosum L.) with electron beam irradiation: Dosimetry and product quality. *J. Food Process Engineer.* **31,** 155–172.

Mukherjee, A., Speh, D., Jones, A. T., Buesing, K. M., and Diez-Gonzalez, F. (2006). Longitudinal microbiological survey of fresh produce grown by farmers in the Upper Midwest. *J. Food Protection* **69,** 1928–1936.

Murrow, L. B., Blake, P., and Kreckman, L. (2002). Outbreak of cyclosporiasis in Fulton County, Georgia. *Georgia Epidemiology Report* **18,** 1–2.

Niu, M. T., Polish, L. B., Robertson, B. H., Khanna, B. K., Woodruff, B. A., Shapiro, C. N. et al. (1992). Multistate outbreak of hepatitis A associated with frozen strawberries. *J. Infect. Dis*. **166,** 518–524.

Noah, N. D. (1981). Foodborne outbreaks of hepatitis A. *Medical Laboratory Science* **38,** 428.

Ortega, Y. R., Sterling, C. R., Gilman, R. H., Cama, V. A., and Diaz, F. (1993). *Cyclospora* species. A new protozoan pathogen of humans. *N. Engl. J. Med*. **328,** 1308–1312.

Ortega, Y. R., Gilman, R. H., and Sterling, C. R. (1994). A new coccidian parasite (Apicomplexa: Eimeriidae) from humans. *J. Parasitol*. **80,** 625–629.

Pieniazek, N. J. and Herwaldt, B. L. (1997). Reevaluating the molecular taxonomy: Is human-associated *Cyclospora* a mammalian *Eimeria* species? *Emerg. Infect. Dis*. **3,** 381–383.

Pirtle, E. C. and Beran, G. W. (1991). Virus survival in the environment. *Rev. Sci. Tech*. **10,** 733–748.

Ponka, A., Maunula, L., von Bonsdorff, C. H., and Lyytikainen, O. (1999). An outbreak of calicivirus associated with consumption of frozen raspberries. *Epidemiol. Infect*. **123,** 469–474.

Popa, J., Hanson, E. J., Todd, E. C. D., Schilder, A. C., and Ryser, E. T. (2007). Efficacy of chlorine dioxide gas sachets for enhancing the microbiological quality and safety of blueberries. *J. Food Protect*. **70,** 2084–2088.

Puupponen-Pimiä, R., Nohynek, L., Alakomi, H. L., and Oksman-Caldentey, K. M. (2005). Bioactive berry compounds—Novel tools against human pathogens. *Appl. Microbiol. Biotech*. **67,** 8–18.

Reid, T. M. S. and Robinson, H. G. (1987). Frozen raspberries and hepatitis A. *Epidemiol. Infect*. **98,** 109–112.

Relman, D. A., Schmidt, T. M., Gajadhar, A., Sogin, M., Cross, J., and Yoder, K. (1996). Molecular phylogenetic analysis of *Cyclospora*, the human intestinal pathogen, suggests that it is closely related to *Eimeria* species. *J. Infect. Dis.* **173,** 440–445.

Rodgers, S. L., Cash, J. N., Siddiq, M., and Ryser, E. T. (2004). A comparison of different chemical sanitizers for inactivating *Escherichia coli* O157:H7 and *Listeria monocytogenes* in solution and on apples, lettuce, strawberries, and cantaloupe. *J. Food Protect.* **67,** 721–731.

Ryan, T., Wilkinson, J. M., and Cavanagh, J. M. A. (2001). Antibacterial activity of raspberry cordial in vitro. *Res. Vet. Sci.* **71,** 155–159.

Sapers, G. M. (2001). Efficacy of washing and sanitizing methods for disinfection of fresh fruit and vegetable products. *Food Tech. Biotech.* **39,** 305–311.

Sapers, G. M. and Simmons, G. F. (1998). Hydrogen peroxide disinfection of minimally processed fruits and vegetables. *Food Tech.* **52,** 48–52.

Sarig, P., Zahavi, T., Zutkhi, Y., Yannai, S., Lisker, N., and Ben-Arie, R. (1996). Ozone for control of post-harvest decay of table grapes caused by *Rhizopus stolonifer*. *Phys. Mol. Plant Path.* **48,** 403–415.

Seymour, I. J. and Appleton, H. (2001). Foodborne viruses and fresh produce. *J. Appl. Microbiol.* **91,** 759–773.

Shields, J. M. and Olson, B. H.(2003). *Cyclospora cayetanensis*: A review of an emerging parasitic coccidian. *Intern. J. Parasitol.* **33,** 371–391.

Smith, H. V., Paton, C. A., Mtambo, M. M. A., and Girdwood, R. W. A. (1997). Sporulation of *Cyclospora* sp. Oocysts. *Appl. Environ. Microbiol.* **63,** 1631–1632.

Solomon, E. B., Yaron, S., and Matthews, K. R. (2002). Transmission of *Escherichia coli* O157:H7 from contaminated manure and irrigation water to lettuce plant tissue and its subsequent internalization. *Appl. Environ. Microbiol.* **68,** 397–400.

Sturbaum, G. D., Ortega, Y. R., Gilman, R. H., Sterling, C. R., Cabrera, L., and Klein, D. A. (1998). Detection of *Cyclospora cayetamensis* in wastewater. *Appl. Environ. Micro.* **64,** 2284–2286.

Suthanthangjai, W., Kajda, P., and Zabetakis, I. (2005). The effect of high hydrostatic pressure on the anthocyanins of raspberry (*Rubus idaeus*). *Food Chem.* **90,** 193–197.

Sy, K. V., McWatters, K. H., and Beuchat. L. R. (2005). Efficacy of gaseous chlorine dioxide as a sanitizer for killing *Salmonella*, yeasts, and molds on blueberries, strawberries, and raspberries. *J. Food Protect.* **68,** 1165–1175.

Timbo, B., Ross, M., Street, D., and Guzewich, J. (2007). FDA's use of epidemiological data, trace-back investigations and farm investigations as regulatory tools during outbreaks of *Cyclospora cayetanensis* infections associated with produce in the U.S. 1995-2005. *IAFP Ann. Mtg. Abstracts Book*, P3–4.

Tournas, V. H. and Katsoudas, E. (2005). Mould and yeast flora in fresh berries, grapes and citrus fruits. *Int. J. Food Microbiol.* **105,** 11–17.

Udompijitkul, P., Daeschel, M. A., and Zhao, Y. (2007). Antimicrobial effect of electrolyzed oxidizing water against *Escherichia coli* O157:H7 and *Listeria monocytogenes* on fresh strawberries (*Fragaria* x *ananassa*). *J. Food Sci.* **72,** M397–M406.

Yuk, H.-G., Bartz, J. A., and Schneider, K. R. (2006). The effectiveness of sanitizer treatments in inactivation of *Salmonella* spp. from bell pepper, cucumber, and strawberry. *J. Food Sci.* **71,** M95–M99.

Wei, K., Zhou, H., Zhou, T., and Gong, J. (2007). Comparison of aqueous ozone and chlorine as sanitizers in the food processing industry: Impact on fresh agricultural produce quality. *Ozone: Sci. Engineer.* **29,** 113–120.

Wu, V. C. H. and Kim, B. (2007). Effect of a simples chlorine dioxide method for controlling five foodborne pathogens, yeasts and molds on blueberries. *Food Microbiol.* **24,** 794–800.

PART 4

Avoidance of Contamination

CHAPTER 13

Produce Contamination Issues in México and Central America

Jorge H. Siller-Cepeda
Desert Glory México S De RL de CV, Guadalajara, Jalisco, México

Cristobal Chaidez-Quiroz and Nohelia Castro-del Campo
Centro de Investigación en Alimentación y Desarrollo, A.C., Culiacán, Sinaloa, México

CHAPTER CONTENTS

Introduction	309
Sources of Contamination	312
Irrigation Water	312
Runoff	315
Inadequate Disinfection Processes at Packinghouses	316
Conditions for Agricultural Workers	318
Good Agricultural Practices	321
Outbreak-Related Cases in México and Central America	323
Conclusions	326

INTRODUCTION

México and Central America (Guatemala, Nicaragua, Honduras, El Salvador, Belize, Panama, and Costa Rica) are important countries exporting considerable volumes of fruits and vegetables to the United States. From 2006 to 2007, the dollar value of exported vegetables, fresh fruits (other than bananas), and bananas and plantains from all these countries increased 9.5%, 29%, and 13.2%, respectively (FAS/USDA, 2008). México was the main exporter by far on vegetables and fruits other than bananas, followed by Costa Rica and Guatemala (Tables 13.1 and 13.2). In the case of bananas and plantains, Guatemala, Costa Rica, and Honduras were, in that order, the main exporters of these crops to the United States (Table 13.3).

Table 13.1 Fresh Vegetables (January–December, values in 1000 dollars) Imported by United States from México and Central American Countries in 2006 and 2007

Country	2006	2007	% Change
México	2,573,341	2,800,221	+ 8.82
Guatemala	23,016	31,146	+ 35.32
Belize	5	2	− 60.0
Honduras	14,679	20,066	+ 36.79
El Salvador	2,285	4,793	+ 109.76
Nicaragua	10,357	14,315	+ 38.22
Costa Rica	51,074	59,518	+ 16.53
Panamá	4,339	3,045	− 29.82

FAS/USDA (2008) www.fas.usda.gov/ustrade/USTImBICO.asp?QI

Table 13.2 Fresh Fruits Other Than Bananas (January–December, values in 1000 dollars) Imported by United States from México and Central American Countries in 2006 and 2007

Country	2006	2007	% Change
México	1,131,857	1,627,415	+ 43.78
Guatemala	109,308	135,552	+ 24.01
Belize	15,658	13,642	− 12.8
Honduras	33,984	36,861	+ 8.47
El Salvador	431	992	+ 130.16
Nicaragua	6,239	9,373	+ 50.23
Costa Rica	443,544	423,262	− 5.43
Panamá	9,733	11,658	+ 19.78

FAS/USDA (2008) www.fas.usda.gov/ustrade/USTImBICO.asp?QI

Table 13.3 Fresh Fruits—Bananas and Plantains (January–December, values in 1000 dollars) Imported by United States from México and Central American Countries in 2006 and 2007

Country	2006	2007	% Change
México	17,270	13,813	− 20.02
Guatemala	248,114	309,029	+ 24.55
Belize	0	0	0
Honduras	116,612	137,069	+ 17.54
El Salvador	0	0	0
Nicaragua	7,831	9,128	+ 16.56
Costa Rica	281,506	295,228	+ 4.89
Panamá	2,679	515	− 80.78

FAS/USDA (2008) www.fas.usda.gov/ustrade/USTImBICO.asp?QI

México grows fruits and vegetables on about 4% of its agricultural land. Climate variation from tropical to temperate allows growers to produce a wide spectrum of fruits and almost any vegetable. About 20% of México's fruit and vegetable production is exported, but the vast majority goes to the large and growing domestic market. The main vegetable crops exported from México are tomatoes, bell peppers, and cucumbers, representing 70% of the total value. The main fruits exported from México to the United States are limes, avocados, and mangos. Production practices in México for the export and the domestic markets are quite different. The export industries grow products to meet foreign-market consumer demand, retail preferences, and governmental restrictions (limits on chemical and pesticide residues, programs to deal with quarantine pests, etc.). The technology is quite similar to that used in the United States, as US firms are active in the Mexican export industries. Producers for México's domestic market tend to be more labor-intensive than in the United States and employ more traditional methods of cultivation and harvesting.

The diversity of the natural environment in highland Central America has influenced the production of fruits and vegetables. From the temperate basins of the Valley of Anahuac, to the tropical forests of the south, different climates and soils have conditioned which fruits and vegetables are grown, and how. Guatemala, Costa Rica, and Honduras grow and export mainly tropical fruits including bananas and plantains, with some participation in vegetable exportation, mainly cantaloupe melons and cucumbers. On the other hand, countries like El Salvador, Belize, and Panama grow most of these crops for their domestic markets. It is interesting how El Salvador has made progress, and from the 2006 season to 2007, this country more than doubled its exportation value, mainly focused on cucumber and some oriental vegetables. Although technological and organizational changes have influenced how crops are grown in Central America, most agriculture in those countries is still not developed, and most of the production is dedicated to domestic markets.

Several large foodborne outbreaks have been linked to fresh produce from these regions, including crops such as cantaloupe, tomatoes, peppers, green onions, and berries. Although contributing factors have not been determined in all cases, quite a few notable causes have been proposed. In particular, cross-contamination with fecal matter of both domestic and wild animals has been suggested. In addition, contact with contaminated water also has been identified as a source of contamination. Moreover, the use of untreated manure or sewage as fertilizer, lack of field sanitary toilet facilities, poorly or unsanitized transportation vehicles, and contamination by handlers are also suggested as potential contributing factors.

Although most fresh fruit and vegetable growers from these countries are implementing programs to prevent and reduce produce contamination, such as GAPs (Good Agricultural Practices) and GMPs (Good Manufacturing Practices), there are still several farms that are at the beginning steps. In 2002, the Mexican Commission for Central American Cooperation conducted a large effort through the SAGARPA (Agricultural, Cattle, Rural Development, Fishing and Food Ministry Office) to develop and distribute three manuals to support all of the growers in México and Central America in implementing these programs. These manuals include a guide for growers, describing Good Agricultural Practices; a manual intended to help growers audit their operations and develop SOPs and quality assurance; and a third manual focused on worker hygiene and pesticide management (Siller et al., 2002). After that, some Central American countries developed their own documents; however, most of them are not harmonized with international regulations and are voluntary programs.

The actual impact of these manuals on produce safety has not been quantified yet, but most of the growers are aware that control of foodborne pathogens in produce must begin before produce is even planted by avoiding fields that have been subjected to flooding, or fields on which animals have been recently grazed, or have otherwise been contaminated with manure. After planting, practices are focusing on using only clean potable water for irrigation, and growers are now aware of a continuous program for cleaning and sanitizing harvesting and packing equipment. Training on proper personal hygiene for all field, packinghouse, and processing plant workers, as well as providing adequate sanitary and handwashing facilities has been an important part of these changes. Use of only approved pesticides along with proper training to workers on management and application of these chemicals has been fundamental to reduction of chemical contamination, reductions in environmental impact, and improvements in workers' health. It will be necessary in the near future to evaluate how these programs are progressing in these countries, especially because there is very little published information available.

SOURCES OF CONTAMINATION

Irrigation Water

When water is in contact with fruits and vegetables, the risk of contamination will depend on the microbiological quality of the water source. Water is used in diverse agricultural activities. Water used to apply fertilizers, as well as pesticides, has to meet both chemical and microbiological requirements before its use (Siller-Cepeda et al., 2002). To avoid contamination, wells and all other sources of water must be submitted periodically to chemical and microbiological examination, with the results recorded, and existing problems corrected. Bathing and grazing of animals near water resources should be prohibited, to prevent fecal contamination of water and reduce risks to human health from consumption of contaminated fresh produce. In the particular case of irrigation, contamination is associated with the irrigation type and the kind of crop. Flooding irrigation represents the greatest possibility of contamination if it is used on creeping crops such as lettuce or strawberries by permitting contact with the soil. With the sprinkler irrigation technique, the spray provides a rapid means to contaminate the product. In both cases, water quality is important. With the drip irrigation technique, the risk of contamination is smaller.

Most of the large areas dedicated to growing and exporting vegetables in the winter in northwest México make use of water stored behind dams for irrigation; however, some areas in Sonora, Baja California, Coahuila, and Nuevo Leon States depend mostly on water extracted from wells. Water from dams is conducted to the fields through irrigation channels. Conditions vary widely across the country; some are protected by concrete to avoid water leaks and growth of weeds, others are just abandoned. Areas in Central and South México, as well as some regions in Central American countries use different water sources for irrigation. Microbiological quality has been always a concern in some of these areas, especially areas in Central México and in Central America that use waste water coming from sewage treatment for irrigation of crops intended for the domestic market. Use of wastewater can contaminate produce with pathogenic microorganisms by direct contact. Sadovski et al. (1978) noticed that the use of waste water was of greater concern if applied immediately before harvest rather than during the early stages of the production cycle.

Among the types of irrigation used in México and Central American countries, drip irrigation is the most common type used for agriculture dedicated to exportation, particularly in Northwest México, reducing risk of contamination. However, this represents only around 15% of the land being irrigated. Overhead irrigation with sprinklers is utilized mostly to irrigate fruit crops, especially citrus and apple crops. However, most tropical crops in these regions will depend on seasonal rainy periods for water availability. Flooding irrigation is the type of irrigation more widely used in countries where water is conducted from sources impounded by dams to fields. This type of irrigation represents higher possibilities of contaminating produce, especially when it is applied to creeping crops and where water is in contact with contaminated soil or runoff from large cities or cattle-raising facilities.

On crops that are staked and raised above the soil, such as those vegetables mostly grown for exportation, the risk of contamination is low.

Relatively few studies in México have examined the presence of protozoan parasites in surface waters used for irrigation or for wash-water applications in the fresh-produce industry. Those few studies have revealed the presence of pathogens such as *Cryptosporidium* oocyst (48%) and *Giardia* cyst (50%) in surface water coming from water impounded by dams. In addition wash-water tanks filled with water coming from rivers or impounded by dams and used in selected packinghouses tested positive for *Cryptosporidium* oocysts and *Giardia* cysts with concentrations ranging from 1 to 133 oocysts and 100 to 533 cysts per 100 liters, respectively. This suggests that there may be a risk of contamination of fresh produce as protozoan oocysts/cysts might come in contact with and attach to produce surfaces, posing a risk of infection to consumers who eat these products (Chaidez et al., 2003).

A quantitative microbial risk assessment (QMRA) was conducted to evaluate the public-health impact of protozoan-laden water used for irrigating produce in Northwest México. Specifically, a QMRA was conducted to address the human health impact associated with consumption of tomatoes, bell peppers, cucumbers, and lettuce irrigated with water contaminated with *Cryptosporidium* and *Giardia*. Yearly infection risks were estimated based on the assumption of a 120-day exposure in a given year. Annual risks range from 9×10^{-6} for *Cryptosporidium* at the lowest concentration associated with bell peppers to almost 2×10^{-1} for exposure to *Giardia* on lettuce at the highest detected concentration. With the relatively high number of illnesses resulting from produce-related outbreaks, addressing pre- and postharvest points of contamination for fruits and vegetables consumed raw should be a food industry priority. This research shows how QMRA can be used to interpret microbial contamination data for public-health significance and subsequently provide the foundation for guideline development (Mota, 2004).

A study performed in the Xochimilco agricultural area in México, located at the central part of the country and irrigated with waste water, showed a variety of microorganisms such as Enterobacteriaceae, *Escherichia coli*, *Enterobacter cloacae*, *Klebsiella pneumoniae*, *K. oxytoca*, *Citrobacter freundii*, and *Salmonella* spp., as well as nonfermenting microorganisms such as *Pseudomonas* spp. and *Acinetobacter* spp. Some species are not native to the natural environment and may represent exogenous microorganisms, further indicating a human or animal fecal source. The observed patterns of irregular urban area settlements and presence of animals such as cows or sheep grazing in some areas provide suggestive evidence of the source of nonnative microorganisms. Mexican guidelines for use of residual water for irrigation (SEMARNAT, 1996) specify 1000 or less CFU/100 ml as the fecal coliform limit for acceptable irrigation water for crops likely to be eaten uncooked and for sport fields and parks; this limit is exceeded in areas of Xochimilco (Mazari-Hiriart et al., 2008).

Agricultural water was assessed to determine the presence of pathogens such as *Salmonella* and *Escherichia coli* from January to May of 2005 in four regions of the Culiacan Valley located in Northwest México, an area known for its vast agricultural production, and as one of the major exporters of fresh produce, worldwide. Samples from what is known as the interphase water-sediment of water used to irrigate agricultural crops were positive for 20 strains of *Salmonella*. Serotyping revealed the presence of 13 strains of *Salmonella* Typhimurium, two of Infantis, and one each of Anatum, Agona, Oranienburg, Minnesota, and Give. Ninety-eight percent of the analyzed water samples were contaminated with *E. coli*, averaging 1.6×10^4 CFU/100 ml (Lopez-Cuevas et al., 2005).

Runoff

Potential Movement of Fecal Matter During Rainy Season

An important environmental factor affecting microbial movement is rainfall. It can result in pathogen spread by runoff from places where manure or biosolids have been applied as fertilizer or by leaching through the soil profile. It is known that bacterial and viral groundwater contamination increase during heavy rainfall. The presence of coliforms was monitored for 9.4-meter and 153.3-meter wells. Coliforms were detected in both shallow and deep wells, with bacterial contamination coinciding with the heaviest rainfalls (Gerba and Bitton, 1984). In Quebec, Canada, human and pig enteroviruses were isolated from 70% of the samples collected from a river. The contamination source was attributed to a massive pig raising activity in the area (Payment, 1989). In contrast, some authors have reported a decrease of pathogen concentrations during the rainy season. Cazarez-Diarte et al. (2004) observed a reduction from 41,493 during the dry season to 8525 CFU/100 mL during the rainy season. Tyrrel and Quinton (2003) stated that coliform transport is mediated by water density and turbidity; hence, water body volumes when pluvial precipitations exist are diluted as well as the coliforms present.

Flooding

Another factor that could affect the bacteriological quality of croplands is a history of flooding. This situation can become a problem when floodwaters cover areas on which farm animals have been grazed or confined, upstream from vegetable production areas. Floodwaters can become polluted with animal waste and carry the contaminants downstream, where they may also flood over croplands. Major flooding has also caused rivers to cover or damage sewage treatment plants. Either the floodwaters or effluents from the plants then become contaminated with human, municipal, and industrial wastes. Again, such events can subsequently contaminate downstream croplands. Microorganisms deposited on flooded croplands may remain for months or years after the flood (Beuchat and Ryu, 1997).

Inadequate Disinfection Processes at Packinghouses

Effects of pH, Temperature, Organic Matter Content, and Disinfectant Concentration

Among the factors affecting disinfection processes are pH, temperature, organic matter content, and disinfectant type and concentration. Chlorination is widely practiced as a disinfection process for microbial control in water used to wash fruits and vegetables at most of the packinghouses dedicated for exportation, but this practice is not used in packing of vegetables for domestic market. When properly applied, chlorine-based products are efficient. However, several drawbacks have been identified, including the protection exerted by the organic and inorganic matter to chlorine disinfection (Karch and Loftis, 1998). Available research has shown that increased resistance to disinfection may result from the attachment of microorganisms to various surfaces, including particles, algae, and carbon fines (LeChevallier et al., 1988). Ridgway and Olson (1982) have shown that the majority of viable bacteria in chlorinated drinking water are attached to particles. Presumably, microorganisms entrapped in particles are shielded from disinfectants. Therefore, the ineffectiveness of chlorine and other disinfectants may depend on whether or not the target organisms are readily accessible (Solomon et al., 2002).

Tomato Production Practices in México

Of the total farm volume of vegetables exported from México, tomatoes account for about 30% or 0.7 million tons (Siller, 1999). Sinaloa State produces about 65% of the commercial winter crop of field-grown fresh tomatoes. Approximately 90% of the state production is exported, mainly to the United States and Canada. Its major production season extends from late October through June. Sinaloa also produces cherry and plum type tomatoes, and handling practices are generally the same as for regular tomatoes. Most Sinaloa tomatoes are harvested at the breaker stage, but a few are harvested at the mature-green stage. The technology for production and handling tomatoes intended for distant markets does not differ from those techniques required for handling when this vegetable is intended for distribution through local markets.

Most Mexican packinghouses that export fruits are large, sophisticated, high-volume operations. Upon transfer to the packing line, tomatoes are washed, presized, waxed, sorted and graded, sized, packed into shipping containers, and unitized for shipment while in the packinghouse. Water dump tanks are used routinely for receiving tomatoes at the packinghouse. Pallet bins are emptied into the dump tank while tomatoes are water flumed from gondolas into the dump tank. In each case, tomatoes in the dump tank are flumed to an elevator where they are spray washed and conveyed to the packing lines. Serious losses due to decay occur periodically in shipments during transit or at destination. Poor dump-tank and wash-water management practices can be major contributors to decay or contamination problems.

There are other ways to receive the product at the initial steps of the fresh produce selection and packing process. Some products are received and maintained dry, whereas others are received by submersion in a chlorinated water tank at concentrations fluctuating from 100 to 300 ppm of total chlorine or from 50 to 75 ppm of free chlorine. When immersing the product, parameters such as turbidity, water temperature, disinfectant concentration, and pH are constantly monitored in order to maintain the optimum conditions for the disinfectant. pH is usually maintained at values between 6.5 to 7.0. Water temperature is one of the critical parameters because if a temperature differential of 10 °F or more between water and pulp exists, the produce will tend to absorb water. Zhuang and coworkers (1995) found that tomatoes took up greater numbers of cells of *Salmonella* spp. from an aqueous environment when placed in water that was 15 °C cooler than tomatoes. Showalter (1979) and Bartz and Showalter (1981) found that when tomatoes were dipped into water that was colder than the fruit, creating a negative temperature differential, tomatoes took up 1 to 4% of the fruit weight in water from the environment; most of the water uptake appeared to be in the vascular area beneath the stem scar. In addition to the temperature differential, Bartz (1982) has shown that, with tomatoes, the amount of water uptake from the environment is partially dependent upon the depth of submersion of the fruit.

In the case of the products maintained dry, the washing and disinfection process occurs further on, while the product is transported on selection rollers. These selection rollers are specially designed to cause the product to spin, ensuring a homogeneous brushing, washing, and disinfection process. Most of these systems use gaseous chlorine directly dispersed in the water through a sparger, wetting and disinfecting produce as it is spinning.

Contact time of produce in chlorinated water is short, usually between 15 and 30 seconds (Siller-Cepeda et al., 2007).

A study was undertaken by Chaidez et al. (2003), to determine the efficacy of three commonly used disinfectants in packinghouses of Sinaloa, México: sodium hypochlorite (NaOCl), trichlor-s-triazinetrione (TST), and trichlormelamine (TCM). Even though TST is approved only for swimming pool area disinfection, sometimes it is used for application to produce contact surfaces in México. Each microbial challenge consisted of water containing approximately 8 $\log_{10}$ bacterial CFU ml^{-1}, and 8 $\log_{10}$ viral PFU ml^{-1} treated with 100 and 300 mg l^{-1} of total chlorine with modified turbidity. Water samples were taken after two minutes of contact. They found that chemical disinfectants inactivate *E. coli* and *S.* Typhimurium in water by greater than 6 $\log_{10}$ at the initial test point. It is known that under conditions of high water-quality, waterborne vegetative bacteria are highly susceptible to relatively low doses of chlorine. Factors such as the amount of organic matter surrounding the target organisms are likely to influence the adhesion characteristics of cells and the lethal effect of sanitizers.

Results of this study showed, however, that the amount of organic material present in the wash water influenced the efficacy of disinfectants. Results also show that similar and significant reductions in populations of *E. coli* and *S.* Typhimurium occur in water used to wash fruits and vegetables at the packinghouses using 100 or 300 mg l^{-1} of NaOCl or TST in both average and worst-case water conditions. TST (300 mg l^{-1}) and NaOCl (300 mg l^{-1}) after a exposure of two minutes were found to effectively reduce the number of bacterial pathogens and viral indicators by 8 $\log_{10}$ and 7 $\log_{10}$, respectively ($P = 0.05$). The highest inactivation rate was observed when the turbidity was low and the disinfectant was applied at 300 mg l^{-1}. TCM did not show effective results when compared with TST and NaOCl ($P \leq 0.05$). Significant reduction of MS2 phage only was achieved using 300 mg l^{-1} of NaOCl or TST in average-case water, whereas in the worst-case water challenge, neither NaOCl nor TST or TCM were effective. These findings suggest that turbidity created by the organic and inorganic material present in the water tanks that was carried in by the fresh produce may affect the efficacy of the chlorine-based products.

Conditions for Agricultural Workers

Prevalence and Incidence of Bacterial, Viral, and Protozoan-Related Gastroenteritis

The overall sanitary quality of crops during production is dependent primarily on the growing environment. However, harvesting introduces human and mechanical contact that has an impact on the microbiological safety of fresh produce. The degree of farm workers' personal hygiene can have an important influence on the transmission of pathogenic bacteria to produce being harvested. Farm workers often come from diverse cultural backgrounds, not all of which stress proper personal hygiene as an important behavioral value.

Infectious diarrhea is an important cause of serious morbidity in developed nations, in hospitalized patients, and in travelers to tropical or subtropical regions of the world. Most important, it is a major public health problem in developing countries, where it is an important cause of morbidity and mortality in children. Bacterial pathogens are responsible for more than 50% of diarrheal diseases in developing countries. In México the gastrointestinal infectious diseases have been the first cause of mortality in children from 1 to 4 years old from 2000 to 2005 (SSA, 2005).

Rotavirus is considered the greatest, worldwide cause of viral gastroenteritis in children. Its prevalence is similar in developed and developing countries (Parashar et al., 2003). The prevalence of rotavirus in Paraguayan children from 2004 to 2005 was seasonal, with the highest prevalence during the coolest and driest months of the year; rotavirus prevalence was 23.8% and 14.9% for children and adults, respectively (Amarilla et al., 2007).

Norovirus (NoV), which belongs to the Caliciviridae family, is now recognized as the leading cause of epidemic and endemic nonbacterial gastroenteritis. In industrialized countries, NoV may be responsible for 68 to 93% of nonbacterial gastroenteritis outbreaks (Fankhauser et al., 2002). Studies in developing countries have shown that NoV is a major nonbacterial pathogen that causes acute diarrhea in children (Talal et al., 2000).

Long et al. (2007) provided unique information about the epidemiology of NoV infection and the effect that vitamin A supplementation has on this infection among children living in peri-urban communities of México City. First, they found a high prevalence of NoV infections during the summer months. NoV was isolated from 114 (30.5%) of 374 stool samples collected during the summer months. NoV GI and NoV GII were found in 62 (54.4%) and 52 (45.6%) of the 114 positive samples, respectively. Twenty-five (21.9%) of the 114 NoV positive samples had coinfections: 7 (6.1%) were coinfected with EPEC, 10 (8.8%) with ETEC, and 7 (6.1%) with *G. lamblia*.

Intestinal parasites remain extremely common worldwide. In developing countries, intestinal protozoans are important causes of childhood diarrhea. Cryptosporidiosis is a common cause of chronic diarrhea in patients with AIDS. With the advent of current active antiretroviral therapy, the prevalence of cryptosporidiosis in AIDS has decreased. By contrast, *Cryptosporidium*, *Cyclospora*, and *Giardia* outbreaks continue to be associated with contamination of food or water (Okhuysen and White, 1999). *Cryptosporidium* spp. and *Giardia* spp. are intestinal protozoan parasites that are recognized as prevalent and widespread pathogens of humans and many species of mammals. They constitute a common cause of gastroenteritis that manifests as a watery diarrhea in humans, and are the third most common protozoan infections in humans worldwide (Paziewska et al., 2007).

Inefficient Field Toilets and Handwashing Stations

Workers involved in farming can have an important impact on the microbial safety of produce they handle. Most of the foodborne illnesses are transmitted by humans. It is therefore very important to have agricultural workers adhere to proper sanitary procedures. Workers who either fail to practice or refuse to apply important hygienic practices, such as handwashing, constitute a risk for contaminating the produce they touch with human pathogens. An outbreak of cholera associated with sliced melon was related to agricultural workers as the source of contamination (Ackers et al., 1997). Several critical points must be considered to reduce the contribution of agricultural workers to pathogen transmission. Among them, adequate sanitary facilities must be provided to workers. Portable toilets and handwashing facilities are the minimum requirements to be implemented. This practice is applied on most operations that export produce; however, there are still a lot of operations that do not follow these practices. It is also important that such facilities be placed in a relatively convenient location, in close proximity to work areas (NACMCF, 1998), but outside the packinghouse, in order to reduce cross-contamination by pathogenic microorganisms. These facilities must exist for both sexes, with at least one toilet for every 15 employees. It is also important to provide instructions explaining the handwashing process. Warm water, soap, paper, and disinfectant must always be available. When sanitary facilities are next to or near packinghouse areas, a sanitary rug must be placed on the bathroom's exit door to disinfect the soles of shoes (Siller-Cepeda et al., 2007). Another important issue is that workers must be trained in the importance of proper personal hygiene, specifically handwashing after using restrooms. Ideally, such training would motivate the workers to willingly conform to the required sanitary practices (NACMCF, 1998). Training should be in the first language of the worker to ensure the real transmission of knowledge (Siller-Cepeda et al., 2007).

A study to demonstrate the importance of good handwashing technique as well as the transfer of pathogens between hands and produce was done by Jimenez et al. (2007); they assessed the effectiveness of hand hygiene techniques and quantified the amount of *Salmonella enterica* serovar Typhimurium transferred from volunteers' hands (bare or gloved) to green bell peppers and vice versa. Their results showed that the efficiency of transmission of *Salmonella* from green bell peppers to hands was high, whereas transfer rate from hands to the fresh produce was low. A combination of handwashing and hand rubbing with alcohol gel significantly reduces the presence of *Salmonella* on hand surfaces, and it should be considered as part of routine packinghouse activity. However, the primary method to avoid the presence of *Salmonella* during packinghouse operations would be the strict adherence to the GAPs, which means that the best strategy, undoubtedly, will be prevention.

GOOD AGRICULTURAL PRACTICES

The concept of Good Agricultural Practices (GAP) has evolved in recent years in the context of a rapidly changing and globalized food economy, and as a result of the concerns and commitments of a wide range of stakeholders regarding food production and security, food safety and quality, and the environmental sustainability of agriculture. These stakeholders have representitives in the supply dimension (farmers, farmers' organizations, workers), the demand dimension (retailers, processors, consumers), and those institutions and services (education, research, extension, input supply) that support and connect demand and supply and who seek to meet specific objectives of food security, food quality, production efficiency, livelihoods, and environmental conservation in both the medium and long term.

Broadly defined, a GAP approach aims at applying available knowledge to addressing environmental, economic, and social sustainability dimensions for on-farm production and postproduction processes, resulting in safe and quality food and nonfood agricultural products. Based on generic sustainability principles, it aims at supporting locally developed optimal practices for a given production system based on a desired outcome, taking into account market demands and farmers' constraints and incentives to apply practices.

However, the term GAP has different meanings and is used in a variety of contexts. For example, it is a recognized terminology used in international regulatory frameworks as well as in reference to private, voluntary, and nonregulatory applications that are being developed and applied by governments, civil society organizations, and the private sector.

For many developing countries, the export of fruits and vegetables accounts for significant income from hard-currency earnings. However, rejection of fresh produce has been related to overall quality, presence of nonauthorized pesticides, pesticide residues, and contaminants exceeding permissible limits. Inadequate labeling and packaging, not having the required nutritional information, and bacterial contamination have also been causes of rejection.

Efforts are underway by governments and industry to develop and apply GAPs, GMPs, and HACCP throughout the food chain. The challenges to the system are the lack of or weak coordination between the public and private sectors, training programs targeting appropriate stakeholders, the needs to harmonize national standards with international standards, and especially, the lack of political concern and incentives for adoption of programs at the farmer's level in Latin American countries. The quality and safety programs and initiatives implemented in the region are targeting mainly production supplying export markets, with little or no emphasis in production supplying domestic markets (national consumers' protection aims). Also, food control systems in some countries do not have a clear distinction regarding responsibilities and roles of the government ministries and institutions involved in quality and safety issues at the production level. There is a clear need to define institutional roles in terms of quality and safety for primary production. There is a need for enforcing pesticide regulations. Preventing the misuse of pesticides and emphasizing the use of approved pesticides, applied to effectively control pests and diseases in conformance with the approved Minimum Risk Levels (MRLs) and the International Code of Conduct for Distribution and Use of Pesticides, are important actions that need to be enforced.

In US and EU markets, private initiatives are implemented (EurepGAP, Safe and Quality Food-SQF Code, BRC Global Standard Packaging, ProSafe Certified Program, GAP Certification); however, in Latin America, initiatives are taken over by the private and public sectors (ChileGAP, PIPAA Program-Guatemala, SENASICA–México, SENASA–Argentina, SENA–Colombia, PRMPEX–Peru, OIRSA–El Salvador).

In México, the Agriculture Department (SAGARPA), a federal institution, supports growers in developing Good Agricultural Practices Manuals for main crops. Within its organic structure, the Servicio Nacional de Sanidad, Inocuidad y Calidad Agroalimentaria (SENASICA) offers a certification of agricultural companies under the Good Agriculture and Manufacturer Practices volunteer program. Its aim is to avoid having food-safety barriers become an obstacle for national produce in international markets. This program consists of a system that minimizes any risk in the production and packing of fruits and vegetables. As an example, federal authorities (SENASICA) and state authorities (Government of Baja California) in conjunction with the green onion export industry and growers developed the Green Onion Protocol (GAPs and GMPs) based on the FDA guidelines.

The México Calidad Suprema program is an official mark of identification that guarantees good sanitation, safety, and a high quality for Mexican products. This label seeks to identify products that comply with the following regulations: Mexican Official Norms (NOMs), Mexican Norms (NMX), and International Rules in a confident and transparent system for the benefit of producers, packers, distributors, and consumers; however, it still has not been harmonized with federal agricultural programs dictated by SENASICA.

OUTBREAK-RELATED CASES IN MÉXICO AND CENTRAL AMERICA

The cantaloupe melon is ranked as the sixth highest cause of fresh produce-related foodborne disease, resulting in nearly 1137 cases during the period of 1990 to 2005 (CSPI, 2007). *Salmonella* infections due to cantaloupe consumption have been reported since at least 1990. In that year, cantaloupes presumably originating in México or Guatemala were found to be contaminated with *S.* Chester and caused 245 cases of infections. In 1991, 400 infections were associated with the presence of *S.* Poona in 23 US and Canadian states. The outbreak was associated with consumption of precut cantaloupe that originated from México (JAMA, 1991; CDC, 2002). However, contaminated fruits were discarded after the outbreaks, and thus *Salmonella* was not isolated. Between 1990 and 1991, FDA personnel isolated several *Salmonella* serotypes from 1% of cantaloupe and watermelon samples collected at the border (CDC, 1991). In 1991, in California, 25 persons were infected with *S.* Saphra due to consumption of cantaloupe imported from Altamirano, Guerrero, México (Mohole-Boetani et al., 1999). Between May and June 1998, again the Mexican cantaloupe was involved in an outbreak in Ontario, Canada where 22 illnesses were reported due to the presence of S. Oranienburg. During this outbreak, cantaloupes were imported from México and Central America (Sewell and Farber, 2001). Between 1999 and 2000, FDA analyzed 151 cantaloupe samples arriving from nine different countries and 5% were contaminated with *Salmonella*. All of them were from four Mexican distributors. The recurring presence of *Salmonella*-contaminated cantaloupes triggered the establishment of a surveillance program at field and packinghouse levels to source track the points of contamination and to implement corrective actions to remedy the problem. After a series of visits and training, four Mexican firms implemented GAP and GMP programs and were allowed to export again

(CDC, 2002; FDA, 2001a). In 2001, 29 fruits were examined, and none presented *Salmonella* spp., *Shigella* spp., or *Escherichia coli* O157:H7 (CDC, 2002).

Between 2000 and 2002, three outbreaks occurred in the United States, and they were associated with the consumption of Mexican cantaloupe melons. In the first outbreak, samples were shown to contain *S*. Poona; in total, 155 cases, 28 hospitalizations (18%), and two deaths were reported. *Salmonella* was confirmed by serotyping and PFGE (CDC, 2002). The second outbreak occurred between April and June 2000 where 47 salmonellosis cases were reported. The cases were, again, associated with cantaloupe consumption. The third outbreak was reported between April and May 2001, and 50 salmonellosis (*S*. Poona) cases were confirmed; 10 patients developed septicemia, and two deaths occurred. Another outbreak occurred between March and May, 2002, and 58 cases were confirmed with 10 hospitalizations (CDC, 2002). Although these three cases were associated with consumption of Mexican cantaloupe, studies conducted by the FDA with a sample of 115 cantaloupes cultivated in United States, showed the presence of *Salmonella* and *Shigella* in 2.6% and 0.9% of the melons, respectively (FDA, 2001b). Castillo and coworkers (2004) collected and analyzed 1735 samples, including cantaloupe melon, irrigation and surface waters from six farms and packinghouses in south Texas, and three in Colima México. A total of 1.8% resulted positive for *Salmonella* spp. However, the levels of contamination were similar in Mexican and US farms. The Mexican cantaloupe melon industry has not yet recovered even to 10% of the volumes exported before these events occurred.

In the spring of 1996, an outbreak occurred in the United States and Canada where a total of 1465 cases of cyclosporiasis were identified in 20 states and two Canadian provinces. Florida initiated an investigation, led by the Florida Department of Health, because the largest number of clusters occurred in this state. The investigation determined the size of the outbreak, identified the vehicle of transmission, and discovered more regarding the morbidity associated with cyclosporiasis. The researchers conducted a case-control study, looking at the clusters of cases associated with a common food item, and attempted to trace that food item back to its country of origin. It was found that the consumption of raspberries was strongly associated with cyclosporiasis, and that Guatemalan raspberries were the source of the cyclosporiasis outbreak. This conclusion was supported by information from 19 other states as well (Calvin et al., 2003).

The occurrence of a second and similar outbreak, described by Herwaldt and colleagues (1999), prompted another look at this foodborne illness and what must be done to prevent it. They confirmed Guatemalan raspberries as the vehicle for *Cyclospora cayetanensis*. Many of their findings are similar to those reported in the 1996 outbreak investigation. One notable finding is that case exposures generally consisted of only a few raspberries, but the median attack rate among persons who ate raspberries was 91.7%. This suggests a very low infectious dose for *C. cayetanensis* and relatively uniform contamination of the implicated raspberry lots. It is unlikely that such contamination of raspberries would result from contact by an infected worker; rather, it seems more likely that an environmental reservoir was responsible. Contaminated water used for irrigation or pesticide spraying continues to be an important consideration. Contamination of raspberries through exposure to bird or insect droppings on packing material stored on open contaminated space also remains a possibility.

It is clear that the control measures instituted after the 1996 outbreak were inadequate, since importation of fresh Guatemalan raspberries into Canada in the spring of 1998 caused another outbreak (CDC, 1998).

In the fall of 2003, large outbreaks of hepatitis A in the United States (Tennessee, North Carolina, and Georgia) were associated with consumption of raw or undercooked green onions from México. The source of the green onions associated with the outbreak in North Carolina was never determined by the FDA. Between October and early November, before the FDA's first announcement regarding contaminated green onions, another very large outbreak of hepatitis A occurred in Pennsylvania among diners at one restaurant. Over 500 people contracted hepatitis A, and three died. Later, the FDA announced that this outbreak was also associated with green onions from México and named the four firms that grew the product associated with the outbreak. Before the 2003 outbreaks of hepatitis A in the United States, many growers in México already used third-party certification for GAPs and GMPs. Despite survey results suggesting that most growers have an interest in food safety, a lack of concern by only a few growers can affect the entire industry (Calvin et al., 2004).

Recently, Mexican tomatoes and hot peppers (serrano and jalapeño type) were associated with the largest produce-related outbreak in US history. With over 1400 cases and 286 hospitalizations in most of the US states, this incident dealt an economic blow to the produce industry in both countries, and challenged the consumers' confidence in the safety of the food supply vegetables. *Salmonella* Saintpaul, the causative agent of the outbreak, was isolated from serrano and jalapeño pepper samples from two packinghouses in Tamaulipas, México, after several unproductive weeks of inspecting and sampling in Mexican tomato packinghouse operations. To date, the mechanism by which these vegetables were contaminated has not yet been determined (CDC, 2008).

CONCLUSIONS

Produce safety in México and Central America requires continued insights and recommendations for achieving desired outcomes including implementation strategies and pilot activities. Private and government agencies need to identify mechanisms and the next steps for the way forward with appropriate partners in the development and implementation of a Good Agriculture Practice approach. Having Good Agricultural and Management practices in place ensures that the process is working correctly. It is important to understand where products are coming from and where they are going. It requires development, implementation, and verification of specifications. Collaborations are needed throughout the entire food chain from farm-to-fork. Other important considerations are recognition of the power of risk perception, and the understanding that there is no zero risk, but that risk assessments and evaluations are needed. Additionally, the need for preparedness planning before a crisis occurs, the need for auditing of suppliers, and the proper use of documents such as letters of guarantee and traceability are important matters that must be considered.

REFERENCES

Ackers, M., Pagaduan, R., Hart, G. et al. (1997). Cholera and sliced fruit: Probably secondary transmission from an asymptomatic carrier in the United States. *Int. J. Infect. Dis.* **1,** 212–214.

Amarilla, A., Espindola, E. E., Galeano, M. E. et al. (2007). Rotavirus infection in the Paraguayan population from 2004 to 2005: High incidence of rotavirus strains with short electropherotype in children and adults. *Med. Sci. Monit.* **13,** 333–337.

Bartz, J. A. (1982). Infiltration of tomatoes immersed at different temperatures to different depths in suspensions of *Erwinia carotovora* subsp. *Carotovora*. *Plant Dis.* **66,** 302–306.

Bartz, J. A. and Showalter, R. K. (1981). Infiltration of tomatoes by aqueous bacterial suspensions. *Phytopathology* **71,** 515–518.

Beuchat, L. R., Ryu, J. H. (1997). Produce handling a processing practice. *Emerg. Infect. Dis.* **3,** 459–465.

Calvin, L., Flores, L., and Foster, W. (2003). Case study: Guatemalan raspberries and *Cyclospora*. *Food Safety in Food Security and Food Trade*. http://ageconsearch.umn.edu/bitstream/123456789/19445/1/fo031007.pdf

Calvin, L., Avendano, B., and Schwentesius, R. (2004). The economics of food safety: The case of green onions and hepatitis A outbreaks. Electronic Outlook Report from the Economic Research Service. www.ers.usda.gov/publications/vgs/nov04/VGS30501/VGS30501

Castillo, A., Mercado, I., Lucia, L. M. et al. (2004). *Salmonella* contamination during production of cantaloupe: A binational study. *J. Food Prot.* **67**(4), 713–720.

Cazarez-Diarte, J. G. (2004). Presencia y supervivencia de coliformes fecales, *Salmonella* spp y *Listeria* spp. en agua de uso agrícola del Valle de Culiacán. Master's Degree Thesis. Centro de Investigación en Alimentación y Desarrollo, A. C. Unidad Culiacán.

CDC. (1991). Multistate outbreaks of *Salmonella* Poona infections—United States and Canada, 1991. *J. Amer. Med. Assoc.* **266**(9), 1189–1190.

CDC. (1998). Outbreak of cyclosporiasis—Ontario, Canada, May 1998. *MMWR* **47**, 806–809.

CDC. (2002). Multistate outbreaks of *Salmonella* serotype Poona infections associated with eating cantaloupe from México—United States and Canada, 2000–2002. *Morb. Mortal. Wkly. Rep.* **51**(46), 1044–1047.

CDC. (2008). Outbreak of *Salmonella* serotype Saintpaul infections associated with multiple raw produce items—United States, 2008. *Morbidity Mortality Weekly Report* **57**, 929–934.

Chaidez, C., Moreno, M., Rubio, W. et al. (2003). Comparison of the disinfection efficacy of chlorine-based products for inactivation of viral indicators and pathogenic bacteria in produce wash water. *International Journal of Environmental Health Research.* **13**, 295–302.

Fankhauser, R. L., Monroe, S. S., Noel, J. S. et al. (2002). Epidemiologic and molecular trends of "Norwalk-like viruses" associated with outbreaks of gastroenteritis in the United States. *J. Infect. Dis.* **186**, 1–7.

FAS/USDA. (2008). US trade imports—Foreign Agricultural Services. www.fas.usda.gov/ustrade/USTImBICO.asp?QI=

FDA. (2001a). FDA survey of imported fresh produce FY 1999 field assignment. www.cfsan.fda.gov/~dms/prodsur6.html.

FDA. (2001b). Survey of domestic fresh produce: Interim results. www.cfsan.fda.gov/~dms/prodsur9.html

Gerba, C. P. and Bitton, G. (1984). Microbial pollutants: Their survival and transport pattern to groundwater. In *Groundwater pollution microbiology* (G. Bitton and C. P. Gerba, Eds.), pp. 39–54. Wiley, New York.

Herwaldt, B. L. and Beach, M. J. (1999). The return of *Cyclospora* in 1997: Another outbreak of cyclosporiasis in North America associated with imported raspberries. The *Cyclospora* Working Group. *Ann. Intern. Med.* **130**, 210–220.

Jiménez, E. M., Siller, J. H., Valdez, J. B., Carrillo, A., and Chaidez, C. (2007). Bidirectional *Salmonella enterica* serovar Typhimurium transfer between bare/glove hands and green bell pepper and its interruption. *International Journal of Environmental Health Research.* **17**, 381–388.

Karch, E., Loftis, D. (1998). Disinfection contact time and kinetics. Environmental Information Management Civil Engineering Dept., Virginia Tech, pp. 1–4.

LeChevallier, M. W., Cawthon, C. D., Lee, R. G. (1988). Factors promoting survival of bacteria in chlorinated water supplies. *Appl. Envir. Microbiol.* **54**, 649–654.

Long, K. Z., Garcia, C., Santos, J. I. et al. (2007). Vitamin A supplementation has divergent effects on norovirus infections and clinical symptoms among Mexican children. *The Journal of Infectious Diseases* **196**, 978–985.

Lopez-Cuevas, O. (2005). Resistencia antimicrobiana de *Escherichia coli* y serotipos de *Salmonella* aisladas de agua y suelo de uso agrícola. Master's Degree Thesis.

Centro de Investigacion en Alimentacion y Desarrollo, A.C. Unidad Culiacan. Culiacan, Sinaloa, Mexico.

Mota, A. (2004). Risk assessment from *Giarda* and *Cryptosporidium* in irrigation water from Culiacan River. University of Texas at El Paso. Master Thesis.

Mazari-Hiriart, M., Ponce-de-Leon, S., Lopez-Vidal, Y. et al. (2008). Microbiological implications of periurban agriculture and water reuse in Mexico city. *Plos one.* **3,** 1–8.

Mohle-Boetani, J., Reporter, R., Werner, S. B. et al. (1999). An outbreak of *Salmonella* serogroup Saphra due to cantaloupe from Mexico. *J. Infect. Dis.* **180,** 1361–1364.

National Committee on the Microbiological Criteria for Foods (NACMCF). (1998). Microbiological safety evaluations and recommendations on fresh produce. *Food Control* **9,** 321–347.

Okhuysen, P. C., White, A. C. (1999). Parasitic infections of the intestines. *Current Opinion in Infectious Disease* **12,** 467–472.

Payment, P. (1989). Presence of human and animal viruses in surface and ground water. *Water Sci Technol.* **21,** 283–285.

Paziewska, A., Bednarska, M., Niewęgłowski, H. et al. (2007). Distribution of *Cryptosporidium* and *Giardia* spp. in selected species of protected and game mammals from north-eastern Poland. *Ann. Agric. Environ. Med.* **14,** 265–270.

Ridgway, H. F. and Olson, B. H. (1982). Chlorine resistance patterns of bacteria from two drinking water distribution systems. *Appl. Environ. Microbiol.* **44,** 972–987.

Sadovski, A., Fattal, Y. B., and Goldberg, D. (1978). Microbial contamination of vegetables irrigated with sewage effluent by the drip method. *J. Food Prot.* **41,** 336–340.

SEMARNAT (Secretaria de Medio Ambiente, Recursos Naturales y Pesca). NORMA Oficial Mexicana NOM-001-ECOL-1996. Límites máximos permisibles de contaminantes en las descargas de aguas residuales en aguas y bienes nacionales.

Sewell, A. M. and Farber, J. M. (2001). Foodborne outbreaks in Canada linked to produce. *J. Food Prot.* **64**(11), 1863–1877.

Secretaria de Salud y Asistencia SSA. (2005). Principales causas de mortalidad en edad pre-escolar (1 a 4 años) 2000–2005.

Showalter, R. K. (1979). Postharvest water intake by tomatoes. *Hort. Sci.* **14**(2), 125.

Siller, C. J. (1999). Importancia económica de la horticultura en México. Proceedings of the 1st International Congress and Exhibition of Horticulture. Mazatlán, Sin., Mexico. 3–11.

Siller-Cepeda, J. H., Baez-Sanudo, M. A., Sanudo-Barajas, A., and Baez Sanudo, R. (2002). Manual de Buenas Prácticas Agrícolas. Guía para el Agricultor. Centro de Investigación en Alimentación y Desarrollo, A.C. SAGARPA. 1ra (ed).

Siller-Cepeda, J. H., Baez-Sanudo, M., Chaidez-Quiroz, C., and Gardea Bejar, A. (2007). Produccion poscosecha en la industria fruticola. In *Buenas Practicas en la Produccion de alimentos.* Trillas. 1st ed.,171–221.

Solomon, E. B., Yaron, S., Matthews, K. R. (2002). Transmission of *Escherichia coli* O157:H7 from contaminated manure and irrigation water to lettuce plant tissue and its subsequent internalization. *Appl. Environ. Microbiol.* **68,** 397–400.

Talal, A. H., Moe, C. L., Lima, A. A. et al. (2000). Seroprevalence and seroincidence of Norwalk-like virus infection among Brazilian infants and children. *J. Med. Virol.* **61,** 117–124.

Tyrrel, S. F., Quinton, J. N. (2003). Overland flow transport of pathogens from agricultural land receiving faecal wastes. *Applied Microbiology* **94,** 87–93.

Zhuang, R. Y., Beuchat, L. R., and Angulo, F. J. (1995). Fate of *Salmonella* Montevideo on and in raw tomatoes as affected by temperature and treatment with chlorine. *Appl. Environ. Microbiol.* **61**(6), 2127–2131.

CHAPTER 14

Regulatory Issues in Europe Regarding Fresh Fruit and Vegetable Safety

Gro S. Johannessen and Kofitsyo S. Cudjoe

Section for Food Bacteriology and GMO, National Veterinary Institute, Oslo, Norway

CHAPTER CONTENTS

Introduction	332
The European Union	333
Basic Facts	333
The European Free Trade Association (EFTA) and The European Economic Area (EEA)	334
European Fruit and Vegetable Production	334
Fresh Produce Contamination Problems in Europe	335
Foodborne Bacteria	336
Parasites	337
Foodborne Human Pathogenic Virus	337
Molds and Mycotoxins	338
European Regulations	339
EU Central Regulations	339
European Food Safety Authority (EFSA)	340
Rapid Alert System for Food and Feed	340
Hygiene and Control Rules	342
GAP and Quality Assurance in Fruit and Vegetable Production	344
Import from Countries Outside EU and EEA (third countries)	345
GlobalGAP (formerly EurepGAP)	346
Differences from US Regulations	346
Funding of Food-Safety Research in Europe	348
Sources for Further Information	349
Acknowledgements	349

INTRODUCTION

In the last decade there appears to have been an increase in the number of outbreaks of foodborne disease associated with fruits and vegetables (Doyle and Erickson, 2008). Several factors may have influenced the notion of these apparent increases in numbers. In outbreak situations there has been an increased awareness that fresh produce could be the cause of disease. An increased consumption of fruits and vegetables due to the expanded import and export trade has led to an increase in choices available to consumers. Due to improved technology, fresh produce can now be transported over long distances within a short period of time, thus leading to import of products from areas far away from the consumption centers. Furthermore, the trendy habit of demanding fresh and "natural" products may also be an important factor.

In 2002, the Scientific Committee on Food in the European Union (EU) published a risk profile on the microbiological contamination of fruits and vegetables eaten raw (the Scientific Committee on Food, 2002). The report concludes that "the most efficient way to improve safety of fruits and vegetables is to rely on a proactive system of reducing risk factors during production and handling. Apart from washing, other methods of decontamination seem to have a limited influence on safety." The report recommends among other things that a more robust traceability system would improve epidemiological investigation of suspected foodborne illness, that there is a need for production measures for fruits and vegetables based on Good Hygienic Practice (GHP), Good Agricultural Practice (GAP), and Hazard Analysis of Critical Control Points (HACCP), and that water and organic fertilizers should be of such quality that they do not contaminate the products with harmful microorganisms.

Early in 2008, the FAO/WHO published a report on microbiological hazards in fresh fruits and vegetables (FAO/WHO, 2008). This report gives leafy green vegetables (including fresh herbs) highest priority as commodities of global concern. This ranking was due to the large volume of production and export, the fact that this product type has been associated with numerous outbreaks with different agents, and that the growing and processing are highly complex. A level 2 priority was given to berries, green onions, melons, sprouted seeds, and tomatoes; level 3 priority was given to a large group, comprising carrots, cucumbers, almonds, baby corn, sesame seeds, onion and garlic, mango, paw paw, and celery.

Although Europe has a large production of fruits and vegetables, the demand for fresh fruits and vegetables during all seasons necessitates import from other areas. This is perhaps particularly important in the northern areas where the growing season is rather short, and greenhouse production is not sufficient to meet all consumer requirements.

This chapter is limited mainly to regulatory issues in Europe regarding fresh fruits and vegetables, but we will also discuss some dried products, especially dried fruits due to problems with mycotoxins. Grain crops will not be discussed. We will also focus on microbiological issues of human concern in fresh produce. In this context mycotoxins are included. We will not discuss issues regarding plant health, which is out of the scope for this chapter.

THE EUROPEAN UNION

Basic Facts

The European Union (EU) began its activities as the European Coal and Steel Union (ECSC) in 1951. The six nations of Belgium, France, Germany, Italy, Luxembourg, and the Netherlands signed the Paris treaty, which entered into force in 1952. In 1953 the first Common Market for coal and iron was set into place. The treaty of Rome, establishing among others the European Economic Community, was signed in 1957 and entered into force on January 1, 1958. On July 1, 1968, the six member states removed customs duties on goods imported from each other, allowing free cross-border trade for the first time. On January 1, 1973, the six member states became nine when Denmark, Ireland, and the United Kingdom formally entered the EU. The expansion of the EU has continued, and the EU currently comprises 27 member states (Table 14.1). Three candidate countries

Table 14.1 Member States in the European Union (October 3, 2008)

Member State (Year of Joining)	
Austria (1995)	Latvia (2004)
Belgium (ECSC,1951)	Lithuania (2004)
Bulgaria (2007)	Luxembourg (ECSC, 1951)
Cyprus (2004)	Malta (2004)
Czech Republic (2004)	The Netherlands (ECSC, 1951)
Denmark (1973)	Poland (2004)
Estonia (2004)	Portugal (1986)
Finland (1995)	Romania (2007)
France (ECSC, 1951)	Slovakia (2004)
Germany (ECSC, 1951)	Slovenia (2004)
Greece (1981)	Spain (1986)
Hungary	Sweden (1995)
Ireland (1973)	United Kingdom (1973)
Italy (ECSC, 1951)	

of Croatia, the Former Yugoslav Republic of Macedonia, and Turkey are awaiting approval of their membership application.

The European Free Trade Association (EFTA) and The European Economic Area (EEA)

The EFTA was founded in 1960 by Austria, Denmark, Norway, Portugal, Sweden, Switzerland, and the United Kingdom. EFTA is an intergovernmental organization set up for the promotion of free trade and economic integration to the benefits of its member states. In 1992 the community, the member states, and the then seven members of EFTA negotiated and signed the agreement that created the European Economic Area (EEA). Currently, the EFTA countries are Iceland, Norway, Liechtenstein, and Switzerland, and these countries, with the exception of Switzerland, are also part of the EEA. Today EFTA maintains the management of the EFTA Convention (intra-EFTA trade), the EEA Agreement (EFTA-EU relations), and the EFTA Free Trade Agreements (third-country relations).

Although Switzerland decided not to take part in the EEA, the Agreement was maintained because the remaining countries wished to take part in EU's Internal Market, while not assuming the full responsibilities of a membership. The EEA countries have the right to be consulted during formulation of community legislation, but they do not have the right to a voice in the decision-making. All the new community legislation that is covered in the EEA is integrated into the Agreement through an EEA Joint Committee decision and subsequently becomes part of the national legislation of the EEA states. The function of the EEA Joint Committee is to adopt decisions extending community regulation and directives to the EEA states. In this process of adopting community legislation, the EEA states have to speak with one voice. As a result of the EEA Agreement, EU legislation, such as the Food Law, will also enter into force in the EEA states, and the same regulation with respect to import from third countries (countries outside EU and the EEA) applies to the EEA states.

European Fruit and Vegetable Production

The average production of fruits and vegetables per year in 2003 to 2005 in the EU was 38.3 and 66 million tons, respectively (Anon., 2006). The fruit and vegetable sector is responsible for 17% of total EU agricultural production in terms of value (Anon., 2007). Typical fruit and vegetable farms in the member states are rather small, on average less than 10 hectares (Martinez-Palou and Rhoner-Thielen, 2008). The production of fruits and vegetables tends to be concentrated in only a few member states. Italy and Spain have

Table 14.2 Noncommunity Countries with Largest Import (Tons) of Fruits and Vegetables into EU-27 in 2000 and 2007 (from Martinez-Palou and Rhoner-Thielen, 2008)

Country	Import (% of Total) 2000	Import (% of Total) 2007
Thailand	20.3	7.4
Canada	6.3	—*
Turkey	6.0	7.3
Ecuador	5.9	6.2
Brazil	5.7	6.2
South Africa	5.2	5.7
Costa Rica	5.0	8.2
United States	5.0	—
Morocco	3.9	4.5
China	—	6.8
Colombia	—	5.3
Others	36.7	42.4

*Not on top-9 list in this year.
(From Martinez-Palou and Rhoner-Thielen, 2008).

the largest total production, followed by France and Greece, the Netherlands, Germany, and Poland. Some member states are quite specialized and produce a large quantity of only one commodity. For example, Italy and Spain accounted for the majority of tomato and orange production in 2006, mostly due to favorable climatic conditions, whereas more than 50% of apple production took place in Poland, France, and Italy. The overall production of the main fruit and vegetable crops in the EU has been remarkably stable in the period from 1996 to 2006.

In 2007, EU-27 imported a total of 22.6 million tons of fruits and vegetables from third countries (countries outside EU and EEA) (Martinez-Palou and Rhoner-Thielen, 2008) (Table 14.2). This is an increase of 16.4% from 2000, and much of this growth is due to a marked increase in the import of fruits and nuts.

FRESH PRODUCE CONTAMINATION PROBLEMS IN EUROPE

As in all other parts of the developed world, Europe has also encountered problems and outbreaks with respect to microbiological contamination of fruits and vegetables. There have been large multinational outbreaks and small national outbreaks involving numerous different commodities, thus reflecting the variety and complexity of production and distribution systems.

It is worthy to note that differences in climatic conditions makes southern Europe self-sufficient, while northern Europe has to import fruits and vegetables to meet their needs.

Foodborne Bacteria

One of the first recognized multinational outbreaks associated with fresh vegetables in Europe was in 1994 when an increasing number of domestic cases of *Shigella sonnei* was observed in May and June (Kapperud et al., 1995). A similar increase was also observed independently in additional European countries, suggesting a multinational outbreak. Epidemiological investigations in Norway, Sweden, and the United Kingdom incriminated Iceberg lettuce imported from Spain as the source. Although *Shigella sonnei* was never isolated from lettuce, and the traceback in Spain was difficult, microbiological analyses of the lettuce from patients' homes revealed a heavy fecal contamination, giving further evidence that the imported lettuce was the actual source. Since then several outbreaks have occurred involving produce imported from third countries (i.e., countries outside the EU) and fruits and vegetables cultivated and sold within EU/EEA.

Several outbreaks of foodborne disease associated with fresh produce were reported in Eurosurveillance in 2007. It seems to be a trend that more outbreaks are multinational. This is probably a result of the worldwide trade and transport of foods. The investigation of an increase in cases of *Salmonella* Senftenberg in the United Kingdom uncovered an international outbreak with cases in Denmark, the Netherlands, and the United States (Pezzoli et al., 2007). *Salmonella* Senftenberg with similar PFGE-types and plasmid profiles were isolated from both the patients and fresh basil. The fresh basil was analyzed as part of a nationwide survey of fresh herbs on retail sale and imported from Israel.

In November 2007, Eurosurveillance, which publishes peer-reviewed information on communicable diseases from a European perspective, published information on an outbreak of *Salmonella* Weltewreden infections in Norway, Denmark, and Sweden associated with alfalfa sprouts (Emberland et al., 2007). It was concluded from this outbreak investigation that alfalfa sprouts grown from contaminated seeds were the source of the outbreak. The seeds used in Denmark and Norway were part of the same batch and were traced, according to invoices, to retailers in Germany and the Netherlands. The seeds used in Finland came from the same Dutch supplier, but were not part of the same batch. Further investigations have shown that the seeds originated from Pakistan (RASFF 2007.0760, RASFF 2007.0760-add01, -add02 and –add03). This outbreak shows that the import and trade routes

Table 14.3 Selected Outbreaks of Foodborne Infections Associated with Fresh Fruits and Vegetables Imported from Third Countries

Country	Food	Agent	Exporting Country if Known	Reference
Denmark, Australia	Baby corn	*Shigella sonnei*	Thailand	(Lewis et al., 2007a, 2007b; Stafford et al., 2007)
Sweden	Lime leaves	*Salmonella*	Thailand	www.slv.se
Sweden	Almonds	*Salmonella* Enteritidis	Probably California	(Ledet et al., 2007)

can be rather complicated and hard to follow. However, European legislation on tracing requirements has proven to be effective and important in ensuring European food safety.

For other outbreaks with noncommunity or third-country sources, see Table 14.3.

Parasites

Fruits and vegetables that are eaten raw and without peeling have been demonstrated to harbor a range of protozoan parasites such as *Giardia*, *Cryptosporidium*, *Cyclospora*, and the helminth parasite *Ascaris*. Some of these organisms—particularly, the protozoan parasites—have caused infections characterized by prolonged diarrhea (Dawson, 2005; Döller et al., 2002; Hoang et al., 2005). These parasites are essentially derived from waste water reuse and are therefore of primary public health concern. Outbreaks involving *Cryptosporidium* have been reported in the United States in unpasteurized apple juice in Maine (1993) and New York (1996) and unwashed salad onions in Washington (1997) (Anon., 1996, 1998; Millard et al., 1994). Outbreaks involving *Giardia* in the United States have been reported in a fruit salad at a party in New Jersey (1986) (Rose and Slifko, 1999). *Cyclospora* outbreaks have been associated with raspberries, mixed baby lettuce, basil, and salads in diverse states in the United States, Canada, and Germany (Brockmann et al., 2001; Chalmers et al., 2000; Herwaldt, 2000).

Foodborne Human Pathogenic Virus

Viruses cannot replicate in or on foods but might sometimes be present on fresh produce as a result of fecal contamination. This contamination can originate in the growing and harvesting area from contact with polluted water and inadequately or untreated sewage sludge used for irrigation and fertilization. Alternatively, fruits or vegetables handled by an infected person might become contaminated with a virus and transmit infection. The most frequently reported foodborne viral infections are viral gastroenteritis and hepatitis A. Several epidemiological studies have associated viral hepatitis A infections with the consumption of fecally contaminated raw vegetables or drinking water (Gaulin et al., 1999; Hernandez et al., 1997; Kuritsky et al., 1985; Long et al., 2002; Nygard et al., 2001; Rosenblum et al., 1990; Warner et al., 1991). Hepatitis A virus (HAV) and norovirus were most frequently associated with foodborne and waterborne outbreaks (Bidawid et al., 2000; Dubois et al., 2002; Leggitt and Jaykus, 2000; Sair et al., 2002; Ward et al., 1982).

Molds and Mycotoxins

The major problems of mold contamination of fruits and vegetables are economic with a significant loss of useful food materials (Moss, 2008). There are a few examples implicating a role for mycotoxins in the safety of fresh fruits and fruit juices. Undoubtedly, the most important is patulin, mainly produced by *Penicillium expansum*. Patulin is especially important in apple juice and apple products. Some members of the black-spored *Aspergillus niger* group, particularly *A. carbonarius*, may cause bunch rot of grapes, and ochratoxin has been detected in both fresh fruits and raisins as well as in wine and grape juice (Belli et al., 2004). The mold *Alternaria alternata* grows on a wide range of fruits and vegetables and is a major pathogen of fresh tomatoes, in which it can produce tenuazonic acid. According to Moss (2008), unlike patulin and ochratoxin A, there are no regulatory limits set for tenuzonic acid or other *Alternaria* metabolites, reflecting the lack of any evidence implicating them in human illness.

According to the Healthy Nut Initiative (1998), nuts should be classified as fruits, and in particular, as so-called shell fruits. This classification thus includes fruits with edible kernels contained in inedible shells. Nut consumption in Europe is on the rise. However, edible nuts, dried figs, and spices can be associated with aflatoxins and other mycotoxins. This contamination can be uneven and spasmodic.

In the United Kingdom, the Contaminants in Food (Amendment) Regulations (United Kingdom, 1999), which were made under sections of the Food Safety Act 1990, set limits for aflatoxins in groundnuts, nuts, dried fruits, and cereals. A higher limit is provided for groundnuts, nuts, and dried fruits intended for further processing before human consumption. The higher limit for these commodities recognizes that processing and sorting can reduce the levels of aflatoxin contamination in consignments below that of the lower limit. The limits are low and were set on the basis that they represent the lowest level technologically achievable, consistent with meeting food-safety objectives. The regulations are targeted at those products that surveillance has indicated may be most highly or most frequently contaminated with aflatoxin. These make the greatest contributions to consumer exposure, and controls targeted at these products are the most effective way of reducing exposure.

EUROPEAN REGULATIONS

In January 2000, the Commission of European Communities launched its White Paper on Food Safety (The European Commission, 2000), in which a radical new approach to food safety was proposed. This process was driven by the need to ensure that food safety was and is a priority of the EU. The key points in the White Paper were the establishment of an independent European Food Authority and the set up of a new legal framework covering the whole food chain, including animal feed production. Focus was also directed at food-safety controls, especially controls of imports at the borders of the community, consumer information, and an international dimension with respect to an effective presentation of the actions to trading partners.

EU Central Regulations

On January 28 2002, the Food Law (Regulation (EC) No 178/2002) was passed in the EU, and became immediately applicable in the Member States (European Union, 2002). The Food law establishes common principles and responsibilities and the means to provide a strong science base, efficient organizational arrangements and procedures to underpin decision-making in matters of food and feed safety. It is stated in Article 1, paragraph 2 that "this regulation lays down the general principles governing food and feed in safety in particular, at Community and national level. This regulation applies to all stages of production, processing and distribution of food and feed, but shall not apply to primary production for private domestic use, domestic preparation, handling or storage of food for private domestic consumption."

Under this legislation, food operators shall not place unsafe food on the market. The key obligations of food business operators are:

- Operators are responsible for the safety of the food and feed that they produce, transport, store, or sell.

- Operators shall be able to rapidly identify any supplier or consignee.

- Operators shall immediately inform the competent authorities if they have a reason to believe that their food or feed is not safe.

- Operators shall immediately withdraw food or feed from the market if they have a reason to believe that it is not safe.

- Operators shall identify and regularly review the critical points in their processes and ensure that controls are applied at these points.

- Operators shall cooperate with the competent authorities in actions taken to reduce risks.

European Food Safety Authority (EFSA)

The European Food Safety Authority (EFSA) was set up in January 2002 as part of a program to improve food safety, ensure a high level of consumer protection, and restore and maintain confidence in the food supply within the EU (www.efsa.europa.eu). In Europe, risk assessments are done independently of the risk management, and EFSA's role is to assess and communicate all risks associated with the food chain. EFSA produces scientific opinions and advice in close cooperation and open consultation with national authorities and other stakeholders. These are important in providing a sound foundation for European policy and legislation making and in supporting the European Commission, European Parliament, and EU member states in taking effective and timely risk management decisions. EFSA consists of a Scientific Committee and Scientific Panels that are composed of highly qualified experts in risk assessment. The Food Law states that EFSA should cooperate closely with the competent bodies (i.e., national food safety authorities) in the member states if it is to operate effectively.

Rapid Alert System for Food and Feed

An important tool for the rapid exchange of information with respect to different contaminants of food and feed is the Rapid Alert System for Food and Feed (RASFF). The purpose of the RASFF is to provide the control authorities with an effective tool for exchange of information on measures taken to ensure food safety. The RASFF has been in operation since 1979. The legal basis for this rapid information exchange system is Article 50 in the Food Law. RASFF systematically informs countries outside the EU (third countries of origin) of notifications concerning products manufactured in, distributed to, or dispatched from these countries through the Commission delegates. Although a country may be mentioned as the origin of a product, that does not necessarily imply that the hazard originated in the country concerned. However, if serious problems are detected several times, a letter is sent to the competent authority of the country. The relevant country is then expected to take appropriate measures to rectify the situation. The member states may also intensify their import checks. In addition, the Food and Veterinary Office (FVO) uses the information provided by RASFF when prioritizing their inspection program. This can be illustrated by an outbreak of *Salmonella* Weltewreden infections associated with alfalfa sprouts in Norway, Denmark, and Sweden, where the origin was traced to Pakistan (see Foodborne bacteria).

The notifications are listed under three headings: Alert notification, which is sent when a food or feed presenting a serious risk is on the market and when immediate action is required; information notification, which is sent when a risk has been identified for food or feed where it is on the market, but the other members of the network do not need to take immediate actions; and border rejection, where food and feed consignments have been tested and rejected at the external borders of the EU and the EEA when a health risk was found.

The RASFF contact points in the member states receive alert notifications and additional information regarding the alert notifications via e-mail. If there are other special situations, notifications of these also are sent via e-mail. Lists of all alert, additional, and information notifications, border rejections, and news are distributed daily to the RASFF contact points. The RASFF contact points go through the messages, and if there is anything that is particularly interesting, further information can be collected from the CIRCA-database of the EU Commission. Weekly overviews of the RASFF notifications are posted on the RASFF Web site (http://ec.europa.eu/food/food/rapidalert/index_en.htm).

An example of how RASFF works is the situation with the occurrence of *Salmonella* in fresh herbs and leafy greens imported from Southeast Asia, in particular Thailand. In 2005 there were more notifications (87) to RASFF than in previous years concerning the presence of unwanted bacteria potentially pathogenic to humans in fresh herbs and leafy greens. The increase observed was probably due to sporadic findings in some countries, followed by more extensive surveillance. The majority of the notifications concerned fresh herbs imported to Europe from Thailand. In Norway, for example, the findings resulted in a temporary sales ban of such products from Thailand. As a consequence of this, the Thai authorities undertook an initiative and introduced an Action Plan where among other things 14 "risk products," later increased to 23, were to be followed by a certificate documenting that the products had been tested for *E. coli* and *Salmonella* before export to the

EU and Norway. There are two types of certificates; one from the Thai Agricultural Department, which, in addition to the bacteriological analyses, also shows that the producer is certified in a quality assurance system ensuring hygienic production. The second type of certificate is from non-certified producers; however, there is still a requirement that the "risk products" have been analyzed for *E. coli* and *Salmonella*.

In 2006 and 2007 the notifications of unwanted bacteria in these types of products decreased again to a level more similar to previous years. As a consequence of the recurrent problem with bacterial contamination, EU's Food and Veterinary Office (FVO) carried out a mission to Thailand in September 2007 in order to "assess the official control system in place to prevent microbiological contamination in fresh herbs and spices intended for export to the European Union" (Food and Veterinary Office, 2007). The results from a small surveillance study in Norway on fresh herbs and leafy greens imported from Southeast Asia indicated that the introduction of the certificate of analysis reduced the risk of contaminated fresh herbs and leafy greens being exported to Europe (Mattilsynet, 2008; Norwegian Scientific Committee for Food Safety, 2008). This shows that findings of unwanted agents in Europe can improve the production system in other countries.

Hygiene and Control Rules

In April 2004 new hygiene rules were adopted, and these became applicable in the member states on January 1, 2006. These hygiene rules comprise three regulations and one directive. The most important for the fruit and vegetable chain is Regulation (EC) No 852/2004 of the European Parliament and of the Council of 29 April 2004 on the hygiene of food stuffs (European Union, 2004a). The other two regulations and the directive, Regulation (EC) No 853/2004, Regulation (EC) No 854/2004, and Directive 2004/41/EC, are more concerned with food of animal origin and are thus not that important in this context. The hygiene rules particularly focus on the following points:

- That the food business operator has primary responsibility for food safety
- That food safety is ensured throughout the food chain starting with primary production
- That general procedures based on HACCP principles must be implemented
- That basic common hygiene requirements must be applied

- That certain food establishments must be registered or approved

- That guides to good practice for hygiene or for the application of HACCP principles should be developed as valuable instruments to aid food business operators at all levels of the food chain to comply with the new rules

- That flexibility is provided for food produced in remote areas and for traditional products and methods

The control rule (Regulation (EC) No 882/2004 of the European Parliament and of the Council on official controls, performed to ensure the verification of compliance with feed and food law, animal health, and animal welfare rules) were also adopted at the same time as the hygiene rules (European Union, 2004b). This regulation states that member states shall ensure that official controls are carried out regularly by the competent authority, on a risk basis and with appropriate frequency. The compliance can be verified in several ways, such as by audits, inspections, monitoring, surveillance, sampling, and testing. The controls shall be carried out at any stage of production, processing and distribution and may be carried out without prior warning except when prior notification is necessary. The competent authorities are also responsible for regular official control on food of nonanimal origin that are imported from third countries.

Another important regulation is Commission Regulation (EC) No 2073/2005 of 15 November on microbiological criteria for foodstuffs (European Union, 2005). This regulation is directed at food operators and gives food-safety criteria and process-hygiene criteria. A food-safety criterion means a "criterion defining the acceptability of a product or a batch of foodstuff applicable to products placed on the market." Food-safety criteria for fresh produce is absence of *Salmonella* in five 25 g samples of sprouted seeds, precut ready-to-eat fruits and vegetables, and unpasteurized fruit and vegetable juices. For the seeded sprouts, samples should be collected either of the seeds prior to sprouting or in the production process when the possibility is greatest to detect *Salmonella*. A process hygiene criterion means "a criterion indicating the acceptable functioning of the production process. Such a criterion is not applicable to products placed on the market. It sets an indicative contamination value above which corrective actions are required in order to maintain the hygiene of the process in compliance with food law." The process hygiene criteria are shown in Table 14.4.

This means that if all sample units have values less than m, the results are satisfactory, if maximum c/n sample units have numbers between m and M and the rest of the sample units are less than M, the results are

Table 14.4 Process Hygiene Criteria

Product	Microorganism	Sampling Plan*		Criteria		When in Process
		n	c	m	M	
Precut ready-to-eat fruits and vegetables	E. coli	5	2	100 CFU/g	1000 CFU/g	During production
Unpasteurized fruit and vegetable juice	E. coli	5	2	100 CFU/g	1000 CFU/g	During production

*n = number of units making up the sample, c = number of units with results that can be between m and M.

acceptable, or they are unsatisfactory if one of the values is greater than M or more than c/n values are between m and M.

In addition, the EU has set limits for maximum levels for certain contaminants in foodstuffs (European Commission, 2002, 2004). These two regulations concern mycotoxins, in particular aflatoxins, ochratoxin A, and patulin in product types such as spices, dried fruits, and apples.

On the basis that apple juice is so widely consumed, the EU set a limit for patulin of 50 µg/kg for all fruit juices, 25 µg/kg for solid apple produce used for direct consumption, and 10 µg/kg in apple juice and apple products for babies and young children (European Commission, 2004).

For example, the EU has established a maximum tolerance for aflatoxin in almonds shipped to its member countries. Handlers who choose to ship almonds to the EU must comply with EU specifications. However, in the United States, there are no mandatory requirements pertaining to aflatoxin. The absence of official, specific outgoing quality requirements for shipments to the EU has forced the hands of the almond industry to develop their own voluntary aflatoxin testing protocol for handlers to follow when shipping almonds to the EU.

GAP and Quality Assurance in Fruit and Vegetable Production

In the EU, primary production, that is, "production, rearing or growing of plant products such as grains, fruits, vegetables and herbs as well as their transport within and storage and handling of products (without substantially changing their nature) at the farm and their further transport to an establishment", is covered by the regulation no. 852/2004 on the hygiene of food stuffs (Anon., 2005). This means that HACCP-based procedures should be implemented.

However, the regulation does not apply to small quantities of primary products. This means that farmers are allowed to sell their products directly to the consumer, to local retail shops for direct sale to the consumer, and to local restaurants. It is up to the member states to define "small quantities" depending on the local situation. The member states must also lay down national rules in order to ensure that the safety of such foods is guaranteed.

According to the 2003 Common Agricultural Policy (CAP) reform, all farmers receiving direct payment from the EU must respect "cross compliance"; that is, farmers must comply with all legislation affecting their business (Anon., 2003). This was made compulsory in 2005 (Anon., 2008a). This means that those who receive direct payment are obliged to keep land in good agricultural and environmental condition. The cross-compliance concept links direct payments to the farmers to their respect of, among other things, environmental requirements at both EU and national levels. Since protection of soil and water from pollution and contamination is imperative, this will also be a positive additional factor for the production of safe fruits and vegetables. The reason for this is that by complying with the agricultural and environmental legislation, both on a EU and national level, the pollution and contamination of soil and water, at least theoretically, will be reduced, thus resulting in, for example, cleaner water that may be used for irrigation.

By respecting cross compliance (agricultural legislation) and implementing and maintaining HACCP procedures (food legislation), complemented by national legislation in the respective fields, a sound basis for the production of safe fresh fruits and vegetables is laid.

Import from Countries Outside EU and EEA (third countries)

In the Food Law (178/2002) it is stated in Article 11 that food imported into the community "shall comply with the relevant requirements of food law or conditions recognised by the Community to be at least equivalent thereto or, where a specific agreement exists between the Community and the exporting country, with requirements contained therein." There is also a demand for traceability; however, the regulations do not have an extraterritorial effect outside the EU. That means that the requirements extend from the importer to the retailer. There are food business's contractual arrangements that exist. The responsibility is placed on the food business operator, and in the case of import, this is the importer. It is important to be aware that food business operators in third countries need to respect the relevant requirements with regard to the hygiene of food as stated in article 3–6 of Regulation (EC)852/2004 (European Union, 2004a). This means that there is a general obligation to monitor food safety of products and processes under operator's responsibility, that there needs to be general hygiene provisions for primary production, that there are detailed requirements after primary production, and that for certain products there are microbiological requirements. It is the responsibility of the importer to ensure compliance with the requirements.

For import of foods of animal origin, registration of food businesses is necessary. This is different for the plant food. Here it is usually sufficient that exporting companies in third countries are known and accepted as suppliers by importers in the EU.

Further, according to the EU rules on food hygiene, food business operators in third countries intending to export foodstuffs into the EU must put in place, implement, and maintain procedures based on the HACCP-principles after primary production (Anon., 2008b). The Commission is responsible for requesting third countries to provide accurate and up-to-date information on the general organization and management of sanitary control systems. The contact point for the Commission in third countries is the competent authority. It is also noticeable that it is incumbent upon the importer to ensure compliance with the relevant requirements or food law or with conditions recognized as equivalent.

GlobalGAP (formerly EurepGAP)

The Good Agricultural Practice (GAP) concept is used in GlobalGAP (formerly known as EurepGAP), which is a private-sector body that sets voluntary standards for the certification of agricultural products around the world (www.globalgap.org). GlobalGAP serves as a practical manual of GAP anywhere in the world. Members of GlobalGAP are both retailers and producers, and the standard is a "pre-farm-gate" standard that covers the production of a certified product from farm inputs to when the product leaves the farm. The GlobalGAP is a single integrated standard with modular applications for different product groups; for example, fruits and vegetables have their own specific module (www.globalgap.org/cms/front_content.phpidcat=3). In the control point and compliance criteria for fruits and vegetables, there are specific points with respect to microbiological quality of irrigation water and hygiene risk analysis at several points during the process, which also include worker hygiene.

DIFFERENCES FROM US REGULATIONS

In a guidance for industry to minimize microbial food safety hazards from the field through distribution of fresh fruits and vegetables, the US Food and Drug Administration (FDA) encouraged compliance with the principle that all applicable local, state, and federal laws and regulations, or corresponding or similar laws, regulations, or standards for operators outside the United States, should be followed for agricultural practices FDA (1998). It was hoped that by identifying basic principles of microbial food safety within the realm of growing, harvesting, packing, and transporting fresh produce, users of this guide would be better prepared to recognize and address the principal elements known to give rise to microbial food safety concerns. The main thrust of the document is based on these eight principles:

Principle 1. Prevention of microbial contamination of fresh produce is favored over reliance on corrective actions once contamination has occurred.

Principle 2. To minimize microbial food safety hazards in fresh produce, growers, packers, or shippers should use Good Agricultural and Management Practices in those areas over which they have control.

Principle 3. Fresh produce can become microbiologically contaminated at any point along the farm-to-table food chain. The major source of microbial contamination with fresh produce is associated with human or animal feces.

Principle 4. Whenever water comes in contact with produce, its source and quality dictates the potential for contamination. Minimize the potential of microbial contamination from water used with fresh fruits and vegetables.

Principle 5. Practices using animal manure or municipal biosolid wastes should be managed closely to minimize the potential for microbial contamination of fresh produce.

Principle 6. Worker hygiene and sanitation practices during production, harvesting, sorting, packing, and transport play a critical role in minimizing the potential for microbial contamination of fresh produce.

Principle 7. Follow all applicable local, state, and federal laws and regulations, or corresponding or similar laws, regulations, or standards for operators outside the United States, for agricultural practices.

Principle 8. Accountability at all levels of the agricultural environment (farm, packing facility, distribution center, and transport operation) is important to a successful food safety program. There must be qualified personnel and effective monitoring to ensure that all elements of the program function correctly and to help track produce back through the distribution channels to the producer.

The issuing of guidance is one major way that the FDA, the Federal Agency responsible for enforcement of food laws, regulates the food industry. For example, in October 1999, the FDA issued two guidance documents to enhance the safety of sprouts, a product that has been implicated in at least 1300 cases of foodborne illness. The guidance advises sprout producers and seed suppliers of steps they should take to reduce microbial hazards common to sprout production. A companion guide provides producers with the latest information about testing spent irrigation water, an important step to ensure the safety of sprouts. Consequently, the FDA then closely monitors the safety of sprouts and the adoption of prevention practices recommended in the guidance, and considers enforcement actions against producers who do not have preventive controls in place (www.cfsan.fda.gov/~lrd/hhsprout.html). Similar guidance documents were issued for the supply chains of melon (2005), fresh tomatoes (2006), and leafy greens (2006).

FUNDING OF FOOD-SAFETY RESEARCH IN EUROPE

The Frame Programmes (FP) have been the main financial tools through which EU supports research and development activities in almost all scientific disciplines, and it is estimated that 5 to 10% of research in the EU is financed through the FPs. The current FP (FP7) was fully operational on January 1, 2007 and will expire in 2013. FP7 is organized into four basic components of European Research, one of which is Cooperation, which is defined as collaboration between industry and academia in key technology areas. By this, European industry is invited to play an active part in European research.

International cooperation between the EU and third countries is an integral part of this action and is encouraged. This can be illustrated by the fact that organizations and researchers from more than 100 countries all over the world are involved in EU research programs. One of the themes identified in FP7 is "Food, agriculture and fisheries and biotechnology." The primary aim of funding in this theme is to build a knowledge-based bioeconomy (KBBE). The EU has earmarked more than 1.9 billion euros for funding of this theme over the duration of FP7, and work-programs for the theme are published annually.

Another organization supporting food-safety research in Europe is SAFEFOODERA, a multinational network with the primary objective of establishing a European platform for protecting consumers against health risks from the consumption of food (see www.safefoodera.net). SAFEFOODERA is a coordinated action ERA-NET of 15 member states, three associated countries, and two regional organizations. According to data collected by SAFEFOODERA, in the Netherlands approximately 25 million euros was allocated in 2008 for food-safety research, and in the United Kingdom, Norway, and Belgium, the figures were approximately 14.3 million, 5.5 million and 3.2 million euros, respectively.

Food-safety research also is financed through national research councils/agencies and other sources, but it is difficult to estimate the extent of this funding.

SOURCES FOR FURTHER INFORMATION

A lot of information on the EU, the common agricultural policy (CAP), and the European Food Law can be found on the following web pages. There is also useful information on the RASFF pages and statistics from the EU can be found on the web pages of Eurostat. Information on European research and development can be found on the Cordis Web site.

> http://ec.europa.eu/food/food/index_en.htm
> http://ec.europa.eu/agriculture/foodqual/index_en.htm
> www.efsa.europa.eu/EFSA/efsa_locale-1178620753812_home.htm
> http://ec.europa.eu/eurostat
> http://ec.europa.eu/agriculture/index_en.htm
> http://ec.europa.eu/external_relations/eea/
> http://cordis.europa.eu/home_en.html
> www.efta.int
> www.globalgap.org
> www.safefoodera.net

ACKNOWLEDGEMENTS

The authors would like to thank Knut G. Berdal, Ellen Christensen, and Johs. Kjosbakken for their contributions to this chapter.

REFERENCES

Anonymous. (1996). Apple cider outbreak in New York. *Government Epidemiology, Research, Business and Technology News: Cryptosporidium Capsule* **2**, 5.

Anonymous. (1998). Outbreak at a banquet in Spokane. *Government Epidemiology, Research, Business and Technology News: Cryptosporidium Capsule* **3**, 4–5.

Anonymous. (2003). Cross compliance.

Anonymous. (2005). Guidance document. Implementation of certain provisions of Regulation (EC) No 852/2004 on the hygiene of foodstuffs, pp. 1–16. Brussels: European Commission. Health & Consumer Protection Directorate-General.

Anonymous. (2006). Major features of the sector of fresh fruits and vegetables in the EU. http://ec.europa.eu/agriculture/capreform/fruitveg/presentations/fresh_en.pdf, accessed on July 15, 2008.

Anonymous. (2007). Fruit and vegetable reform. Brussels. http://europa.eu/rapid/pressReleaseAction.do?reference=MEMO07/28, accessed on July 16, 2008.

Anonymous. (2008a). Agriculture and the environment: Cross compliance. http://ec.europa.eu/agriculture/envir/index_en.htm, accessed on July 8, 2008.

Anonymous. (2008b). Guidance document. Key questions related to import requirements and the new rules on food hygiene and official controls. European Commission: Health & Consumer Protection Directorate General.

Belli, N., Ramos, A. J., Sanchis, V., and Marin, S. (2004). Incubation time and water activity effects on ochratoxin A production by Aspergillus section Nigri strains isolated from grapes. *Lett. Appl. Microbiol.* **38**, 72–77.

Bidawid, S., Farber, J. M., and Sattar, S. A. (2000). Rapid concentration and detection of hepatitis A virus from lettuce and strawberries. *J. Virol. Methods.* **88**, 175–185.

Brockmann, S., Dietrich, K., Döller, P. C., Dreweck, C., Filipp, N., Wagner-Wiening, C. et al. (2001). Cyclosporiasis in Germany. *Eurosurveillance Weekly* **5**.

Chalmers, R. M., Nichols, G., and Rooney, R. (2000). Foodborne outbreaks of cyclosporiasis have arisen in North America. Is the United Kingdom at risk? *Commun. Dis. Public Health* **3**, 50–55.

Dawson, D. (2005). Foodborne protozoan parasites. *Int. J. Food Microbiol.* **103**, 207–227.

Döller, P. C., Dietrich, K., Filipp, N., Brockmann, S., Dreweck, C., Vonthein, R. et al. (2002). Cyclosporiasis outbreak in Germany associated with the consumption of salad. *Emerg. Infect. Dis.* **8**, 992–994.

Doyle, M. P. and Erickson, M. C. (2008). Summer meeting 2007—The problems with fresh produce: An overview. *J. Appl. Microbiol.* **105**, 317–330.

Dubois, E., Agier, C., Traore, O., Hennechart, C., Merle, G., Cruciere, C., and Laveran, H. (2002). Modified concentration method for the detection of enteric viruses on fruits and vegetables by reverse transcriptase-polymerase chain reaction or cell culture. *J. Food Prot.* **65**, 1962–1969.

Emberland, K. E., Ethelberg, S., Kuusi, M., Vold, L., Jensvoll, L., Lindstedt, B. A. et al. (2007). Outbreak of *Salmonella* Weltevreden infections in Norway, Denmark and Finland associated with alfalfa sprouts, July–October 2007. *Eurosurveill.* **12**, E071129.

European Commission. (2002). Commission Regulation (EC) No 472/2002 of 12 March 2002 amending Regulation (EC) No 466/2001 setting maximum levels for certain contaminants in foodstuffs.

European Commission. (2004). Commission Regulation (EC) No 455/2004 of 11 March 2004 amending Regulation (EC) No 466/2001 as regards patulin.

European Union, H. &. C. P. D. G. (2002). Regulation (EC) No 178/2002 of the European Parliament and of the Council of 28 January 2002 laying down the general principles and requirements of food law, establishing the European Food Safety Authority and laying down procedures in matters of food safety.

European Union, H. &. C. P. D. G. (2004a). Regulation (EC) No 852/2004 of the European Parliament and of the Council of 29 April 2004 on the hygiene of food stuffs.

European Union, H. &. C. P. D. G. (2004b). Regulation (EC) No 882/2004 of the European Parliament and of the Council of 29 April 2004 on official controls performed to ensure the verification of compliance with feed and food law, animal health and animal welfare rules.

European Union, H. &. C. P. D. G. (2005). Commission Regulation (EC) No 2073/2005 of 15 November 2005 on microbiological criteria for foodstuffs.

FAO/WHO. (2008). Microbiological hazards in fresh fruits and vegetables. ed. FAO/WHO, pp. 1–28. Rome/Geneva: FAO/WHO.

Food and Veterinary Office. (2007). Final report of a mission carried out in Thailand from 18–27 September 2007 in order to assess the official control system in place to prevent microbiological contamination in fresh herbs and spices intended for export to the European Union. Food and Veterinary Office, pp. 1–23. European Commission, Health & Consumer Directorate–General.

Gaulin, C. D., Ramsay, D., Cardinal, P., and D'Halevyn, M. A. (1999). Epidemic of gastroenteritis of viral origin associated with eating imported raspberries. *Can. J. Public Health* **90,** 37–40.

Healthy Nut Initiative. (1998). Nuts and aflatoxins. Data, facts, background. Hamburg, Germany: Healthy Nut Initiative.

Hernandez, F., Monge, R., Jimenez, C., and Taylor, L. (1997). Rotavirus and hepatitis A virus in market lettuce (Latuca sativa) in Costa Rica. *Int. J. Food Microbiol.* **37,** 221–223.

Herwaldt, B. L. (2000). *Cyclospora cayetanensis*: A review, focusing on the outbreaks of cyclosporiasis in the 1990s. *Clin. Infect. Dis.* **31,** 1040–1057.

Hoang, L. M., Fyfe, M., Ong, C., Harb, J., Champagne, S., Dixon, B., and Isaac-Renton, J. (2005). Outbreak of cyclosporiasis in British Columbia associated with imported Thai basil. *Epidemiol. Infect.* **133,** 23–27.

Kuritsky, J. N., Osterholm, M. T., Korlath, J. A., White, K. E., and Kaplan, J. E. (1985). A statewide assessment of the role of Norwalk virus in outbreaks of food-borne gastroenteritis. *J. Infect. Dis.* **151,** 568.

Ledet, M. L., Hjertqvist, M., Payne, L., Pettersson, H., Olsson, A., Plym, F. L., and Andersson, Y. (2007). Cluster of *Salmonella* Enteritidis in Sweden 2005–2006. Suspected source: Almonds. *Eurosurveill.* **12,** E9–10.

Leggitt, P. R. and Jaykus, L. A. (2000). Detection methods for human enteric viruses in representative foods. *J. Food Prot.* **63,** 1738–1744.

Lewis, H. C., Ethelberg, S., Lisby, M., Madsen, S. B., Olsen, K. E., Rasmussen, P. et al. (2007a). Outbreak of shigellosis in Denmark associated with imported baby corn, August 2007. *Eurosurveill.* **12,** E070830.

Lewis, H. C., Kirk, M., Ethelberg, S., Stafford, R., Olsen, K., Nielsen, E. M. et al. (2007b). Outbreaks of shigellosis in Denmark and Australia associated with imported baby corn, August 2007—Final summary. *Eurosurveill.* **12,** E071004.

Long, S. M., Adak, G. K., O'Brien, S. J., and Gillespie, I. A. (2002). General outbreaks of infectious intestinal disease linked with salad vegetables and fruit, England and Wales, 1992–2000. *Commun. Dis. Public Health* **5,** 101–105.

Martinez-Palou, A. and Rhoner-Thielen, E. (2008). Fruit and vegetables: Fresh and healthy on European tables. Eurostat, pp. 1–8. European Commission.

Mattilsynet. (2008). Tilsyn med ferske bladgrønnsaker og krydderurter importert fra Asia med hensyn på *Salmonella* og *Campylobacter*. Mattilsynet, pp. 1–25. Oslo, Norway: Mattilsynet.

Millard, P. S., Gensheimer, K. F., Addiss, D. G., Sosin, D. M., Beckett, G. A., Houck-Jankoski, A., and Hudson, A. (1994). An outbreak of cryptosporidiosis from fresh-pressed apple cider. *JAMA* **272,** 1592–1596.

Moss, M. O. (2008). Fungi, quality and safety issues in fresh fruits and vegetables. *J. Appl. Microbiol.* **104,** 1239–1243.

Norwegian Scientific Committee for Food Safety. (2008). Risk assessment of import and dissemination of intestinal pathogenic bacteria via fresh herbs and leafy vegetables from South-East Asia, pp. 1–36. Oslo, Norway: VKM–Norwegian Committee for Food Safety.

Nygard, K., Andersson, Y., Lindkvist, P., Ancker, C., Asteberg, I., Dannetun, E. et al. (2001). Imported rocket salad partly responsible for increased incidence of hepatitis A cases in Sweden, 2000–2001. *Eurosurveill.* **6,** 151–153.

Pezzoli, L., Elson, R., Little, C., Fisher, I. S., Yip, H., Peters, T. M. et al. (2007). International outbreak of *Salmonella* Senftenberg in 2007. *Eurosurveillance Weekly* **12**.

Rose, J. B. and Slifko, T. R. (1999). *Giardia, Cryptosporidium,* and *Cyclospora* and their impact on foods: A review. *J. Food Prot.* **62,** 1059–1070.

Rosenblum, L. S., Mirkin, I. R., Allen, D. T., Safford, S., and Hadler, S. C. (1990). A multifocal outbreak of hepatitis A traced to commercially distributed lettuce. *Am. J. Public Health* **80,** 1075–1079.

Sair, A. I., D'Souza, D. H., Moe, C. L., and Jaykus, L. A. (2002). Improved detection of human enteric viruses in foods by RT-PCR. *J. Virol. Methods.* **100,** 57–69.

Stafford, R., Kirk, M., Selvey, C., Staines, D., Smith, H., Towner, C. and Salter, M. (2007). An outbreak of multi-resistant *Shigella* sonnei in Australia: Possible link to the outbreak of shigellosis in Denmark associated with imported baby corn from Thailand. *Eurosurveill.* **12,** E070913.

The European Commission. (2000). White paper on Food Safety.

The Scientific Committee on Food. (2002). Risk profile on the microbiological contamination of fruits and vegetables eaten raw. Brussels: European Commission Health and Consumer Protection Directorate-General.

United Kingdom. (1999). United Kingdom (Great Britain): Contaminants in Food (Amendment) Regulations 1999.

Ward, B. K., Chenoweth, C. M., and Irving, L. G. (1982). Recovery of viruses from vegetable surfaces. *Appl. Environ. Microbiol.* **44,** 1389–1394.

Warner, R. D., Carr, R. W., McCleskey, F. K., Johnson, P. C., Elmer, L. M., and Davison, V. E. (1991). A large nontypical outbreak of Norwalk virus. Gastroenteritis associated with exposing celery to nonpotable water and with *Citrobacter freundii*. *Arch. Intern. Med.* **151,** 2419–2424.

CHAPTER 15

Regulatory Issues in Japan Regarding Produce Safety

Kenji Isshiki

Division of Marine Life Science, Research Faculty of Fisheries Science, Hokkaido University, Hokkaido, Japan

Md. Latiful Bari and Shinichi Kawamoto

Food Hygiene Laboratory, National Food Research Institute, Tsukuba, Japan

Takeo Shiina

Distribution Engineering Laboratory, National Food Research Institute, Tsukuba, Japan

CHAPTER CONTENTS

Introduction	354
Domestic Fresh Produce Production	357
Domestic Consumption of Fresh Produce	361
Fresh Produce-Related Outbreaks in Japan	361
Domestic Food Chain Approach from Farm to Table	364
Good Agricultural Practices (GAP) in Japan	364
Outline of Japan GAP	365
Dealing with GAP in the Private Sector and Producer	366
Government Initiatives for Ensuring Produce Safety and Gaining Consumer Confidence	367
Initiatives in Ensuring Produce Safety and Stable Supply	367
Initiatives for Gaining Consumer Confidence	367
Imports and Distribution of Fresh Produce	368
Increasing Agricultural and Food Imports	368
Distribution Route of Fresh Produce	373
Vegetable Imports and Compliance	374
Safety Regulations and Their Enforcement in Japan	376
Scandals and a Major Change of Attitude	376
New Food Safety Policy and Public Standards	377

The Produce Contamination Problem: Causes and Solutions
© 2009, Elsevier, Inc. All rights of reproduction in any form reserved.

Main Role of The Food Safety Commission	378
Applicable Laws and Regulations	380
Plant Protection Law	380
Food Sanitation Law	380
Product Liability Law	381
Enforcement of Regulations and Standards in Practice	381
Private Sector View on Standards	384
Company Strategies and Company-Specific Quality Standards	384
Company-Specific Quality Standards as a Differentiation Strategy	385
Traceability	385
Conclusions	386

INTRODUCTION

Food safety encompasses many kinds of potential hazards in food. Examples include foodborne pathogens such as *Escherichia coli* O157:H7, *Salmonella*, *Listeria monocytogenes*, among others; naturally occurring mycotoxins, such as aflatoxin; and pesticide residues. These hazards can pose acute risks (consumers become ill immediately) or chronic risks (consumers' risk of chronic illness is enhanced). Some hazards are easily controlled or detected whereas others occur naturally and may be difficult for producers to see or eliminate.

In Japan, most food-safety hazards pose only small risks due to the quality of Japanese food production and the strong standards in place. However, food-safety issues are receiving more attention now for several reasons. First, science is now better able to trace many foodborne illnesses and their outcomes to specific pathogens found in food. Second, as consumers live longer and become more affluent, they demand higher levels of quality and safety in their food. Third, changes in production practices and new sources of food, such as imports, introduce new kinds of risks into the food system. Finally, as more foods are purchased away from home or purchased in prepared form, consumers exercise less control over food safety.

Public policy sets standards for food safety. Such standards reflect policy decisions about acceptable risks and costs of risk avoidance. For many food-safety hazards, consumers cannot detect the hazard at the time of purchase, and producers may also be unable to measure or guarantee a particular level of safety. Therefore, consumers cannot always make their demand for safer food known through purchase decisions, and producers cannot always supply what consumers demand. Public policies attempt to address this market failure by setting standards that ensure some level of acceptable safety for all consumers.

It is now possible for scientists to trace specific foodborne pathogens to their food production origin through genetic fingerprinting. Some foodborne pathogens have only recently been identified, and have evolved to pose new threats. *Escherichia coli* O157:H7, which appeared in the 1980s, poses a new potential threat to consumers of food products. All these trends in scientific and public awareness increase the attention to food safety and the potential for this issue to impact the farm sector.

There are three different government agencies with authority over different aspects of food safety in Japan. Food safety is primarily the responsibility of the Ministry of Health, Labor, and Welfare (MHLW), the Ministry of Agriculture, Forestry, and Fisheries (MAFF), and the Food Safety Commission (FSC). However, new legislation needs to be introduced to unify responsibility into a single agency, in order to use public resources more efficiently to address the most important risks. With increased scientific and public awareness, there have been changes in the way that public agencies approach certain food hazards. The Food Safety Commission has advocated a risk assessment approach to the design of food safety regulations. This means looking at how hazards enter food during production, and where it is easiest to control them. A related idea is that the benefits of a regulation should exceed its costs. The risk assessment framework should help to identify whether and how regulation can provide the greatest benefits (higher safety) for the lowest costs.

A related trend in food-safety regulation is the voluntary use of the Hazard Analysis Critical Control Point (HACCP) systems of safety management. In 1996, the MHLW introduced the use of HACCP in production plants, in order to reduce microbial pathogens in foods. The use of HACCP reflects a growing recognition that it is important to prevent and control hazards before they reach the consumer. HACCP requires identification of critical control points and the development of procedures for monitoring controls and addressing any failures in control. Another important development in food safety policy was the passage of the Agricultural Chemicals Restriction Law in May 2006. This legislation set a consistent standard for risks from pesticide residues in food. The standard requires reasonable certainty that no harm will result to infants and children from aggregate exposure to all residues. Farmers growing many different crops have used these chemicals for many years.

All these changes in food-safety regulation influence farm production. New regulations requiring control of pathogens may also lead processors to place greater emphasis on hazard control in contracts with farm producers. Tracing food-safety problems to their source helps both industry and regulators to find the best control methods, but it can place additional responsibilities on farm producers. Increased attention to management of food safety and quality at all points in the supply chain is often seen as one cause of increased vertical integration (i.e., processor control) in food production.

One approach to food safety is that responsibility is shared by all those involved in food production and consumption. Yet, even acceptance of shared responsibility does not preclude controversy over who will bear specific risks or the costs of risk avoidance. Changes in regulation and in food production, processing, and consumption may alter who bears food-safety risks and costs. What should be the roles of producers, processors, distributors, consumers, and government agencies in assuring food safety? What kinds of information do consumers need to make informed choices about the safety of foods that they buy? To what degree can we rely on the food industry to respond to consumer concerns about food safety? What kinds of new information or research does the food industry need to respond to increased food-safety regulation, increased consumer concern, and growing competition from international trade?

The use of cost/benefit analysis and risk assessment to set standards is still an imperfect science, at best. Scientific certainty about risks and costs will never be possible. Furthermore, consumers do not view different kinds of risks in the same way. Risks that are manmade, unfamiliar, undetectable, and involuntary are viewed with greater fear than risks that are natural, familiar, detectable, and voluntary. What levels of safety are desired, and what risk standards should be applied to foods? Should standards be based primarily on expert risk assessments, consumer risk perceptions, or a combination of the two? Should risk standards be consistent across foodborne risk sources (e.g., risks from pesticide residues and foodborne pathogens)? How should risks to consumers be compared with costs to industry of reducing risks? Should standards be flexible to adapt to new technologies and new scientific information?

Some risks have greater consequences for particular groups of consumers, but not for everyone. Some foodborne pathogens lead to more serious infections in the old and the young. It is also the case that some farms or firms will have greater costs of compliance with food-safety standards. For example, small processing farms have higher costs of adopting HACCP; on the other hand, large farms can easily adapt to comply with the HACCP regulation. Should standards be set to protect the most vulnerable consumers, or should they be set to protect the "average" consumer? Should standards be enforced for all firms equally, or should special consideration be given to small businesses and farms?

One approach to incorporate food safety into farm programs would treat food safety at the farm level as analogous to conservation efforts. Japanese farmers currently receive payments to cover the costs of certain conservation activities. Similar payments could be designed for the costs of improving food safety, such as documented procedures to reduce microbial pathogens. The advantage of this approach is to make farm income policy consistent with consumer protection goals. The disadvantage is that it would address production only at the farm level, which is only one point in the food chain and not necessarily the source of many food-safety hazards. Furthermore, production practices that improve safety are not well defined for many hazards, and compliance would be difficult to monitor.

The approach of placing more reliance on consumers and industry for food-safety assurance would place more responsibility for food safety on consumers and industry, and would mean reduced government involvement in setting standards. This might be achieved through following a stricter rule for comparing benefits and costs of intervention. In other words, new regulation would be justified only by a very large gap between benefits and costs. Even with reduced regulations, a government role in providing information might still be retained, which would assist market forces in assuring food safety. The government could mandate that the food industry provide certain kinds of food safety information to consumers, in order to help them make the most informed choices about food purchases and preparation. Two examples are the required labels on unpasteurized fruit juices and the safe handling labels on fresh vegetables or processed products.

DOMESTIC FRESH PRODUCE PRODUCTION

Japan is a large market for fresh and processed vegetables—the wholesale value of the market in 2000 was about 3 trillion yen ($23 billion). The high value of Japan's vegetable consumption reflects both high consumption per person and high prices for vegetables. Japan's vegetable production includes almost all the vegetables commonly used in North America and Europe, as well as Asian vegetables. Vegetable production has been one of the dynamic sectors of Japan's agriculture, and is one of the few sectors that support widespread full-time farming. As a source of aggregate Japanese farm income, vegetable production is as important as rice or livestock production (Ito and Dyck, 2004).

Vegetables are in a declining trend in both cultivated area and production volumes. The total domestic production was 15.67 million tons in 1975 and it gradually decreased to 12.28 million tons in 2004 (Table 15.1). Conversely, the volume of imports is increasing sharply owing to the tardiness of Japanese producing areas in responding to demand for processing and commercial use.

Table 15.1 The Annual Production, Quantity of Imports and Consumption of Vegetables in Japan

	Domestic Production (1000 t)			Quantity of Imports (1000 t)				Domestic Consumption (1000 t)
Year	Green & Yellow Vegetables	Other Vegetables	Total	Frozen Vegetables	Fresh Vegetables	Other Vegetables	Total	
1975	2750	12,924	15,674	25	39	166	230	15,896
1980	2956	13,514	16,470	141	107	247	495	16,964
1985	2933	13,522	16,455	180	124	562	866	17,320
1990	2848	12,892	15,740	345	261	945	1551	17,289
1995	2854	11,754	14,608	578	738	1312	2628	17,236
2000	2743	10,927	13,670	773	971	1258	3002	16,670
2004	2645	9641	12,286	759	968	1324	3051	15,333

Source: Homepage of the Ministry of Agriculture, Forestry and Fisheries of Japan www.kanbou.maff.go.jp/www/fbs/fbs-top.htm.

Vegetable production tends to be small-scale and specialized. Japan's vegetable operations are typically set up to absorb the full-time labor of one, two, or three family members; the size of the operation is limited to what these workers can do, with part-time or occasional help from other family members or hired nonfamily members. A farm household will produce a few vegetables, or even only one type. Rice production is a common sideline activity or income source and often is contracted out by vegetable farmers, who reserve their labor for their vegetable crops.

For many vegetables, covered production is important. The most common coverings are vinyl houses, followed by glass houses and plastic tunnels. In 2000, 72% of tomatoes and sweet peppers, 69% of cucumbers, 45% of eggplants, and 34% of lettuce crops were grown in covered facilities (MAFF). Vinyl and glass houses usually include heating/ventilation machinery for climate control and systems to control fertilizer and pesticide application. Covered facilities typically produce higher yields than open-field vegetable production and provide the opportunity to raise crops over a longer season. Because Japan's main islands stretch almost as far from north to south as the continental United States, the nation's effective growing season for a vegetable is already long; with covered production, it is extended even more. Nevertheless, in the coldest winter months Japan's production of tender vegetables shrinks dramatically, creating an opportunity for imports from Southern Hemisphere and tropical countries.

The domestic production of widely consumed vegetables is also declining, and the production of lettuce, cucumber, tomato, cabbage, and radish has declined gradually between 2000 and 2004 (Figure 15.1).

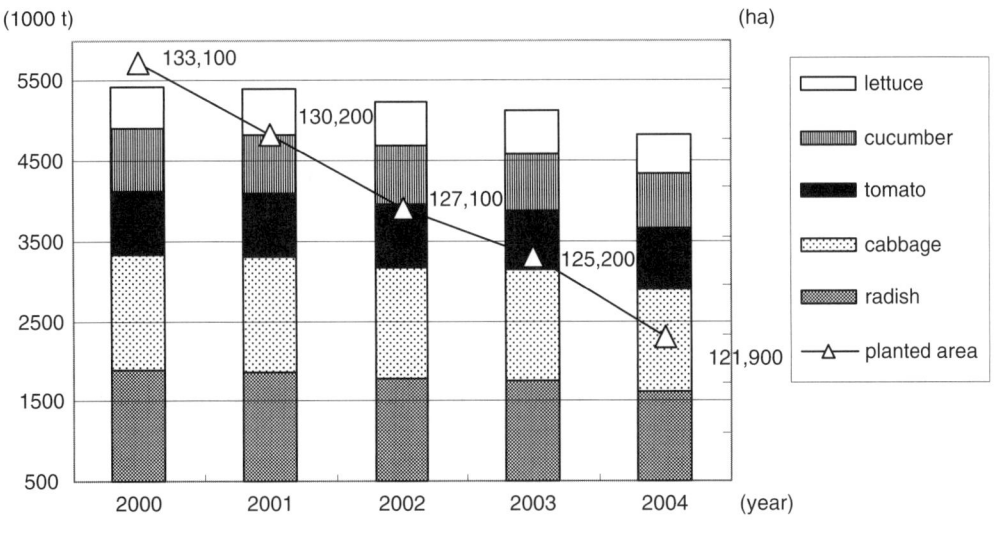

FIGURE 15.1 Domestic production of selected vegetables and the planted area.

The planted area for the production has declined too, from 133,100 ha in 2000 to 121,900 ha in 2004 (Figure 15.1).

Fruit production in Japan benefits from abundant water and a relatively mild climate, but suffers from high humidity, which encourages plant diseases. Temperate fruits such as apples and pears are grown in large volumes. Citrus fruit production is significant in and around the western part of Japan, such as the island of Shikoku. The main citrus product is the unshu mandarin orange. Japan's climate does not support significant commercial production of bananas and other tropical fruits, and pineapple production is small. Like vegetable farmers, fruit farmers tend to be specialized. The multiyear planning horizon for orchard production reduces year-to-year flexibility (Ito and Dyck, 2004).

Greenhouse production of certain fruits, such as melons and strawberries, has increased over the last decades and requires a large fixed investment. Local fruit cooperatives, tied to regional and national federations, have been very important. The cooperatives provide packing, distribution, and marketing functions for member farmers.

Stabilizing prices is a major goal of Japan's vegetable and fruit sectors. Farmers fear volatile prices that could depress their main source of income.

The government, through the MAFF, wants to avoid price swings that would hurt farmers, and seeks stable prices for consumers. Retail firms, while welcoming lower prices, also place considerable weight on price stability and do not like to risk sudden price hikes (Ito and Dyck, 2004).

Several mechanisms operate to stabilize prices, or to correct the effects of volatile prices if they cannot be avoided. Each year the MAFF surveys supply-and-demand conditions for four major vegetables—onions, cabbage, Chinese cabbage, and Japanese radishes (daikon)—and sets a target for the planted area of each. Given historical yields, the target area is expected to produce a volume that will satisfy domestic consumption without significant changes in prices. The target planting area is then divided up regionally and passed on to cooperative federations, which make prefectural targets. Finally, each local cooperative is assigned a target area and works with its farmers to achieve, but not exceed, that area. For 10 other vegetables, national producer groups are entrusted with the responsibility of stabilizing prices by coordinating planting decisions of their members. These associations (in addition to the producing groups of the four major vegetables) are supported by the Vegetable Supply Stabilization Fund (VSSF) when prices or harvests are disappointing (Ito and Dyck, 2002).

Price compensation guarantees payments of a portion of the difference between current season wholesale prices and a moving average of prices in previous seasons, depending on a variety of factors (Ito and Dyck, 2002). For onions, potatoes, and cabbages, the VSSF makes advance purchases for stockholding, releasing stocks in case of market price spikes. The MAFF also has the authority to subsidize cooperatives for shipping low-graded vegetables that are not usually shipped in order to dampen price increases.

Producer groups for vegetables not included among the 14 handled by the VSSF receive government support for undertaking similar supply management plans. The Fruit Supply Stabilization Fund operates to plan production and stabilize prices for certain fruits, currently for citrus, apples, peaches (for processing), and pineapples (OECD, 1995). Some large cooperative units also do autonomous planning, especially Hokuren, the Hokkaido cooperative federation. Hokuren tries to reach a targeted onion production level, set with regard to the MAFF area target and the prospective planting in other major Japanese production areas. In addition to volume, timing the release of onion stocks is a critical factor in Hokuren's planning. To maintain its onion supply to Japan's markets, Hokuren purchases imports from outside Japan when its own supplies are short of its targets. In recent years, subsidies for planting vegetables have been less than those for some other crops (Ito and Dyck, 2002). Recent diversification for fruits has been less important than for vegetables.

DOMESTIC CONSUMPTION OF FRESH PRODUCE

Japan is a large market for vegetables and fruits. One indicator is the value of consumption; the total wholesale value of vegetables in 2000 was about 3 trillion yen (about $23 billion) (MAFF). For the United States in the same year, the value of the 25 leading vegetables (shipping point basis) was $9.27 billion (NASS, 2002). Fruit and nut wholesale value in Japan was 1.58 trillion yen (about $14 billion) (MAFF). The high value of Japan's vegetable and fruit consumption reflects both high consumption per person and high prices for vegetables and fruits. In 2000, Japan's consumers each ate about 101.9 kg of vegetables (MAFF). US consumption per person in the same year was about 137 kg (potatoes excluded for both countries) (NASS, 2002). Japan's consumption has declined over the last quarter-century when measured in kilograms. However, the decline appears to reflect a move away from heavy vegetables (such as Japanese radishes) toward lighter ones (Ito and Dyck, 2004). On a caloric basis, consumption per person appears to have remained stable. Japan's leading vegetables by value are tomatoes, cucumbers, cabbages, Welsh onions (which resemble leeks), lettuce, and bulb onions. In addition, potato consumption, at 16.2 kg per person, is quite important (MAFF). Besides vegetables commonly used in the United States, Japan consumes those associated with Northeast Asian diets in substantial amounts: Japanese radishes, burdock roots, bamboo shoots, lotus roots, Chinese cabbages, fresh soybeans, taros, and shiitake and enokidake mushrooms (MAFF).

Japan's fruit consumption was 41.5 kg per person in 2000. The volume of fruit per person has hovered around 40 kg over the last 25 years. The caloric value of fruit consumption appears to have increased slightly. Leading fruits by wholesale value are mandarin oranges, strawberries, apples, grapes, bananas, watermelons, pears, persimmons, and peaches (MAFF). Retail marketing of vegetables and fruits in Japan emphasizes freshness and quality. Appearance and size are important characteristics. Produce is commonly packaged with labeling that advertises its origin. Japan's cooperatives, for example, usually highlight their names and locations on their produce packages, so that a consumer knows not just the prefecture, but also even the town where the produce was grown. Since April 2000, fresh fruits and vegetables must be labeled with the country of origin (or prefecture, if the produce is domestic) (FAS, USDA, GAIN #JA9022, 1999; #JA1049, 2001).

FRESH PRODUCE-RELATED OUTBREAKS IN JAPAN

The potential for widespread outbreaks of human infection caused by consumption of raw produce was dramatically realized during the summer of 1996 in Japan. More than 9000 cases of *E. coli* O157:H7 infections were reported.

The largest outbreak resulted in three deaths and affected more than 9000 school children in and around Sakai City. Raw radish sprouts, which had been grown on one farm, appeared to have transmitted the pathogen, although the mechanism of sprout contamination was not determined. During 2000 to 2002, again *E. coli* O157:H7 outbreaks occurred with lightly fermented turnip, kimchi, and lightly fermented cucumber, respectively, in different prefectures of Japan. An outbreak of norovirus involving 244 cases in Yamagata and Hokkaido prefectures, associated with potato salad, occurred in 2003 and 2004. In 2005, a *Salmonella* associated outbreak occurred in Oita prefecture due to consumption of sweet potato. In the same year, a large outbreak of *E. coli* due to consumption of Japanese styled kimchi occurred in Chiba prefecture, and one foodborne illness, linked with *Salmonella* Montevideo in white radish sprouts has been reported in Miyagi prefecture. Several major outbreaks related to the consumption of fresh produce are listed in Table 15.2. An outbreak linked with

Table 15.2 Outbreak Cases of Foodborne Illness on Vegetables in Japan

Year	Foods	Microorganisms Involved	Patients	Reference
1996	Radish sprouts (?)	*E. coli* O157:H7	9451	①
2000	Lightly fermented turnip	*E. coli* O157	7	②
2001	Japanese styled kimchi	*E. coli* O157:H7	29	②
2002	Bean sprouts dressed with vinegar	*E. coli*	204	③
2002	Lightly fermented cucumber	*E. coli* O157	112	②
2003	Potato salad	Norovirus	42	③
2004	Potato salad	Norovirus	202	③
2005	Sweet potato	*Salmonella* spp	67	③
2005	Japanese styled kimchi	*E. coli* O6:H16	401	②
2005	White radish sprouts	*Salmonella* Montevideo	12	④
2006	Boiled green vegetables, dressed with grated sesame seed	*Campylobacter jejuni*	40	③
2006	Boiled bracken shoot	*Bacillus cereus*	22	③
2007	Cabbage (?)	*Salmonella* spp	5	③
2007	Potato salad	*E. coli*	35	③

①*Hideshi Michino et al. (1999).* Am. J. Epidemiol. *150(8):786–796.*
②*Homepage of Infectious Disease Surveillance Center, Infectious Agents Surveillance Report* http://idsc.nih.go.jp/iasr/index-cj.html.
③*Homepage of the Ministry of Health, Labour and Welfare of Japan* www.mhlw.go.jp/topics/syokuchu/index.html.
④*N. Saito et al. (2006). PFGE analysis of Salmonella* Montevideo *isolated from an outbreak of food poisoning and the case of sporadic salmonellosis,* Jpn. J. Food Microbiol., *23(3), 143–148.*

Campylobacter in boiled green vegetables, dressed with grated sesame seed, occurred in Kyoto prefecture in 2006. Another outbreak involving 22 cases in Nagano prefecture occurred in June 2006 due to consumption of boiled bracken shoot containing *Bacillus cereus* (MHLW). In April 2007, a *Salmonella* outbreak from cabbage involving five cases occurred in Kyoto prefecture, and in June 2007 an outbreak involving 35 cases in Miyazaki prefecture occurred due to consumption of potato salad containing a pathogenic strain of *E. coli* (Table 15.2).

Mushroom poisoning is caused by the consumption of raw or cooked fruiting bodies of a group of higher fungi that have evolved contemporaneously with plants. Mushrooms are widely distributed in nature, and thousands of species have been identified. About 100 species of mushrooms are poisonous to humans, and 15 to 20 mushroom species are lethal when ingested. No simple rule exists for distinguishing edible mushrooms from poisonous mushrooms. In more than 95% of mushroom toxicity cases, poisoning occurs as a result of misidentification of the mushroom by individuals who are not experts in mushroom identification. In fewer than 5% of the cases, poisoning occurs after the mushroom is consumed for its mind-altering properties (Rania Habal, 2006).

The severity of mushroom poisoning may vary depending on where the mushroom is grown, growth conditions, the amount of toxin delivered, and the genetic characteristics of the mushroom. Boiling, cooking, freezing, or processing may not alter the mushroom's toxicity. Variations in clinical effects may depend on an individual's susceptibility. In general, children, older persons, and persons with disabilities are at a higher risk of developing serious complications than are healthy young adults. The toxins involved in mushroom poisoning such as Amanitin, Gyromitrin, Orellanine, Muscarine, Ibotenic Acid, Muscimol, Psilocybin, and Coprine are produced naturally by the fungi themselves. In 2004 at Niigata Prefecture, an outbreak of acute encephalopathy was noted among patients with renal dysfunction after eating autumn mushrooms (MHLW). *Pleurocybella porrigens* was thought to be a possible causative agent of this outbreak in which 11 persons became ill. The number of fresh-produce–related outbreaks and the ratio of mushroom outbreaks to outbreaks from other vegetables, and the percentage of vegetable-related outbreaks to outbreaks from all other foods in Japan are shown in Figure 15.2. The number of outbreaks was based on the public surveillance conducted by the MHLW. It can be seen that the percentage of mushroom-related outbreaks was greater than that of other vegetables between 1996 and 2005.

FIGURE 15.2

The number of outbreak cases of foodborne illness on fresh and processed vegetables.

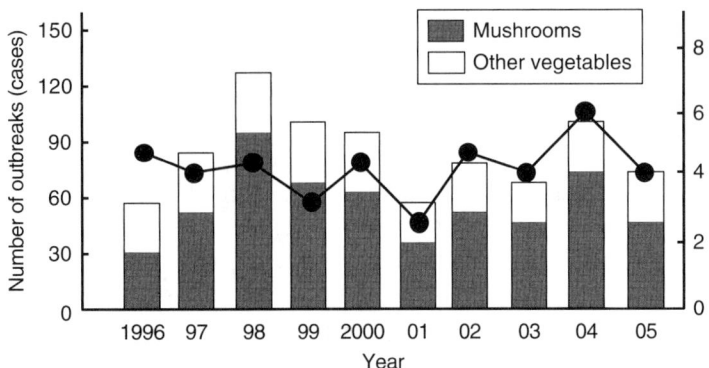

Percentage of vegetable-related outbreaks to outbreaks from all other foods (●)
Source: MHLW, 2006

DOMESTIC FOOD CHAIN APPROACH FROM FARM TO TABLE

Good Agricultural Practices (GAP) in Japan

A manual entitled *The Hygienic Management of Radish Sprout Production*, in which the general concept of HACCP was presented, was introduced after the massive radish-sprout–associated outbreaks in December 1996, and the manual worked well for the improvement of hygiene level in radish sprout production. This manual calls attention to the issue of the hygienic practices for fresh fruits and vegetables. Next, in March, 1999, *The Hygienic Management Guide of the Hydroponics* was issued for the improvement of hygienic management of production and shipping of green vegetables. In March 2002, *Advanced Hygienic Management Guide for Fresh Vegetables—From Production to Consumption* was issued. Thereafter, *The Hygienic Management Guide of Fresh Vegetables—From Production to Consumption (A Simple Edition)* appeared in February 2003. The material covered in the simple edition is shown diagrammatically in Figure 15.3.

The *Hygienic Management Guide for Fresh Vegetables—From Production to Consumption (A Complete Edition)*, which is recognized as Japan's GAP document, was issued in March 2003. The contents of the guide including a simple edition were posted in the homepage of the MAFF. The MAFF is actively introducing the guide to the public, making an effort at widespread dissemination of information in the guide.

The introduction and dissemination of GAP were promoted by the grant-in-aids from the MAFF such as *Fresh Agricultural Produce Safety*

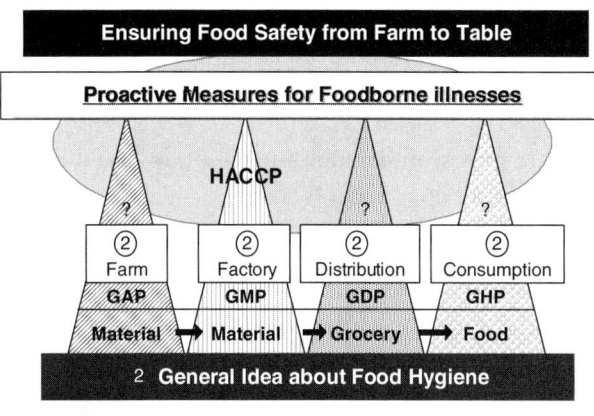

FIGURE 15.3 *General idea about food hygiene.*

Securing Countermeasure Business and *Securing Food Safety and Reliability* in 2004 and 2005, respectively. The Plant Products Safety Division under the Food and Consumer Affairs Bureau has been assigned with the task. This task is composed of two subjects: (1) establishment and dissemination of GAP and (2) introduction and implementation of GAP in produce production (Figure 15.3). The Japanese Agricultural Standards Association was entrusted with the establishment and dissemination of GAP, and the local government and private sectors were responsible for the introduction and implementation of GAP in the production locations.

Outline of Japan GAP

The Hygienic Management Guide for Fresh Vegetables—From Production to Consumption (A Complete Edition) consists of four parts: the common edition, the production edition, the distribution edition, and the consumption edition. As for the common edition, the purpose, scope of the guide, definition of terms, workers' hygiene, education, training, and labeling on the product for consumer's information, packaging, and transportation are mentioned as general remarks. The pathogens, especially *E. coli* O157:H7 and *Salmonella*, are listed as hazards, and countermeasures for minimizing the risk of hazards are mentioned. A guide and implementation manual entitled *The GAP for the Food Safety* was issued in March 2005. In this manual, GAPs for each crop were set as a goal, and GAPs for fruits, cereals, mushrooms, and to some extent, vegetables, were established. As for vegetables, agricultural chemical residues, pathogens, patulin (mycotoxin), and mold are listed among the hazards (Shiina, 2006).

The GAP introduction system is composed of 10 processes, as follows, based on the 12 processes and seven principles in HACCP.

Process 1: Organize the GAP promotion team or promotion committee.

Process 2: Develop proper understanding of conditions during harvesting and shipping.

Process 3: Develop proper understanding of production fields and facilities such as conditions, location, and so on.

Process 4: Prepare a flow diagram for production process.

Process 5: Execute the hazard analysis.

Process 6: Establish countermeasure actions.

Process 7: Identify checklist items and prepare the checklist.

Process 8: Establish verification procedures.

Process 9: Establish the improvement action.

Process 10: Establish record-keeping and documentation procedures.

Dealing with GAP in the Private Sector and Producer

Private large enterprises, such as AEON Group, promote "AEON Produce Suppliers–Quality Management Standards," which introduce a traceability system that allows customers to see the history of fresh produce stretching right back to the area where they were produced. With the TOPVALU Green Eye brand, customers with mobile phones are able to access the production information for individual products simply by scanning the QR code on the package.

The Japanese Consumers' Cooperative Union (COOP) adopted an Agricultural Produce Quality Assurance System in 2004 after false labeling incidents in 2001 and 2002. This system is composed of three fundamental principles: standardization, product management and shipping, and certification. All the COOP chain stores have engaged in this system.

Some agriculture associations and corporations are engaged in GAP at the production level and got the EurepGAP certification. However, these associations were concerned about fulfilling the increased documentation and deskwork requirements of different safety standards. Therefore, recently, the Japan Good Agricultural Initiative (JGAI) was established for achieving a safe and sustainable agriculture system by developing and introducing GAP to produce. The name of this association was changed to Japan

Good Agriculture Practice (JGAP) Association in May, 2006. The activities of the JGAP Association are:

1. Support introduction of JGAP to the members, and establishment and dissemination of GAP.

2. Collect and provide information and cooperate in produce distribution and marketing.

3. Provide technical information and support for the members concerning the GAP management tools (Shiina, 2006).

GOVERNMENT INITIATIVES FOR ENSURING PRODUCE SAFETY AND GAINING CONSUMER CONFIDENCE

Initiatives in Ensuring Produce Safety and Stable Supply

There are increased public concerns over food safety, triggered by incidences of BSE in Japan and other countries, problems of food poisoning, outbreaks of highly pathogenic avian influenza, and fraudulent food labeling, among other issues. In order to ensure safe produce for the consumers, it is important to ensure safety by taking into account the whole produce chain, from production to consumption. The MAFF is promoting introduction of GAP for produce safety, which highlights the appropriate use of agricultural materials and such, implementation of risk communication among stakeholders, and development of standard operating procedure for risk management.

Initiatives for Gaining Consumer Confidence

Traceability Systems

Traceability systems for food have been introduced into the food products industry, and linkages between these systems and those of shipping destinations have also advanced. The rate of introduction of a traceability system in foods other than beef has increased up to 40% (FY2005). According to the results from the general inspection of fresh foods, the percentage of inappropriate labeling has decreased from 25.3% (FY2003) to 14.8% (FY2005) (MAFF, 2006). The MAFF has been making continuous efforts to improve food labeling.

Food Labeling

As for food labeling, the labeling of place of origin of fresh food products has been steadily implemented. Processed foods subject to the mandatory labeling of place of origin of principal ingredients have grown in number.

Future issues include further deliberation on the improved food labeling system, dissemination of information to business entities, and strengthening administrative monitoring and guidance. For the food-service industry, guidelines for the labeling of place of origin have been formulated, and accordingly, business entities in the food-service industry are voluntarily taking initiatives.

Legal Compliance

There has been a spate of illegal rice distribution by agricultural groups, illegal labeling by food product companies, and other such incidents. In the future, legal compliance by agricultural groups and food product companies will be required, and efforts aimed at information disclosure will be strengthened.

Cooperativeness

In the process of foods being processed, distributed, and consumed, the real and perceived distance in understanding between consumers, food industry businesses, and producers has widened, and this is thought to have an impact on various problems concerning food. In the future, it will be important to build relationships of trust such that consumers, producers, and others all fulfill their social roles and remain "visible" to each other. Figure 15.4 illustrates MAFF measures in ensuring food safety in order to regain consumer confidence.

IMPORTS AND DISTRIBUTION OF FRESH PRODUCE

Increasing Agricultural and Food Imports

Despite high levels of protection of domestic farmers, Japan is the world's largest net importer of agricultural and food products. From 1965 to the present, the food self-sufficiency ratio in Japan showed a sharp decrease from 73% to 40% (on a calorie supply basis). At present, the import ratio is approximately 55% of the consumption of fruits and 20% of the consumption of vegetables. The domestic production of vegetables decreased from 15.7 million tons in 1975 to 12.2 million tons in 2004 (Table 15.1). The increase of imports has caused a decrease in domestic vegetable production because the market of vegetables expanded little. The volume of vegetables has increased steadily from 0.23 million tons in 1975 to 3.0 million tons in 2000 and remained constant since then (Table 15.1).

The value of Japan's vegetable imports was approximately 450 billion Yen in 2006. The rise in volume coincided with a decline in prices for most of Japan's vegetable imports and a decrease in the aggregate value of vegetable imports (Figure 15.5).

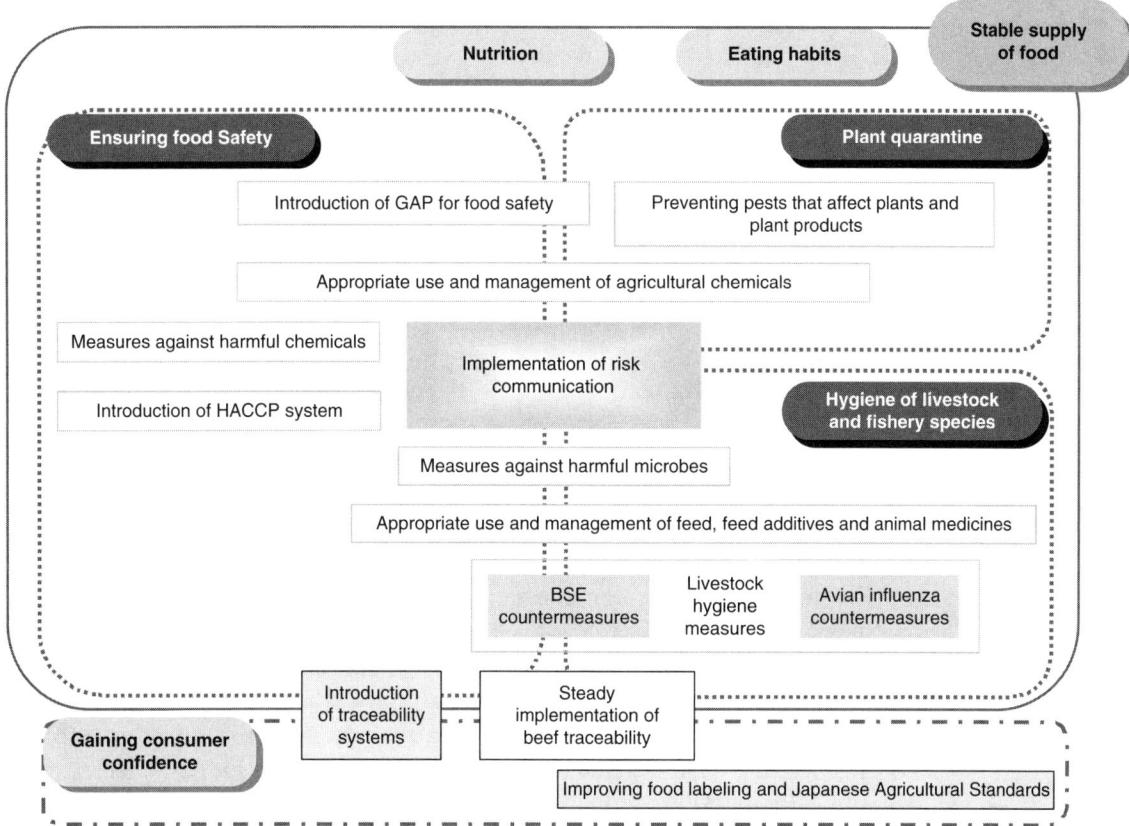

Source: MAFF, 2006

FIGURE 15.4 MAFF measures in ensuring food safety and gaining consumer confidence.

Imports are distributed among four main categories: dried vegetables and beans, vegetables that are provisionally preserved so that they can be further processed, frozen vegetables, and fresh vegetables (Ito and Dyck, 2004). Fresh vegetable imports have shown some growth in recent years. Unit values for fresh vegetable imports (as an aggregate) increased by one-third from 2000 to 2006, and volumes rose by 58% (Figure 15.5).

The import volume of frozen, dried, and provisionally preserved vegetables also increased in 2006 compared to previous years. For frozen vegetables, a 17% rise in unit values coincided with a 19% rise in the volume of imports from 2000 to 2006. The 42% increase of frozen vegetable import unit value over the same years was accompanied by a 21% gain in volume (Table 15.4).

China is the largest source of Japan's vegetable imports, supplying virtually all the provisionally preserved vegetables, most of the mushrooms, half

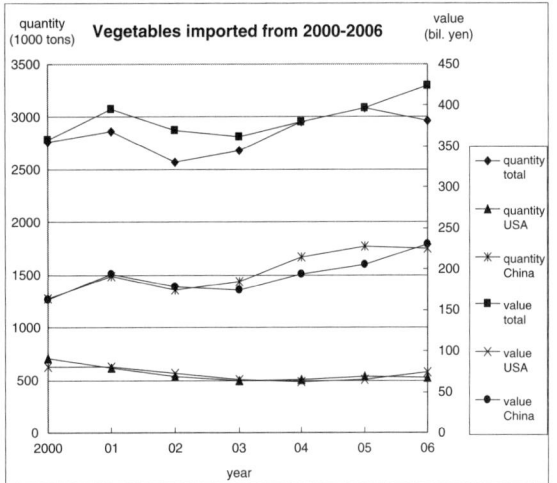 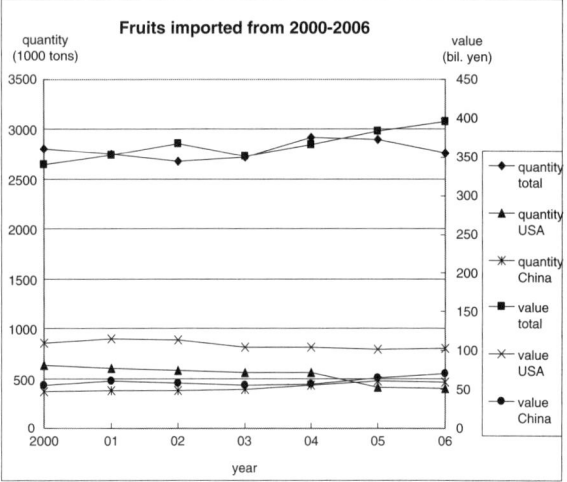

FIGURE 15.5 *Japan's total vegetable and fruit imports, 2000–2006 average volume and value.*

the dried vegetables, and substantial shares of the fresh and frozen vegetables (Tables 15.3 and 15.4). China's share of Japan's imports has been rising, growing from 50.7% in 2001 to 60% in 2006. Shares of other major exporters to Japan (except for Thailand) have fallen. China's leading frozen exports to Japan are green soybeans, taros, spinach, and mixed vegetables. The leading fresh vegetables from China are Japanese radishes, leeks, peas, and garlic (MAFF). China's rising vegetable exports to Japan were bolstered by many factors. With its low production costs and geographic proximity to Japan, China attracted foreign investment, especially from Japanese trading

Table 15.3 Top Major Supplying Countries of Fresh or Chilled Vegetables to Japan, 2000–2006

	China		USA		Korea		New Zealand		Thailand	
	Quantity	Value	Quantity	Value	Quantity	Value	Quantity	Value	Quantity	Value
Year	1000 t	Billion yen	1000 t	Billion yen	1000 t	Billion yen	1000 t	Billion yen	1000 t	Billion yen
2000	294.3	35.6	260.8	21.8	27.3	10.1	159.8	9.9	10.2	2.4
2001	403.5	42.5	194.9	20.7	38.8	11.1	147.5	10.0	12.0	3.0
2002	319.9	30.2	150.9	19.0	24.3	8.9	95.3	8.7	11.9	3.8
2003	388.1	34.8	140.7	17.2	23.2	8.9	139.7	10.3	7.7	3.2
2004	503.9	38.5	138.0	16.2	36.2	9.3	104.5	8.2	16.1	4.1
2005	589.7	40.9	142.0	13.8	33.0	9.6	129.7	9.4	16.2	4.0
2006	541.8	44.7	107.7	13.5	19.8	7.1	85.2	8.5	13.9	4.3

Source: MOF, 2006.

Table 15.4 Top Major Supplying Countries of Frozen Vegetables to Japan, 2000–2006

	China		USA		Thailand		Taiwan		Canada	
	Quantity	Value	Quantity	Value	Quantity	Value	Quantity	Value	Quantity	Value
Year	1000 t	Billion yen	1000 t	Billion yen	1000 t	Billion yen	1000 t	Billion yen	1000 t	Billion yen
2000	222.4	29.6	260.9	27.2	19.7	3.3	25.1	4.9	34.7	3.3
2001	249.4	37.0	252.8	28.7	17.1	3.0	23.3	5.0	33.2	3.5
2002	207.6	31.1	236.5	27.4	18.6	3.3	23.8	5.2	36.6	3.8
2003	182.6	25.2	204.7	22.8	22.1	3.6	27.8	5.7	30.3	2.9
2004	222.5	28.9	205.9	21.4	24.9	3.8	30.2	5.8	43.6	4.0
2005	237.9	32.1	220.0	23.1	24.5	3.9	26.4	5.2	37.9	3.8
2006	258.4	37.2	237.5	26.4	26.7	4.5	24.4	5.3	35.3	3.8

Source: MOF, 2006.

companies. These businesses provided the seeds, spores, and production/packing techniques, and imported the harvest for Japanese retailers. Improved ocean freight service from major Chinese ports to Japan also increased China's competitiveness. Mushrooms are the leading imports, comprising 14 to 18% of the total value of Japan's vegetable imports. Frozen potato products, chiefly french fries, are the next largest import item, making up 9 to 11% of imports. Other imports are distributed over a wide range of vegetables (MAFF).

Japan's imports from the United States, the second largest source of its vegetable supply, are concentrated in the fresh and frozen categories. Frozen potato products, fresh broccoli, fresh and dried onions, frozen and dried sweet corn, and asparagus are the leading commodities (MAFF, 2006).

Thailand is the third most important supplier, exporting fresh vegetables and tropical fruits, mushrooms, fresh onions, frozen sweet corn, and fresh peppers.

Seasonal differences are a factor in Japan's vegetable imports, especially of asparagus, with large Southern Hemisphere and tropical shipments from Oceania and Southeast Asia. Among the 10 largest suppliers to Japan, New Zealand, Thailand, and Australia have seasonal advantages. However, their market shares, like that of most countries except China, tended to decline slightly over the 2000–2006 period.

Australia was the major supplying country of fresh asparagus in 2002. The Philippines and Thailand were second and third, respectively. The Philippines held the number two position in 2002, but fell to ninth place in 2005. Imports of fresh or chilled vegetables from Thailand showed a rapid rise to fifth place in 2005 and remain constant (Table 15.5).

Table 15.5 Ranking of Major Supplying Countries of Fresh or Chilled and Frozen Vegetables to Japan, 2005

Fresh or Chilled

	Country	Quantity 1000 t	Value Billion yen
1	China	589.7	40.9
2	United States	142.0	13.8
3	Korea	33.0	9.6
4	New Zealand	129.7	9.4
5	Thailand	16.2	4.0
6	Mexico	24.6	3.4
7	Australia	16.6	3.0
8	Netherlands	5.4	2.5
9	Philippines	5.6	1.9
10	North Korea	0.8	1.7
11	Taiwan	13.8	1.2
	World	996.8	95.5

Frozen

	Country	Quantity 1000 t	Value Billion yen
1	China	297.6	45.0
2	United States	267.4	29.5
3	Thailand	31.5	6.1
4	Taiwan	26.4	5.2
5	Canada	46.4	5.1
6	New Zealand	28.9	3.8
7	Vietnam	9.0	1.5
8	Indonesia	6.0	1.1
9	Ecuador	5.7	1.1
	World	737.3	102.0

Source: MAFF, 2006.

The quantity of fresh green soybeans (edamame) is small in comparison with frozen green soybeans. Taiwan has been the major supplier for the last five years, followed by mainland China. Taiwan supplied 88% of the imported quantity in 2005, and this share has not changed much since 1998. Since the product is fresh, the transportation distance is important. Therefore, China and Taiwan are the main suppliers because of their proximity to Japan. About 50% of frozen green soybeans come from China.

The Philippines has a very strong position in the Japanese market for pineapples and mango, supplying 98% of the former and 63% of the latter.

Mexico supplies 25% of the mango imports. The Philippines has an advantage over Mexico in shipping time and cost.

Distribution Route of Fresh Produce

There is not one single distribution route for imported fresh fruits and vegetables. Figure 15.6 summarizes the different routes schematically. The main distribution route for imports goes from foreign producers via trading companies (importers), wholesalers, and intermediary wholesalers to the retailers, whereas the distribution route for domestic fruits and vegetables goes from wholesalers, and intermediary wholesalers to the retailers. However, processed products, such as frozen green soybeans, are not handled in wholesale markets; they go directly from foreign producers via trading companies to retailers or commercial users. Supermarket chains also directly import the main fruits and vegetables, but not 100% of their required amount. The remainder is purchased from trading companies and wholesalers. Fruits and vegetables that are not sold in large amounts are purchased from smaller trading companies that specialize in those specific products (Jonker et al., 2005).

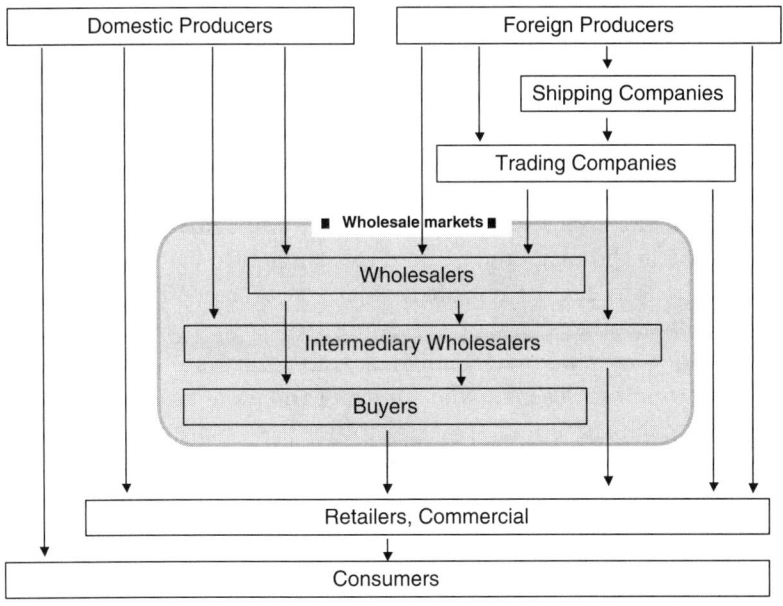

FIGURE 15.6

Distribution route of domestic and imported fruits and vegetables.

Source: modified from Jonker et al., 2005

Moreover, fruits and vegetables that are not sold in large amounts are often are transported together with large-volume products because of savings in logistics and transport. A very large share of the imported fruits and vegetables in the Japanese market, especially pineapples and bananas, is handled by a relatively small number of big companies.

Vegetable Imports and Compliance

The 1990s saw a rush of foreign investments in China. This influx of foreign investments led to the six-fold increase in the value of imported frozen vegetables from 1990 to 1992 through 1998 to 2000 (Wu Huang, 2002). The factors important to Japanese consumers are freshness (thus, countries geographically close to Japan have an edge), quality, visual perfection, taste, stability of supply, and low prices (Ito and Dyck, 2002).

The quality of China's products is improving and is "reportedly meeting customer preferences" (Shields and Huang, 2004). However, the fruit and vegetable industry does not, in general, use grade standards (such as for uniform product size and appearance); thus, a number of private firms based their criteria on customer specifications. Moreover, there is no widespread use of modern packing and packaging techniques. The abundant supply of low-cost labor deters the adoption of such improvements (Shields and Huang, 2004).

In 2002, excess levels of pesticides were detected in frozen spinach from China (JETRO, 2001a, 2002b, 2002c, 2002d). Japan's MHLW called on related industries to voluntarily suspend imports of frozen spinach produced in China. The negative impact spread to green soybeans and other Chinese frozen vegetables, and all other imported frozen vegetables as well. The Enforcement Order for the Food Sanitation Law was amended to add frozen vegetables to the list of products subject to ordered inspection. The amended law also authorized the Japanese government to ban imports from particular countries in case of repeated violations of agricultural chemical residue standards or other standards (promulgated September 2002). This incident spurred the trend to establish a traceability system to assure consumers and to inform them of the source of the produce.

Talks between Japanese and Chinese authorities resulted in measures such as the cessation of use of the "offending" agricultural chemical, the institution of export inspections and the issuance of health certificates, and the issuance of export permits only to products of registered farms. Japanese frozen food companies have agreements with the Chinese producers to directly control the type, quantity, and number of applications of agricultural chemicals, and to maintain traceable production records.

The suspension was lifted in February 2003, and imports resumed in the middle of 2004 (Jonker et al., 2005).

Compliance concerning microbiological standards in fresh and frozen vegetables and violations of the microbiological standard by country is shown in Table 15.6. China with 128 violations was the highest (37% of the gross number of violations of microbiological standards), followed by Thailand with 58 (16.8%), and Vietnam, 46 (13.3%). Data listed by product and type show that violations concerning the microbe standard of frozen food (number of natural microflora, coliform bacteria, and *E. coli*) were the most frequent violations occurring in these areas (MHLW, 2006a).

The MHLW and the prefectural governments have been engaged in surveillance for pathogens in fresh produce after the radish sprout outbreaks in 1996. Almost all the fruits and vegetables are under the testing program for the presence of pathogens. Only selected fresh produce is listed in

Table 15.6 Violation Cases against Component Standard by Nations, Products, and Events in FY2005

Nation	Products	Events	Number of Cases*
China	Frozen food (vegetables)	Viable count (7), coliform bacteria (5), *E. coli* (3)	15
	Frozen food (beans)	Viable count (3), coliform bacteria (1), *E. coli* (1)	5
	Frozen food (fruits)	Coliform bacteria (1)	1
Thailand	Frozen food (fruits)	Viable count (2), coliform bacteria (1)	3
	Frozen food (vegetables)	Viable count (2), coliform bacteria (1)	3
Korea	Frozen food (vegetables)	Viable count (1), coliform bacteria (1)	2
France	Frozen food (fruits)	Viable count (1)	1
Philippines	Frozen food (fruits)	Coliform bacteria (1)	1
Indonesia	Frozen food (fruits)	Coliform bacteria (1), viable count (1)	2
Taiwan	Frozen food (fruits)	Viable count (2), coliform bacteria (1)	3
Italy	Frozen food (vegetables)	Viable count (1)	1
Brazil	Frozen food (fruits)	Viable count (2), coliform bacteria (1)	3
Canada	Frozen food (vegetables)	Coliform bacteria (1)	1
Australia	Frozen food (vegetables)	Viable count (2)	2
Peru	Frozen food (fruits)	Coliform bacteria (1), viable count (1)	2

*Number of cases indicates the total number of violation events.
Source: MHLW, 2006.

Table 15.7 Surveillance on Pathogens in Specific Fresh Produce (FY2000–2006)

Samples	Number of Samples							E. coli Positive Samples							Salmonella Positive Samples							E. coli O157 Positive Samples						
	00	01	02	03	04	05	06	00	01	02	03	04	05	06	00	01	02	03	04	05	06	00	01	02	03	04	05	06
White radish sprouts	200	171	149	135	121	114	97	20	23	19	10	9	10	11	-	-	-	-	-	-	-	-	-	-	-	-	-	-
Alfalfa	55	27	32	22	20	35	22	7	6	8	1	1	-	4	1	-	-	-	-	-	-	-	-	-	-	-	-	-
Lettuce	214	204	178	139	123	116	110	18	12	11	8	6	7	3	-	-	-	-	1	-	-	-	-	-	-	-	-	-
Japanese honewort	166	149	133	102	95	92	66	51	45	27	28	25	24	23	-	-	-	-	1	-	-	-	-	-	-	-	-	-
Bean sprouts	230	213	148	135	147	122	109	55	78	35	28	41	33	36	-	-	-	-	-	-	-	-	-	-	-	-	-	-
Cucumbers	-	206	178	130	125	124	101	-	12	4	9	4	11	5	-	-	-	-	-	2	-	-	-	-	-	-	-	-
Pre-cut vegetables	178	187	155	107	177	137	160	7	8	3	8	8	13	11	-	-	-	-	-	-	-	-	-	-	-	-	-	-
Vegetables used for pickles	-	-	105	86	101	117	74	-	-	6	7	5	3	-	-	-	-	1	-	-	-	-	-	-	-	-	-	-

Source: MHLW

Table 15.7. In 2000, one sample of alfalfa was found positive for *Salmonella* out of 55 samples tested. In 2005, two samples of cucumber were found positive for *Salmonella* out of 124 samples tested. However, no *Escherichia coli* O157:H7 was found in the samples tested during the survey from 2000 to 2006.

SAFETY REGULATIONS AND THEIR ENFORCEMENT IN JAPAN

Scandals and a Major Change of Attitude

The attitude of Japanese consumers, government officials, producers, processors, and retailers toward food safety has changed greatly in the last few years, following a series of food-related accidents and scandals, which include the outbreak of *Escherichia coli* O157:H7 in sprouts; BSE; food poisoning; the use of bad ingredients in dairy production; and mislabeling origin of production areas for beef, vegetables, and other food. Numerous scandals occurred, but five incidents greatly affected Japanese consumers' consciousness of food safety and distrust in food manufacturers.

1. In July 1996, a sprout-associated outbreak occurred in Sakai City, Japan, affecting approximately 9000 people and causing three deaths.

2. In August 2000, the subsidiary of a reputable dairy company caused massive food poisoning by enterotoxin-producing *Staphylococcus aureus* in low-fat milk that affected 14,700 people.

3. In September 2001, BSE was detected. The government promptly introduced very severe countermeasures, including BSE prion testing on all slaughtered cattle. The general public considered the investigation into the cause of the outbreak insufficient. Dissatisfaction against the MAFF spread, and anxiety about BSE grew among Japanese consumers.

4. In 2002, the Japanese government announced that an agency of the MAFF would buy out the domestic beef stored from before the BSE outbreak and incinerate it because the MAFF could not check all stored beef for BSE infection. One large and well-known food manufacturer committed fraud by making the agency buy an old stock of imported beef.

5. In 2002, excessive pesticide residues were detected in frozen vegetables (mainly spinach) from China on several occasions, which received great attention from the media. Also illegal pesticides were detected in domestic fruits and vegetables.

For a long time, Japanese government agencies claimed that the Japanese regulatory system guaranteed the safest food in the world. Most Japanese people believed this claim. However, the food-related accidents and scandals that involved reputable Japanese food companies have badly affected consumers' trust in regulatory authorities and undermined public trust in food safety. Food safety has become an important factor for governmental policy and company strategies.

New Food Safety Policy and Public Standards

To overcome such situations and to reinforce countermeasures to secure food safety, reliability, and public trust, the Japanese government reviewed the food-safety system. The Food Sanitation Law was amended, and a new Food Safety Basic Law was established, which became effective in July, 2003. Under this law, an independent advisory committee under the cabinet office, the Food Safety Commission, was established. The mission of this independent commission is to (1) conduct risk assessment on food in a scientific, independent, and fair manner, making recommendations to relevant ministries based upon the results from the risk assessment, (2) implement risk communication among stakeholders such as consumers and food-related business operators, and (3) respond to food safety incidents and emergencies. An overview of the new food safety policy is shown in Figure 15.7.

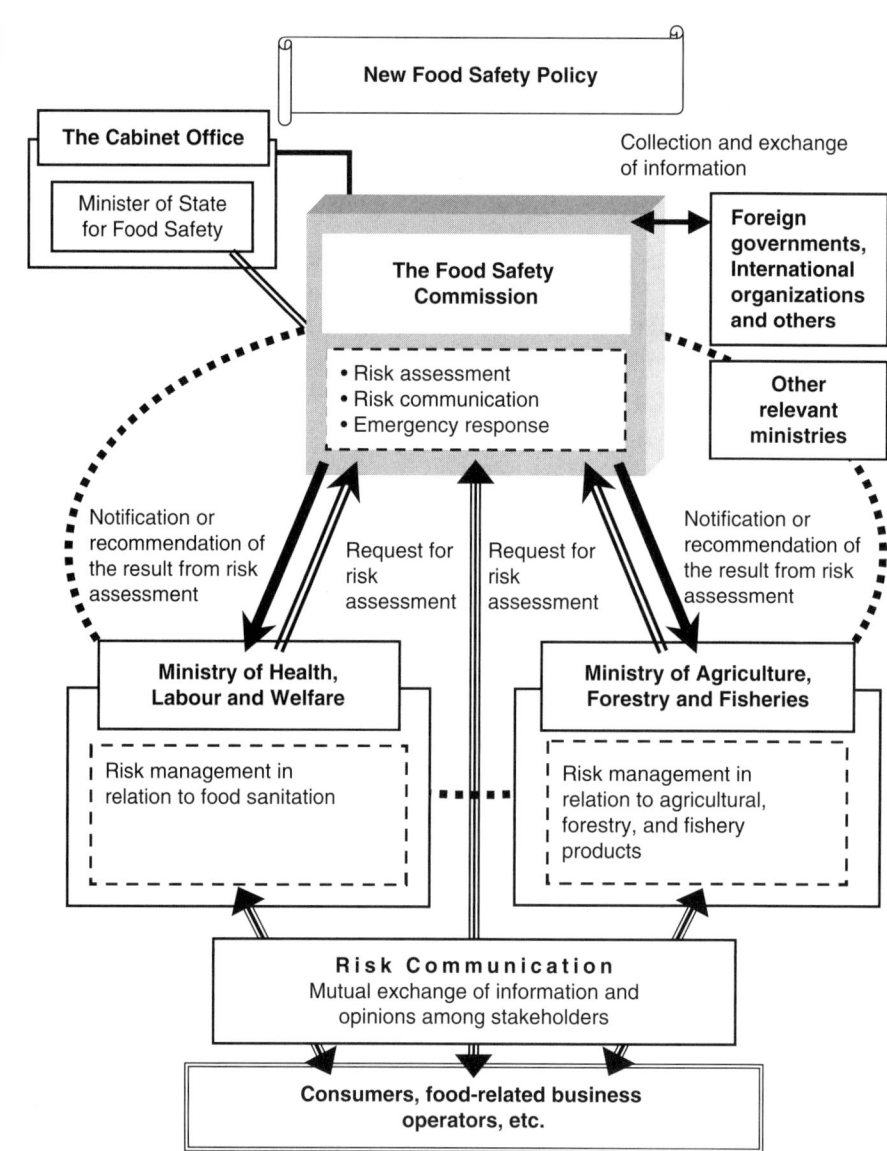

FIGURE 15.7

Overview of new food safety policy.

Source: FSC

Main Role of The Food Safety Commission

Implementation of Risk Assessments

"Risk assessment" denotes the assessment of the effect of food on human health, as referred to in the Food Safety Basic Law, so that after eating certain food, the probability of risk, and the degree of the adverse effects of harmful factors on health, can be appropriately controlled. Risk assessment is implemented by conducting assessments in an objective, neutral, and fair manner, and on the basis of scientific knowledge.

The Food Safety Commission implements risk assessments, either based on the requests for assessment from risk management organizations, or from the Food Safety Commission itself, performing a risk assessment for a particular food item if it believes that this food item may cause hazards to human health. Based on the results from these risk assessments, the Food Safety Commission makes recommendations to the relevant ministries through the prime minister to implement policies/recommendations for ensuring food safety.

Promotion of Risk Communication

The Food Safety Commission plans and implements risk communication, the mutual exchange of information, and opinions or recommendations generated by risk assessment among stakeholders, including consumers, with various forms of communication methods such as public meetings, use of the media, and the commission's Web site.

Responses to Emergency Situations

The Food Safety Commission routinely collects and analyzes domestic and international information on hazards pertaining to food safety. Therefore, in case of any major food-related incident or any emergency situation, the commission could take rapid countermeasures and appropriate governmental responses by issuing requests to the relevant ministries for their prompt action, or by providing easy-to-understand information to the public to prevent the expansion and reoccurrence of the hazard.

Other Roles

Under the Food Safety Basic Law, the government is responsible for determining the Basic Matters related to the implementation of measures, which are to be taken in accordance with the provisions contained within the same law. Another role of the Food Safety Commission is to accept requests and advice from the Prime Minister, and to provide views on the items that should be incorporated into the Basic Matters. In January 2004, the Cabinet endorsed the Basic Matters that were recommended by the Food Safety Commission. Since then, the risk management organizations, including the Food Safety Commission, the Ministry of Health, Labor and Welfare, and the Ministry of Agriculture, Forestry and Fisheries, have been working together to implement measures to ensure food safety.

Applicable Laws and Regulations

At the time of importation, the most important regulations are those set in the Plant Protection Law (under the jurisdiction of the MAFF), and in the Food Sanitation Law (under the jurisdiction of the MHLW). The Plant Protection Law deals with plant quarantine and is applicable to fruits and vegetables only. Food sanitation inspections are applicable to fruits and vegetables and fishery products, and they apply at the time of sale. All food products distributed and marketed in Japan are subject to the labeling regulations. The main law concerning labeling is the JAS (Japanese Agricultural Standard) Law, which explains the details of mandatory and voluntary labeling.

Plant Protection Law

Fruits and vegetables from a certain country or region are either allowed or not allowed to be exported to Japan, so there is no confusion or uncertainty. This either/or scenario differs from the pesticides requirement, which sets a maximum level of pesticide residue that cannot be exceeded. Even if the importation of certain plants or plant products is prohibited according to the Plant Protection Law, it may be allowed under prescribed conditions of quarantine and after the completion of specified procedures. For example, mangoes from certain tropical areas disallowed according to the Plant Protection Law, among other fruits and vegetables, receive a fumigation treatment (with chemicals) or a vapor heat treatment (with steam) in the exporting country that will enable them to enter Japan. Fruits and vegetables need to have a phytosanitary certificate issued by the exporting country. Nonetheless, if quarantine pests are found during the import inspection in Japan, the fruits and vegetables have to be treated (disinfected) or discarded, depending on the particular conditions.

Food Sanitation Law

The Food Sanitation Law is to prevent the occurrence of health hazards arising from human consumption of food, by making necessary regulations and taking effective measures to protect the health of the people. Food sanitation inspections are applicable to fruits and vegetables and fishery products at the time of importation, and can also be applied at the time of sale. For fresh fruits and vegetables, residual pesticides, additives, component standards, fungal toxin, presence of pathogens and microbiological conditions are assessed before the food can be entered into the markets. In May 2006, Japan adopted a positive list with maximum residue levels (MRLs) for specific pesticides. If the residue exceeds the maximum limit, the product cannot be imported into Japan.

The impact of the Food Sanitation Law will greatly depend on the way it is implemented. The maximum residue levels were set lower, making it more difficult for suppliers to comply. According to an inspection company, increasing the precision of analysis by using more advanced equipment means increasing costs, because the equipment costs more. However, the equipment for the zero-tolerance test is not considered very expensive, and the tests are not considered very difficult, although this may depend on the particular pesticide, food additive, and microbiological safety standard.

Product Liability Law

The Product Liability Law states that the producer or importer is liable in case of a problem with a product, but the retailer is exempt, except if the retailer causes the problem. Consequently, if a food-poisoning incident occurs, the manufacturer has a tremendous financial loss. Among other financial setbacks, the products have to be withdrawn from the shelves of numerous retail shops, the products are destroyed, and compensation has to be paid to the victims. Besides that, the company will be the target of public criticism, and if the incident is very severe, it will be hard for the company to survive. In comparison to this, the loss for retailers will be limited to not receiving the expected profit of the withdrawn product. The situation is similar for imported products. The importer has to bear the liability for problems with the products, unless the wholesaler or retailer causes the accident. Although a retail company may not be liable in case of an incident, it can suffer damage to its reputation (Jonker et al., 2005).

Enforcement of Regulations and Standards in Practice

At the time of importation, three inspections related to the Food Sanitation Law are conducted:

1. **Document.** Examination of the results of the inspections conducted by the public organizations of the exporting country. Wherever parties agree that the inspection will take place, inspectors dispatched from Japan will conduct document inspections.
2. **Monitoring.** Random sampling inspections including microbiological evaluation have been conducted by the MHLW. The costs are borne by the MHLW.

3. **Order.** Inspections conducted by importers based on the order issued by the MHLW.

Order inspections are conducted after three violation cases have been reported for an import item. The numbers of samples for order inspections are greater than those for monitoring inspections. As the number of violation cases increases, so does the number of samples. Whereas it is allowed for importers to sell imported goods before the results of monitoring inspections are known, it is not possible to sell imported goods in the case of order inspections until it is confirmed that there are no violations. Since residual pesticides found in Chinese vegetables became a big problem in the beginning of 2002, more stringent measures were taken for both monitoring and order inspections, and the frequency of sampling for both increased.

Under the Plant Protection Law, fresh foods and agricultural products are subject to visual observation inspections for harmful pests, mold, and other undesired characteristics. These inspections are conducted per import item in the form of sample inspections. Both food sanitation and quarantine inspections are conducted on each shipment entering the country. However, if the exporting country conducts food sanitation inspections and the government of the exporting country certifies the safety of the products, food sanitation inspection (in Japan) may require only document inspections. In strict contrast, quarantine inspections are conducted without exception.

In addition to the food sanitation inspections at the port of entry (controlled by the MHLW), the Japanese government conducts two other kinds of food sanitation inspections: inspections at wholesale markets (controlled by the health stations of prefectural governments) and inspections on the sales floors of retailers (controlled by the health centers of prefectural governments). The sampling inspection conducted by prefectures is in accordance with the Prefectures' Monitoring and Guidance Plan.

Apart from these inspections conducted by the government, private companies (importers and retailers) regularly conduct their own inspections. Staff of big retail chains visit production sites and inspect the farming conditions of the produce they purchase before the import season begins. They do this not only for the goods they import on their own, but also for goods purchased through trading companies. If they purchase goods from unspecified dealers, traceability cannot be established, and it is impossible to trace back information to the production sites (MHLW, 2006a). The overview of monitoring and guidance system of imported food is shown in Figure 15.8.

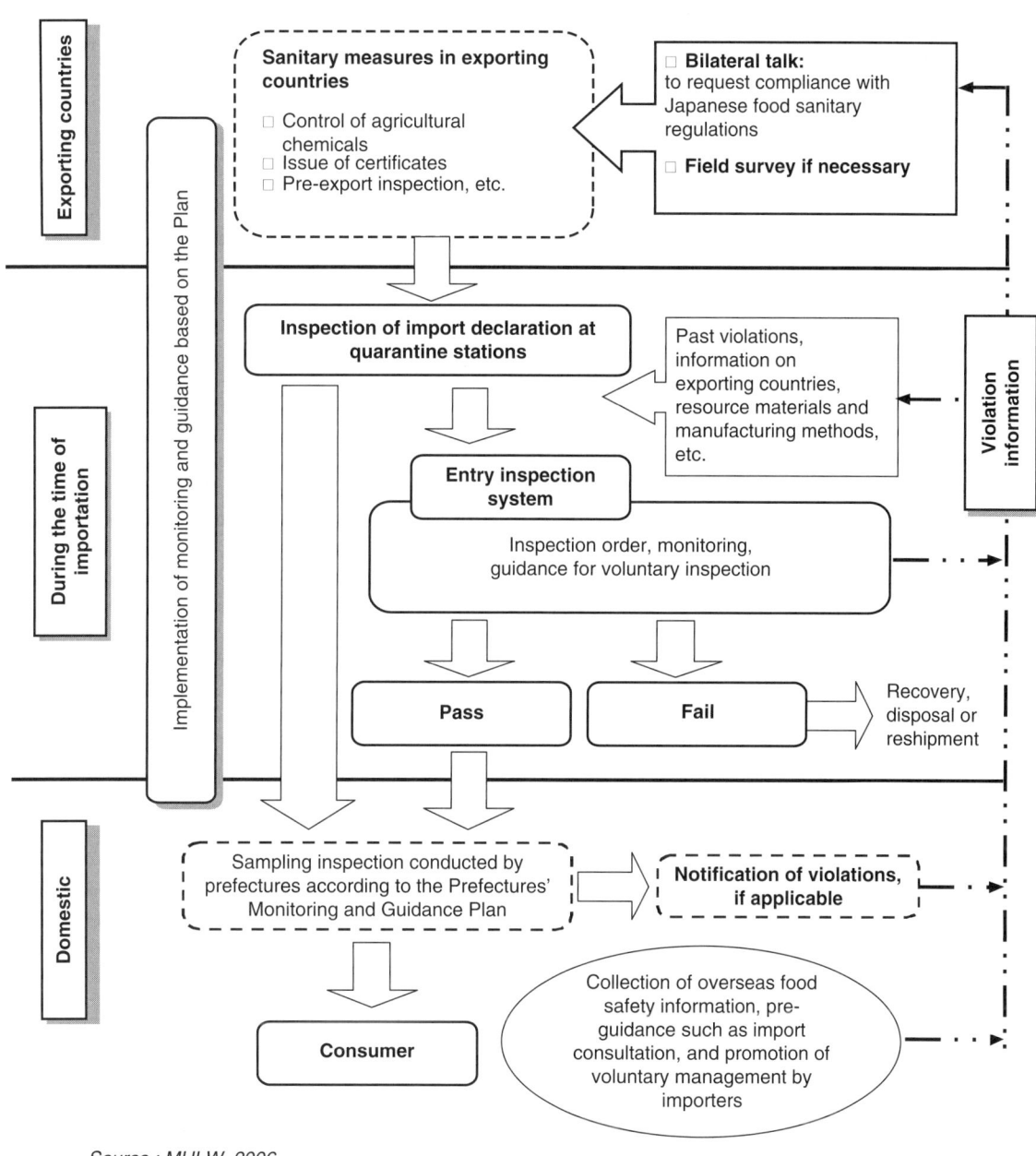

FIGURE 15.8 Overview of monitoring and guidance system on imported foods.

Private Sector View on Standards

For fruits and vegetables, the relative priority of sanitary and phytosanitary standards and other standards in the overall trade relations for importers (trading companies and retailers) are somewhat different among the products and among the companies. However, in general, the most important points are "overall product quality" and "compliance with food-safety requirements," followed by "price" and "reliability of supply." "Transport cost" is also considered important. "Phytosanitary regulations" are not mentioned as a priority issue, because the regulations are clear and do not cause any confusion. The origin of the product plays a significant role in addition to the factors just mentioned. Most Japanese believe that domestic food is safer than imported food, and imported food from other OECD countries is trusted more than that from developing countries. Chinese products are believed to be the least safe. Increased attention has been given to the safety of imported vegetables from China because they often have pesticide residues above the standards. After providing warnings to the Chinese authorities, Japan, on several occasions, has put a ban on spinach imports (Jonker et al., 2005).

Traceability information is becoming the standard for fruits and vegetables, in particular, to enhance consumer confidence. The ban on spinach imports, although it may be justifiable on the basis of frequent violations, is an example of one of the governmental actions to restore public confidence in regulators.

Company Strategies and Company-Specific Quality Standards

In Japan, the most important driving force in defining standards was consumers' concern about food safety after the food-related incidents and scandals. These incidents compelled the food processors and producers to face the reality that they can no longer survive unless they put strenuous efforts into securing food safety. Food-handling companies, especially the leading companies with a high profile, are increasingly focusing on how to secure safety and relieve consumers' anxiety. Although the government sets the public regulatory safety standards, these companies have become nervous and have set up stricter measures to protect themselves from food-safety incidents. They are fully aware that if they mishandle the safety matter, they will have to pay the price because they will be forced out of business far more quickly than before by the government, business partners, and consumers—not only because of the product liability law, but mainly because of the loss of their reputations. They have learned that the companies affected suffer social ostracism. For instance, a case of mislabeling seriously damages the trust of the company's business partners and consumers, provoking many partners to become angry and cancel contracts. Consumers, too, will easily lose their trust in companies that have deceived them and will not buy these companies' products anymore (Jonker et al., 2005).

Company-Specific Quality Standards as a Differentiation Strategy

Food safety has become an important factor in commercial strategies. The attitude of each individual company may be different, but the issue undoubtedly is an element of its strategy. The leading companies are usually aggressively promoting their food-safety standards and quality standards as differentiating elements. The sensitivity of consumers offers opportunities because a manufacturer can distinguish its products from its competitors by meeting certain standards and providing traceability information. In addition, those retailers that have successfully incorporated safety steps into their management systems are enhancing their brand values. It means that safety-conscious retailers and manufacturers can be more successful in their businesses. In addition to the regulatory food safety standards, Japanese supermarket chains pay attention to HACCP and the International Organization for Standardization (ISO) and set their own original and company-specific quality standards. The number of retailers that make their own quality standards has been increasing recently. Company quality standards relate mainly to issues including size, uniformity of size, appearance, and freshness. In addition, retailers might set minimum residue levels lower than what the governmental regulations require (Jonker et al., 2005).

Another factor enhancing retailers' incentives to define their own standards is related to demographics. The average life expectancy of Japanese people is the highest in the world. The over-65 population total is approximately 25.6 million and accounts for 20.1% of the total population in 2005. The senior citizens are very interested in keeping their health. Many of them are wealthy so they prefer to buy food products with a good health image, even if the prices are a little higher. Strict standards can distinguish such a company's products and indicate that they are better for the consumers' health than the competitors' products.

Traceability

In a recent trend, retailers are presenting information about domestic products and the background of their producers in the store, for example, above or near the shelves. Such information appeals to the consumers' demand for traceability. It shows that the products are produced in a safe way and that the producer takes all responsibility for the products. Increasingly, posting background information is also being done for imported food products. Securing food safety is considered important by almost all companies in the supply chain. Supply-chain management for controlling and managing the entire supply chain has become an essential means for retail chains to guarantee food safety. The word "traceability" is more often referred to, and means, in practice, providing the consumer with information about the origin and background of the product. It is becoming a requirement and in some cases, especially for vegetable products from China, information can be obtained from the Internet after typing a number printed on the package (Jonker et al., 2005).

CONCLUSIONS

The food-safety–related incidents and scandals of the last few years have greatly affected Japanese consumers' consciousness of food safety. For a long time, the Japanese government agencies claimed that the Japanese regulatory system guaranteed the safest food in the world. Most Japanese people believed this claim. However, the food safety incidents and scandals, which involved reputable Japanese food companies, have badly affected consumers' trust in regulatory authorities and undermined public trust in food safety. Food safety has become an important factor for governmental policy and company strategies. In the policy field, the Japanese government has reviewed the food-safety system. The Food Sanitation Law was amended, and the new Food Safety Basic Law took effect by the summer of 2003. Under the latter, the Food Safety Commission, an independent advisory commission under the cabinet office, was established.

The sensitivity of consumers to food safety made food handling companies, especially the leading companies with high profiles, increasingly keen to achieve food safety. These companies are fully aware that if they mishandle food safety requirements, for example, by mislabeling, they will be forced out of business far more quickly than before, by the government, business partners, and consumers. Therefore, tracking and tracing have become a requirement. The sensitivity of consumers also offers opportunities, because manufacturers can distinguish their products from their competitors by meeting certain standards or providing traceability information. It means that safety-conscious retailers and manufacturers can be more successful in their businesses. In addition to the public food safety standards, Japanese supermarket chains set their own company-specific quality standards. Recently, the number of retailers with their own standards has been increasing. Company quality standards mainly relate to issues such as size, uniformity of size, appearance, and freshness. For suppliers in developing countries, these standards are hard to meet. Furthermore, they often differ per retailer, and retailers are usually very strict. Furthermore, it is not easy for companies from developing countries to become new suppliers to Japanese buyers, particularly with regard to the quality aspects. The Japanese buyers prefer bigger suppliers with good access to capital and technology.

The key problems encountered in sanitary, phytosanitary, and other standards relate to the maximum pesticide and antibiotic residue limits, sanitary issues, packing, labeling, and company quality standards. The two most important criteria for the compliance of suppliers in exporting countries with the standards are company-specific factors and country-specific factors. The company-specific factors include the size of the company, experiences with sanitation standards, and the culture and training of the staff. Country-specific factors include the degree to which public authorities have already implemented their own food sanitation regulations, the number of inspectors, authorities' understanding of the necessity for meeting sanitation standards, and public sector assistance to producers and exporters.

Nearly all Japanese importers (trading companies and retailers) send technical staff to suppliers' factories to supervise production and advise them about improvements. The estimate is that the total cost for a buyer to control safety and quality of foreign suppliers' products does not exceed several percent of the total sales of fruits and vegetables. Often the Japanese buyers of fruits and vegetables visit the exporting country before they import the produce. They investigate the actual conditions of the production and transportation systems of the particular supplier and the particular country. If they find any problem, the products will not be imported.

At the time of importation, there are, in principle, three requirements for fruits and vegetables: plant quarantine (according to the Plant Protection Law), food sanitation inspection (according to the Food Sanitation Law), and customs inspection. The importation of certain plants or plant products is prohibited according to the Plant Protection Law, but importation may be allowed under prescribed conditions of quarantine and after the completion of specified procedures. With regard to phytosanitary regulations, fruits and vegetables from a certain county or region are either allowed or not allowed to be exported to Japan, so there is little confusion or uncertainty.

The Food Sanitation Law applies at the time of importation, as well as at the time of sale. It has been established that absence of any pathogenic microorganisms is a prerequisite to the import of fresh fruits and vegetables into the Japanese market as well as at the time of sale. Therefore, regular testing and inspection are conducted to ensure the safety of the fresh produce. Besides this, if a product contains a chemical for which there is no specified MRL, the product cannot be imported into Japan. The improvement of traceability in the supply chain, assistance in equipping packing and processing facilities to meet sanitary requirements, education and training to increase awareness and knowledge of food safety, reinforcement of both public and private inspection systems, and improvements in infrastructure are required to assist suppliers in complying with the Japanese standards.

REFERENCES

Anonymous. (2005). "A guide to GAP" (tentative proposal). The Ministry of Agriculture, Forestry and Fisheries, Food and Consumer Affairs Bureau, Plant Products Safety Division.

FAS, Foreign Agricultural Service, USDA. (2001). Report JA9022 in the GAIN series. www.fas.usda.gov/scriptsw/attacherep/default.asp

FAS, Foreign Agricultural Service, USDA. (June 4, 2001). Report JA1049 in the GAIN series. www.fas.usda.gov/scriptsw/attacherep/default.asp

FSC, Food Safety Commission. (2004). www.fsc.go.jp/english/index.html

Ito, K. and Dyck, J. (2002). Vegetable policies in Japan. Electronic outlook report from the economic research service. USDA VGS–293-01, November 2002.

Ito, K. and Dyck, J. (2004). Japan's fruit and vegetable market. In *Global trade patterns in fruits and vegetables*, Chapter 7. Agriculture and Trade Report No. WRS-04-06. USDA. Washington, DC.

JETRO. (2001a). "Fresh and Frozen Vegetables." In *Marketing Guidebook for Major Imported Products*. Tokyo: Japan External Trade Organization.

JETRO. (2002b). *The Overview of Plant Quarantine: Based on Plant Protection Law*, Tokyo: Japan.

JETRO. (2002c). "Fresh and Frozen Vegetables." In *Marketing Guidebook for Major Imported Products*. Tokyo: Japan External Trade Organization.

JETRO. (2002d). "Fresh Fruits." In *Marketing Guidebook for Major Imported Products*. Tokyo: Japan External Trade Organization.

Jonker, H. T., Ito, H., and Fujishima, H. (2005). Food safety and quality standards in Japan—Compliance of suppliers from developing countries. Agriculture and Rural Development Discussion paper, pp. 1–50. The World Bank.

MAFF. (2006). Annual Report on Food, Agriculture and Rural Areas in Japan FY 2005. Part 1, Trend in Food, Agriculture and Rural Areas Summary.

MAFF. (2007). The Ministry of Agriculture, Forestry and Fisheries of Japan. www.maff.go.jp/e/index.html. Accessed on July 2007.

MHLW. (2006a). Results of monitoring and guidance based on the imported foods monitoring guidance plan for FY 2005. Department of Food Safety, Pharmaceutical and Food Safety Bureau, MHLW.

MHLW. (2006b). Introduction of the positive list system for agricultural chemical residues in foods. Department of Food Safety, Ministry of Health, Labour and Welfare. www.mhlw.go.jp/english/topics/foodsafety/positivelist060228/introduction.html. Accessed on June 2007.

MHLW. (2007). *Administration of Food Safety*. Tokyo: Department of Food Safety, Pharmaceutical and Food Safety Bureau, Ministry of Health, Labor and Welfare. www.mhlw.go.jp/english/topics/foodsafety/index.html. Accessed on July 2007.

Michino, H., Araki, K., Minami, S., Takaya, S., Sakai, N., Miyazaki, M. et al. (1999). Massive outbreak of *Escherichia coli* O157:H7 infection in schoolchildren in Sakai City, Japan, associated with consumption of white radish sprouts. *Am. J. Epidemiol.* **150**(8), 787–796.

MOF. (2006). *Japan exports and imports*. Tokyo: Ministry of Finance. Trade Statistics of Japan, home page www.customs.go.jp/toukei/info/index_e.htm. Accessed on July 2007.

NAAS. (2002). National Agricultural Statistics Services.

Noriyuki, S., Hiratsuka, M., Sugawara, N., Kobayashi, T., Watanabe, S., Yamada, W. et al. (2006). PFGE analysis of *Salmonella* Montevideo isolated from an outbreak of food poisoning and the case of sporadic Salmonellosis. *Jpn. J. Food Microbiol.* **23**(3), 143–148. (In Japanese)

OECD, Organization for Economic Cooperation and Development. (1995). *Agricultural reform and its impact on the fruit and vegetables sector in OECD countries*. Paris.

Rania, H. (2006). Toxicity, mushroom. Health On the Net Foundation, National Center for Medical Informatics. www.emedicine.com/med/topic1527.htm. Accessed on October 15, 2007.

Shields, D. and Wu Huang, S. (2004). China's Fruit and Vegetable Market. In *Global trade patterns in fruits and vegetables*, Chapter 6. Agriculture and Trade Report No. WRS-04-06, USDA. Washington, DC.

Shiina, T. (2006). Trend in good agricultural practices in Japan. *Nippon Shokuhin Kagaku Kogaku Kaishi* **53**(12), 659–664. (In Japanese)

Tanaka, H., Maruyama, T., Shinohara Y., Someya, T., Konuma, K., Isshiki, K. et al. (2003a). The hygienic management guide to fresh vegetables—From production to consumption (a complete edition). Japan Greenhouse Horticulture Association. www.maff.go.jp/soshiki/seisan/yasai/4.pdf. Accessed on July 2007.

Tanaka, H., Maruyama, T., Shinohara Y., Someya, T., Konuma, K., Isshiki, K. et al. (2003b). The hygienic management guide of fresh vegetables—From production to consumption (a simple edition). Japan Greenhouse Horticulture Association. www.maff.go.jp/soshiki/seisan/yasai/yasai.html. Accessed on July 2007.

World Trade Organization. (2001). Notifications by Japan concerning its commitments http://docsonline.wto.org/gen_search.asp. Accessed on July 2007.

Wu Huang, S. (2002). *China Increases Exports of Fresh and Frozen Vegetables to Japan*, ERS, VGS-292-01. USDA, Washington, DC.

PART 5

Technology for Reduction of Human Pathogens in Fresh Produce

CHAPTER 16

Disinfection of Contaminated Produce with Conventional Washing and Sanitizing Technology

Gerald M. Sapers, Ph.D. (Emeritus)
Eastern Regional Research Center, Agricultural Research Service, US Department of Agriculture

CHAPTER CONTENTS

Introduction	393
Washing and Sanitizing Agents	395
Detergent Products	395
Chlorine	395
Alternatives to Chlorine	397
Other Approved Sanitizing Agents for Produce	401
Sanitizing Agents for Organic Crops	403
Expectations for Sanitizing Agents	403
Washing Equipment	403
Types of Washers	403
Efficacy of Commercial Washers	404
Produce Washes for Food Service and Home Use	405
Efficacy of Washing and Sanitizing Methods for Problem Commodities	407
Leafy Vegetables	407
Tomatoes	409
Cantaloupe	410
Apples	412
Conclusions	414

INTRODUCTION

Prepackaged fresh and fresh-cut fruits and salad vegetables represent a major segment of the fresh produce industry. In preceding chapters, we have seen how such products can become contaminated with human pathogens, resulting in outbreaks of foodborne illness. With the exception of vegetable products normally cooked by the consumer, these items are not subjected to a final inactivation step prior to consumption. However, they are washed with water or sanitizing agents, primarily to remove soil and pesticide residues, but also to remove or inactivate human pathogens and spoilage-causing microorganisms. In this chapter we will examine the efficacy of produce disinfection treatments that are based on washing and sanitizing technology.

In addition to populations of epiphytes, freshly harvested produce may contain localized, heavy loads of microbial contaminants, including plant and human pathogens, often associated with soil, decay, and mechanical injury. Washing such produce can transfer microbial contaminants to the wash water and thence to other, uncontaminated raw material as well as the conveying, packing, and processing equipment. Addition of a sanitizing agent to the wash water can greatly reduce the population of planktonic bacterial cells and thus lower the risk of cross-contamination. Such reductions can improve product safety and shelf-life, thereby enabling grower/packers to ship their products nationwide or to overseas markets.

However, cleaning and sanitizing agents are much less effective in removing or inactivating human pathogens and other microorganisms that have attached to produce surfaces. This is a consequence of strong microbial attachment and attachment to inaccessible sites such as pores, punctures, surface irregularities, and cut surfaces, which limits contact between sanitizer solutions and the targeted microorganisms. Also, bacterial incorporation within biofilms will confer greater resistance to microcidal agents (see Chapter 2 for more information on microbial attachment). Thus, the level of pathogen reduction obtained by washing and sanitizing may be inadequate to assure food safety.

In this chapter we will review the characteristics of conventional and novel washing and sanitizing agents used in produce packing and fresh-cut processing, the characteristics of commercial equipment used for washing and disinfecting produce, disinfection treatments suitable for food service and consumer use, and regulatory restrictions regarding sanitizing agents. we will examine the efficacy of such disinfection treatments in reducing pathogen levels on commodities that have a history of association with outbreaks of foodborne illness such as leafy vegetables, tomatoes, cantaloupes, and apples. Previously, these topics have been reviewed by Beuchat (1998), Parish et al. (2003), and Sapers (2003, 2005). Also, recommendations regarding washing and sanitizing appear in the FDA's *Guide to Minimize Microbial Food Safety Hazards of Fresh-cut Fruits and Vegetables* (FDA, 2007a).

WASHING AND SANITIZING AGENTS

Detergent Products

A number of surfactants, including sodium n-alkylbenzene sulfonate, sodium dodecylbenzene sulfonate, sodium mono- and dimethyl naphthalene sulfonates, sodium 2-ethylhexyl sulfate, and others are permitted by the FDA for washing fruits and vegetables (21CFR173.315). Various detergent formulations for washing fresh produce are commercially available including products prepared at a neutral pH, acidified with citric or phosphoric acid, or made alkaline with sodium or potassium hydroxide (see www.cerexagri.com—source of Decco products, www.microcideine.com, www.stepan.com, and www.afcocare.com for details).

Detergents reduce microbial populations on produce surfaces by detachment rather than inactivation. Studies of the efficacy of various commercial detergent formulations, compared to other treatments, in reducing populations of human pathogens on inoculated fruits and vegetables have been reported for apples (Sapers et al., 1999; Wright et al., 2000; Kenney and Beuchat, 2002), strawberries (Raiden et al., 2003), cantaloupe (Sapers et al., 2001), tomatoes (Raiden et al., 2003; Sapers and Jones, 2006), and lettuce (Raiden et al., 2003). The results of these studies indicate that detergent washes sometimes achieve population reductions as great as 2 to 3 logs, equaling or surpassing sodium hypochlorite, but in other cases showed no greater efficacy than water (Raiden et al., 2003).

Chlorine

Because of its microbicidal activity and low cost, chlorine (as sodium or calcium hypochlorite or Cl_2 gas) is the agent most widely used to sanitize fresh produce (Suslow, 2000). Typically, sodium hypochlorite concentrations of 50 to 200 ppm are used. Concentrations may be expressed as total available chlorine (the calculated amount of chlorine present in the sanitizer solution, which includes both free and combined forms of chlorine) or as free chlorine, which depends on the actual chlorine, hypochlorous acid and hypochlorite ion concentrations. At the pH used in packinghouse water systems, the elemental chlorine concentration is near 0, and free available chlorine is the sum of hypochlorous acid and hypochlorite ion. The chlorine concentration (total available or free) can be monitored using test kits based on colorimetric measurements (www.chemetrics.com; www.emscience.com; www.hach.com), or by measurement of the oxidation-reduction potential (ORP).

Depending on the pH, hypochlorite solutions contain varying proportions of hypochlorite ion and hypochlorous acid, the latter having the most bactericidal activity. To enhance the antimicrobial activity of hypochlorite solutions, the solution pH may be reduced from the alkaline range (about pH 9) to the slightly acidic range of 6 to 7 by addition of citric acid, a mineral acid or buffer (available commercially from www.cerexagri.com). Additionally, some benefits may be realized by adding a surfactant to the hypochlorite solution to improve contact with the microbial surface (Segall, 1968; Spotts, 1982). Chlorine is highly reactive with organic species originating in soil and debris or leached from damaged produce into the process water, resulting in rapid chlorine depletion and greater survival of the targeted microflora when the organic load is high (Suslow, 2000). Hence, the chlorine level in wash water should be monitored continuously by measuring the ORP and replenished to maintain the desired concentration using automated commercial systems (see www.pulseinstruments.net or www.globalspec.com).

Although chlorine has a broad spectrum of antimicrobial activity and is highly efficacious in inactivating planktonic microorganisms in wash water (depending on the organic load), it is far less effective against bacteria attached to produce surfaces. Population reductions reported in the literature for indigenous microflora and for human pathogens on inoculated samples rarely exceed 2 logs (99%) (Brackett, 1987; Zhuang et al., 1995; Beuchat et al., 1998; Garcia et al., 2003). Such reductions may have a large impact on the incidence of spoilage and will significantly reduce the risk of foodborne illness by reducing the load of attached pathogens and protecting against cross-contamination. However, because of the low infectious dose of some pathogens, they cannot assure safety.

To provide an acceptable level of safety for fresh juice, the FDA has mandated a 5-log reduction (99.999%) of pathogens in the juice product. It is clear that this cannot be accomplished solely by use of chlorine (or any other sanitizers); in the case of apple juice, this must be accomplished by pasteurization or UV-treatment of the finished juice. With citrus juices, the 5-log reduction can be apportioned between surface treatment of the fruit and juice treatment (FDA, 2001).

Chlorine solutions are considered to be highly corrosive, especially at low pH, and will shorten the life of tanks and other stainless steel equipment used in produce packing/processing operations. Also, because of reports in the literature indicating potential mutagenicity and carcinogenicity from exposure to reaction products of chlorine with food constituents, there is some concern in the food industry regarding future regulatory restrictions on the use of this sanitizer (Chang et al., 1988; Hidaka et al., 1992). Consequently, a number of alternatives to chlorine have been developed or are under study for use by the food industry.

Alternatives to Chlorine

Electrolyzed Water

Electrolyzed water (also known as electrolyzed oxidizing or EO water) has received much attention as a replacement for chlorine in sanitizing produce, although in principle, it represents an alternative means of generating hypochlorous acid (Izumi, 1999). Hypochlorous acid is formed at the anode during electrolysis of water containing some sodium chloride. Depending on the sodium chloride concentration, the available chlorine level can reach or exceed 100 ppm. If the electrolyzed water generator has a membrane separating the electrodes, highly acidic (pH < 3.0) water will be produced at the anode, and alkaline water (pH ≥ 11.0) will be produced at the cathode. Electrolyzed water is considered to be an effective sanitizing agent at low pH with an oxidation-reduction potential greater than 1000 mV, and it can be used as a cleaning agent at high pH with a redox potential less than 800 mV (Deza et al., 2003; Yang et al., 2003; Ozer and Demirci, 2006).

Electrolyzed water is highly effective in reducing the population of planktonic cells (Venkitanarayanan et al., 1999a), but like chlorine, its efficacy in reducing bacterial populations attached to produce surfaces is generally limited to 1 to 3 logs (Izumi, 1999; Park et al., 2001). Other studies have yielded population reductions between 1 and 7 logs, depending on the commodity, method of inoculation and recovery, inoculation site, interval between inoculation and treatment, method of treatment, and strength of electrolyzed water (Koseki et al., 2003, 2004; Yang et al., 2003; Deza et al., 2003; Bari et al., 2003). It is questionable whether some of the larger reductions in attached microbial populations can be realized in a packing or processing plant situation.

Aqueous Chlorine Dioxide and Acidified Sodium Chlorite

Solutions of chloride dioxide (ClO_2), at residual concentrations not to exceed 3 ppm, are permitted by the FDA for sanitizing fresh fruits and vegetables (21CFR173.300). Such treatment shall be followed by a potable water rinse. ClO_2 must be generated on-site by such means as reaction of sodium chlorite with either chlorine gas or a mixture of sodium hypochlorite and hydrochloric acid. Information on ClO_2 generators can be obtained from the many vendors of this equipment (see Vulcan Chemical (800–873–4898); CH2O Inc. (Fresh-Pak™; www.ch2o.com); Rio Linda Chemical Co., Inc. (916–443–4939); Bio-Cide International, Inc. (Oxine®; www.bio-cide.com); International Dioxcide (www.idiclo2.com); Alex C. Fergusson, Inc. (www.afcocare.com; 800–345–1329); CDG Technology, Inc. (www.cdgtechnology.com); and others). Chlorine dioxide gas also can be generated from sachets containing a dry mixture of sodium chlorite or sodium chlorate and an activator (www.icatrinova.com). In contrast to hypochlorite, ClO_2 is claimed to be more effective at neutral pH, less reactive with organics, less corrosive, and form fewer chlorinated byproducts (Anon., 2001). However, ClO_2 is unstable, and at partial pressures greater than 120 mm Hg (15.8% by volume at atmospheric pressure), becomes explosive (see CDG Technology Chlorine Dioxide Safety Fact Sheet at www.cdgtechnology.com).

Pathogen population reductions obtained with ClO_2 solutions vary widely from study to study, depending on the target organism, the commodity, and to a lesser extent, on the ClO_2 concentration (1–5 ppm), but are in the range of less than 1 to 3 logs (Zhang and Farber, 1996; Wisniewsky et al., 2000; Han et al., 2001; Huang et al., 2006). However, Rogers et al. (2004) reported much larger reductions in the populations of *E. coli* O157: H7 and *L. monocytogenes* on inoculated apples, lettuce, strawberries, and cantaloupe. Using ClO_2 gas generated from dry reactant sachets, Lee et al. (2006) obtained 4.5-log reductions in *Alicyclobacillus acidoterrestris* spores, inoculated on apple surfaces; this organism is responsible for spoilage of apple juice.

Acidified sodium chlorite (ASC) is produced by mixing a solution of sodium chlorite with any GRAS acid and is considered to be a source of chlorous acid ($HClO_2$), the primary antimicrobial agent, although some ClO_2 is produced gradually as ASC decomposes (Warf, 2001). ASC is permitted by the FDA for use as an antimicrobial agent in the water applied to processed fruits and vegetables at concentrations of 500 to 1200 ppm and pH 2.3 to 2.9 by spraying or dipping, provided that the treatment is followed by a potable water rinse and a 24-hour waiting period prior to consumption; with leafy vegetables, only application by dipping is permitted (21CFR173.325). Most efficacy studies indicate that treatment of fresh produce with up to 1200 ppm ASC can reduce the natural microflora and *Salmonella*, *Staphylococcus aureus*, *E. coli* O157:H7, and *L. monocytogenes* on inoculated produce (lettuce, cucumbers, bell peppers, tomatoes, Chinese cabbage, cantaloupes, strawberries, apples, alfalfa sprouts) by about 1 to 3 logs (Park and Beuchat, 1999; Conner, 2001; Fett, 2002; Caldwell et al., 2003; Yuk et al., 2005, 2006; Inatsu et al., 2007). In side-by-side comparisons, ASC was at least as efficacious if not superior to hypochlorite. However, Gonzalez et al. (2004) obtained reductions in the population of *E. coli* O157:H7 as great as 5.25 logs on inoculated shredded carrot by treatment with 1000 ppm ASC.

Aqueous Ozone

Ozone is a highly effective, broad spectrum antimicrobial agent, effective at low concentrations and short contact times (Wickramanayake, 1991;

Restaino et al., 1995). Ozone is highly unstable and decomposes to non-toxic products. However, it is corrosive to equipment and can cause physiological injury to produce and degrade product color and flavor. Ozone is toxic and an irritant to workers at concentrations in air greater than 0.1 ppm (29CFR1910.1000) and must be adequately vented to avoid worker exposure (Anon., 2001). Ozone is approved for food use by the FDA (21CFR173.368). Food applications of ozone have been reviewed by Graham (1997), Kim et al. (1999), Xu (1999), Khadre et al. (2001), Smilanick (2003), Suslow (2004), Sharma (2005), and Karaca and Velioglu (2007). Ozone must be generated on-site by passing air or oxygen through a corona discharge or UV light (Xu, 1999). Information about commercial ozone generators is available on-line from Air Liquide (www.airliquide.com), Praxair, Inc. (www.praxair.com), Novazone (www.novazone.net), Pure Ox (www.pureox.com), GE Water & Process Technologies (www.gewater.com), Ozonia North America, Inc. (www.ozonia.com), Lynntech, Inc. (www.lynntech.com), Clean Air & Water Systems, LLC (www.caawsystems.com), and others.

Ozonation can reduce bacterial populations in flume and wash water; typical use rates for disinfection of postharvest water are 2 to 3 ppm (Suslow, 2004). The efficacy of ozone treatment of fresh produce is generally similar to that of chlorine and other chlorine alternatives (Kim et al., 1999; Garcia et al., 2003). However, ozone treatment was ineffective in reducing populations of *E. coli* O157:H7 in the stem and calyx regions of inoculated apples (Achen and Yousef, 2001), reducing postharvest fungal decay of pears (Spotts and Cervantes, 1992), and decontaminating alfalfa seeds inoculated with *E. coli* O157:H7 (Sharma et al., 2002) and *L. monocytogenes* (Wade et al., 2003), probably because of the difficulty in contacting and inactivating bacteria attached in inaccessible sites (see Chapter 2). In contrast, Rodgers et al. (2004) reported much higher population reductions with several commodities. Conditions for obtaining a 5-log reduction of *E. coli* O157:H7 in apple cider by treatment with ozone gas were described by Steenstrup and Floros (2004). However, an outbreak of cryptosporidiosis associated with ozonated apple cider suggests that this application is not feasible, perhaps because of the inherent inadequacy of ozone in inactivating *Cryptosporidium* or improper application of ozone. The FDA advised juice processors not to use ozone unless they can demonstrate a 5-log reduction through ozonation (FDA, 2004; Blackburn et al., 2006).

Bialka and Demerci (2007) obtained large reductions in the populations of *E. coli* O157:H7 and *Salmonella* on inoculated raspberries and strawberries, commodities that are difficult to treat because of their fragility, by treatment with aqueous ozone (1.7 to 8.9 mg/liter), but treatment times were as long as 64 minutes. This study did not report treatment effects on sample shelf-life. One might expect some fungal spoilage during storage of washed small fruits unless the fungal population was greatly reduced by exposure to ozone.

Peroxyacetic Acid

Peroxyacetic acid (PAA), a highly effective antimicrobial agent (Block, 1991), is actually an equilibrium mixture of PAA, hydrogen peroxide, and acetic acid. This product is approved by the FDA (21CFR173.315) for addition to wash water at concentrations not to exceed 80 ppm, and, under EPA regulations, is exempt from the requirements of a tolerance for residues resulting from treatment of fruits and vegetables with PAA solutions at concentrations up to 100 ppm (40CFR180.1196). PAA is a strong oxidizing agent, and handling may be hazardous at high concentrations. PAA is available at various strengths from Ecolab, Inc. (www.ecolab.com), FMC Corp. (www.fmcchemicals.com), and Solvay Interox (www.solvayinterox.com). PAA formulations are recommended for treating process water and are also claimed to substantially reduce microbial populations on fruit and vegetable surfaces (www.ecolab.com/initiatives/foodsafety). Lower concentrations of PAA are effective in killing pathogenic bacteria in aqueous suspension than would be required with chlorine (Block, 1991). However, population reductions for the indigenous microflora and human pathogens on inoculated produce are generally no greater than 1 or 2 logs (Sapers et al., 1999; Wisniewsky et al., 2000; Wright et al., 2000; Lukasik et al., 2003; Caldwell et al., 2003; Nascimento et al., 2003; Beuchat et al., 2004; Oh et al., 2005; Yuk et al., 2005, 2006) with few exceptions (Park and Beuchat, 1999; Rodgers et al, 2004; Allwood et al., 2004). Formulations of PAA containing octanoic acid were more effective in killing yeasts and molds in fresh-cut vegetable process waters but had little effect on population on fresh-cut vegetables (Hilgren and Salverda, 2000).

Efficacy of Combination of Treatments

Certain combinations or sequences of treatments may show synergism or an additive effect in reducing populations of microbial contaminants on produce. Combinations of lactic acid with chlorine (Zhang and Farber, 1996; Escudero et al., 1999; Materon, 2003) or hydrogen peroxide (Venkitanarayanan et al., 1999c, 2002; Lin et al., 2002; Rupasinghe et al., 2006), acetic acid with hydrogen peroxide (Liao et al., 2003), and ozone with chlorine (Garcia et al., 2003) show promise. Huang et al. (2006) reported enhancement of ClO_2 treatment by sonication. Sequential washing of whole cantaloupes with an acidic detergent, followed by a 2000 ppm chlorine wash, reduced the total aerobic plate count of the fresh-cut melon pieces initially and delayed outgrowth of survivors during storage at 4 °C, suggestive of injury to survivors (Sapers et al., 2001). On the other hand, addition of surfactants did not enhance efficacy of chlorine (Zhang and Farber, 1996; Escudero et al., 1999) or hydrogen peroxide (Sapers and Jones, 2006). Further research in this area may yield treatment combinations that show greater efficacy.

Other Approved Sanitizing Agents for Produce

Hydrogen Peroxide

Hydrogen peroxide (HP), a highly effective antimicrobial agent (Block, 1991), may be a potential alternative to chlorine for sanitizing fresh produce (Sapers, 2003), although HP's regulatory status in the United States requires clarification. Use of HP as a wash for raw agricultural commodities is covered under regulations of the US Environmental Protection Agency (EPA), and such applications are exempt from the requirements of a tolerance if the concentration used is 1% or less per application (40CFR180.1197). However, use of HP in fresh-cut processing operations would be regulated by the FDA, and although HP is considered GRAS for certain specified applications, its use as a produce wash is not addressed by FDA regulations (21CFR184.1366).

Numerous studies have demonstrated the efficacy of dilute hydrogen peroxide in sanitizing fresh produce including mushrooms (Sapers et al., 2001), apples (Sapers et al., 1999, 2000, 2002), melons (Sapers et al., 2001; Ukuku et al., 2004), and eggplant and sweet red pepper (Fallik et al., 1994). Hydrogen peroxide treatments were ineffective in decontaminating sprouts (Fett, 2002) or the seeds used to produce sprouts (Weissinger and Beuchat, 2000). Contact with stainless steel and aluminum alloy equipment can destabilize hydrogen peroxide solutions, and such equipment must be passivated by treatment with nitric or citric acid solution prior to exposure to H_2O_2 to render it less reactive (Sapers, 2003). Information on hydrogen peroxide applications can be obtained from FMC Corp. (www.fmcchemicals.com), Solvay Interox (www.solvayinterox.com), US Peroxide (www.h2o2.com), Degussa Corp. (www.degussa.com), and BiosSafe Systems (www.biosafesystems.com).

Organic Acids

Organic acids such as citric, lactic, and acetic acids are effective antibacterial agents (Doores, 2005) and have been classified by the FDA as GRAS (21CFR184.1005, 1033, 1061). Information about applications of lactic acid and lactates can be obtained from Purac America, Inc. (www.purac.com). Numerous studies have demonstrated the efficacy of organic acids, used in combination with other sanitizing agents in reducing pathogen levels on fresh produce (see page 401). However, use of organic acid alone in wash water has been less effective, resulting in pathogen reductions of 1 log or less for *L. monocytogenes* in shredded lettuce with acetic and lactic acids (Zhang and Farber, 1996), 1 log or less for *E. coli* O157:H7 and *Salmonella* in apples with vinegar (Lukasik et al., 2003; Liao et al., 2003), 1.6 to 2 log *E. coli* O157:H7 in cantaloupe with 2% lactic acid at 55 °C (Alvarado-Casillas et al., 2007), and less than 1 log *S.* Typhimurium on tomatoes with 2% lactic acid (Ibarra-Sánchez et al., 2004). However, greater reductions were reported for organic acid treatment of *Yersinia enterocolitica* on lettuce (Escudero et al., 1999), *E. coli* O157:H7 on cantaloupes (Materon, 2003) and apples (Wright et al., 2000), *E. coli* (CDC1932) on iceberg lettuce (Vijayakumar and Wolf-Hall, 2002), and mesophilic aerobes on lettuce (Nascimento et al., 2003). Differences in results between studies on the same commodity using similar treatments probably reflect differences in methodology.

Alkaline Products

Sodium metasilicate (AvGard® XP) has been marketed by Danisco A/S (www.danisco.com) as an antimicrobial rinse to reduce human pathogen populations on processed beef and poultry, and this product has been approved by FDA for produce washing (21CFR184.1769a). The antimicrobial activity of alkaline products such as sodium metasilicate and trisodium phosphate (TSP, AvGard®) is probably due to their high pH (11–12), which disrupts the cytoplasmic membrane (Mendonca et al., 1994; Sampathkumar et al., 2003). Population reductions of 1 to 3 logs have been reported with alkaline produce washes (Zhuang and Beuchat, 1996; Pao et al., 2000; Lukasik et al., 2003), although treatment of shredded lettuce with TSP was ineffective in killing *L. monocytogenes* (Zhang and Farber, 1996). TSP was highly effective in inactivating *E. coli* O157:H7 in biofilms but less effective against biofilms of *S. typhimurium* and *L. monocytogenes* (Somers et al., 1994). TSP has fallen out of favor because of phosphate disposal issues (http://meatupdate.csiro.au/new/Trisodium%20Phosphate.pdf).

Iodine

An iodine-based system (Isan® system) for treatment of fruits and vegetables is claimed to provide a high kill rate, require no pH adjustment, and be less corrosive than other sanitizers (www.ioteq.com; Klein and Morris, 2004). However, data demonstrating efficacy of this treatment against human pathogens have not been published. This system is approved in Australia and New Zealand.

Sanitizing Agents for Organic Crops

Packers and processors of organic crops must conform to special USDA regulations regarding use of nonagricultural substances for washing and sanitizing processed organic products, if these products are to be labeled and marketed as organic. Approved antimicrobial agents, identified in the USDA National Organic Program List of Allowed and Prohibited Substances (7CFR205.605), include chlorine materials (calcium hypochlorite, sodium hypochlorite, ClO_2) and PAA for disinfecting and sanitizing food contact surfaces, and HP, ozone, and PAA for use in wash or rinse water according to FDA limitations. Additional restrictions placed on chlorine materials state that "residual chlorine levels in the water shall not exceed the maximum residual disinfectant limit under the Safe Drinking Water Act." According to Suslow (2000), this is interpreted to be "10 ppm residual chlorine measured downstream of the wash step."

Expectations for Sanitizing Agents

Numerous studies have demonstrated that use of chlorine and other sanitizing agents permitted by the FDA and EPA cannot achieve better than 1- to 3-log reductions in microbial populations attached to fresh produce. Some incremental improvements in efficacy may be possible. It is clear that washing and sanitizing treatments represent a hurdle, accomplishing some good by reducing the microbial load, but not enough to assure safety. When infectious doses are small, as few as 10 cells for *E. coli* O157:H7 (FDA, 2003), a 1-, 2-, or even 3-log reduction may not be enough to prevent significant numbers of people from getting sick.

WASHING EQUIPMENT

Types of Washers

Many types of washers are available to the produce industry, designed according to the characteristics (shape, size, and fragility) and special requirements of specific commodities, for removal of soil, debris, and pesticide residues from harvested produce. Such equipment is not generally designed specifically to remove microorganisms attached to fruit and vegetable surfaces. Design criteria are reviewed by Saravacos and Kostaropoulos (2002). Types of commercial washers for fresh produce include dump tanks, flatbed and U-bed brush washers, reel washers, pressure washers, hydro air agitation wash tanks, and immersion pipeline washers. Suppliers of such equipment are listed in buyers' guides published on-line at sites such as the Postharvest Resources Web site of the University of Florida (http://postharvest.ifas.ufl.edu) and the United Fresh Produce Association (www.unitedfresh.org/programs/special).

Efficacy of Commercial Washers

Studies conducted by Annous et al. (2001) with dip-inoculated apples demonstrated that the population of attached *E. coli* (strain K12) could be reduced by about 1 log (90%) by passage of the apples through a dump tank containing water with minimal agitation. However, further cleaning of the apples in a flat-bed brush washer (rotating brushes in a horizontal plane under spray) had little additional effect on the remaining *E. coli* population, irrespective of whether the washing agent used was water, a detergent, or a biocide (Table 16.1). Similar results were obtained in experiments with a U-bed brush washer (rotating brushes in U-shaped configuration causing tumbling action under spray) (Sapers, 2002). Survival of *E. coli* was attributed to attachment on inaccessible surfaces in the stem and blossom ends of the apples, infiltration within the latter region, or incorporation into resistant biofilms. Greater efficacy was obtained when the apples were washed by full immersion in a sanitizing solution with vigorous agitation (Sapers et al., 2002).

Garcia et al. (2006) identified the washing step in commercial apple cider production as a potential source of contamination, possibly because of excessive microbial build-up in dump tanks and improper cleaning and sanitizing of washing equipment. In a study of commercial washing practices in the Rio Grande River Valley of Texas, Gagliardi et al. (2003)

Table 16.1 Decontamination of Apples Inoculated with *E. coli* (Strain K12) with Sanitizing Washes Applied in a Flat-Bed Brush Washer[a]

Wash Treatment	Temp. (°C)	*E. coli* ($\log_{10}$ CFU/g)[b]		
		Before Dump Tank	After Dump Tank	After Brush Washer
Water	20	5.49 ± 0.09	4.92 ± 0.37	4.81 ± 0.26
	50	5.49 ± 0.09	5.03 ± 0.15	4.59 ± 0.08
200 ppm Cl_2	20	5.87 ± 0.07	5.45 ± 0.05	5.64 ± 0.23
8% Na_3PO_4	20	5.49 ± 0.09	5.02 ± 0.43	4.98 ± 0.02
	50	5.49 ± 0.09	5.02 ± 0.08	4.75 ± 0.45
1% acidic detergent	50	5.87 ± 0.07	5.49 ± 0.03	5.42 ± 0.50
5% H_2O_2	20	5.87 ± 0.07	5.46 ± 0.40	5.27 ± 0.09
	50	5.87 ± 0.07	5.54 ± 0.31	5.49 ± 0.10

[a] From Annous, B.A. et al. (2001). Reprinted with permission from the *Journal of Food Protection*. Copyright held by the International Association for Food Protection, Des Moines Iowa.
[b] Mean of 4 determinations ± standard deviation.

reported little or no reduction and some significant increases in populations of coliforms and enterococci in cantaloupes cleaned in a "spray-propulsioned" immersion wash tank, hydrocooled, and then spray rinsed on a conveyor line. Much of the contamination was traced to the wash tank or hydrocooler, perhaps resulting from soil accumulation and chlorine depletion. Cantaloupes may be especially difficult to disinfect, even if fully immersed in the sanitizing solution, because of microbial attachment within inaccessible sites in the netting and stem scar (Richards and Beuchat, 2004). Hassenberg et al. (2007) reported only a small decrease in the population of microorganisms in lettuce washed with ozonated water in a commercial lettuce washing facility.

PRODUCE WASHES FOR FOOD SERVICE AND HOME USE

Many washing and sanitizing agents that can achieve pathogen reduction levels as great as 3 logs under commercial treatment conditions are not suitable for food service or home use because the users lack the technical skills, knowledge, and equipment to apply treatments safely and effectively. However, because of greater awareness by food-service managers and many consumers of the increasing risk of produce-associated foodborne illness, there has been an explosion of interest in produce washes that can be used safely by food-service workers or in the kitchen. A sampling of Web sites describing such products is provided in Table 16.2. Most of these fruit and vegetable washes contain mixtures of surfactants, and in some cases, are combined with chelating agents, buffers, and antioxidants. Product descriptions on their Web sites claim that the washes are capable of removing dirt, pesticide residues, waxes, animal waste, and bacteria from fruit and vegetable surfaces. However, only two of these products, PRO-SAN (www.microcideinc.com) and Victory (www.ecolab.com) are claimed to be bactericidal and appear to be suitable for the institutional market. Victory is a peroxyacetic acid-based antimicrobial produce wash designed specifically for the food-service industry.

There are few scientific studies validating the use of produce washes marketed for home use. Lukasik et al. (2003) obtained population reductions of 1 to 2 logs on strawberries inoculated with *E. coli* O157:H7, *S.* Montevideo, and several viruses, by treatment with Fit® and Healthy Harvest (a nonionic surfactant product). Much larger reductions were reported for tomatoes inoculated with *Salmonella* serotypes and washed with Fit®, using a standardized method of testing (Harris et al., 2001). Smith et al. (2003) reported reductions of only 1 log in the microbial load on lettuce by treatment with Victory. Other studies have evaluated diluted

Table 16.2 Fruit and Vegetable Wash Products Available on the Internet[a]

Product Name	Composition	Web Site
Clean Greens	Surfactants, chelating agents, buffers, antioxidants	www.cleangreensinc.com
Earth Friendly Fruit & Vegetable Wash	Surfactant, citric acid	www.kalyx.com
Fit Fruit & Vegetable Wash	Citric acid, oleic acid, glycerol, ethyl alcohol, baking soda, potassium hydrate, distilled grapefruit oil	www.tryfit.com
Fruit and Vegetable Wash	Surfactants, chelating agents, antioxidants	www.organicandnature.com
Fruit & Vegetable Wash	Surfactant blend	www.vegiwash.com
Fruit-Vegetable Wash	Contains grapefruit seed extract	www.naturescrib.com
Healthy Harvest Vegetable	Nonionic surfactant	www.kidsorganics.com
Mom's Veggie Wash	Surfactant blend	www.veggiewash.com
PRO-SAN Fruit and Vegetable Wash	Unspecified	www.microcideinc.com
Sprout Spray Fruit & Veggie Wash	Unspecified	www.handypantry.com
Veggie Wash	Unspecified ingredients from citrus, corn, and coconut	www.citrusmagic.com
Veggi Wash Fruit Too	Plantaren, sucrose esters, cocoyl glutamate, trisodium citrate, glycerine	www.goodnessdirect.co.uk
Veg'n'Fru Wash	Sorbitol-based, all natural	www.vegnfruwash.com
Watkins Fresh Wash	Sodium cocoyl glutamate, polysorbate 20 and 80, glycerine	www.watkinsonline.com

[a] Listing of products in Table 16.2 does not constitute an endorsement by the author, and the products listed therein are not recommended over other products of a similar nature not identified by the author.

vinegar (Vijayakumar and Wolf-Hall, 2002; Parnell and Harris, 2003; Nascimento et al., 2003) as a produce wash for consumer use. Neither the FDA (www.cfsan.fda.gov/~dms/prodsafe.html) nor the USDA (www.fsis.usda.gov/Fact_Sheets/Does_Washing_Food_Promote_Food_Safety/index.asp) recommend that consumers wash fruits and vegetables with soap, detergents, or commercial produce washes. They *do* recommend, however, washing under cold running tap water to remove any lingering dirt and scrubbing with a clean produce brush if the produce has a firm surface. However, reductions in the bacterial load on cantaloupe obtained by washing with water and scrubbing were poor, only 70%, but not much worse than the 90% reduction obtained by dipping in 150 ppm sodium hypochlorite (Barak et al., 2003).

Table 16.3 Equipment for Small Scale Application of Commercial Sanitizing Agents for Fresh Produce

Sanitizing Agent	Product Name	Website
Electrolyzed water	ElectroCide System	www.electrolyzercorp.com
	Models BTM 3000, Watrix 10G, AC 2.0L, AM 2.0L, AL-5GR	http://heartspring.net/electrolyzedwater.html
	Hoshizaki ROX 20TA-U Water Electrolyzer	www.hoshizakiamerica.com
	Sterilox Food Safety Generator	www.puricore.com
Ozone	Applied Ozone Systems	www.appliedozone.com
	The Aqua Clean AQ-20	www.ozonesafefood.com
	ClearWater Tech HDO3 Dissolved Ozone Systems	www.cwtozone.com
	Lotus Sanitizing System	www.sharperimage.com
	Pacific Ozone Element Ozone FS2180	www.pacificozone.com
Chlorine Dioxide (sachets and tablets)	Engelhard Aseptrol	www.engelhard.com
	Highland Fresh Technologies Selectrocide	www.highcor.com
	Quiplabs MB 10 Tablet	www.quiplabs.com

Small-scale systems for applying electrolyzed water, ozone, and chlorine dioxide are now being marketed (Table 16.3). Some of these may have application for food service use, but treatment control and safety issues must be addressed before such equipment can be recommended. Venkitanarayanan et al. (1999b) reported that an electrolyzed water treatment was effective in inactivating foodborne pathogens on plastic cutting boards.

EFFICACY OF WASHING AND SANITIZING METHODS FOR PROBLEM COMMODITIES

Leafy Vegetables

Leafy vegetables and herbs, including lettuce (romaine, iceberg, mesclun), spinach, parsley, and cilantro have been implicated in numerous outbreaks of food poisoning caused by *E. coli* O157:H7, *Salmonella*, Norwalk-like virus, hepatitis A, and other human pathogens in recent years (DeWaal and Barlow, 2002). Fresh-cut salad vegetables are subjected to triple-wash treatments and, since the inception of this industry, have been claimed to be safe. Yet bagged or fresh-cut spinach, and romaine and iceberg lettuce have been associated with major outbreaks. In view of these outbreaks, we must question whether washing and sanitizing treatments are capable of disinfecting contaminated leafy vegetables.

Numerous studies have indicated the limited ability of chlorine to reduce populations of human pathogens on inoculated lettuce, typically reporting 1- to 2-log reductions, with regrowth during posttreatment storage (Beuchat and Brackett 1990; Zhang and Farber, 1996; Beuchat et al., 1998; Delaquis et al., 2002; Lang et al., 2004; Beuchat et al., 2004; Hellström et al., 2006). Lang et al. (2004) reported somewhat greater reductions with parsley. Francis and O'Beirne (1997) found greater regrowth of *L. innocua* at 8 °C in shredded lettuce that had been dipped in 100 ppm chlorine than in water-dipped controls. They suggested that this might be due to a reduction in the population of indigenous microflora, thereby giving *L. innocua* a competitive advantage. Beuchat et al. (2004) reported large reductions in the concentration of free chlorine as the ratio of lettuce to solution and treatment time increased, the largest decreases occurring with shredded iceberg lettuce and iceberg pieces. These losses were attributed to release of chlorine-consuming tissue juices from the cut lettuce. Pirovani et al. (2001) modeled depletion of chlorine during washing of fresh-cut spinach with chlorinated water at different chlorine concentrations, water-to-produce ratios, and treatment times, all of which affected extent of depletion. Reductions in the total microbial populations were only 2 to 3 logs.

Electrolyzed water has been evaluated as a sanitizing agent for lettuce with mixed results. This may be explained in part by differences in methodology used by investigators, especially the method of inoculation. Population reductions for *E. coli* O157:H7 and *Salmonella* spp. on inoculated head lettuce treated with acidic electrolyzed water (200 ppm free available chlorine) were only about 1 log for dip inoculation, compared to reductions of about 2.5 logs for samples spot inoculated on the inner surface, and 4.5 logs for samples spot inoculated on the outer surface (Koseki et al., 2003). These results may provide some insight into the survival of human pathogens on contaminated lettuce leaves cleaned and sanitized with commercial washing equipment. Similar reductions were reported previously by Park et al. (2001), with iceberg lettuce spot inoculated with *E. coli* O157:H7 and *L. monocytogenes*. Other studies with inoculated romaine and iceberg lettuce reported 2-log reductions (Yang et al., 2003; Koseki et al., 2004).

Various other FDA-approved alternatives to chlorine including ozone, ClO_2, PAA, and detergents have been evaluated for sanitizing lettuce. Treatment of shredded or fresh-cut iceberg lettuce with ozone resulted in 1- to 2-log reductions in counts of the natural microflora (Kim et al., 1999; Garcia et al., 2003); ozone-chlorine combinations were more effective than the individual treatments (Garcia et al., 2003). Treatment of inoculated shredded lettuce with 1 to 5 ppm ClO_2 resulted in minimal reductions in

L. monocytogenes population (≤ 1 log) (Zhang and Farber, 1996), but treatment of inoculated romaine lettuce leaves with 5 to 40 ppm ClO_2 in combination with ultrasonification reduced *Salmonella* and *E. coli* O157:H7 populations by 2 to 3 logs (Huang et al., 2006). Application of 80 ppm PAA (Tsunami 100) to lettuce reduced the population of mesophilic aerobes and total coliforms by 1.85 and 1.44 logs, respectively (Nascimento et al., 2003), and reduced the population of *L. monocytogenes* on inoculated iceberg and romaine lettuce by 0.7 to 1.8 logs (Beuchat et al., 2004). Application of 40 ppm Tsunami 200 as an aerosol to spot inoculated iceberg lettuce leaves reduced populations of *E. coli* O157:H7, *S.* Typhimurium, or *L. monocytogenes* by 2.2, 3.3, and 2.7 logs after 30-minute exposure (Oh et al., 2005).

It is apparent that the efficacy of commercially available, FDA-approved sanitizing agents against human pathogens on contaminated lettuce and other leafy vegetables is limited to population reductions of 1 to 3 logs at best. Presumably, this is a consequence of the strong attachment of contaminants to inaccessible sites on the leaf surface and cut edges, internalization of bacterial cells within the leaf (Solomon et al., 2002), and incorporation of cells within resistant biofilms (Carmichael et al., 1999) so that contact between the human pathogens and sanitizing agent is insufficient for inactivation (see Chapter 2). Additionally, depletion of chlorine and other sanitizing agents by reaction with tissue juices at cut edges of leaves during washing (Pirovani et al., 2001, Beuchat et al., 2004) may contribute to their limited efficacy.

Tomatoes

Tomatoes have a history of association with outbreaks of *Salmonella* food poisoning (Cummings et al., 2001; CDC 2002, 2005, 2006). Research conducted in the 1990s demonstrated that *Salmonella* could be inactivated on the unbroken skin of tomatoes by washing with chlorinated water (100 ppm free Cl_2) but could survive if attached in the stem scar, in core tissue, or within growth cracks (Wei et al., 1995; Zhuang et al., 1995). Furthermore, internalization of bacteria by infiltration through the stem scar into the core could be driven by a temperature differential between the tomato fruit (warm) and the wash water (cold) or by a hydrostatic pressure differential depending on the depth of immersion of tomatoes in a wash tank (Bartz and Showalter, 1981; Bartz, 1982). Guo et al. (2002) demonstrated infiltration of *Salmonella* through the stem scar when tomatoes were placed stem-scar-down in contact with inoculated moist soil. However, this condition would not be likely to occur under field conditions unless accidentally detached or dropped fruits were subsequently harvested from the ground.

Various sanitizing agents and methods of application have been employed in attempts to improve the disinfection of tomatoes inoculated with *Salmonella*. However, population reductions generally were in the range of 1 to 3 logs for a spray application of 2000 ppm chlorine (Beuchat et al., 1998), a dip treatment with 15% trisodium phosphate (Zhuang and Beuchat, 1996), a dip or spray treatment with lactic acid solution at 55 °C (Ibarra-Sanchez et al., 2004), a dip treatment with 5% HP at 60 °C (Sapers and Jones, 2006), treatment with 1200 ppm acidified sodium chlorite, 87 ppm PAA (Yuk et al., 2005), and treatment of air-dried tomatoes with 5 ppm aqueous ClO_2 (Pao et al., 2007). Treatments in which the tomatoes were spot inoculated on smooth skin away from the stem scar area (Harris et al., 2001; Venkitanarayanan et al., 2002; Bari et al., 2003) tended to yield greater population reductions (4–7.5 logs) than when the tomatoes were spot inoculated at the stem scar or dip inoculated (Raiden et al., 2003; Ibarra-Sanchez et al., 2004; Yuk et al., 2005; Pao et al., 2007).

Some of the larger population reductions reported in studies where the interval between inoculation and treatment was brief (< 60 min) may be indicative of treatment efficacy when contamination occurs on the packing or processing line; for example, in dump tanks, hydrocoolers, or flumes. However, such treatments might be substantially less effective when the interval between inoculation and treatment is longer (days), as would be the case with preharvest contamination where pathogens might be protected by attachment in protected sites (growth cracks, punctures, or other surface irregularities) or by biofilm formation. In experiments with dip inoculated tomatoes, Sapers and Jones (2006) obtained 1.8- and 2.6-log reductions in the *Salmonella* population when the inoculated tomatoes were held at 20 °C for one hour prior to treatment with 150 ppm chlorine at 20 °C or 5% hydrogen peroxide at 60 °C, respectively, but reductions were less than 1.5 log when the tomatoes were held for 24 hours at 20 °C prior to treatment (Table 16.4).

We can conclude from these studies that the efficacy of washing and sanitizing treatments in decontaminating tomatoes is greatly limited when the site of contamination is in punctures, cracks, or the stem scar area, and if contamination occurred preharvest. More emphasis must be placed on avoidance of contamination. The FDA has initiated a collaborative effort to "identify practices or conditions that potentially lead to product contamination" (FDA, 2007b).

Cantaloupe

Because of the association of cantaloupes with large *Salmonella* outbreaks (CDC, 1991, 2002; DeWaal and Barlow, 2002), much attention has been given to the efficacy of washing and sanitizing treatments in reducing

Table 16.4 Efficacy of Wash Treatments in Reducing Population of *Salmonella* on Dip-Inoculated Tomatoes[a,b]

Treatment	Storage at 20 °C (h)	Population Reduction ($\log_{10}$ CFU/g)[c]
Rinsed control at 20 °C for 2 min	1	1.11 ± 0.18 C
	24	0.40 ± 0.32 D
150 ppm Cl_2 at 20 °C for 2 min	1	1.78 ± 0.49 B
	24	1.34 ± 0.39 BC
5% H_2O_2 at 60 °C for 2 min	1	2.59 ± 0.74 A
	24	1.45 ± 0.33 BC
H_2O at 60 °C for 2 min	1	1.75 ± 0.11 B
	24	0.99 ± 1.00 CD

[a] From Sapers and Jones (2006). Reprinted with permission from the Journal of Food Science. Copyright held by the Institute of Food Technologists, Chicago, IL.
[b] Inoculum prepared from cocktail containing S. Montevideo (G4639) and S. Baildon (61–99); mean inoculum population was 10.13 ± 0.04 for the treatment comparisons.
[c] Mean population reductions ± standard deviations based on corresponding control means for 2 or 3 independent experiments, each with duplicate trials; control means were 5.61 ± 0.27 and 5.42 ± 0.26 $\log_{10}$ CFU/g for the 0-time and 24 h treatment time comparisons, respectively. Means not followed by the same letters are significantly different ($P < 0.05$).

pathogen populations on cantaloupe surfaces. It is believed that contamination of fresh-cut melon results from transfer of human pathogens on the surface to the interior flesh during cutting (Ukuku and Sapers, 2001; Ukuku and Fett, 2002). As with lettuce and tomatoes, efforts to decontaminate cantaloupes by application of sanitizers have achieved limited success. Typically, 1- to 3-log reductions have been reported for cantaloupes inoculated with *Salmonella* spp. and treated with 150 to 200 ppm chlorine (Park and Beuchat, 1999; Barak et al., 2003), 80 ppm PAA (Park and Beuchat, 1999), and 1 to 5% HP (Park and Beuchat, 1999; Ukuku and Sapers, 2001; Ukuku et al., 2004). Other studies have reported similar reductions in the populations of the native microflora (Sapers et al., 2001; Ukuku et al., 2001) and in populations of a nonpathogenic *E. coli* (Ukuku et al., 2001) and *L. monocytogenes* (Ukuku and Fett, 2002) on inoculated cantaloupes washed with 1000 ppm chlorine and 5% HP. However, larger reductions in *Salmonella* were reported with 850 ppm acidified sodium chlorite (Park and Beuchat, 1999) and in *E. coli* O157:H7 and *L. monocytogenes* with 80 ppm PAA, 100 and 200 ppm chlorinated trisodium phosphate, 3 and 5 ppm ClO_2, and 3 ppm ozone (Rodgers et al., 2004). Materon (2003) also reported large population reductions for *E. coli* O157:H7 on inoculated cantaloupes treated with combinations of 200 ppm chlorine + 1.5% lactic acid or 1.5% lactic acid + 1.5% HP.

It is not clear why population reductions were so much greater in some studies than in others. However, studies with *Salmonella* and a nonpathogenic *E. coli* showed that the efficacy of sanitizer treatments for cantaloupe disinfection decreased as the interval between inoculation and treatment increased from 24 to 72 hours (Ukuku and Sapers, 2001; Ukuku et al., 2001). Differences in methodology for melon inoculation, storage, treatment application, and recovery and enumeration of the targeted pathogen can all influence the experimental results (Beuchat and Scouten, 2004; Ukuku and Fett, 2004; Annous et al., 2005). The demonstration of rapid biofilm formation by *Salmonella* on the surface of spot inoculated cantaloupes may explain storage effects on the efficacy of disinfection treatments discussed earlier (Annous et al., 2005).

The efficacy of sanitizer treatments in inactivating *Salmonella* on cantaloupe surfaces can be enhanced by application at elevated temperatures, for example, 2% lactic acid at 55 to 60 °C (Alvarado-Casillas et al., 2007) and 5% HP at 70 °C (Ukuku et al., 2004; Ukuku, 2006). Hot water surface pasteurization of cantaloupes, inoculated with *Salmonella*, achieved population reductions approaching 5 logs with no detrimental effects on melon quality (Ukuku et al., 2004; Ukuku, 2006; Annous et al., 2004). The rind apparently has sufficient insulating ability to protect the flesh from thermal injury.

In a study of cantaloupe contamination conducted at four packinghouses located in the Rio Grande Valley of South Texas, Materon (2003) reported that washing resulted in significant reductions (2–3 logs) in populations of aerobic bacteria, total coliforms, and fecal coliforms in a study. In contrast, Gagliardi et al. (2003) reported elevations in bacterial counts of cantaloupes from packinghouses in the Rio Grande Valley, sampled before and after washing and packing. Contamination during processing was traced to a primary wash tank or hydrocooler and may be a reflection of poor water quality, the organic load, and chlorine depletion. It is evident that implementation of Good Agricultural and Manufacturing Practices as well as improvements in disinfection technology are needed to correct such deficiencies.

Apples

Because of the history of *E. coli* O157:H7 outbreaks associated with unpasteurized apple cider, there have been numerous studies of the efficacy of washing and sanitizing agents in disinfecting contaminated apples (Beuchat et al., 1998; Sapers et al., 1999; Wisniewsky et al., 2000; Wright et al., 2000; Achen and Yousef, 2001; Kenney and Beuchat, 2002; Venkitanarayanan et al., 2002; Sapers et al., 2002; Parnell and Harris, 2003; Rodgers et al., 2004). Population reductions on apples inoculated with *E. coli* O157:H7 were generally in the range of 1 to 3 logs for most disinfection treatments. In studies of the efficacy of disinfection treatments for apples, Sapers et al. (2002) identified rapid attachment of bacterial cells to apple surfaces, attachment to inaccessible sites in the stem and calyx areas, and attachment and growth in skin punctures as factors limiting efficacy. The attachment of *E. coli* O157:H7 to surface and internal structures of apples, including discontinuities in the waxy cuticle, lenticels, punctures, the floral tube, seeds, cartilaginous pericarp, and internal trichomes was demonstrated by confocal scanning laser microscopy (Burnett et al., 2000). Rubbing dip-inoculated apples was reported to seal attached cells of *E. coli* O157:H7 within the waxy cutin platelets, thereby protecting them from disinfection (Kenney et al., 2001). Treatment efficacy was poor when apples were treated by spraying in a commercial brush washer (Annous et al., 2001), and efficacy could be improved by immersing apples in the sanitizing solution with good agitation, by heating sanitizer solutions, and by removal of calyx and stem tissue (Sapers et al., 2002). Other studies have demonstrated the localized concentration and survival of microbial contaminants in punctures and the stem and calyx regions of apples (Riordan et al., 2001; Fatemi et al., 2006). Fleischman et al. (2001) and Sapers et al. (2002) demonstrated the efficacy of hot water in disinfecting apple surfaces; however, such treatments will result in discoloration and softening of surface tissues. Five-log pathogen reductions are now mandated for apple cider (and other fresh juices), but for cider, these reductions must be obtained entirely on the juice, not part on the whole fruit and the remainder on the juice (FDA, 2001). Such reductions can be achieved by heat pasteurization or UV treatment of cider.

Although the cider safety issue is no longer dependent on fruit disinfection, potential safety problems remain with fresh-cut apple slices that are vulnerable to contamination with human pathogens in the processing environment; such products may have an extended shelf-life, providing an opportunity for outgrowth of such contaminants. Detection of *L. monocytogenes* in fresh-cut apples resulted in a recall (FDA, 2001). None of the sanitizing treatments currently available for whole or cut apples can consistently achieve a 5-log reduction in pathogen population without altering product identity. However, conditions allowing contamination of the product in the processing plant environment can be corrected, and survival and growth of human pathogens in the product can be suppressed (Karaibrahimoglu et al., 2004; Pilizota and Sapers, 2004).

CONCLUSIONS

In the absence of a final disinfection step to eliminate spoilage organisms and human pathogens from fresh and fresh-cut fruits and vegetables, the produce industry has depended on avoidance of contamination, providing good plant sanitation, and employing presumably effective washing and sanitizing agents and methods of application to provide consumers with safe products. Yet, in spite of such measures, outbreaks remain a problem. Conventional washing and sanitizing treatments can achieve reductions in the microbial load on fresh produce of 1- to 3-log reductions, good enough to improve quality, but not enough to assure safety.

If the surviving microbial populations on washed and sanitized produce represent cells attached in inaccessible sites, incorporated into resistant biofilms, or internalized within the produce interior, then incremental improvements in washing technology will not solve the problem. Some more potent, penetrating treatment, such as application of ionizing radiation, is required. Since there may be commodities that cannot be irradiated or subjected to other high energy, nonthermal treatments without significant loss of quality, other approaches are required as well. This shifts the burden of solving the produce contamination problem from the area of postharvest interventions to preharvest interventions, the avoidance of contamination by human pathogens. Unless this fact is fully appreciated and efforts to reduce produce contamination are comprehensive, not just restricted to washing and sanitizing technology, produce safety cannot be assured.

REFERENCES

Achen, M. and Yousef, A. E. (2001). Efficacy of ozone against *Escherichia coli* O157: H7 on apples. *J. Food Sci.* **66,** 1380–1384.

Allwood, P. B., Malik, Y. S., Hedberg, C. W. et al. (2004). Effect of temperature and sanitizers on the survival of feline calicivirus, *Escherichia coli*, and F-specific coliphage MS2 on leafy salad vegetables. *J. Food Prot.* **67,** 1451–1456.

Alvarado-Casillas, S., Ibarra-Sánchez, S., Rodríguez-García, O. et al. (2007). Comparison of rinsing and sanitizing procedures for reducing bacterial pathogens on fresh cantaloupes and bell peppers. *J. Food Prot.* **70,** 655–660.

Annous, B. A., Burke, A., and Sites, J. E. (2004). Surface pasteurization of whole fresh cantaloupes inoculated with *Salmonella* Poona or *Escherichia coli*. *J. Food Prot.* **67,** 1876–1885.

Annous, B. A., Sapers, G. M., Jones, D. M. et al. (2005). Improved recovery procedure for evaluation of sanitizer efficacy in disinfecting contaminated cantaloupes. *J. Food Sci.* **70,** M242–M247.

References

Annous, B. A., Sapers, G. M., Mattrazzo, A. M. et al. (2001). Efficacy of washing with a commercial flat-bed brush washer, using conventional and experimental washing agents, in reducing populations of *Escherichia coli* on artificially inoculated apples. *J. Food Prot.* **64**, 159–163.

Annous, B. A., Solomon, E. B., Cooke, P. H. et al. (2005). Biofilm formation by *Salmonella* spp. on cantaloupe melons. *J. Food Safety* **25**, 276–287.

Anon. (2001). Methods to reduce/eliminate pathogens from fresh and fresh-cut produce. Chapter V in *Analysis and evaluation of preventive control measures for the control and reduction/elimination of microbial hazards on fresh and fresh-cut produce.* Table V-1. U.S. Food and Drug Administration, Center for Food Safety and Applied Nutrition, September 30, 2001. www.cfsan.fda.gov/~comm/ift3-5.html

Barak, J. D., Chue, B., and Mills, D. C. (2003). Recovery of surface bacteria from and surface sanitization of cantaloupes. *J. Food Prot.* **66**, 1805–1810.

Bari, M. L., Sabina, Y., Isobe, S. et al. (2003). Effectiveness of electrolyzed acidic water in killing *Escherichia coli* O157:H7, *Salmonella* enteritidis, and *Listeria monocytogenes* on the surface of tomatoes. *J. Food Prot.* **66**, 542–548.

Bartz, J. A. (1982). Infiltration of tomatoes immersed at different temperatures to different depths in suspensions of *Erwinia carotovora* subsp. *carotovora. Plant Dis.* **66**, 302–306.

Bartz, J. A. and Showalter, R. K. (1981). Infiltration of tomatoes by aqueous bacterial suspensions. *Phytopathol.* **71**, 515–518.

Beuchat, L. R. (1998). Surface decontamination of fruits and vegetables eaten raw: A review. Food Safety Unit, World Health Organization, WHO/FSF/FOS/98.2. www.who.int/foodsafety/publications/fs_management/en/surface_decon.pdf

Beuchat, L. R. and Brackett, R. E. (1990). Survival and growth of *Listeria monocytogenes* on lettuce influenced by shredding, chlorine treatment, modified atmosphere packaging and temperature. *J. Food Sci.* **55**, 755–758, 870.

Beuchat, L. R., Nail, B. V., Adler, B. B. et al. (1998). Efficacy of spray application of chlorinated water in killing pathogenic bacteria on raw apples, tomatoes, and lettuce. *J. Food Prot.* **61**, 1305–1311.

Beuchat, L. R., Adler, N. B., and Lang, M. M. (2004). Efficacy of chlorine and peroxyacetic acid sanitizer in killing *Listeria monocytogenes* on iceberg and Romaine lettuce using simulated commercial processing conditions. *J. Food Prot.* **67**, 1238–1242.

Beuchat, L. R. and Scouten, A. J. (2004). Factors affecting survival, growth, and retrieval of *Salmonella* Poona on intact and wounded cantaloupe rind and stem scar tissue. *Food Microbiol.* **21**, 683–694.

Bialka, K. L. and Demirci, A. (2007). Efficacy of aqueous ozone for the decontamination of *Escherichia coli* O157:H7 and *Salmonella* on raspberries and strawberries. *J. Food Prot.* **70**, 1088–1092.

Blackburn, B. G., Mazurek, J. M., Hlavsa, M. et al. (2006). Cryptosporidiosis associated with ozonated apple cider. *Emerg. Infect. Dis.* **12**, April. Available from www.cdc.gov/ncidod/EID/vol12no04/05-0796.htm.

Block, S. S. (1991). Peroxygen compounds. Chapter 9 in *Disinfection, sterilization, and preservation*, 4th ed. (Block, S. S., Ed.). Lea & Febiger, Philadelphia.

Brackett, R. E. (1987). Antimicrobial effect of chlorine on *Listeria monocytogenes*. *J. Food Prot.* **50,** 999–1003.

Burnett, S. L., Chen, J., and Beuchat, L. R. (2000). Attachment of *Escherichia coli* O157:H7 to the surfaces and internal structures of apples as detected by confocal scanning laser microscopy. *Appl. Environ. Microbiol.* **66,** 4679–4687.

Caldwell, K. N., Adler, B. B., Anderson, G. L. et al. (2003). Ingestion of *Salmonella enterica* Serotype Poona by a free-living nematode, *Caenorhabditis elegans*, and protection against inactivation by produce sanitizers. *Appl. Environ. Microbiol.* **69,** 4103–4110.

Carmichael, I., Harper, I. S., Coventry, M. J. et al. (1999). Bacterial colonization and biofilm development on minimally processed vegetables. *J. Applied Microbiol. Symposium Suppl.* **85,** 45S–51S.

CDC. (1991). Multistate outbreak of *Salmonella* Poona infections—United States and Canada. MMWR **40,** 549–552.

CDC. (2002a). Outbreak of *Salmonella* serotype Javiana infections—Orlando, Florida, June 2002. MMWR **51,** 683–684.

CDC. (2002b). Multistate outbreaks of *Salmonella* serotype Poona infections associated with eating cantaloupe from Mexico—United States and Canada, 2000–2002. MMWR **51,** 1044–1047.

CDC. (2005). Outbreaks of *Salmonella* infections associated with eating Roma tomatoes—United States and Canada, 2004. MMWR **54,** 325–328.

CDC. (2006). Salmonellosis—Outbreak investigation, October 2006. Division of Bacterial and Mycotic Diseases, Centers for Disease Control and Prevention, Press Release Nov. 23, 2006. www.cdc.gov/ncidod/dbmd/diseaseinfo/salmonellosis2006/outbreak_notice.htm

Chang, T.-L., Streicher, R., and Zimmer, H. (1988). The interaction of aqueous solutions of chlorine with malic acid, tartaric acid, and various fruit juices, a source of mutagens. *Anal. Lett.* **21,** 2049–2067.

Cummings, K., Barrett, E., Mohle-Boetani, J. C. et al. (2001). A multistate outbreak of *Salmonella enterica* serotype Baildon associated with domestic raw tomatoes 1999. *Emer. Infect. Dis.* **7,** 1046–1048.

Delaquis, P., Stewart, S., Cazaux, S. et al. (2003). Survival and growth of *Listeria monocytogenes* and *Escherichia coli* O157:H7 in ready-to-eat iceberg lettuce. *J. Food Prot.* **65,** 459–464.

DeWaal, C. S. and Barlow, K. Outbreak Alert! 2002, 7th ed. Center for Science in the Public Interest. Washington, DC.

Deza, M. A., Araujo, M., and Garrido, M. J. (2003). Inactivation of *Escherichia coli* O157:H7, *Salmonella* enteritidis and *Listeria monocytogenes* on the surface of tomatoes by neutral electrolyzed water. *Letters in Appl. Microbiol.* **37,** 482–487.

D'Lima, C. B. and Linton, R. H. (2002). Inactivation of *Listeria monocytogenes* on lettuce by gaseous and aqueous chlorine dioxide gas and chlorinated water. Abstract 15D-4. IFT 2002 Annual Meeting and Food Expo, Anaheim, CA.

Doores, S. (2005). Organic acids. Chapter 4 in *Antimicrobials in food* (P. M. Davidson, J. N. Sofos, and A. L. Branen, Eds.), pp. 91–142. CRC Press, Boca Raton.

Escudero, M. E., Velázquez, L., Di Genaro, M. S. et al. (1999). Effectiveness of various disinfectants in the elimination of *Yersinia enterocolitica* on fresh lettuce. *J. Food Prot.* **62,** 665–669.

Fallik, E., Aharoni, Y., Grinberg, S. et al. (1994). Postharvest hydrogen peroxide treatment inhibits decay in eggplant and sweet red pepper. *Crop Protection* **13,** 451–454.

Fatemi, P., LaBorde, L. F., Patton, J. et al. (2006). Influence of punctures, cuts, and surface morphologies of Golden Delicious apples on penetration and growth of *Escherichia coli* O157:H7. *J. Food Prot.* **69,** 267–275.

FDA. (2001). Hazard Analysis and Critical Control Point (HAACP); Procedures for the Safe and Sanitary Processing and Importing of Juice; Final Rule. Federal Register, **66**(13), 6137–6202.

FDA. (2001). Enforcement report. Recalls and field corrections: Foods—class I. Recall number F-535–1. 20 August 2001. Sliced apples in poly bags. www.fda.gov/bbs/topics/ENFORCE/2001/ENF00708.html

FDA. (2003). *The bad bug book. Foodborne pathogenic microorganisms and natural toxins handbook.* US Department of Health and Human Services, Food and Drug Administration, Center for Food Safety and Applied Nutrition. Updated: 01-30-2003. www.cfsan.fda.gov/~mow/intro.html

FDA. (2004). Guidance for industry: Recommendations to processors of apple juice or cider on the use of ozone for pathogen reduction purposes. US Department of Health and Human Services, Food and Drug Administration/CFSAN, August 2004. www.cfsan.fda.gov/~dms/juicgu12.html

FDA. (2007a). Guidance for Industry. Guide to minimize microbial food safety hazards of fresh-cut fruits and vegetables. Draft Final Guidance. US Department of Health and Human Services, Food and Drug Administration, Center for Food Safety and Applied Nutrition, March 2007. www.cfsan.fda.gov/~tdms/prodgui3.html

FDA. (2007b). FDA News. FDA implementing initiative to reduce tomato-related foodborne illnesses. US Department of Health and Human Services, Food and Drug Administration. June 12, 2007. www.fda.gov/bbs/topics/NEWS/2007/NEW01651.html

Fett, W. F. (2002). Reduction of native microflora on alfalfa sprouts during propagation by addition of antimicrobial compounds to the irrigation water. *Int. J. Food Microbiol.* **72,** 13–18.

Fleischman, G. J., Batore, C., Merker, R. et al. (2001). Hot water immersion to eliminate *Escherichia coli* O157:H7 on the surface of whole apples: Thermal effects and efficacy. *J. Food Prot.* **64,** 451–455.

Francis, G. A. and O'Beirne, D. (1997). Effects of gas atmosphere, antimicrobial dip and temperature on the fate of *Listeria innocua* and *Listeria monocytogenes* on minimally processed lettuce. *Int. J. Food Sci. and Technol.* **32,** 141–151.

Gagliardi, J. V., Millner, P. D., Lester, G. et al. (2003). On-farm and postharvest processing sources of bacterial contamination to melon rinds. *J. Food Prot.* **66,** 82–87.

Garcia, A., Mount, J. R., and Davidson, P. M. (2003). Ozone and chlorine treatment of minimally processed lettuce. *J. Food Sci.* **68,** 2747–2751.

Garcia, L., Henderson, J., Fabri, M. et al. (2006). Potential sources of microbial contamination in unpasteurized apple cider. *J. Food Prot.* **69**, 137–144.

Gonzalez, R. J., Luo, Y., Ruiz-Cruz, S. et al. (2004). Efficacy of sanitizers to inactivate *Escherichia coli* O157:H7 on fresh-cut carrot shreds under simulated process conditions. *J. Food Prot.* **67**, 2375–2380.

Graham, D. M. (1997). Use of ozone for food processing. *Food Technol.* **51**(6), 72–75.

Guo, X., Chen, J., Brackett, R. E. et al. (2002). Survival of *Salmonella* on tomatoes stored at high relative humidity, in soil, and on tomatoes in contact with soil. *J. Food Prot.* **65**, 274–279.

Han, Y., Linton, R. H., Nielson, S. S. et al. (2001). Reduction of *Listeria monocytogenes* on green peppers (*Capsicum annuum* L.) by gaseous and aqueous chlorine dioxide and water washing and its growth at 7 °C. *J. Food Prot.* **64**, 1730–1738.

Harris, L. J., Beuchat, L. J., Kajs, T. M. et al. (2001). Efficacy and reproducibility of a produce wash in killing *Salmonella* on the surface of tomatoes assessed with a proposed standard method for produce sanitizers. *J. Food Prot.* **64**, 1477–1482.

Hassenberg, K., Idler, C., Molloy, E. et al. (2007). Use of ozone in a lettuce-washing process: An industrial trial. *J. Sci. Food Agric.* **87**, 914–919.

Hellström, S., Kervinen, R., Lyly, M. et al. (2006). Efficacy of disinfectants to reduce *Listeria monocytogenes* on precut iceberg lettuce. *J. Food Prot.* **69**, 1565–1570.

Hidaka, T., Kirigaya, T., Kamijo, M. et al. (1992). Disappearance of residual chlorine and formation of chloroform in vegetables treated with sodium hypochlorite. *J. Food Hygiene Soc. Japan.* **33**, 267–273.

Hilgren, J. D. and Salverda, J. A. (2000). Antimicrobial efficacy of a peroxyacetic/octanoic acid mixture in fresh-cut vegetable process waters. *J. Food Sci.* **65**, 1376–1379.

Huang, T-S., Xu, C., Walker, K. et al. (2006). Decontamination efficacy of combined chlorine dioxide with ultrasonification on apples and lettuce. *J. Food Sci.* **71**, M134–M139.

Ibarra-Sánchez, L. S, Alvarado-Casillas, S., Rodríguez-García, M. O. et al. (2004). Internalization of bacterial pathogens and their control by selected chemicals. *J. Food Prot.* **67**, 1353–1358.

Inatsu, Y., Bari, M. L., and Kawamoto, S. (2007). Application of acidified sodium chlorite prewashing treatment to improve the food hygiene of lightly fermented vegetables. *JARQ* **41**, 17–23.

Izumi, H. (1999). Electrolyzed water as a disinfectant for fresh-cut vegetables. *J. Food Sci.* **64**, 536–539.

Karaca, H. and Velioglu, Y. S. (2007). Ozone applications in fruit and vegetable processing. *Food Reviews Intern.* **23**, 91–106.

Karaibrahimoglu, Y., Fan, X., Sapers, G. M. et al. (2004). Effect of pH on the survival of *Listeria innocua* in calcium ascorbate solutions and on quality of fresh-cut apples. *J. Food Prot.* **67**, 751–757.

Kenney, S. J. and Beuchat, L. R. (2002). Comparison of aqueous commercial cleaners for effectiveness in removing *Escherichia coli* O157:H7 and *Salmonella muenchen* from the surface of apples. *Int. J. Food Microbiol.* **74**, 47–55.

Kenney, S. J., Burnett, S. L., and Beuchat, L. R. (2001). Location of *Escherichia coli* O157:H7 on and in apples as affected by bruising, washing, and rubbing. *J. Food Prot.* **64,** 1328–1333.

Kim, J.-G., Yousef, A. E., and Chism, G. W. (1999). Use of ozone to inactivate microorganisms on lettuce. *J. Food Safety* **19,** 17–34.

Klein, P. and Morris, S. C. (2004). Technological breakthrough in postharvest sanitation treatment with the iodine based Isan Process. ISHS Acta Horticulturae 687: International Conference Postharvest Unlimited Downunder 2004. www.actahort.org/members/showpdf?booknrarnr=687_15

Koseki, S., Isobe, S., and Itoh, K. (2004). Efficacy of acidic electrolyzed water ice for pathogen control on lettuce. *J. Food Prot.* **67,** 2544–2549.

Koseki, S., Yoshida, K., Kamitami, Y. et al. (2003). Influence of inoculation method, spot inoculation site, and inoculation size on the efficacy of acidic electrolyzed water against pathogens on lettuce. *J. Food Prot.* **66,** 2010–2016.

Lang, M. M., Harris, L. J., and Beuchat, L. R. (2004). Survival and recovery of *Escherichia coli* O157:H7, *Salmonella*, and *Listeria monocytogenes* on lettuce and parsley as affected by method of inoculation, time between inoculation and analysis, and treatment with chlorinated water. *J. Food Prot.* **67,** 1092–1103.

Lee, S.-Y., Dancer, G. I., Chang, S.-S. et al. (2006). Efficacy of chlorine dioxide gas against *Alicyclobacillus acidoterrestris* spores on apple surfaces. *Int. J. Food Microbiol.* **108,** 364–368.

Liao, C.-H., Shollenberger, L. M., and Phillips, J. G. (2003). Lethal and sublethal action of acetic acid on *Salmonella* in vitro and on cut surfaces of apple slices. *J. Food Sci.* **68,** 2793–2798.

Lin, C.-M., Moon, S. S., Doyle, M. P. et al. (2002). Inactivation of *Escherichia coli* O157:H7, *Salmonella enterica* Serotype Enteritidis, and *Listeria monocytogenes* by hydrogen peroxide and lactic acid and by hydrogen peroxide with mild heat. *J. Food Prot.* **65,** 1215–1220.

Lukasik, J., Bradley, M. L., Scott, T. M. et al. (2003). Reduction of poliovirus 1, bacteriophages, *Salmonella* Montevideo, and *Escherichia coli* O157:H7 on strawberries by physical and disinfectant washes. *J. Food Prot.* **66,** 188–193.

Materon, L. A. (2003). Survival of *Escherichia coli* O157:H7 applied to cantaloupes and the effectiveness of chlorinated water and lactic acid as disinfectants. *World J. Microbiol. Biotechnol.* **19,** 867–873.

Mendonca, A. F., Amoroso, T. L., and Knabel, S. J. (1994). Destruction of Gram-negative food-borne pathogens by high pH involves disruption of the cytoplasmic membrane. *Appl. Environ. Microbiol.* **60,** 4009–4014.

Nascimento, M. S., Silva, N., Catanozi, M. P. L. M. et al. (2003). Effects of different disinfection treatments on the natural microbiota of lettuce. *J. Food Prot.* **66,** 1697–1700.

Oh, S.-W., Dancer, G. I., and Kang, D.-H. (2005). Efficacy of aerosolized peroxyacetic acid as a sanitizer of lettuce leaves. *J. Food Prot.* **68,** 1743–1747.

Ozer, N. P. and Demirci, A. (2006). Electrolyzed oxidizing water treatment for decontamination of raw salmon inoculated with *Escherichia coli* O157:H7 and *Listeria monocytogenes* Scott A and response surface modeling. *J. Food Eng.* **72,** 234–241.

Pao, S., Davis, C. L., and Kelsey, D. F. (2000). Efficacy of alkaline washing for the decontamination of orange fruit surfaces inoculated with *Escherichia coli*. *J. Food Prot.* **63,** 961–964.

Pao, S., Kelsey, D. F., Khalid, M. F. et al. (2007). Using aqueous chlorine dioxide to prevent contamination of tomatoes with *Salmonella enterica* and *Erwinia carotovora* during fruit washing. *J. Food Prot.* **70,** 629–634.

Parish, M. E., Beuchat, L. R., Suslow, T. V. et al. (2003). Methods to reduce/eliminate pathogens from fresh and fresh-cut produce. Chapter 5 in *Comprehensive Reviews in Food Science and Food Safety* **2** (Supplement), 161–173.

Park, C. M. and Beuchat, L. R. (1999). Evaluation of sanitizers for killing *Escherichia coli* O157:H7, *Salmonella*; and naturally occurring microorganisms on cantaloupes, honeydew melons and asparagus. *Dairy, Food and Environmental Sanitation* **19,** 842–847.

Park, C.-M., Hung, Y.-C., Doyle, M. P. et al. (2001). Pathogen reduction and quality of lettuce treated with electrolyzed oxidizing and acidified chlorinated water. *J. Food Sci.* **66,** 1368–1372.

Parnell, T. L. and Harris, L. J. (2003). Reducing *Salmonella* on apples with wash practices commonly used by consumers. *J. Food Prot.* **66,** 741–747.

Pilizota, V. and Sapers, G. M. (2004). Novel browning inhibitor formulation for fresh-cut apples. *J. Food Sci.* **69,** 140–143.

Pirovani, M. E., Guemes, D. R., and Piagnetini, A. M. (2001). Predictive models for available chlorine depletion and total microbial count reduction during washing of fresh-cut spinach. *J. Food Sci.* **66,** 860–864.

Raiden, R. M., Sumner, S. S., Eifert, J. D. et al. (2003). Efficacy of detergents in removing *Salmonella* and *Shigella* spp. from the surface of fresh produce. *J. Food Prot.* **66,** 2210–2215.

Restaino, L. et al. (1995). Efficacy of ozonated water against various food-related microorganisms. *Appl. Environ. Microbiol.* **61,** 3471–3475.

Richards, G. M. and Beuchat, L. R. (2004). Attachment of *Salmonella* Poona to cantaloupe rind and stem scar tissues as affected by temperature of fruit and inoculum. *J. Food Prot.* **67,** 1359–1364.

Riordan, D. C. R., Sapers, G. M., and Annous, B. A. (2000). The survival of *E. coli* O157:H7 in the presence of *Penicillium expansum* and *Glomerella cingulata* in wounds on apple surfaces. *J. Food Prot.* **63,** 1637–1642.

Riordan, D. C., Sapers, G. M., Hankinson, T. H. et al. (2001). A study of U.S. orchards to identify potential sources of *Escherichia coli* O157:H7. *J. Food Prot.* **64,** 1320–1327.

Rodgers, S. L., Cash, J. N., Siddiq, M. et al. (2004). A comparison of different chemical sanitizers for inactivating *Escherichia coli* O157:H7 and *Listeria monocytogenes* in solution and on apples, lettuce, strawberries, and cantaloupe. *J. Food Prot.* **67,** 721–731.

Rupasinghe, H. P. V., Boulter-Bitzer, J., and Odumeru, J. A. (2006). Lactic acid improves the efficacy of anti-microbia washing solutions for apples. *J. Food Agr. Environ.* **4,** 44–48.

Sampathkumar, B., Khachatourians, G. G., and Korber, D. R. (2003). High pH during trisodium phosphate treatment causes membrane damage and destruction of *Salmonella enterica* Serovar Enteriditis. *Appl. Environ. Microbiol.* **69,** 122–129.

Sapers, G. M. (2002). Washing and sanitizing raw materials for minimally processed fruit and vegetable products. Chapter 11 in *Microbial Safety of Minimally Processed Foods* (J. S. Novak, G. M. Sapers, V. K. Juneja, Eds.), pp. 221–253. CRC Press, Boca Raton.

Sapers, G. M. (2003). Hydrogen peroxide as an alternative to chlorine for sanitizing fruits and vegetables, FoodInfo Online Features, IFIS Publishing, 23 July 2003. http://foodsciencecentral.com/library.html#ifis/12433

Sapers, G. M. (2005). Washing and sanitizing treatments for fruits and vegetables. Chapter 17 in *Microbiology of Fruits and Vegetables* (G. M. Sapers, J. R. Gorny, and A. E. Yousef, Eds.), pp. 375–400. CRC Press, Boca Raton.

Sapers, G. M. and Jones, D. M. (2006). Improved sanitizing treatments for fresh tomatoes. *J. Food Sci.* **71,** M252–M256.

Sapers, G. M., Miller, R. L., Annous, B. A. et al. (2002). Improved anti-microbial wash treatments for decontamination of apples. *J. Food Sci.* **67,** 1886–1891.

Sapers, G. M., Miller, R. L., and Mattrazzo, A. M. (1999). Effectiveness of sanitizing agents in inactivating *Escherichia coli* in Golden Delicious apples. *J. Food Sci.* **64,** 734–737.

Sapers, G. M., Miller, R. L., Jantschke, M. et al. (2000). Factors limiting the efficacy of hydrogen peroxide washes for decontamination of apples containing *Escherichia coli*. *J. Food Sci.* **65,** 529–532.

Sapers, G. M., Miller, R. L., Pilizota, V. et al. (2001). Shelf-life extension of fresh mushrooms (*Agaricus bisporus*) by application of hydrogen peroxide and browning inhibitors. *J. Food Sci.* **66,** 362–366.

Sapers, G. M., Miller, R. L., Pilizota, V. et al. (2001). Antimicrobial treatments for minimally processed cantaloupe melon. *J. Food Sci.* **66,** 345–349.

Sapers, G. M. and Sites, J. E. (2003). Efficacy of 1% hydrogen peroxide wash in decontaminating apples and cantaloupe melons. *J. Food Sci.* **68,** 1793–1797.

Saravacos, G. D. and Kostaropoulos, A. E. (2002). *Handbook of Food Processing Equipment*. Kluwer Boston, Norwell. pp. 256–258. Available on-line at www.Google.com.

Segall, R. H. (1968). Reducing postharvest decay of tomatoes by adding a chlorine source and the surfactant Santormerse F85 to water in field washers. *Fla. State Hort. Soc. Proc.* **81,** 212–214.

Sharma, R. (2005). Ozone decontamination of fresh fruit and vegetables. In *Improving the safety of fresh fruit and vegetables* (W. Jongen, Ed.). Woodhead Publishing, Cambridge, UK.

Sharma, R. R., Demirci, A., Beuchat, L. R. et al. (2002). Inactivation of *Escherichia coli* O157:H7 on inoculated alfalfa seeds with ozonated water and heat treatment. *J. Food Prot.* **65,** 447–451.

Smilanick, J. L. (2003). Use of ozone in storage and packing facilities. Presented at Washington Tree Fruit Postharvest Conference, December 2–3, 2003, Wenatchee, WA.

Smith, S., Dunbar, M., Tucker, D. et al. (2003). Efficacy of a commercial produce wash on bacterial contamination of lettuce in a food service setting. *J. Food Prot.* **66,** 2359–2361.

Solomon, E. B., Yaron, S., and Matthews, K. R. (2002). Transmission of *Escherichia coli* O157:H7 from contaminated manure and irrigation water to lettuce plant tissue and its subsequent internalization. *Applied Environ. Microbiol.* **68,** 397–400.

Somers, E. B., Schoeni, J. L., and Wong, A. C. L. (1994). Effect of trisodium phosphate on biofilm and planktonic cells of *Campylobacter jejuni, Escherichia coli* O157:H7, *Listeria monocytogenes* and *Salmonella* typhimurium. *Int. J. Food Microbiol.* **22,** 269–276.

Spotts, R. A. (1982). Use of surfactants with chlorine to improve pear decay control. *Plant Disease* **66,** 725, 1982.

Spotts, R. A. and Cervantes, L. A. (1992). Effect of ozonated water on postharvest pathogens of pear in laboratory and packinghouse tests. *Plant Dis.* **76,** 256–259.

Steenstrup, L. D. and Floros, J. D. (2004). Inactivation of *E. coli* O157:H7 in apple cider by ozone at various temperatures and concentrations. *J. Food Proc. Preserv.* **28,** 103–116.

Suslow, T. (2000). Chlorination in the production and postharvest handling of fresh fruits and vegetables. Chap. 6 in *Fruit and vegetable processing* (D. McLaren, Ed.), pp. 2–15. Published by the Food Processing Center at the University of Nebraska, Lincoln, NE. http://ucce.ucdavis.edu/files/filelibrary/5453/4369.pdf.

Suslow, T. (2000). Postharvest handling for organic crops. University of California ANR Communication Services. Publication 7254. http://anrcatalog.ucdavis.edu/pdf/7254.pdf

Suslow, T. V. (2004). Ozone application for postharvest disinfection of edible horticultural crops. Publication 8133. University of California Division of Agriculture and Natural Resources. http://anrcatalog.ucdavis.edu/pdf/8133.pdf

Ukuku, D. O. (2006). Effect of sanitizing treatments on removal of bacteria from cantaloupe surface, and recontamination with *Salmonella. Food Microbiol.* **23,** 289–293.

Ukuku, D. O. and Fett, W. (2002). Behavior of *Listeria monocytogenes* inoculated on cantaloupe surfaces and efficacy of washing treatments to reduce transfer from rind to fresh-cut pieces. *J. Food Prot.* **65,** 924–930.

Ukuku, D. O. and Fett, W. (2004). Method of applying sanitizers and sample preparation affects recovery of native microflora and *Salmonella* on whole cantaloupe surfaces. *J. Food Prot.* **67,** 999–1004.

Ukuku, D. O., Pilizota, V., and Sapers, G. M. (2001). Influence of washing treatment on native microflora and *Escherichia coli* population of inoculated cantaloupes. *J. Food Safety* **21,** 31–47.

Ukuku, D. O., Pilizota, V., and Sapers, G. M. (2004). Effect of hot water and hydrogen peroxide treatments on survival of *Salmonella* and microbial quality of whole and fresh-cut cantaloupe. *J. Food Prot.* **67,** 432–437.

Ukuku, D. O. and Sapers, G. M. (2001). Effect of sanitizer treatments on *Salmonella* Stanley attached to the surface of cantaloupe and cell transfer to fresh-cut tissues during cutting practices. *J. Food Prot.* **64,** 1286–1291.

Venkitanarayanan, K. S., Ezeike, G. O, Hung, Y-C. et al. (1999a). Efficacy of electrolyzed oxidizing water for inactivating *Escherichia coli* O157:H7, *Salmonella* Enteritidis, and *Listeria monocytogenes*. *Appl. Environ. Microbiol.* **65**, 4276–4279.

Venkitanarayanan, K. S., Ezeike, G. O. I., Hung, Y.-C. et al. (1999b). Inactivation of *Escherichia coli* O157:H7 and *Listeria monocytogenes* on plastic kitchen cutting boards by electrolyzed oxidizing water. *J. Food Prot.* **62**, 857–860.

Venkitanarayanan, K. S., Lin, C.-M., Bailey, H. et al. (2002). Inactivation of *E. coli* O157:H7, *Salmonella* Enteritidis and *Listeria monocytogenes* on apples, oranges, and tomatoes by lactic acid with hydrogen peroxide. *J. Food Protect.* **65**, 100–105.

Venkitanarayanan, K., Zhao, T., and Doyle, M. P. (1999c). Inactivation of *E. coli* O157:H7 by combination of GRAS chemicals and temperatures. *Food Microbiol.* **16**, 75–82.

Vijayakumar, C. and Wolf-Hall, C. E. (2002). Evaluation of household sanitizers for reducing levels of *Escherichia coli* on iceberg lettuce. *J. Food Prot.* **65**, 1646–1650.

Wade, W. N., Scouten A. J., McWatters, K. H. et al. (2003). Efficacy of ozone in killing *Listeria monocytogenes* on alfalfa seeds and sprouts and effects on sensory quality of sprouts. *J. Food Prot.* **66**, 44–51.

Warf, C. C., Jr. (2001). The chemistry & modes of action of acidified sodium chlorite. Abstract 91–1. Acidified sodium chlorite—An antimicrobial intervention for the food industry. 2001 IFT Annual Meeting – New Orleans, LA.

Wei, C. I, Huang, T. S, Kim, J. M. et al. (1995). Growth and survival of *Salmonella* Montevideo on tomatoes and disinfection with chlorinated water. *J. Food Prot.* **58**, 829–836.

Weissinger, W. R. and Beuchat, L. R. (2000). Comparison of aqueous chemical treatments to eliminate *Salmonella* on alfalfa seeds. *J. Food Protect.* **63**, 1475–1482.

Wickramanayake, G. B. (1991). Disinfection and sterilization by ozone. Chapter 10 in *Disinfection, Sterilization, and Preservation*, 4th ed. (Block, S. S., Ed.). Lea & Febiger, Philadelphia.

Wisniewsky, M. A, Glatz, B. A, Gleason, M. L. et al. (2000). Reduction of *Escherichia coli* O157:H7 counts on whole fresh apples by treatment with sanitizers. *J. Food Prot.* **63**, 703–708.

Wright, J. R., Sumner, S. S., Hackney, C. R. et al. (2000). Reduction of *Escherichia coli* O157:H7 on apples using wash and chemical sanitizer treatments, *Dairy, Food and Environ. Sanit.* **20**, 120–126.

Xu, L. (1999). Use of ozone to improve the safety of fresh fruits and vegetables. *Food Technol.* **53**(10), 58–63.

Yang, H., Swem, B. L., and Li, Y. (2003). The effect of pH on inactivation of pathogenic bacteria on fresh-cut lettuce by dipping treatment with electrolyzed water. *J. Food Sci.* **68**, 1013–1017.

Yuk, H-G., Bartz, J. A., and Schneider, K. R. (2005). Effectiveness of individual or combined sanitizer treatments for inactivating *Salmonella* spp. on smooth surface, stem scar, and wounds of tomatoes. *J. Food Sci.* **70**, M409–M414.

Yuk, H-G., Bartz, J. A., and Schneider, K. R. (2006). The effectiveness of sanitizer treatments in inactivation of *Salmonella* spp. from bell pepper, cucumber, and strawberry. *J. Food Sci.* **71**, M95–M99.

Zhang, S. and Farber, J. M. (1996). The effects of various disinfectants against *Listeria monocytogenes* on fresh-cut vegetables. *Food Microbiol.* **13,** 311–321.

Zhuang R-Y and Beuchat, L. R. (1996). Effectiveness of trisodium phosphate for killing *Salmonella* Montevideo on tomatoes. *Letters Appl. Microbiol.* **232,** 97–100.

Zhuang, R-Y, Beuchat, L. R, and Angulo, F. J. (1995). Fate of *Salmonella* Montevideo on and in raw tomatoes as affected by temperature and treatment with chlorine. *Appl. Environ. Microbiol.* **61,** 2127–2131.

7CFR205.605. Nonagricultural (nonorganic) substances allowed as ingredients in or on processed products labeled as "organic" or "made with organic (specified ingredients or food group(s))." *Code of Federal Regulations*, Title 7, Part 205, Section 205.605. Current as of May 14, 2007.

21CFR173.300. Chlorine dioxide. *Code of Federal Regulations*, Title 21, Part 173, Section 173.300. Revised as of April 1, 2006.

21CFR173.315. Chemicals used in washing or to assist in the peeling of fruits and vegetables. *Code of Federal Regulations*, Title 21, Part 173, Section 173.315. Revised as of April 1, 2006.

21CFR173.325. Acidified sodium chlorite solutions. *Code of Federal Regulations*, Title 21, Part 173, Section 173.315. Revised as of April 1, 2006.

21CFR173.368. Ozone. *Code of Federal Regulations*, Title 21, Part 173, Section 173.368. Revised as of April 1, 2006.

21CFR184.1005. Direct food substances affirmed as generally recognized as safe. Listing of specific substances affirmed as GRAS. Acetic Acid. *Code of Federal Regulations*, Title 21, Part 184, Subpart B, Section 184.1005. Revised as of April 1, 2006.

21CFR184.1033. Direct food substances affirmed as generally recognized as safe. Listing of specific substances affirmed as GRAS. Citric Acid. *Code of Federal Regulations*, Title 21, Part 184, Subpart B, Section 184.1033. Revised as of April 1, 2006.

21CFR184.1061. Direct food substances affirmed as generally recognized as safe. Listing of specific substances affirmed as GRAS. Lactic Acid. *Code of Federal Regulations*, Title 21, Part 184, Subpart B, Section 184.1061. Revised as of April 1, 2006.

21CFR184.1366. Hydrogen peroxide, *Code of Federal Regulations*. Title 21, Part 184, Section 184.1366.

21CFR184.1769a. Direct food substances affirmed as generally recognized as safe. Listing of specific substances affirmed as GRAS. Sodium metasilicate. *Code of Federal Regulations*, Title 21, Part 184, Subpart B, Section 184.1769a. Revised as of April 1, 2006.

29CFR 1910.1000 TABLE Z-1 Limits for air contaminants. *Code of Federal Regulations*, Title 29, Part 1910. Section 1910.1000.

40CFR180.1196. Peroxyacetic acid; exemption from the requirement of a tolerance. *Code of Federal Regulations*, Title 40, Part 180, Section 180.1196.

40CFR180.1197. Hydrogen peroxide; exemption from the requirement of a tolerance, *Code of Federal Regulations*, Title 40, Part 180, Section 180.1197.

CHAPTER 17

Advanced Technologies for Detection and Elimination of Pathogens

Brendan A. Niemira and Howard Q. Zhang

USDA Agricultural Research Service, Eastern Regional Research Center, Wyndmoor, PA

CHAPTER CONTENTS

Introduction	425
Detection Methods	426
Immunomagnetic Beads and Biosensors: Separation and Concentration	427
PCR-Based Methods	428
Computer/AI Optical Scanning	429
Antimicrobial Intervention Technologies	430
Cold Plasma	431
Irradiation	433
Pulsed Light	434
High-Pressure Processing	436
Sonication	436
Biological Controls	437
The Challenge of Technology Development for Organic Foods	438
Acknowledgements	439

INTRODUCTION

In recent years, the incidence of foodborne illness (FBI) outbreaks associated with contaminated fruits, vegetables, salads, and juices has increased notably (Sivapalasingam et al., 2004). Adherence to established industry standards for Good Agricultural Practices (GAP), Good Manufacturing Practices (GMP), and Good Handling Practices (GHP) can serve to reduce risk. However, by themselves, these practices have not been able to prevent repeated product recalls and illnesses of exposed consumers. Although various food-safety interventions have been proposed as antimicrobial processes that are broadly applicable for produce (a produce "kill step"), barriers to their widespread implementation have hampered the food-safety efforts of the fresh-produce industry (UFPA, 2007; JIFSAN, 2007). Leafy green vegetables such as lettuce and spinach, tomatoes, melons, sprouts, and other fresh produce thus remain vectors for *Escherichia coli* O157:H7, *Salmonella*, *Listeria*, and other pathogens.

When taken as part of a unified production approach, the GAP, GMP, and GHP protocols in the preharvest, postharvest, and postprocessing environments constitute the components of a food-safety–oriented HACCP plan. Verification of the efficacy and consistency of the HACCP plan requires detection techniques that are scientifically valid and suitable for commercial production of fresh and fresh-cut produce. This chapter will discuss the latest research in developing rapid, sensitive, and accurate detection technologies, present a summary of the latest research on advanced intervention technologies to inactivate pathogens on produce, and briefly discuss the applicability of new technologies for the production and processing of organic fruits and vegetables.

DETECTION METHODS

The benchmarks and standards (e.g., the applicable Good Agricultural and Good Manufacturing Practices and guidance documents) of produce safety are the underpinning of the specific actions taken by growers, processors, shippers, and retailers. In the field, during harvest and processing, after packaging and shipping, and in the retail or foodservice environment, the steps taken to ensure the safety of fresh and fresh-cut fruits and vegetables fall into three general categories. These are protocols to *exclude* pathogens from the plants, produce, or packages, and protocols to *contain* or to *eradicate* pathogens (e.g., by application of treatments to suppress growth or to reduce populations sufficiently to assure safety). If completely effective, excluding pathogens will serve to prevent contamination in the first place. Improving these protocols has been the primary focus of much of the industry's effort to date. Although these improvements are achieving positive results, reliance on only one type of control is unlikely to be completely or optimally effective. Hence, the focus of this section is on scientific efforts to improve the industry's ability to detect pathogens and subsequently contain contaminated produce.

Standard microbiological testing for quality purposes is common in the produce industry. However, regular testing for contamination by pathogens, including *E. coli* O157:H7, *Salmonella*, *Shigella*, and *L. monocytogenes*, historically has been less common, although this is becoming more prevalent in the wake of changes in the industry. Relatively slow traditional sampling, enrichment, and enumeration can take 48 to 72 hours, an extremely long time for a commodity class with a shelf-life of only 7 to 14 days. Tests that result in an unacceptable percentage of false positives result in needless product recall and expense, whereas false negatives put the health of the consumer and the viability of the company name and reputation at risk. For tests to be useful for the produce industry, they need to be rapid, sensitive, and accurate, identifying critical limits of viable or potentially viable pathogens in a timely manner.

Immunomagnetic Beads and Biosensors: Separation and Concentration

In order to screen high volumes of food material, either from pooled samples or from a flow-through system, a means of concentrating the samples must be employed. Air and water sampling has traditionally used flow-through filtration, with analysis of the filter membrane as the diagnostic step. For bulky, fibrous, or otherwise difficult-to-filter produce material, this is a less than optimal approach. Recently, advances in the use of immunomagnetic beads have improved the sensitivity and reliability of detection (Tu et al., 2008).

As the name suggests, this technique involves the use of specialized iron beads, coated with plastic. These micron-sized beads are surface-treated with antibodies that are specific to the pathogen of interest. The food sample is prepared as a slurry, and the beads are added to the suspension. During mixing, the entire population of bacteria in the suspension comes in contact with the beads. Nontarget organisms remain in suspension, while the target pathogens bind to the antibodies on the bead surface. A powerful magnetic field, either from an electromagnet or from a neodymium boron iron magnet, is used to collect the beads for further analysis. This type of magnetic collection allows for flow-through collection of a large volume of material, and is a significant improvement over older methods that relied on centrifugation to concentrate and collect the immunoattractant material. Since the primary basis for the recognition and capture is based on the immunoattraction of the bead surface, this technology can be adapted for a range of purposes. In addition to concentrating pathogens such as *Salmonella* (Tu et al., 2008), this technology can be used to concentrate and analyze chemical contaminants of interest, such as toxins and adulterants (Gessler et al., 2006).

An adaptation of this type of immunoattraction is in the use of active biosensors (Nugen and Baeumer, 2008). These are probes for detecting an analyte characteristic of a pathogen or other organism, that integrates a biological component, such as a whole bacterium or a biological product (e.g., an enzyme or antibody) with an electronic component to yield a measurable signal. Rather than free-floating beads collected using a physical process, such as centrifugation or magnetic separation, comparably-sized nanofabricated sensor tips are connected to a sensor bank. Upon binding of a bacterium to the immunoattractive sensor tip, the configuration of the sensor is altered, inducing a signal in the attached biosensor. This may be a change in electrical conductivity, in optical properties (absorption, reflection, wavelength shift, etc.), or in physical conformation as in a cantilever design. Varying types of biosensors are employed in flow-through or "dipstick" type systems. As with immunomagnetic beads, sample preparation is critical. Sample viscosity, concentration of extraneous or contaminating material, and so on can influence the efficiency and accuracy of the testing.

PCR-Based Methods

Polymerase chain reaction (PCR) based detection systems are extremely sensitive to the presence of select DNA sequences associated with particular pathogens. Customized probes will bind only to the target sequence, which acts as a template for DNA synthesis by DNA polymerase, during multiple cycles of heating and cooling, leading to rapid amplification. Samples processed using traditional PCR methods would undergo many cycles of replication and denaturation. Once the cycling was complete and the target sequence had been fully amplified, the sample would be run on an agarose gel. Using a binding marker such as ethidium bromide, the DNA banding pattern from the samples would be read to establish presence/absence, and, to a more limited extent, a quantification of the original copy number and prevalence of a specific organism. A number of tools were developed in association with this basic procedure to speed up, automate, and increase the sensitivity and accuracy of the process. For example, digital scanners were applied to read the gels and attempt to quantify the size, position, and intensity of the bands in different lanes.

Traditional PCR methods read the sample after the cycling reactions had been completed. In contrast, real-time PCR methods draw the sample earlier in the reaction process (ABS, 2002; La Paz et al., 2007). This allows for discrimination of the samples during the exponential amplification phase, before the rate of signal increase slows down as all the polymerase becomes saturated. This improves the sensitivity of the process. Also, real-time PCR systems do not use a gel to read the signal, but analyze the PCR product directly as it is produced. This is done with a photodetector using fluorescent binding dyes that adhere to the DNA. Real-time PCR assays are thus sensitive and specific, and allow detection and accurate quantitation (La Paz et al., 2007). A number of commercial versions of real-time PCR detection technologies are available.

Recently, advances in thermocycler design and improvements in double-labeled probes have increased the utility of this approach. By running multiple fast real-time PCR amplifications simultaneously in conventional 96-well plates, total throughput of samples is improved. Since differential signal chemistries can be used for the different reactions, these reaction products can be analyzed simultaneously.

The conventional and real-time PCR approaches were originally used to isolate and amplify a single DNA or RNA target sequence. Reverse transcriptase-PCR, which amplifies RNA, can be used to differentiate between viable and dead organisms, unlike standard PCR, which amplifies DNA from both. Thus, the analysis and detection focused on a single gene or gene product. Multiplex PCR, however, amplifies multiple genes or gene products in a single reaction, broadening the scope and efficiency of the process (Chang et al., 2008). Combining the multiwell thermocycler design with the techniques of multiplex PCR leads to even greater increases in total throughput and accuracy. When used for molecular epidemiology, multiplex PCR can be used to cross-verify multiple genes against one another, thereby improving the accuracy of detection and identification.

A pre-enrichment step can enhance the ability of detection systems to pick up indications of live pathogens. However, in addition to the time required for preenrichment, information regarding the original concentration of bacteria is degraded or lost entirely. The selection and application of real-time PCR protocols, either single target or multiplex, will be driven in part by the intentions of the detection scheme, and the real-world limitations of cost and complexity. Although the more advanced techniques are becoming simpler, more commoditized and more widely available, there is an unavoidable relationship between complexity, technical difficulty, and cost (La Paz et al., 2007; Chang et al., 2008). Care must be shown in matching the capabilities of the detection system with the actual needs of the industrial and commercial environment.

Computer/AI Optical Scanning

Computerized optical scanning is technology that has been under development for some years. Using digital images of produce, it has been possible for some time to use computer software to measure the extent of damage caused by plant pathogens (Niemira et al., 1999). This quantification is based on differences in reflection and light absorption caused by the different physical properties of healthy versus diseased tissue. More recently, optical scanning in the visible and near-infrared has shown promise in identifying fecal contamination on poultry carcasses (Chao et al., 2008). Using this system, images of the carcass are taken at a number of different wavelengths. Since contamination spots have different spectral characteristics than the underlying skin tissue, subtractive analysis of the various images can enhance the ability of software to identify problem areas.

When applying these techniques to identification of human pathogen contamination on produce, a number of specific hurdles remain. First, it is known that areas of damaged or diseased tissue can increase the harborage of human pathogens (Wells and Butterfield, 1997). Therefore, machine vision, which can be used to identify physical defects (Lee et al., 2008), also serves to reduce risk of contamination with human pathogens. However, *Salmonella*, *E. coli* O157:H7, and other human pathogens are often resident on the surfaces of fruits and vegetables without being associated with gross physical defects. Also, while machine vision has proven useful in detecting identifiable fecal contamination, it has been less effective at detecting independent human pathogens not associated with fecal material. As biofilms represent a potential area of enhanced risk, recent research is developing the ability of machine vision tools to locate and identify biofilm material. Work using *in vitro* biofilms of *E. coli*, *Pseudomonas pertucinogena*, *Erwinia chrysanthemi*, and *L. inocula* has shown that excitation with UV-A at 320 to 400 nm enhances the signal from biofilms (Jun et al., 2008). Subtractive analysis of the resulting images collected at various wavelengths provides an actionable level of contrast between biofilm-containing and clear areas. Expansion of the work to *in vivo* biofilms on the surfaces of fruits and vegetables is required. This type of detection technology remains an area of active technological development.

ANTIMICROBIAL INTERVENTION TECHNOLOGIES

The focus of this section is on the third possible response to pathogen contamination, eradication. Chemical sanitizers, including a variety of chlorine-based sanitizers, are a standard feature of produce processing. These are intended primarily as a means to prevent cross-contamination, rather than as a true kill-step that would eliminate pathogens where they are present in fruits and vegetables. The limited efficacy of conventional liquid-based sanitizing solutions has led to an investigation of a variety of alternatives. Treatments that rely on novel or precision application of chemical sanitizers, such as gas-phase treatments using volatile essential oils (Matan et al., 2005), chlorine dioxide, or ozone (Linton et al., 2006) are the subject of widespread research in order to optimize these treatments for industrial application. One of the most challenging aspects of antimicrobial gas phase treatments such as chlorine dioxide is maintaining a uniform level of treatment at a high enough concentration to be efficacious. Modified approaches have investigated using lower concentrations renewed over extended treatment times, hours or even days, such as would be available during shipping. As with shorter duration treatments, uniformity and process control remain obstacles to full implementation.

A number of studies have investigated other promising technologies for sanitizing produce, including use of electrolyzed water (Ayebah et al., 2006), ozonated water (Koseki and Isobe, 2006) and advanced thermal treatments (Annous et al., 2004). These technologies are finding use in a number of different areas in the fresh and fresh-cut produce industry. The remainder of this section will present an overview of several key technologies that are not yet widely implemented in commercial settings. For some of these, there are important technological hurdles to be overcome. For others, the barriers to adoption are regulatory or cost engineering in nature. These technologies will be key areas for research in the future.

Cold Plasma

Cold plasma is a promising new sanitizing technology for fresh produce. A number of technological challenges are being addressed in ongoing research. As energy is added to materials, they change state, going from solid to liquid to gas, with large-scale intermolecular structure breaking down. As additional energy is added, the intra-atomic structures of the components of the gas break down, yielding plasmas—concentrated collections of ions, radical species, and free electrons (Birmingham and Hammerstrom, 2000; Fridman et al., 2005; Gadri et al., 2000; Niemira and Sites, 2008). Therefore, although it is technically a distinct state of matter, cold plasma for all practical purposes may be regarded as an energetic form of gas. Cold plasma technologies used to treat foods have been grouped into three general categories (Niemira and Sites, 2008): *electrode contact* (in which the target is in contact with or between electrodes), *direct treatment* (in which active plasma is deposited directly on the target), and *remote treatment* (in which active plasma is generated at some distance, and plasma is moved to the target).

Electrode contact systems have been shown to achieve reductions as great as 5 logs of *E. coli*, *Staphylococcus aureus*, *Bacillus subtillis*, and

Saccharomyces cerevisiae on foods and inert surfaces (Deng et al., 2007; Kelly-Wintenberg et al., 1999). Cultures of *E. coli* placed within the 1 mm gap spacing of the plasma reactors were reduced by 4.6 and 5.1 log CFU/ml after treatments of 10 and 60 seconds, respectively (Sladek and Stoeffels, 2005). As the space between the plasma emitter and the treated culture was increased from 1 mm, antimicrobial efficacy was reduced, until at 10 mm spacing, no reductions were observed at any power level tested. Using the remote treatment reactor, Montie et al. (2000) reduced *E. coli* and *S. aureus* inoculated on polypropylene by 4 or 2 log CFU/ml, respectively, after a 10-second treatment. D-values of 22 seconds (*Shigella flexneri* and *Vibrio parahaemolyticus*) to 51 seconds (*E. coli* O157:H7) for pathogens on agar were obtained using the one atmosphere uniform glow discharge plasma system (OAUGDP) (Kayes et al., 2007). Treatment with the OAUGDP for two minutes reduced *E. coli* O157:H7 on Red Delicious apples by approximately 3 log CFU, reduced *S*. Enteritidis on cantaloupe by approximately 3 log CFU, and reduced *L. monocytogenes* on iceberg lettuce by approximately 2 log CFU (Critzer et al., 2007).

A gliding arc cold plasma system effectively inactivated *E. coli* O157:H7 and *Salmonella* on agar plates and on the surface of golden delicious apples (Niemira and Sites, 2008). In that study, higher flow rates of plasma (30 or 40 L/min) were more effective than lower flow rates (10 or 20 L/min) in inactivating these pathogens on inoculated apples, and longer exposures were more effective than shorter. At the highest flow rate, treatments of three minutes reduced *Salmonella* by 3.4 log CFU, and reduced *E. coli* O157:H7 by 3.5 log CFU. An important area for future research is evaluation of cold plasma treatments to porous surfaces, such as stem scars and fresh-cut surfaces. Niemira et al. (2005) applied gliding arc plasma to the rough, porous surface of cantaloupes inoculated with *E. coli* 25922. Treatments of one or three minutes with a 260 mA gliding arc reduced the population by 1.0 or 1.3 $\log_{10}$ CFU. These results suggest that cold plasma may hold additional promise with respect to difficult-to-sanitize surfaces.

A modified approach to the design of cold plasma emitters offers a potential in-package treatment process. By applying electrically conductive labels to the inside of a container's surface, cold plasma may be generated by induction, thereby leading to the generation of ozone and other sanitizing plasma species inside the package. This approach, called PlasmaLabel by Schwabedissen et al. (2007), resulted in a 4 log CFU reduction of *B. subtillis* on agar following a 10-minute treatment. This 10-minute treatment also increased the ozone concentration inside the package to approximately 2000 ppm. Although the authors did not report sensory impact of this level of ozone, preserving the quality of the treated produce is an essential element for any antimicrobial treatment. The development of the process will be determined by the optimization of the shape of the applied electrodes, and their method of application (screen-printed, applied, bonded, etc.). Cold plasma is a developing field; research is ongoing to advance the state of the art in cold plasma emitter design, and to improve the operational application of the technology to fruits and vegetables. Research on sensory impact of the process on the treated fruits and vegetables will be critical for establishing protocols for commercial use.

Irradiation

Irradiation is a nonthermal process in which high-energy electrons or photons are applied to foods, resulting in the inactivation of associated pathogens (Niemira and Fan, 2006). Until recently, irradiation of produce was limited to disinfestation and storage-life extension. In 2008, the FDA approved the use of irradiation up to 4.0 kGy on fresh lettuce and fresh spinach to kill human pathogens such as *E. coli* O157:H7 and *Salmonella* (FDA, 2008). This intended use, to improve food safety and shelf-life, opens new opportunities for implementation of the technology in the arena of lettuce and spinach safety. However, protocols to use the technology effectively in the industrial setting must address matters of cost, consumer acceptance, and retail marketing.

An extensive body of research has demonstrated that this technology is safe and effective. Recently, research has focused on the ability of irradiation to address contamination of pathogens within the interior spaces of a leaf, fruit, or vegetable, which are inaccessible to conventional antimicrobial treatments. Microbiological analysis is made problematic by the inefficient uptake of bacteria via roots and vasculature, complicating the development of a clear risk analysis for this kind of contamination. However, it is clear that penetrating processes such as irradiation may be uniquely suited for dealing with this type of contamination. Nthenge et al. (2007) showed that irradiation eliminated pathogenic bacteria that were internalized within leaf tissues as a result of root uptake. The lettuce plants, grown in hydroponic solutions inoculated with *E. coli* O157:H7, contained the pathogen in the leaf tissue. In that study, irradiation effectively killed the pathogen whereas a treatment with 200 ppm aqueous chlorine was ineffective.

Irradiation was shown to be similarly effective in eliminating internalized *E. coli* O157:H7 from baby spinach and various types of lettuce (Romaine, Iceberg, Boston, green leaf, red leaf), whereas 300 or 600 ppm sodium hypochlorite was generally ineffective (Niemira, 2007, 2008). D_{10}

values for internalized cells (0.30–0.45 kGy) were two- to three-fold higher than for surface associated cells (0.12–0.14 kGy) (Niemira, 2007). The mechanism for this increase in D_{10} value has not been fully described. Pathogen populations within the leaf are generally expected to be very low in a commercial setting, due to the poor efficiency of uptake via the roots. Therefore, near-complete elimination of internalized pathogens may potentially be a practical goal using irradiation doses that do not cause undue sensory damage.

Irradiation is a penetrating process. Along with heat, it is one of the few treatments than can be applied to foods after they are already in the package. Once conventional washes and similar treatments have removed gross contamination, such as adherent residues or foreign matter (insect parts, manure flecks, etc.), more advanced treatments can be used to further reduce risk. These advanced treatments are intended to complement, not to replace, conventional controls and antimicrobial treatments (Fan et al., 2008). Once the produce is as clean as the conventional processes can make it, it is packaged to avoid potential cross-contamination. Irradiation would then be applied to further reduce microbial load. Hence, the packaged produce would then remain untouched by hand or machine during distribution, wholesale and retail, until it is opened at point-of-consumption by the consumer or by foodservice workers. It is expected that irradiation would therefore be used as a terminal process, incorporated into a processing line to be applied postpackaging (Fan et al., 2008).

Pulsed Light

Recent advances in electronics and lighting technology have renewed interest in pulsed light as an antimicrobial process. In addition to improvements in xenon flash lamps and related technologies that produce intense flashes of broad-spectrum light, narrow spectrum light emitting diode (LED) sources are also of interest. These are being explored for their applicability to fresh produce, and to food-contact surfaces in a produce processing environment.

Applications of broad-spectrum pulsed light for decontamination of surfaces were recently reviewed by Gomez-Lopez et al. (2007). The antimicrobial mode of action of pulsed light is based on the activity of UV-C. Across a range of wavelengths tested, maximum inactivation of *E. coli* was achieved around 270 nm, with antimicrobial efficacy dropping off to zero above 300 nm (Wang et al., 2005). The authors ascribed the majority of antimicrobial efficacy to the 220 to 290 nm range. Total aerobic counts of white cabbage, leek, paprika, carrots, and kale were reduced by 1.6 to 2.6 log CFU/cm^2 following treatment with wide-spectrum pulsed light (Hoornstra et al., 2002). The total luminance was 0.30 J/cm^2, delivered in two pulses. Working with inoculated raspberries exposed to pulsed UV light from a xenon flash lamp, Bialka and Demirci (2008) reduced *E. coli* O157:H7 and *Salmonella* by 3.9 and 3.4 log CFU/g, respectively. A parallel study using inoculated strawberries reported reductions of 2.1 and 2.9 log CFU/g for *E. coli* O157:H7 and *Salmonella*. Total luminance for these studies was reported as 25.7 to 72.0 J/cm^2. It should be noted that ensuring uniformity of treatment in a commercial-scale system will be a critical factor in scaling up this technology. The engineering challenges in effectively treating a line throughput of hundreds or thousands of pounds per minute must be thoroughly considered.

Photothermal effects from exposure to pulsed light are a known factor associated with this technology (Gomez-Lopez et al., 2007). In a study of alfalfa seeds treated with pulsed light, the authors noted that treatments were limited by excessive heating of the seeds caused by the intensity of the flash lamp (Sharma and Demirci, 2003). In the case of small fruits treated with pulsed light (Bialka and Demirci, 2008), the temperature of strawberries and raspberries increased to 69 and 79 °F, respectively, during the 60-second treatment time. However, the authors of that study reported that there was no significant effect on the sensory properties of the treated fruits. It may be that future applications of pulsed light could operate in a cold room, or use a posttreatment stream of sterile cold air to remove excessive heat from the treated product.

Recent research has examined a variation on pulsed light technology that uses narrow-spectrum illumination to treat surfaces with light in the visible spectrum. This research derives from the field of medical applications of pulsed light, where an intense UV spectrum cannot be employed without damage to soft tissues of the eye, mouth, or skin. Using selective wavelength filters on a xenon lamp system, Maclean et al. (2008a) demonstrated maximum inactivation of *Staphylococcus aureus*, including methicillin-resistant strains, at 405 nm. A combination treatment of narrow-band light at 405 nm and 880 nm produced by banks of specialized LEDs reduced *S. aureus* and *Pseudomonas aeruginosa* by as much as 1.3 log CFU. Research into the mode of action of blue light has identified oxygen availability as a key factor (Maclean et al., 2008b). The specific chemistry proposed by the authors is related to photoactivation of intracellular porphyrins. These results suggest that narrow-band pulsed blue light may be a promising area of future research.

High-Pressure Processing

High pressure processing (HPP) has been successfully applied to a number of liquid and semisolid vegetable-derived foods such as juices, sauces, and guacamole, as well as to meats and seafood products. The process causes a number of physical effects on foods that can extend shelf-life, as with guacamole, or simplify subsequent processing and preparation at point-of-consumption, as with oysters (Considine et al., 2008). It also can effectively reduce the microbial load of contaminating microflora and is therefore the subject of consideration from a food-safety perspective. Chemical analysis of treated juices, pulps, and similar processed products indicates an acceptable level of sugar retention and enzyme inactivation (Butz et al., 2002). However, the process induces significant changes in protein conformation and cellular structure. For whole fruits or vegetables, these changes can alter the cellular electrolyte balance and enzyme activity, leading to discoloration and off-aromas (Considine et al., 2008). Fruits and vegetables with voids of any kind are unlikely to be good candidates for HPP, as the compression and re-expansion of the membranes around the void cause changes in texture and color. Also, extremely high pressures involved in the process shift the critical point of water, leading to glass transitions within the cells of the treated product. The freezing of cellular and intercellular water caused by these transitions can rupture membranes. Solid frozen fruit products, such as whole berries or sliced fruits, have minimal internal voids and meet different quality criteria than fresh produce products. Therefore, HPP may hold promise as a process for juices, pulps, and purees, as well as for quick-frozen fruits and vegetables.

Sonication

Ultrasonication is the process of exposing contaminated foods and food contact surfaces to high frequency sound waves. Recent research has focused on using ultrasound to enhance the antimicrobial efficacy of applied antimicrobial compounds. Huang et al. (2006) found that a 170-kHz ultrasound treatment increased the efficacy of chlorine dioxide treatments when applied to apples inoculated with *Salmonella* or *E. coli* O157:H7. However, that same study showed that the treatment was generally ineffective at enhancing chemical efficacy for contaminated lettuce. Other studies have also shown that ultrasound does not enhance sanitizer efficacy when applied to lettuce, and can cause sensory damage to leaves with longer treatment times (Ajlouni et al., 2006). Therefore, although ultrasound appears to show potential for improving the treatments applied to some commodities, defining the commercial protocols of its use will require additional information. One possibility is to contribute to the sanitization of wash-water tanks and transport-water flumes, to prevent cross-contamination. Further research will define the role that ultrasonication can play in an overall produce processing system.

Biological Controls

Fresh produce supports a varied and complex microflora. Total aerobic populations can range from 10^2 to 10^9 CFU/g and can include bacteria, yeasts, and fungi (Fett, 2006). Complex interactions among the microflora can enhance or detract from the establishment and growth of enteric pathogens (Lund, 1992; Liao et al., 2003). Utilizing the suppressive and antagonistic effects of native microflora has been a topic of research for a number of years (Beuchat and Bracket, 1990; Nguyen-the and Carlin, 1994; Matos and Garland, 2005).

Liao (2007) demonstrated inhibition of enteric pathogens with native microflora derived from alfalfa seeds and from baby carrot. An isolate of the antagonist *P. fluorescens* (isolate Pf 2-79) suppressed *Salmonella*, *E. coli*, and *L. monocytogenes* on bell pepper disks. Efficacy of suppression was related to the population ratio of antagonist to pathogen. Where this ratio is 100:1, the pathogen is most effectively suppressed. This antagonist Pf 2-79 was also effective in suppressing *Salmonella* on sprouting seeds (Liao, 2008). Pretreatment of seeds with Pf 2-79 before sprouting suppressed *Salmonella* growth by 2 to 3 log CFU. These results suggest a role for cultures of compatible antagonists as a dip treatment.

An emerging area of research is the use on produce of bacteriophages as a targeted antimicrobial tool. Anti-*Listeria* bacteriophage treatments for packaged ready-to-eat meat and poultry products were recently approved by the FDA (FDA, 2006). In this application, the phage is applied as a liquid preparation to the surfaces of the food product immediately prior to packaging. It is expected that, much like treatments that use conventional bacterial antagonists, phage-based treatments for fruits and vegetables would be applied as a dip or spray. The specific and most optimal means of usage are still being determined, for example in field application as a preharvest treatment, during postharvest processing, and so on. Initial populations of *E. coli* O157:H7 attached to a coupon of stainless steel, a common food contact material for up to four days at 4 °C, were reduced by 1 to 2 log CFU by bacteriophage KH1; however, populations enmeshed in biofilms were protected (Sharma et al., 2005). When combined with nisin, a phage treatment reduced *L. monocytogenes* on apples and melons by 2.3 and 5.7 log CFU, respectively (Leverentz et al., 2003).

THE CHALLENGE OF TECHNOLOGY DEVELOPMENT FOR ORGANIC FOODS

The regulations governing organic fruits and vegetables (National Organic Program) limit which technologies can be used during production and processing (CFR, 2005). These regulations establish science-based limits on what additives and processes can be used and applied, consistent with the tenets and philosophy of organic production. For example, although chlorine-based sanitizers can be used, they must be limited in use and application so as to result in residues of no more than 4 ppm. High pressure processing, ohmic heating, and pulsed electric field processing of juices is permitted for organic products. In contrast, irradiation is a prohibited process for organic foods, and cannot be used under any circumstances for fresh and fresh-cut fruits and vegetables to be labeled organic.

New technologies may not be specifically addressed by existing governing regulations. The guiding tenets of organic production are applied on a case-by-case basis for approval of new technologies or adaptation of existing technologies. When considering new interventions for which no previous context exists, either organic or conventional, science-based decision making is part of the overall approach to regulation. For example, the FDA has not yet issued a ruling with respect to the applicability of cold plasma for organic foods. In such a circumstance, should the specific constraints on organic produce be taken as constraints on the technology? This is a question with important implications for economics as well as for food technology. Although the needs of a specific commodity or market can often be a driver for innovation and technology development, this type of a targeted approach can limit the advancement of an otherwise promising technology. An awareness of the governing regulations for organic or other specialty foods can help to guide development of appropriate technology, or in the adaptation of existing technologies to organic implementation. The most useful approach is to strike a balance of allowing scientific innovation to generate a host of new tools, and refining their implementation for the specific needs of important commodities or markets.

ACKNOWLEDGEMENTS

The authors would like to thank Drs. Lindsey Keskinen and Christopher H. Sommers for their critical reviews of this chapter. Mention of trade names and commercial products in this publication is solely for the purpose of providing specific information and does not imply recommendation or endorsement by the US Department of Agriculture.

REFERENCES

ABS, AppliedBioSystems (2002). Real-time PCR vs. traditional PCR. www.appliedbiosystems.com/support/tutorials/pdf/rtpcr_vs_tradpcr.pdf. Accessed Oct 17, 2008.

Ajlouni, S., Sibrani, H., Premier, R., and Tomkins, B. (2006). Ultrasonication and fresh produce (Cos lettuce) preservation. *J. Food Sci.* **71,** M62–M68.

Annous, B. A., Burke, A., and Sites, J. E. (2004). Surface pasteurization of whole fresh cantaloupes inoculated with *Salmonella* Poona or *Escherichia coli*. *J. Food Prot.* **67,** 1876–1885.

Ayebah, B., Hung, Y. C., Kim C, and Frank, J. F. (2006). Efficacy of electrolyzed water in the inactivation of planktonic and biofilm *Listeria monocytogenes* in the presence of organic matter. *J. Food Prot.* **69,** 2143–2150.

Beuchat, L. R. and Brackett, R. E. (1990). Inhibitory effect of raw carrots on *Listeria monocytogenes*. *Appl. Environ. Microbiol.* **56,** 1734–1742.

Bialka, K. L. and Demirci, A. (2008). Efficacy of pulsed UV-light for the decontamination of *Escherichia coli* O157:H7 and *Salmonella* spp. on raspberries and strawberries. *J. Food Sci.* **73,** M201–M207.

Birmingham J. G. and Hammerstrom, D. J. (2000). Bacterial decontamination using ambient pressure nonthermal discharges. *IEEE Trans. Plasma Sci.* **28,** 51–55.

Butz, P., Fernández García, A., Lindauer, R., Dieterich, S., Bognár, A., and Tauscher, B. (2003). Influence of ultra high pressure processing on fruit and vegetable products. *J. Food Engineering* **56,** 233–236.

CFR, Code of Federal Regulations. (2005). Title 7—Agriculture, Chapter I—Agricultural marketing service (standards, inspections, marketing practices), Department of Agriculture (continued) Part 205—National Organic Program. www.access.gpo.gov/nara/cfr/waisidx_05/7cfr205_05.html. Accessed Oct 15, 2008.

Chang, Y. C., Wang, J. Y., Selvam, A., Kao, S. C., Yang, S. S., and Shih, D. Y. C. (2008). Multiplex PCR detection of enterotoxin genes in Aeromonas spp. from suspect food samples in Northern Taiwan. *J. Food Prot.* **71,** 2094–2099.

Chao, K., Nou, X., Liu, Y., Kim, M. S., Chan, D. E., Yang, C., Patel, J. R., and Sharma, M. (2008). Detection of fecal/ingesta contaminants on poultry processing equipment surfaces by visible and near-infrared reflectance spectroscopy. *Applied Engineering in Agriculture* **24,** 49–55.

Critzer, F. A., Kelly-Wintenberg, K., South, S. L., and Golden, D. A. (2007). Atmospheric plasma inactivation of foodborne pathogens on fresh produce surfaces. *J. Food Prot.* **70,** 2290–2296.

Considine, K. M., Kelly, A. L., Fitzgerald, G. F., Hill, C., and Sleator, R. D. (2008). High-pressure processing—Effects on microbial food safety and food quality. *FEMS Microbiology Letters* **281,** 1–9.

Deng, S. R., Ruan, C. Y., Mok, G., Huang, X. L., and Chen, P. (2007). Inactivation of *Escherichia coli* on almonds using nonthermal plasma. *J. Food Sci.* **72,** M62–M66.

Fan, X., Niemira, B. A., and Prakash, A. (2008). Ionizing irradiation of fresh and fresh-cut fruits and vegetables. *Food Technology* **62,** 36–43.

FDA (US Food and Drug Administration). (2006). FDA approval of *Listeria*-specific bacteriophage preparation on ready-to-eat (RTE) meat and poultry products. www.cfsan.fda.gov/~dms/opabacqa.html. Accessed Oct 15, 2008.

FDA. (2008). U.S. Food and Drug Administration—Final Rule (73 FR 49593). Irradiation in the production, processing and handling of food. 21 CFR Part 179. www.cfsan.fda.gov/~lrd/fr080822.html. Accessed September 2, 2008.

Fett, W. F. (2006). Inhibition of *Salmonella enterica* by plant-associated pseudomonads in vitro and on sprouting alfalfa seed. *J. Food Prot.* **69,** 719–728.

Fridman, A., Chirokov, A., and Gutsol, A. (2005). Non-thermal atmospheric pressure discharges. *J. Phys. D: Appl. Phys.* **38,** R1–R24.

Gadri, R. B., Roth, J. R., Montie, T. C., Kelly-Wintenberg, K., Tsai, P., Helfritch, D. J. et al. (2000). Sterilization and plasma processing of room temperature surfaces with a one atmosphere uniform glow discharge plasma (OAUGDP). *Surface Coatings Technol.* **131,** 528–542.

Gessler, F., Hampe, K., Schmidt, M., and Bohnel, H. (2006). Immunomagnetic beads assay for the detection of botulinum neurotoxin types C and D. *Diagnostic Microbiol. Inf. Dis.* **56,** 225–232.

Gomez-Lopez, V. M., Ragaert, P., Debevere, J., and Devlieghere, F. (2007). Pulsed light for food decontamination: A review. *Food Sci. Technol.* **18,** 464–473.

Hoornstra, E., de Jong, G., and Notermans, S. (2002). Preservation of vegetables by light. In *Frontiers in microbial fermentation and preservation* (Society for Applied Microbiology, Ed.), pp. 75–77. Wageningen, The Netherlands.

Huang, T-S., Xu, C., Walker, K., West, P., Zhang, S., and Weese, J. (2006). Decontamination efficacy of combined chlorine dioxide with ultrasonication on apples and lettuce. *J. Food Sci.* **71,** M134–M139.

JIFSAN (Joint Institute for Food Safety and Applied Nutrition). (2007). Tomato Safety Research Needs Workshop. www.jifsan.umd.edu/tomato_wkp2007.htm. Accessed Feb 6, 2008.

Jun, W., Lee, K., Millner, P. D., Sharma, M., Chao, K., and Kim, M. S. (2008). Portable hyperspectral fluorescence imaging system for detection of biofilms on stainless steel surfaces. *Proceedings of SPIE*, March 17–20, 2008, Orlando, FL.

Kayes, M. M., Critzer, F. J., Kelly-Wintenberg, K., Roth, J. R., Montie, T. C., and Golden, D. A. (2007). Inactivation of foodborne pathogens using a one atmosphere uniform glow discharge plasma. *Foodborne Path. Dis.* **4,** 50–59.

Kelly-Wintenberg, K., Hodge, A., Montie, T. C., Eleanu, L. D., Sherman, D., Roth, J. R. et al. (1999). Use of a one atmosphere uniform glow discharge plasma to kill a broad spectrum of microorganisms. *J. Vac. Sci. Technol. A* **17,** 1539–1544.

Koseki, S. and Isobe, S. (2006). Effect of ozonated water treatment on microbial control and on browning of iceberg lettuce (Lactuca sativa L.). *J. Food Prot.* **69**, 154–160.

La Paz, J. L., Esteve, T., and Pla, M. (2007). Comparison of real-time PCR detection chemistries and cycling modes using Mon810 event-specific assays as model. *J. Agric. Food Chem.* **55**, 4312–4318.

Lee, K., Kang, S., Delwiche, S. R., Kim, M. S., and Noh, S. (2008). Correlation analysis of hyperspectral imagery for multispectral wavelength selection for detection of defects on apples. *Sensing and Instrumentation for Food Quality and Safety* **2**, 90–96.

Leverentz, B., Conway, W. S., Camp, M. J., Janisiewicz, W. J., Abuladze, T., Yang, M. et al. (2003). Biocontrol of *Listeria monocytogenes* on fresh-cut produce by treatment with lytic bacteriophages and a bacteriocin. *Appl. Env. Microbiol.* **69**, 4519–4526.

Liao, C.-H., McEvoy, J. L., and Smith, J. L. (2003). Control of bacterial soft rot and foodborne human pathogens on fresh fruits and vegetables. In *Advances in plant disease management* (H. C. Huang and S. N. Aharya, Eds.), pp. 165–193. Research Signpost, Kerala, India.

Liao, C.-H. (2007). Inhibition of foodborne pathogens by native microflora recovered from fresh peeled baby carrot and propagated in cultures. *J. Food Sci.* **72**, M134–M139.

Liao, C.-H. (2008). Growth of *Salmonella* on sprouting seeds as affected by the inoculum size, native microbial load and *Pseudomonas fluorescens* 2–79. *Lett. Appl. Microbiol.* **46**, 232–236.

Linton, R. H., Han, Y., Selby, T. L., and Nelson, P. E. (2006). Gas-/vapor-phase sanitation (decontamination) treatments. In *Microbiology of fruits and vegetables* (Sapers, G. M., Gorny, J. R., and Yousef, A. E., Eds.), pp. 401–436. Taylor and Francis, New York.

Lund, B. M. (1992). Ecosystems in vegetable foods. *J. Appl. Baceriol. (Symp. Suppl.)* **73**, 115S–126S.

Maclean, M., MacGregor, S. J., Anderson, J. G., and Woolsey, G. (2008a). High-intensity narrow-spectrum light inactivation and wavelength sensitivity of *Staphylococcus aureus*. *FEMS Microbiology Letters* **285**, 227–232.

Maclean, M., MacGregor, S. J., Anderson, J. G., and Woolsey, G. (2008b). The role of oxygen in the visible-light inactivation of *Staphylococcus aureus*. *J. Photochemistry Photobiology B: Biology* **92**, 180–184.

Matan, N., Rimkeereea, H., Mawsonb, A. J., Chompreedaa, P., Haruthaithanasana, V., and Parker, M. (2005). Antimicrobial activity of cinnamon and clove oils under modified atmosphere conditions. *Int. J. Food Microbiol.* **107**, 180–185.

Matos, A. and Garland, J. L. (2005). Effects of community versus single strain of inoculants on the biocontrol of *Salmonella* and microbial community dynamics in alfalfa sprouts. *J. Food Prot.* **68**, 40–48.

Montie, T. C., Kelly-Wintenberg, K., and Roth, J. R. (2000). An overview of research using the one atmosphere uniform flow discharge plasma (OAUGDP) for sterilization of surfaces and materials. *IEEE Trans. Plasma Sci.* **28**, 41–50.

Nguyen-the, C. and Carlin, F. (1994). The microbiology of minimally processed fresh fruits and vegetables. *Crit. Rev. Food Sci. Nutr.* **34**, 371–401.

Niemira, B. A., Kirk, W. W., and Stein, J. M. (1999). Screening for late blight susceptibility in potato tubers by digital analysis of cut tuber surfaces. *Plant Disease* **83**, 469–473.

Niemira, B. A., Alvarez, I., Annous, B. A., Gutsol, A., and Fridman, A. (2005). Antimicrobial efficacy of cold atmospheric pressure plasma applied to inoculated food surfaces. P2. Institute of Food Technologists Nonthermal Processing Division Meeting, Wyndmoor, PA Sept 2005.

Niemira, B. A. and Fan, X. (2006). Low dose irradiation of fresh produce: Safety, sensory and shelf life. In *Food irradiation research and technology* (C. H. Sommers and X. Fan, Eds.), pp. 169–184. Blackwell Publishing, Ames, IA.

Niemira, B. A. (2007). Relative efficacy of sodium hypochlorite wash vs. irradiation to inactivate *Escherichia coli* O157:H7 internalized in leaves of romaine lettuce and baby spinach. *J. Food Prot.* **70**, 2526–2532.

Niemira, B. A. (2008). Irradiation vs. chlorination for elimination of *Escherichia Coli* O157:H7 internalized in lettuce leaves: Influence of lettuce variety. *J. Food Sci.* **73**, M208–M213.

Niemira, B. A. and Sites, J. (2008). Cold plasma inactivates *Salmonella* Stanley and *Escherichia coli* O157:H7 inoculated on golden delicious apples. *J. Food Prot.* **71**, 1357–1365.

Nthenge, A. K., Weese, J. S., Carter, M., Wei, C. I., and Huang, T. S. (2007). Efficacy of gamma radiation and aqueous chlorine on *Escherichia coli* O157:H7 in hydroponically grown lettuce plants. *J. Food Prot.* **70**, 748–752.

Nugen, S. R. and Baeumner, A. J. (2008). Trends and opportunities in food pathogen detection. *Anal. Bioanal. Chem.* **391**, 451–454.

Schwabedissen, A., Lacinski, P., Chen, X., and Engemann, J. (2007). PlasmaLabel—A new method to disinfect goods inside a closed package using dielectric barrier discharges. *Contrib. Plasma Phys.* **47**, 551–558.

Sivapalasingam, S., Friedman, C. R., Cohen, L., and Tauxe, R. V. (2004). Fresh produce: A growing cause of outbreaks of foodborne illness in the United States, 1973 through 1997. *J. Food Prot.* **67**, 2342–2353.

Sharma, M., Ryu, J.-H., and Beuchat, L. R. (2005). Inactivation of *Escherichia coli* O157:H7 in biofilm on stainless steel by treatment with an alkaline cleaner and a bacteriophage. *J. Appl. Microbiol.* **99**, 449–459.

Sharma, R. R. and Demirci, A. (2003). Inactivation of *Escherichia coli* O157:H7 on inoculated alfalfa seeds with pulsed ultraviolet light and response surface modeling. *J. Food Sci.* **68**, 1448–1453.

Sladek, R. E. J. and Stoeffels, E. (2005). Deactivation of *Escherichia coli* by the plasma needle. *J. Phys. D: Appl. Phys.* **38**, 1716–1721.

Tu, S., Reed, S. A., Gehring, A. G., and He, Y. (2008). Detection of *Salmonella* Enteriditis from egg components using different immunomagnetic beads and time-resolved fluorescence. *Food Analytical Methods* 10.1007/s12161–008-9033-4.

UFPA (United Fresh Produce Association). (2007). Leafy greens food safety research conference. www.unitedfresh.org/newsviews/leafy_greens_food_safety_research. Accessed Feb 6, 2008.

Wang, T., MacGregor, S. J., Anderson, J. G., and Woolsey, G. A. (2005). Pulsed ultra-violet inactivation spectrum of *Escherichia coli*. *Water Research* **39**, 2921–2925.

Wells, J. M. and Butterfield, J. E. (1997). *Salmonella* contamination associated with bacterial soft rot of fresh fruits and vegetables in the marketplace. *Plant Disease* **81**, 867–872.

CHAPTER 18

Conclusions and Recommendations

Casey J. Jacob
Diagnostic Medicine/Pathobiology, Kansas State University, Manhattan, KS

Benjamin J. Chapman
4-H Youth Development and Family & Consumer Sciences, NC Cooperative Extension Service, North Carolina State University, Raleigh, NC

Douglas A. Powell
Associate Professor, Food Safety, Diagnostic Medicine/Pathobiology, Kansas State University, Manhattan, KS

CHAPTER CONTENTS

Introduction	445
Sources of Contamination	446
Commodities at Risk	447
Challenges of Produce Disinfection	449
Investigating Contamination on the Farm	449
Pre-emptive Food Safety Programs	450
The Farm-to-Fork Approach	451

INTRODUCTION

From the local market to the local megalomart, the year-round availability of fresh fruits and vegetables has never been greater. Produce is a nutritional superstar, but because many fruits and vegetables are not cooked, anything with which they come into contact is a possible source of contamination. Is the water used for irrigation or rinsing spinach clean, or is it contaminated with human pathogens? Do the workers who harvest the produce follow strict hygienic practices such as thorough handwashing? What happens to that spinach when it gets chopped up? The possibilities for contamination, as documented in the preceding chapters, are seemingly endless.

So what are consumers to think? A diet rich in fresh fruits and vegetables is actively promoted as the cornerstone of a healthy lifestyle. But, public awareness about produce-associated foodborne illness has reached a tipping point as a consequence of the outbreak of *E. coli* O157:H7 in spinach in the fall of 2006 and other recent outbreaks involving many hundreds of cases of foodborne illness.

The social and economic impacts of these outbreaks are far-reaching and visible. The challenge lies in how to maximize the benefits of a diet rich in fresh fruits and vegetables while minimizing known risks. The produce contamination problem must be fixed.

SOURCES OF CONTAMINATION

Contamination begins on the farm in the soil, water, and amendments used to nurture safe, nutritious crops. As noted by Millner (Chapter 4), on-farm conditions and practices are critical determinants of the sanitary condition of fresh produce from both organic and conventional sources. Identification of on-farm pathogen reservoirs and vectors can aid development and use of farm-specific pathogen reduction programs. Research is needed to evaluate the effectiveness of various manure management practices designed to reduce pathogen loading on-site and minimize pathogen accumulation off-site (via run off or transport). Some animal viruses, while not zoonotic pathogens, may be suitable indicators of manure treatment efficacy, although research is required to validate their use. Additional research is needed to determine appropriate field management strategies for land areas adjacent to fresh produce crop fields to reduce fugitive enteric pathogen contamination. Many approaches to biological processing of manure exist, however thermophilic composting remains one of the most cost-effective treatment technologies for manure solids that functions well in a variety of environments. A science-based quality control program (such as a HACCP plan with verification) is required to ensure the safety of compost produced for use on fresh-produce crops (http://lubbockonline.com/stories/011609/loc_377888884.shtml).

Millner rather astutely stated that "the non-preferential contamination in fresh-produce–related outbreaks across organic and conventional sources suggests that actual on-site conditions and practices, rather than marketing-based labels, are the critical determinants of the sanitary condition of fresh produce." In other words, microorganisms tend not to preferentially associate with a politically favorable growing regime. The sooner the public discussion and buying patterns returns to a basis of biological safety, the better.

Additional microbial risks are found outside of human practices. Although it is thought that wildlife can be reservoirs for human diseases transmitted by contaminated fresh produce, there is, as noted by Rice (Chapter 7), a lack of evidence directly linking contamination by wildlife to outbreaks of human illness. However, awareness—particularly of the surrounding environment—along with some controls can reduce the risk. Clark (Chapter 6) details circumstantial evidence that incriminates wild birds in the contamination of produce at several points throughout production and processing, but can all the birds be killed? This would likely not be possible, nor would it be desirable. There is a need for economically feasible mechanisms to mitigate wildlife-produce interactions.

COMMODITIES AT RISK

Not all fresh fruits and vegetables are equally susceptible to microbial contamination. Certain commodities—leafy greens, tomatoes, cantaloupes, green onions, herbs, and sprouts—are linked to notably more outbreaks of foodborne disease than others. Tree fruits and nuts are rarely associated with such outbreaks.

Matthews (Chapter 8) notes that the economically advantageous practice of processing leafy greens in the field (including harvesting, washing, and packaging) should be examined for potential exposure to microbial contaminants. In addition, hydrocooling processes could have a significant impact on the microbial safety of leafy greens, but research to that end has focused largely on retail environments over processing environments. Matthews also states that guidelines are in place for the microbiological testing of water, soil amendments, and equipment used to grow and harvest leafy greens, as well as the commodities themselves.

The discussion of raw tomatoes and *Salmonella* in Chapter 10 raises some important questions. Any association between consumption of food or water—and especially fresh produce, given the short shelf-life—is tenuous and based on the best available evidence at the time. However the public-health goal of fewer illnesses is best served by offering evidence-based suggestions to improve the microbiological safety of fresh produce. To that end, Bartz recommends the following:

- Preventing contact between crops and environmental sources of *Salmonella*

- Requiring use of GAPs by producers and GMPs by postharvest handlers
- Requiring a sanitation step in processing
- Evaluating packed tomatoes for cleanliness and a safe dry pack
- Identifying grower, date of harvest, packer, shipper, and such on each package
- Requiring use of HACCP by fresh-cut processors
- Excluding fruit salvaged from defective shipments or culled from regular shipments
- Refusing fruit from noncertified growers-or packers-or from uncontrolled markets
- Providing a clear use-by date and embedding sensors that would indicate temperature abuse or anoxic conditions
- Subjecting food-service suppliers and distributors to periodic, unannounced inspections
- Focusing FDA inspections on the microbiology of fresh products and away from chemical contamination

However, it must be understood that pointing fingers in a farm-to-fork food-safety system is of little use. Everyone, including producers, processors, retailers, and consumers, needs to practice good food-safety behavior, based on the best available evidence at the time, and should expend resources to manage their own responsibilities. Therefore, additional measures may include subjecting producers to periodic, unannounced inspections; verifying that GAPs were being used appropriately by growers; and encouraging proper handwashing—a sanitation step—by everyone handling fresh produce or equipment.

In discussing outbreaks associated with tree fruits and nuts, Keller (Chapter 11) says that the occasional association of human pathogens with tree fruits and nuts is facilitated by particular pathogens' tolerance to some extreme conditions (such as desiccation and acidity). The mechanism of contamination of tree fruits involved in foodborne illness outbreaks is often unknown, though poor sanitation and hygiene practices, unsafe harvest or processing methods, internalization of contaminated wash/rinse water facilitated by a temperature gradient, and changes in normal procedures without appropriate attention to food safety (i.e., the lack of a culture of food safety) can lead to growth and survival of human pathogens in or on such fruits, as well as nuts.

CHALLENGES OF PRODUCE DISINFECTION

When human pathogens are introduced into the production or processing environment, are there ways to minimize the potential for colonization? Solomon and Sharma (Chapter 2) explain that several bacterial pathogens attach rapidly to produce surfaces and cannot be removed with current washing or agitation regimens. Sapers (Chapter 16) says the efficacy of washing and sanitizing agents for produce is often limited by attachment of microorganisms to inaccessible surfaces, infiltration of microorganisms, or incorporation of microorganisms into resistant biofilms. Cut produce is at even greater risk for bacterial colonization, and sanitizing solutions may be rendered ineffective by organic matter released from cut tissues of produce or by biofilms present on the produce. Some bacterial foodborne pathogens can be internalized by produce when a temperature differential exists between a fruit or vegetable and the fluid it is immersed in, such as in a dump tank. Further, washing equipment can also introduce contamination through the accumulation of soil and microbes in wash water or sanitizing solution.

Despite the proclamations of various consumer advisory groups, washing of produce is of limited use. Solomon and Sharma suggest that gaseous sanitizers appear promising, but at this point, as Sapers notes, the best defense is to minimize contact between human pathogens and fresh produce.

INVESTIGATING CONTAMINATION ON THE FARM

The produce problem needs to be addressed on the farm first. Managing risks on an individual farm basis may yield the most significant reductions in produce-related foodborne illness. As Sapers and Doyle explained in the introduction (Chapter 1), the prevalence of produce contamination by enteric pathogens is generally too low for broadly focused testing, and the understanding of human pathogens in the farm environment is limited. Better detection may help allocate risk reduction resources in the most cost-effective manner. But as described by Niemira and Zhang (Chapter 17), there are lots of technologies but not many near practical application.

And when problems do happen, the human factor, as pointed out by Farrar and Guzewich (Chapter 3), is often overlooked. Their chapter provided an outstanding overview to improve the investigative process when an outbreak of produce-related foodborne illness occurs, and an urgent call to pay attention to human behavior. For example, Farrar and Guzewich state,

Where possible, investigators are strongly encouraged to visually observe 'routine' food preparation/food processing procedures and record objective measurements of preparation and processing practices, as well as routine cleaning and sanitation procedures. Often, what is written in procedures manuals is not what actually occurs in the kitchen or in the food processing facility. Even with the inevitable bias introduced during physical observation by investigators, valuable clues are often obtained by simply observing and documenting a process from start to finish.

The same advice applies to the farm.

PRE-EMPTIVE FOOD SAFETY PROGRAMS

Government regulation of microbial food safety generally focuses on employing vague good agricultural practices and do not specify measurable outcomes. Neither European Union regulatory requirements described by Johannessen and Cudjoe (Chapter 14), nor Japanese requirements described by Isshiki et al. (Chapter 15) mandate microbial standards for fresh produce. Some organizations have implemented their own food-safety programs. Isshiki et al. note that Japanese retailers are requiring that suppliers meet food safety and quality standards to protect themselves from food safety incidents and enhance their products for consumers who increasingly value food safety. The California Leafy Greens Marketing Agreement was put in place to demonstrate that industry could control the problem of contamination of leafy greens, and that there was no need for government regulation.

Retailers and consumers are currently driving on-farm food safety program implementation. Clear expectations by these groups can lead producers to reduce the likelihood of illnesses associated with their products through on-farm food-safety programs.

There are several factors that contribute to the successful implementation of scientifically validated risk-reduction practices on-farm. To begin, successful on-farm food-safety programs include ongoing support in the form of workshops, documents, and individuals with expertise in food safety to advise on potential risks, implementation of practices, standards to be met, and methods to evaluate any new risks. It is important that any implemented program be constantly revisited and updated with new science, practice developments, and discoveries of risk. An ideal program would also provide rewards to participating producers and be translatable to buyers. Finally, a successful program of on-farm food safety promotes a change in culture on a farm, as opposed to a change of practices in relation to specific risks.

THE FARM-TO-FORK APPROACH

The real challenge for food-safety professionals is to garner support for safe food practices in the absence of an outbreak; to create a culture that values microbiologically safe food from farm-to-fork at all times, and not just in response to the glare of the media spotlight. A farm-to-fork approach must be used to target food-safety practices to all food handlers at each stage of food production during typical day-to-day operation. Primary production, processing, distribution/retail and consumers are the four sectors that comprise the farm-to-fork food continuum. Pathogens with the potential to cause foodborne illness can contaminate food at any point along the continuum. Food safety is, therefore, a shared responsibility among all involved in the food continuum, from producer to consumer, and across all levels of government.

A farm-to-fork approach to fresh produce food safety involves marketing food safety to those involved in the production of safe produce, as well as consumers. Coordination among packers, shippers, and retailers provides opportunities to establish relationships and explore other value-chain initiatives, where program costs can be recouped and profits increased. The purchase of microbiologically safe fresh fruits and vegetables by consumers translates into fewer sick people.

Index

Note: Page numbers followed by *f* or *t* indicate figures or tables, respectively.

A

Advanced disinfection technologies
  biological controls
    application of bacteriophages as, 437-438
    effects of native microflora in, 437
    Pseudomonas fluorescens isolate as, 437
  cold plasma
    in-package treatment with, 432-433
    microbial reductions with, 431-432
    principle of treatment with, 431
  high pressure processing
    applicable produce items for, 436
    induction of changes by, 436
  irradiation
    FDA approval of, 433
    inactivation of internalized bacteria by, 433-434
    treatment of packaged foods by, 434
  novel applications of chemical sanitizers as, 431
  other promising technologies as, 431
  pulsed light
    broad-spectrum applications of, 434-435
    photothermal effects of, 435
    narrow spectrum applications of, 435
  sonication
    commodity differences in response to, 436
    enhancement of antimicrobial treatments with, 436
  further research needs of, 437
  technology requirements for organic foods
    regulations applicable to, 438
    constraints on new technology development by, 438

B

Berry contamination
  Cyclospora
    characteristics, 279
    outbreaks, 273, 279-280
    transmission, 280-281
  impact of outbreaks
    economic effects, 273-274
    need better diagnostic methods, 273
    need to react to strong epidemiological data, 273
  mitigation strategies
    chemical sanitizing treatments, 292
    contamination reduction strategies, 286t-291t
    disinfection efficacy and microbial inaccessibility, 293
    high hydrostatic pressure, 295
    irradiation, 296-297
    refrigeration and frozen storage, 294-295
    thermal processing, 285
    UV light, 295-296
    washing, 285
    washing and decay, 278
  overview
    berries linked to 22 outbreaks, 272-273
    berry production and consumption, 272
    incidence of contamination in berries, 284
    surface characteristics of berries, 282
  outbreaks associated with berries
    Cyclospora, in raspberries and blackberries, 272-274
    hepatitis A in raspberries and frozen strawberries, 274-276
    implications of outbreak history, 284
    norovirus in raspberries, 277-279
    Salmonella in blueberries/strawberries and grapes, 283
    Staphylococcus aureus in strawberries, 283
  reduction of contamination
    GAPs and GMPs, 272, 298
    Guide to Minimize Microbial Food Safety Hazards for Fresh Fruits and Vegetables, 297
    Produce Safety Action Plan of 2004, 297
  sources of contamination
    commingling in processing facility, 274
    contaminated water for irrigation and pesticide application, 281
    hand destemming strawberries in field, 275
    handling during harvesting, 274-275
    infected farm workers, 274-275
    limited access to latrines and handwashing facilities, 276
    no gloves, 276
    poor personal hygiene, 275-276
Biofilm formation
  biofilm formation by human pathogens on produce, 32

453

Biofilm formation (*Continued*)
  colonization of plant surfaces by bacteria, 31
  efficacy of washing and, 394, 409, 412
  location by optical scanning, 430
  on leaves, 178
  on melons, 208
  strategies of biofilm formation, 31-32

C

Conclusions and recommendations
  challenges of produce disinfection
    bacterial attachment and infiltration, 449
    limited efficacy of washing and sanitizing agents, 449
  consumer attitudes
    benefits of a diet high in fruits and vegetables, 445-446
    risk of foodborne illness, 446
    farm as primary source of contamination, 446
    high risk commodities, 447-448
    investigation of contamination on the farm, 449
  pre-emptive food-safety programs
    driven by retailers and consumers, 450
    implementation, 450
    lack of microbial standards for produce, 450
    need for farm to fork approach, 451
    on-farm food safety programs, 450
  research needs
    evidence of wildlife in contamination, 447
    field and land management practices, 446
    identification of reservoirs and vectors, 446
    manure management practices, 446
Contamination sources
  fresh-cut processing
    bacterial attachment and growth on cut surfaces during, 11, 37
    bacterial transfer from rind to flesh during, 11, 194, 203-204
    contamination of processing aids used for, 11, 257
    cleaning and sanitation during, 11-12, 256-257
    temperature abuse following, 11
  produce packing
    cleaning and sanitation during, 11, 226, 243, 256-257
    equipment for, 10, 173-174, 202-203, 232
    field packing in, 13, 210, 228, 275
    produce handling during, 10, 171, 202-203, 227
    hydrocooling in, 13, 114
    process water for, 10-11, 114, 201-202, 226, 256, 317
    produce washing in, 11, 394, 412
  preharvest contamination
    feces of wildlife as source of, 10, 13, 124-131, 146-155
    irrigation water and, 10, 13, 107-113
    manure as source of, 10, 85-89
    produce handling and, 10, 274-275
    runoff and, 10, 315-316
    specific sources of, 10

D

Defining the contamination problem
  attachment and survival of pathogens on produce
    factors affecting, 9
    conventional disinfection technology and, 11
    interval between contamination and washing and, 9
    outbreak investigations in, 9
  prevalence of produce contamination in
    published studies of, 8
    FDA and USDA surveys of, 8
  produce-associated outbreaks
    as a new problem, 3-4
    economic consequences of, 5
    human pathogens involved in, 5, 6t
    prevalence of, 5, 6t
    public health consequences of, 4
    specific commodities involved in, 7-8, 7t
Detection of pathogens by rapid methods
  computerized optical scanning
    detection of fecal contamination by, 430
    detection of plant tissue damage by, 429-430
    location and identification of bacterial biofilms with, 430
  limitations of traditional methods for, 426-427
  polymerase chain reaction (PCR)-based methods
    description of, 428
    multiplex PCR as, 429
    real-time PCR as, 428-429
    reverse transcriptase PCR as, 429
  separation and concentration
    active biosensors in, 427-428
    immunomagnetic beads in, 427

E

Effective interventions and recommendations
  effective regulatory actions as, 13
  for berries, 286t-291t, 292-297
  for leafy vegetables, 169, 172
  for manure management, 95-97
  for melons, 204-206, 210-214
  for outbreak investigations, 75-76
  for tomatoes, 243-244
  for tree fruits and nuts, 253-255, 259-260
  for wild bird management, 132
  for wildlife management, weakness of, 156
  Good Agricultural Practices (GAPs) as, 13

Good Manufacturing Practices
(GMPs) as, 13
guidance documents and
HACCP plans as, 13
improved disinfection technology
as, 13
new technologies as, 173,
174-176
overview of, 446-451
rapid detection needed for, 13
washing and sanitizing,
limitations as, 414
water quality, 108-109
Efficacy of sanitizing agents
limited efficacy of conventional
decontamination
methodologies
physical location of bacteria on
plant surfaces, 36
survival of aggregates of
bacteria on leaf surfaces, 37
preferential attachment of
pathogens to cut surfaces, 37
resistance of attached
pathogens to disinfection
using chlorine, 37-38
other decontamination methods
chlorine dioxide, 38
gaseous ozone, 38-39
plasma, 39

F

Fresh-cut processing
antimicrobial treatments for,
212-214
contamination during, 11-12
outbreak investigations of, 72-73
tomatoes, 229

G

Gaps in understanding/research
needs
airborne transmission of
pathogens in aerosols and
dust, 12
application of bacteriophages to
disinfection, 175
application of sonication to
disinfection treatments, 437

biofilm formation on
produce, 12
behavior of enteric pathogens in
phylosphere, 39, 167, 177
development of new disinfection
technologies, 173,
174-176, 431
experimental designs more
reflective of real world
conditions, 182
field management strategies for
land adjacent to produce
fields, 91, 446
improvements in disinfection
technology, 13, 412, 414
improvements in diagnostic
methods, 13, 273
indicators of efficacy of manure
treatments, 89, 446
internalization of pathogens, 12,
32-36
knowledge of specific source of
contamination in, 12
manure management, 446
mechanisms of contamination,
in, 12-13, 214
risky agricultural and
manufacturing practices
affecting, 10, 13
role of water in produce
contamination, 114
seasonality of prevalence of
infections and outbreaks for
berries, 282
vulnerability of specific
commodities in, 13
wildlife management, 446-447
wildlife reservoirs of human
pathogens, 143-144
Good Agricultural Practices (GAPs)
as effective interventions, 13
for berries, 272, 298
for leafy vegetables, 169, 172
for melons, 196-197, 214
for tomatoes, 224, 243-244
for tree fruits and nuts, 256,
258-259
in Europe, 332, 342-346
in Japan, 366-367

in Mexico and Central America,
321-323
Good Manufacturing Practices
(GMPs)
as effective interventions, 13
for berries, 272, 298
for leafy vegetables, 169, 172
for manure mitigation, 95-96
in Japan, 366-367
in Mexico and Central America,
321-323
Guidance documents
as effective interventions, 13
in Japan, 364-365, 365f
in Mexico and Central
America, 312
mitigating wildlife-crop
interactions, 155

H

Hazard Analysis and Critical
Control Points (HACCP)
as effective interventions, 13
for leafy vegetables, 169, 172
for manure mitigation, 95-96
for tomatoes, 243-244
in Japan, 355
in Mexico and Central America,
322

I

Internalization
dependence on temperature
differential, 33
efficacy of washing and, 409,
413
immersion of produce into
bacterial inoculum, 33
in leafy vegetables, 180
in melons, 209
internalization through cut or
damaged tissue, 32
in tomatoes, 240-241
uptake through plant roots
bacterial inoculum level and,
34-35
co-inoculation of human
pathogens with
phytopathogens and, 34

Internalization (Continued)
   differences between *E. coli* O157:H7 and *Salmonella*, 35-36
   growing medium (soil, hydroponic etc.) and, 33-35
   plant age (stage of development) and, 33-34

L

Leafy vegetables
   human pathogenic bacteria on
     characteristics of, 179-180
     Escherichia coli O157:H7, 161, 169, 171, 181
     Listeria monocytogenes, 162, 171, 181
     Salmonella, 161, 169, 171, 181
     Shigella, 162, 181
     surveys for foodborne pathogens, 167
   internalization, leaves and roots, 180
   microbial localization in, 180
   leaves
     biofilms on, 178
     characteristics of, 177-178
     microbial interactions with, 178
     native microflora on, 178
   new technologies and mitigation strategies
     acid electrolyzed water, 174
     bacteriophage, 175
     chlorine dioxide gas, 176
     GAPs, GMPs and HACCP, 169, 172
     radio frequency identification devices, 172
     ultrasound, 175
   outbreaks from
     cabbage, 166-167
     lettuce, 166
     spinach, 166, 168
   packaging
     active packaging materials, 177
     antimicrobial films, 177
     edible films, 177
   pre-harvest contamination
     geographical region, growing conditions and, 168
     harvesting practices and, 169-170
     irrigation water and, 169
     soil amendments and, 169
   postharvest contamination
     cooling and, 171
     cutting and, 171, 181
     handling and, 171
     microbiological screening for processing equipment and, 173-174
     sanitizing agents and, 174
     triple wash and, 173

M

Manure as contamination source
   bacterial pathogens in
     E. coli O157:H7, 85
     Enterococcus species, 85
     Listeria monocytogenes, 85
     Salmonella, 85-86
     survival in, 84
     transmission of bacteria in, 84
   manure handling guidelines
     Federal agencies, 82
     organic production, 82-83
     pathogen testing, 81
   mitigation strategies
     Critical Control Points, 96-97
     implementation of GMPs and HACCP, 95-96
     USEPA requirements, 95-96
   overview
     manure treatment technologies, 92
     microorganisms associated with manure, 80*t*
     pathogen spread in the environment, 90
     production practices influencing spread of microorganisms, 90-92
     regulatory guidelines for manure management, 81-82
     survival of microorganisms in the environment, 85-87
   parasites in
     Cryptosporidium, 87-88
     domestic and wildlife sources, 88
     survival in the environment, 88
   treatment of
     composting, 93-97
     composting formats, 94-95
     thermophilic composting, 94
   viruses in
     bovine enterovirus, 89
     hepatitis, 89
     sources, 89
Melons
   antimicrobial treatments for fresh-cut
     chlorine, 212-213
     edible coatings, 213
     irradiation, 213-214
   antimicrobial treatments for whole
     chlorine, 205, 211
     field washing and water washing, 210
     hot water treatment, 211-212
     lactic acid, 211
     peroxide, 211
     surfactants/detergents, 211
   cantaloupe netting
     attachment of bacteria on, 207
     bacterial detachment from cantaloupe and honeydew, 207
     differences between Western and Eastern cantaloupes, 208
   cutting practices
     sources of pathogens, 203
     mechanism for contamination, 203-204
     preventing and minimizing contamination, 205
   factors contributing to outbreaks from
     bacterial growth in cut, 194, 206-207
     contamination of rind, 194
     transfer of bacteria from rind to cut pieces, 194, 203-204
   outbreaks linked to
     norovirus, 194
     Salmonella, 191-194
     Escherichia coli O157:H7, 194
     Campylobacter, 194

Index **457**

postharvest sources of contamination of
  equipment and design of handling facilities, 202-203
  water for cooling, rinsing, washing, disinfection, 201-202
  worker activities, 202
preharvest sources of contamination of
  animal or human feces, 199-200
  facilities and equipment, 200-201
  organic fertilizer, 198-199
  water for irrigation and production, 197-198
  worker activities, 198
prevalence of human pathogens on
  prevalence in U.S. and Mexico, 191
  sampling of Mexican melons, 190-191
regulatory activities
  FDA ban on melon imports from Mexico, 195
  Good Agricultural Practices, 196-197, 214
  memorandum of understanding, 195
  producers allowed to import into the U.S., 196
  SENASICA, 195-196
structural characteristics affecting bacterial survival and growth on
  biofilm formation on, 208
  growth of *Shigella* within whole melons, 206
  growth of *Salmonella* on cut pieces of, 206-207
  growth of *E. coli* O157:H7 and *Salmonella* on rind of, 207
  infiltration and internalization through ground spot and stem scar of, 209
Microbial attachment to fresh produce
  attachment of *E. coli* O157:H7 to lettuce
    cellular and molecular processes involved in attachment, 27-28
    greenhouse or field level experiments, 27
    laboratory experiments with cut lettuce, 26-27
  attachment of *Salmonella* to cantaloupes
    characteristics of cantaloupe and, 30
    efficacy of sanitizers and transfer to cut pieces, 31
    greenhouse level experiments and, 31
    laboratory experiments and, 30-31
    relation of bacterial cell surface charge/hydrophobicity to attachment, 30
  attachment of *Salmonella* to tomatoes
    greenhouse or field level experiments, 29
    laboratory experiments with tomatoes, 28-29
    molecular determinants of *Salmonella* attachment to plant tissue, 29-30
  characteristics of plant surfaces, 25
  ecological niches of human pathogens and introduction into the plant environment
    reservoirs of *E. coli* and *Salmonella*, 23
    survival of *E. coli* in manure, 23-24
    survival of *Salmonella* in manure, 24
    survival of pathogens in water, 24
    sources and persistence of pathogens, 24-25
    2006 *E. coli* O157:H7 spinach outbreak, 25
  overview
    incidence of fresh produce outbreaks, 21-22
    survival and behavior of pathogens in plant environment, 22
    survival of pathogens through the food chain, 25
Microbial survival
  aggregates of bacteria on leaf surfaces, 37
  bacteria on tree fruits and nuts, 250-252
  bacterial pathogens in manure, 23-24, 84
  microorganisms in the environment, 85-87
  parasites in the environment, 88
  pathogens in plant environment, 22
  pathogens in water, 24
  pathogens on berries, 273, 287, 293-294
  pathogens on melons, 206
  pathogens on produce, 9, 113
  pathogens through food chain, 5
  pesticides and, 198
  Salmonella on tomatoes, 238-240

O

Outbreak investigation
  environmental investigations
    chain of custody procedures in, 59, 60f
    environmental samples in, 58-59
    equipment needs for, 57, 59
    farm-to-fork continuum for, 56, 56f
    interviews in environmental investigations and tracebacks, 59, 63
    investigative approach for, 53-56
    physical observation in, 57-58
    team requirements of, 57
  epidemiological investigation
    characteristics of ill individuals in, 51
    identification of specific vehicle in, 51
    interviews in, 52
    limitations of, 54
    methodology for, 52-53
    types of exposure sources in, 52

Outbreak investigation (*Continued*)
  use of questionnaires in, 53
  farm investigations
    collection of environmental samples for, 70
    determining movement of pathogen from reservoir to produce in, 69
    examination of potential contamination sources in, 69-70
    expertise required for, 68-69
    inventory of farm operations in, 69
  fresh-cut processor investigations
    collection of library and water samples in, 73
    commingling in, 73
    washing and fluming steps in, 72
    water quality in, 73
    disinfection procedures and monitoring in, 73
    guidelines for resuming operations, 67-68
  intentional contamination
    communication with law enforcement agencies regarding, 73
    indications of, 73
  overview, 50
  packinghouse investigations
    assessment of water quality and temperature in, 72
    assessment of disinfection procedures and monitoring system in, 72
    collection of environmental samples for, 72
    evaluation of vacuum cooler/hydrocooler operations in, 72
    evaluation of worker health, hygiene and training in, 71
    examination of potential contamination sources in, 71, 71f, 72f
  prevention research phase
    applied research in, 66
    quick, prioritized research in, 66
  recommendations
    guidelines as, 75-76
    lessons learned in investigations affecting, 73-75
  regulatory/enforcement phase
    regulatory or enforcement action needed in, 66
    understanding product distribution and preventing further exposures in, 65
  surveillance and detection
    national surveillance systems and PulseNet in, 51
    reports to local health jurisdictions in, 50-51
  tracebacks
    coordination of, 63-64
    flow diagrams and spreadsheets for, 63, 64f, 65t
    records in, 62
    timeline and triangulation in, 61-62, 61t
    time period of interest, 62
  training needs of investigators for, 66-67

P

Produce-associated outbreaks
  from berries, 272-283
  from leafy vegetables, 166-168
  from melons, 191-194
  from tomatoes, 229-236
  from tree fruits and nuts, 253-257
  in Europe, 335-339
  in Japan, 362-363, 362t
  in Mexico and Central America, 311, 323-325
  prevalence of, 5-8, 6t, 7t
  zoonotic bacterial species linked to, 146-147, 152
Produce safety in Mexico and Central America
  agricultural workers as contamination source
    prevalence of infectious diarrhea in, 319-320
    portable toilets and handwashing facilities for, 320
    training in personal hygiene and use of sanitary facilities, 320
  disinfection processes at packinghouses
    chlorination as, 316
    effect of suspended solids on, 316
  good agricultural practices
    aims of approach in, 321
    enforcement of pesticide regulations and, 322
    European Union program, 332, 342-346
    implementation of, 322-323
    implementation of GMPs and HACCP, 322
    lack of political concern and incentives for, 322
    Mexico Calidad Suprema program and, 323
    responsibilities of government ministries for, 322
  irrigation water as contamination source
    avoidance of contamination by, 313
    bacterial species in waste water and, 314-315
    Cryptosporidium and *Giardia* and, 314
    impounded water and irrigation channels as, 313
    irrigation methods and, 313
    irrigation with waste water, 313
    risk with flooding irrigation, 313-314
    Salmonella and *E. coli* in agricultural water, 315
    use of drip irrigation for export crops and, 313
  major outbreaks of foodborne disease
    cyclosporiasis outbreak from Guatemalan raspberries, 324-325
    hepatitis A outbreak from Mexican green onions, 325
    Salmonella infections from Mexican cantaloupes, 323-324

Salmonella Saintpaul outbreak from Mexican peppers, 325
overview
  actions to improve food safety, 326
  contributing factors, 311
  GAPs and GMPs, 312
  guidance manuals and their impact, 312
  outbreaks linked to fresh produce, 311
  produce exports to US, 309-311, 310*t*
  production practices in Mexico for export and domestic markets, 311
  state of agriculture in Central America, 311
runoff and flooding during rainy season
  contamination of flood waters by animal waste, 315-316
  contamination of surface and ground water, 315
tomato production in Mexico
  crop production practices, 316-318
  dump tank and wash water as sources of contamination, 317
  turbidity effects on efficacy of common disinfectants, 318
  water temperature and avoidance of infiltration, 317

R

Regulatory issues in Europe
  European fruit and vegetable production and imports
    domestic production, 334
    imports from non-E.U. countries, 334-335, 335*t*
  European Union (E.U.)
    European Economic Area, 334
    European Free Trade Association, 334
    member states, 333, 333*t*
  European Union food safety regulation
    Good Hygienic Practice (GHP), GAPs and HACCP principles, 332, 342-346
    European Food Safety Authority (EFSA), 340
    Food Law (EC 178/2002), 339-340
    GAP and quality assurance in produce production, 344-345
    GlobalGAP, 346
    hygiene regulations for produce, 342-344, 344*t*
    Rapid Alert System for Food and Feed (RASFF), 340-342
    regulation on imported foods, 345-346
    White Paper on Food Safety (European Commission), 339
  funding of food safety research in Europe, 348-349
  produce-associated outbreaks of foodborne disease in Europe
    foodborne bacteria (*Shigella* and *Salmonella*), 336-337, 337*t*
    molds and mycotoxins, 338-339
    parasites, 337
    viruses, 337-338
  recognition of produce safety problem
    FAO/WHO report on microbiological hazards, 332
    import of fruits and vegetables to EU, 332-333
    risk profile by EU Scientific Committee on Food, 332
  similarities/differences between E.U. and U.S. food safety regulation, 346-348
  sources for further information, 349
Regulatory issues in Japan
  domestic fresh produce production
    covered production of vegetables, 358
    fruit production, 359
    greenhouse production of fruits, 359
    price stabilization and, 359-360
    vegetable production, 357-359, 358*t*, 359*f*
  fresh produce-related outbreaks
    contamination with human pathogens and, 362-363, 362*t*
    mushroom poisoning and, 363-364, 364*f*
  Good Agricultural Practices (GAPs) in Japan
    GAP introduction system, 366
    guidance manuals for vegetable production and, 364-365, 365*f*
    introduction and dissemination of, 364-365
    private sector adoption of, 366-367
  government initiatives for ensuring produce safety
    building relationships of trust and, 368, 369*f*
    mandatory labeling of place of origin as, 367-368
    public concerns over food safety and, 367
    required legal compliance as, 368
    traceability systems as, 367
  imports and distribution of fresh produce
    China as major source of vegetables, 369-371, 370*t*, 371*t*
    compliance with microbiological standards, 375, 375*t*
    distribution routes of fresh produce and, 373-374, 373*f*
    fruit imports and, 372-373
    increase in imports and, 368
    pesticide compliance of vegetable imports, 374-375
    surveillance for pathogens and, 375-376, 376*t*

Regulatory issues in Japan (*Continued*)
  value of imports and, 368-369, 370*f*
  vegetable imports and, 371-372, 372*t*
  overview
    conclusions, 386-388
    food safety issues in Japan, 354
    government agencies with food safety authority, 355
    hazard control at all points in supply chain, 356
    risks and risk assessment, 356
    role of public policies, 354
    shared responsibility for food safety, 356-357
    use of HACCP, 355
  safety regulations and their enforcement
    applicable laws and regulations for, 380-381
    company strategies and quality standards as, 384-385
    enforcement of, 381-382, 383*f*
    Food Safety Basic Law, 377
    implementation of risk assessment and, 378-379
    new food safety policy and Food Safety Commission, 377, 378*f*
    private sector view of, 384
    risk communication and, 379
    responses to emergency situation, 379
    scandals and a major change in attitudes, 376-377
    traceability and, 385-386

T

Tomatoes
  commercial production of
    harvesting of, 225-226
    types and sizes of, 225
    yield of, 225
  deficiencies in agricultural practices at implicated farms
    contamination of water for irrigation or pesticide application as, 232-233, 235, 236
    proximity of farms to flood-prone drainage ditches as, 235
    proximity of farms to wildlife and cattle as, 234, 235
  deficiencies at implicated packinghouses
    bird feces in packinghouse and on equipment as, 232
    lack of records of chlorine concentration, pH, temperature as, 231
    tomatoes soaked in cold water enabling infiltration as, 233, 235
  factors complicating outbreak investigations
    complexity of marketing chain as, 236
    diversity of *Salmonella* spp. causing enteritis as, 238-239
    insufficient information about menu items as, 237
  fresh-cut tomatoes
    advantages of, 229
    outbreaks linked to, 229
    sourcing raw material for, 229
  marketing of tomatoes
    field packs in, 228
    levels of handling after packinghouse in, 227
    repacker operations in, 227-228
    terminal markets in, 228-229
  overview
    infectious dose of *Salmonella*, 224
    isolation of *Salmonella* from, 223-224
    links to outbreaks from, 223
    multiplication of *Salmonella* on, 224
    violations of good agricultural practices, 224
  packing of
    culling of, 226
    inspection of, 227
    packing of, 226
    ripening of, 227
    sanitation and, 226
    sorting of, 226
    washing, drying and waxing of, 226
    water quality in, 226
  recommendations for
    continuous temperature recording as, 244
    inspection and grading as, 243
    lot identification as, 244
    package labeling as, 244
    inspections of food service suppliers and distributors as, 244
    record keeping as, 244
    rejection of culls, salvaged fruit as, 244
    sanitation steps as, 243
    sensors to detect temperature abuse and anoxia as, 244
    use of GAPs, GMPs and HACCP as, 243-244
  tomato-linked *Salmonella* outbreaks 1990-2006
    1990 S. Carolina, *S. Javiana*, 229-230
    1993 S. Carolina, *S. montevideo*, 230
    1998 Florida, *S. Baildon*, 231
    2002 Florida diced, *S. Javiana*, 232
    2002 S. Carolina sliced, multiserotype, 233-234
    2002 Virginia, *S. Newport*, 232-233
    2005 Delmarva Peninsula, *S. Newport*, 234-235
    2005 Florida diced, *S. Braenderup*, 235
    2006 Delmarva Peninsula, *S. Newport*, 235
    2006 Ohio, *S. Typhimurium*, 236
  unrealistic experimental models affecting validity of recommendations for
    infiltration of *Salmonella* as, 241-242
    internalization of *Salmonella* as, 240-241

Index **461**

survival and growth of *Salmonella* on tomatoes as, 238-240
Tree fruits and nuts
  foodborne outbreaks linked to tree fruits
    Cryptosporidium in unpasteurized apple juice, 253
    Norwalk virus in fruit salad, 254, 257
    Odwalla *E. coli* O157:H7 outbreak in unpasteurized apple juice, 253
    Salmonella in orange juice, 254
    Salmonella in mangoes, 257
    Shigella in orange juice, 257
  foodborne outbreaks linked to nuts
    aflatoxins and other mycotoxins in tree nuts, 255
    Salmonella in almonds, 255
  human pathogens associated with outbreaks from
    acid tolerance of pathogens, 251
    Cryptosporidium, 252, 254, 256
    E. coli O157:H7, 250, 256
    Listeria monocytogenes, 250-251
    Norwalk virus, 254, 257
    reservoirs of, 256
    Salmonella, 251-252, 255-258
    Shigella, 250-251, 257
    surveys of pathogen prevalence, 256-257
    survival characteristics of, 250-252
  measures to reduce contamination of the crop, 258
    exclusion of domestic animals and wildlife, 259
    implementation of GAPs and FDA guidelines, 256, 258-259
    pathogen-free irrigation and pesticide make-up water, 259
    tree-picked and blemish free fruit, 259
  mitigation options
    FDA-mandated 5-log reduction for juices, 253, 260
    hot water surface pasteurization, 260
    irradiation, 260
    limitations of surface treatments, 259
    USDA-mandated program for *Salmonella* in almonds, 255
    UV light, 260
    validation of alternative treatments, 254
  overview
    low incidence of illness from tree fruits and nuts, 250
    tree fruits and nuts provide hostile environment, 250
    tree fruits and nuts consumed raw, 250
  risk factors for contamination
    contamination of processing aids, 257
    contamination of wash water, 256
    infiltration of pathogens during cooling, 257
    poor sanitation and hygienic practices, 256-257
    utilization of ground harvested tree fruit and nuts, 256, 258

V

Vulnerable commodities
  berries, 272-273
  efficacy of washing and sanitizing treatments for, 407-413, 411t
  fresh-cut, 11, 37
  high risk, 447-448
  lettuce and spinach, 166, 168
  melons, 191-194
  prevalence of association with outbreaks, 7-8, 7t, 13
  tomatoes, 229-236
  tree fruits and nuts, 253-257

W

Washing and sanitizing fresh produce
  contamination during washing, 412
  efficacy of sanitizing agents
    bacterial attachment sites and, 36-37, 394, 408-410, 413
    bacterial internalization and, 409, 413
    biofilms and, 394, 409, 412
    depletion of sanitizing agent and, 408
    enhancement of efficacy, 396, 409, 412-413
    interval between contamination and washing and, 410, 412
    pathogen resistance to disinfection, 37-39
  overview
    benefits of washing and sanitizing, 394
    cross contamination during washing, 394
    importance of preharvest interventions, 414
    objectives of washing and sanitizing, 394
    need for penetrating disinfection treatments, 414
  problem commodities
    apples as, 412-413
    cantaloupes as, 410-412
    leafy vegetables as, 407-409
    tomatoes as, 409-410, 411t
  produce washes for food service and home use
    commercial products for, 405, 406t
    equipment for, 407, 407t
    validation studies of, 405-406
  regulatory restrictions on use of washing and sanitizing agents, 395-403
  washing and sanitizing agents
    acidified sodium chlorite, 398, 410-411
    alkaline products, 402, 410-411
    chlorine, 395-396, 408-411
    chlorine dioxide, aqueous, 397-398, 408-409
    combinations of agents, 400-401, 408-409, 411
    detergent products, 395

Washing and sanitizing fresh produce (*Continued*)
  electrolyzed water, 397, 408
  hydrogen peroxide, 401, 411
  iodine, 402
  organic acids, 401-402, 410-411
  ozone, aqueous, 398-400, 408
  peroxyacetic acid, 400, 409-411
  sanitizing agents for organic crops, 403
  washing equipment
    efficacy of, 404-405, 404t, 413
    types of, 403
Water quality and produce production
  contamination of produce during irrigation
    contamination by enteric bacteria, 112
    contamination by viruses, 111-113
    factors affecting contamination of edible plant portions, 111
    types of irrigation, 111
  irrigation water
    canal/channel systems for, 107-108
    factors influencing quality of, 107
    global usage of, 106
    microbial quality of, 107-108
  occurrence of pathogens in irrigation water
    parasites, 109
    Salmonella, *E. coli*, *Campylobacter*, 109-110
    viruses, 109
  other water sources of contamination
    hand washing, 114
    hydrocooling, 114
    water for pesticide application, 113
  quality standards for irrigation water
    California/Arizona treatment standards, 108
    Environmental Protection Agency guidelines, 108
    fecal coliforms, 108-109
  research needs
    ecology of pathogens and bacterial indicators in irrigation delivery systems, 114
    meaningful standards for indicator bacteria, 114
    occurrence of indicator bacterial, 114
    role of water in produce contamination, 114
    survival of foodborne parasites on produce, 114
    survival of pathogens on produce under field conditions
      Cryptosporidium, 113
      Giardia, 113
      viruses, 113
  uses of water in produce production
    irrigation, 105-108
    link between water and produce outbreaks, 106
    washing and hygiene, 105
    water treatment, 105-106
Wild birds as contamination source
  pathogenic bacteria carried by
    Campylobacter, 124-125
    Chlamydia, 125
    Escherichia coli, 125-127
    Listeria, 127
    Salmonella, 127-128
  pathogenic fungi carried by
    Aspergillus, 128
    Cryptococcus, 128-129
    Histoplasma, 129
  mitigation options
    chemical repellents as, 132
    frightening devices as, 132
    lethal vs. non-lethal, 132
  overview
    circumstantial evidence for, 120, 132
    documenting role in produce contamination by, 120
    ecological perspective, 133
    empirical and risk modeling studies, 133
  parasites carried by wild birds
    Cryptosporidia, 130
    Microsporidia, 130
    Toxoplasma, 130, 131t
  wild bird species associated with agriculture
    gulls, 123
    passerines, 123-124
    pigeons, 121-122, 122f
    waterfowl, 123
Wildlife as contamination source
  bacterial pathogens
    deer and apple cider contamination, 147-148
    environmental reservoirs of, 148
    feral swine and spinach contamination, 147
    insects as mechanical vectors of, 150-151
    nematodes as mechanical vectors of, 151
    surveys of wildlife for bacterial human pathogens, 148-150
    wildlife as direct source of produce contamination, 147
    zoonotic bacterial species linked to outbreaks, 146-147
  mitigating wildlife-crop interactions
    guidance documents for, 155
    management strategies for, 155-156
    physical barriers for, 156
    rejection of potentially contaminated produce, 155
    removal of habitat and fencing for, 155
  overview
    animal species involved, 145
    crop production losses from wild animals, 144
    FAO-WHO assessment, 143-144
    implication of wildlife in produce-borne illness, 144-145

increased risk with imports from developing countries, 156-157
limitations of wildlife control programs, 144, 156
weakness of mitigation strategies, 156
zoonotic pathogens, 145-146, 145t

protozoa
foodborne outbreaks of cryptosporidiosis, 152
foodborne transmission of, 152
mechanical transport of *C. parvum* by flies, 153
presence in wastewater and irrigation water, 152
presence of *C. parvum* in fresh produce, 152
primary foodborne zoonotic protozoa, 151-152
role of wild animals in transmission, 152-153

viral pathogens
hepatitis E carriers, 146
host specificity of viruses, 146
routes of infection by, 146

zoonotic helminthes
involvement of wildlife in transmission, 153-155
linkage of *Fasciola hepatica* infection to wild animals, 153
potential foodborne helminthes, 153-155
rarity of foodborne outbreaks from pathogenic nematodes, 153

Food Science and Technology
International Series

Maynard A. Amerine, Rose Marie Pangborn, and Edward B. Roessler, *Principles of Sensory Evaluation of Food*. 1965.
Martin Glicksman, *Gum Technology in the Food Industry*. 1970.
Maynard A. Joslyn, *Methods in Food Analysis*, second edition. 1970.
C. R. Stumbo, *Thermobacteriology in Food Processing*, second edition. 1973.
Aaron M. Altschul (ed.), *New Protein Foods*: Volume 1, *Technology, Part A*—1974. Volume 2, *Technology, Part B*—1976. Volume 3, *Animal Protein Supplies, Part A*—1978. Volume 4, *Animal Protein Supplies, Part B*—1981. Volume 5, *Seed Storage Proteins*—1985.
S. A. Goldblith, L. Rey, and W. W. Rothmayr (eds), *Freeze Drying and Advanced Food Technology*. 1975.
R. B. Duckworth (ed.), *Water Relations of Food*. 1975.
John A. Troller and J. H. B. Christian, *Water Activity and Food*. 1978.
A. E. Bender, *Food Processing and Nutrition*. 1978.
D. R. Osborne and P. Voogt, *The Analysis of Nutrients in Foods*. 1978.
Marcel Loncin and R. L. Merson, *Food Engineering: Principles and Selected Applications*. 1979.
J. G. Vaughan (ed.), *Food Microscopy*. 1979.
J. R. A. Pollock (ed.), *Brewing Science*, Volume 1—1979. Volume 2—1980. Volume 3—1987.
J. Christopher Bauernfeind (ed.), *Carotenoids as Colorants and Vitamin A Precursors: Technological and Nutritional Applications*. 1981.
Pericles Markakis (ed.), *Anthocyanins as Food Colors*. 1982.
George F. Stewart and Maynard A. Amerine (eds), *Introduction to Food Science and Technology*, second edition. 1982.
Hector A. Iglesias and Jorge Chirife, *Handbook of Food Isotherms: Water Sorption Parameters for Food and Food Components*. 1982.
Colin Dennis (ed.), *Post-Harvest Pathology of Fruits and Vegetables*. 1983.
P. J. Barnes (ed.), *Lipids in Cereal Technology*. 1983.
David Pimentel and Carl W. Hall (eds), *Food and Energy Resources*. 1984.
Joe M. Regenstein and Carrie E. Regenstein, *Food Protein Chemistry: An Introduction for Food Scientists*. 1984.
Maximo C. Gacula, Jr. and Jagbir Singh, *Statistical Methods in Food and Consumer Research*. 1984.
Fergus M. Clydesdale and Kathryn L. Wiemer (eds), *Iron Fortification of Foods*. 1985.
Robert V. Decareau (ed.), *Microwaves in the Food Processing Industry*. 1985.

S. M. Herschdoerfer (ed.), *Quality Control in the Food Industry*, second edition. Volume 1—1985. Volume 2—1985. Volume 3—1986. Volume 4—1987.
Walter M. Urbain, *Food Irradiation*. 1986.
Peter J. Bechtel, *Muscle as Food*. 1986.
H. W.-S. Chan, *Autoxidation of Unsaturated Lipids*. 1986.
Chester O. McCorkle, Jr., (ed.), *Economics of Food Processing in the United States*. 1987.
F. E. Cunningham and N. A. Cox (eds), *Microbiology of Poultry Meat Products*. 1987.
Jethro Japtiani, Harvey T. Chan, Jr., and William S. Sakai, *Tropical Fruit Processing*. 1987.
J. Solms, D. A. Booth, R. M. Dangborn, and O. Raunhardt (eds), *Food Acceptance and Nutrition*. 1987.
R. Macrae (ed.), *HPLC in Food Analysis*, second edition. 1988.
A. M. Pearson and R. B. Young, *Muscle and Meat Biochemistry*. 1989.
Marjorie P. Penfield and Ada Marie Campbell, *Experimental Food Science*, third edition. 1990.
Leroy C. Blankenship (ed.), *Colonization Control of Human Bacterial Enteropathogens in Poultry*. 1991.
Yeshajahu Pomeranz, *Functional Properties of Food Components*, second edition. 1991.
Reginald H. Walter (ed.), *The Chemistry and Technology of Pectin*. 1991.
Herbert Stone and Joel L. Sidel, *Sensory Evaluation Practices*, second edition. 1993.
Robert L. Shewfelt and Stanley E. Prussia (eds), *Postharvest Handling: A Systems Approach*. 1993.
Tilak Nagodawithana and Gerald Reed (eds), *Enzymes in Food Processing*, third edition. 1993.
Dallas G. Hoover and Larry R. Steenson, *Bacteriocins*. 1993.
Takayaki Shibamoto and Leonard Bjeldanes, *Introduction to Food Toxicology*. 2009.
John A. Troller, *Sanitation in Food Processing*, second edition. 1993.
Harold D. Hafs and Robert G. Zimbelman (eds), *Low-fat Meats*. 1994.
Lance G. Phillips, Dana M. Whitehead, and John Kinsella, *Structure-Function Properties of Food Proteins*. 1994.
Robert G. Jensen (ed.), *Handbook of Milk Composition*. 1995.
Yrjö H. Roos, *Phase Transitions in Foods*. 1995.
Reginald H. Walter, *Polysaccharide Dispersions*. 1997.
Gustavo V. Barbosa-Cánovas, M. Marcela Góngora-Nieto, Usha R. Pothakamury, and Barry G. Swanson, *Preservation of Foods with Pulsed Electric Fields*. 1999.
Ronald S. Jackson, *Wine Tasting: A Professional Handbook*. 2009.

Malcolm C. Bourne, *Food Texture and Viscosity: Concept and Measurement*, second edition. 2002.
Benjamin Caballero and Barry M. Popkin (eds), *The Nutrition Transition: Diet and Disease in the Developing World*. 2002.
Dean O. Cliver and Hans P. Riemann (eds), *Foodborne Diseases*, second edition. 2002.
Martin Kohlmeier, *Nutrient Metabolism*, 2003.
Herbert Stone and Joel L. Sidel, *Sensory Evaluation Practices*, third edition. 2004.
Jung H. Han (ed.), *Innovations in Food Packaging*. 2005.
Da-Wen Sun (ed.), *Emerging Technologies for Food Processing*. 2005.
Hans Riemann and Dean Cliver (eds) *Foodborne Infections and Intoxications*, third edition. 2006.
Ioannis S. Arvanitoyannis (ed.), *Waste Management for the Food Industries*. 2008.
Ronald S. Jackson, *Wine Science: Principles and Applications*, third edition. 2008.
Da-Wen Sun (ed.), *Computer Vision Technology for Food Quality Evaluation*. 2008.
Kenneth David and Paul Thompson (eds), *What Can Nanotechnology Learn From Biotechnology?* 2008.
Elke K. Arendt and Fabio Dal Bello (eds), *Gluten-Free Cereal Products and Beverages*. 2008.
Debasis Bagchi (ed.), *Nutraceutical and Functional Food Regulations in the United States and Around the World*, 2008.
R. Paul Singh and Dennis R. Heldman, *Introduction to Food Engineering*, fourth edition. 2008.
Zeki Berk, *Food Process Engineering and Technology*. 2009.
Abby Thompson (ed.), Mike Boland and Harjinder Singh, *Milk Proteins: From Expression to Food*. 2009.
Wojciech J. Florkowski (ed.), Stanley E. Prussia, Robert L. Shewfelt and Bernhard Brueckner (eds) *Postharvest Handling*, second edition. 2009.
James N.BeMiller and Roy L. Whistler (eds) *Starch: Chemistry and Technology*, Third Edition, 2009.